T0122313

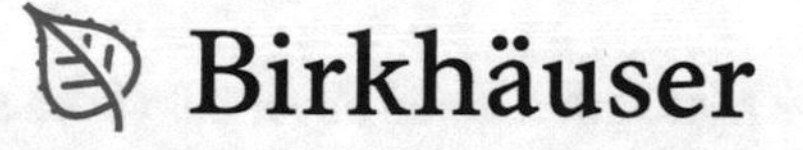

Operator Theory: Advances and Applications
Volume 269

Founded in 1979 by Israel Gohberg

More information about this series at http://www.springer.com/series/4850

Zhongwei Shen

Periodic Homogenization of Elliptic Systems

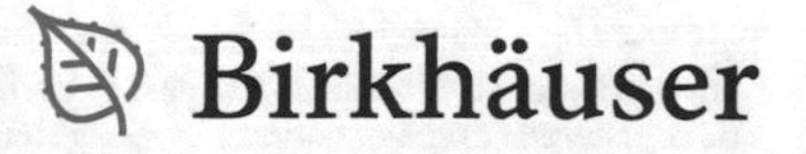

Zhongwei Shen
Department of Mathematics
University of Kentucky
Lexington, Kentucky, USA

ISSN 0255-0156 ISSN 2296-4878 (electronic)
Operator Theory: Advances and Applications
ISBN 978-3-030-08199-7 ISBN 978-3-319-91214-1 (eBook)
https://doi.org/10.1007/978-3-319-91214-1

Mathematics Subject Classification (2010): 35B27, 35J57, 74Q05

This book is published under the imprint Birkhäuser, www.birkhauser-science.com by the registered company Springer Nature Switzerland AG part of Springer Nature.
The registered company address is: Gewerbestrasse 11, 6330 Cham, Switzerland

To my wife Liping
and my sons
Michael and Jeffrey

Contents

Preface ix

1 Introduction

1.1 Homogenization theory 1

1.2 General presentation of the monograph 3

1.3 Readership 5

1.4 Notation 5

2 Elliptic Systems with Periodic Coefficients

2.1 Weak solutions 8

2.2 Two-scale asymptotic expansions and the homogenized operator 14

2.3 Homogenization of elliptic systems 21

2.4 Elliptic systems of linear elasticity 26

2.5 Notes 31

3 Convergence Rates, Part I

3.1 Flux correctors and ε-smoothing 34

3.2 Convergence rates in H^1 for the Dirichlet problem 40

3.3 Convergence rates in H^1 for the Neumann problem 49

3.4 Convergence rates in L^p for the Dirichlet problem 52

3.5 Convergence rates in L^p for the Neumann problem 57

3.6 Convergence rates for elliptic systems of elasticity 59

3.7 Notes 63

4 Interior Estimates

4.1 Interior Lipschitz estimates 66

4.2 A real-variable method 74

4.3 Interior $W^{1,p}$ estimates 82

4.4 Asymptotic expansions of fundamental solutions 88

4.5 Notes 97

5 Regularity for the Dirichlet Problem
5.1 Boundary localization in the periodic setting . . . 100
5.2 Boundary Hölder estimates . . . 103
5.3 Boundary $W^{1,p}$ estimates . . . 110
5.4 Green functions and Dirichlet correctors . . . 114
5.5 Boundary Lipschitz estimates . . . 121
5.6 The Dirichlet problem in C^1 and $C^{1,\alpha}$ domains . . . 129
5.7 Notes . . . 134

6 Regularity for the Neumann Problem
6.1 Approximation of solutions at the large scale . . . 136
6.2 Boundary Hölder estimates . . . 141
6.3 Boundary $W^{1,p}$ estimates . . . 144
6.4 Boundary Lipschitz estimates . . . 150
6.5 Matrix of Neumann functions . . . 159
6.6 Elliptic systems of linear elasticity . . . 161
6.7 Notes . . . 168

7 Convergence Rates, Part II
7.1 Convergence rates in H^1 and L^2 . . . 169
7.2 Convergence rates of eigenvalues . . . 174
7.3 Asymptotic expansions of Green functions . . . 177
7.4 Asymptotic expansions of Neumann functions . . . 187
7.5 Convergence rates in L^p and $W^{1,p}$ . . . 199
7.6 Notes . . . 202

8 L^2 Estimates in Lipschitz Domains
8.1 Lipschitz domains and nontangential convergence . . . 207
8.2 Estimates of fundamental solutions . . . 214
8.3 Estimates of singular integrals . . . 219
8.4 The method of layer potentials . . . 225
8.5 Laplace's equation . . . 231
8.6 The Rellich property . . . 246
8.7 The well-posedness for the small scale . . . 252
8.8 Rellich estimates for the large scale . . . 262
8.9 L^2 boundary value problems . . . 265
8.10 L^2 estimates in arbitrary Lipschitz domains . . . 269
8.11 Square function and $H^{1/2}$ estimates . . . 275
8.12 Notes . . . 281

Bibliography . . . 283

Index . . . 291

Preface

In recent years considerable advances have been made in quantitative homogenization of partial differential equations in the periodic and non-periodic settings. This monograph surveys the theory of quantitative homogenization for second-order linear elliptic systems in divergence form with rapidly oscillating periodic coefficients,

$$\mathcal{L}_\varepsilon = -\text{div}\big(A(x/\varepsilon)\nabla\big),$$

in a bounded domain Ω in $\mathbb{R}^d$. It begins with a review of the classical theory of qualitative homogenization, and addresses the problem of convergence rates of solutions. The main body of the monograph investigates various interior and boundary regularity estimates (Hölder, Lipschitz, $W^{1,p}$, nontangential-maximal-function) that are uniform in the small parameter $\varepsilon > 0$. Additional topics include convergence rates for Dirichlet eigenvalues and asymptotic expansions of fundamental solutions, Green functions, and Neumann functions.

Part of this monograph is based on lecture notes for courses I taught at several summer schools and at the University of Kentucky. Much of the material in Chapters 7 and 8 is taken from my joint papers [64, 66] with Carlos Kenig and Fang-Hua Lin, and from [68] with Carlos Kenig. I would like to express my deep gratitude to Carlos Kenig and Fang-Hua Lin for introducing me to the research area of homogenization and for their important contribution to our joint work.

I would like to thank Jun Geng, Weisheng Niu, B. Chase Russell, Qiang Xu, Yao Xu, and Jinping Zhuge, who read parts of the manuscript, for their helpful comments and suggestions.

This work was supported in part by the National Science Foundation.

Lexington, Kentucky, Fall 2017 Zhongwei Shen

Chapter 1

Introduction

1.1 Homogenization theory

Partial differential equations and systems with rapidly oscillating coefficients are used to model various physical phenomena in inhomogeneous or heterogeneous media, such as composite and perforated materials. Let $\varepsilon > 0$ be a small parameter, representing the inhomogeneity scale – the scale of the microstructure of an inhomogeneous medium. The local characteristics of the medium are described by functions of the form $A(x/\varepsilon)$, which vary rapidly with respect to the space variables. Since ε is much smaller than the linear size of the domain where the physical process takes place, solving the corresponding boundary value problems for the partial differential equations directly by numerical methods may be costly.

To describe the basic idea of homogenization, we consider the Dirichlet problem,

$$\begin{cases} -\text{div}(A(x/\varepsilon)\nabla u_\varepsilon) = F & \text{in } \Omega, \\ u_\varepsilon = f & \text{on } \partial\Omega, \end{cases} \tag{1.1.1}$$

where Ω is a bounded domain in $\mathbb{R}^d$. The $d \times d$ coefficient matrix $A = A(y)$ is assumed to be real, bounded and measurable, and to satisfy the ellipticity condition

$$A\xi \cdot \xi \geq \mu|\xi|^2 \quad \text{for any } \xi \in \mathbb{R}^d, \tag{1.1.2}$$

where $\mu > 0$. Given $F \in H^{-1}(\Omega)$ and $f \in H^{1/2}(\partial\Omega)$, the Dirichlet problem (1.1.1) has a unique weak solution u_ε in $H^1(\Omega)$. Under the additional structural condition that A is periodic with respect to some lattice in $\mathbb{R}^d$, it can be shown that as $\varepsilon \to 0$, the solution u_ε of (1.1.1) converges to u_0 weakly in $H^1(\Omega)$ and thus strongly in $L^2(\Omega)$. Moreover, the limit u_0 is the weak solution of the Dirichlet problem,

$$\begin{cases} -\text{div}(\widehat{A}\,\nabla u_0) = F & \text{in } \Omega, \\ u_0 = f & \text{on } \partial\Omega, \end{cases} \tag{1.1.3}$$

Z. Shen, *Periodic Homogenization of Elliptic Systems*, Operator Theory: Advances and Applications 269, https://doi.org/10.1007/978-3-319-91214-1_1

where the coefficient matrix $\widehat{A}$ is constant and satisfies the ellipticity condition (1.1.2). As a result, one may use the function u_0 as an approximation of u_ε. Since $\widehat{A}$ is constant, the Dirichlet problem (1.1.3) is much easier to handle both analytically and numerically than (1.1.1). Similar results may be proved for Neumann type boundary value problems. More precisely, under the same ellipticity and periodicity assumptions on A, the weak solution of the Neumann problem,

$$\begin{cases} -\operatorname{div}(A(x/\varepsilon)\nabla u_\varepsilon) = F & \text{in } \Omega, \\ n \cdot A(x/\varepsilon)\nabla u_\varepsilon = g & \text{on } \partial\Omega, \end{cases} \tag{1.1.4}$$

with $\int_\Omega u_\varepsilon \, dx = 0$, converges weakly in $H^1(\Omega)$ to the weak solution of the Neumann problem

$$\begin{cases} -\operatorname{div}(\widehat{A}\nabla u_0) = F & \text{in } \Omega, \\ n \cdot \widehat{A}\nabla u_0 = g & \text{on } \partial\Omega, \end{cases} \tag{1.1.5}$$

with $\int_\Omega u_0 \, dx = 0$, where n denotes the outward unit normal to $\partial\Omega$.

The entries of the constant matrix $\widehat{A}$, which are uniquely determined by A, are given by solving some auxiliary problems in a periodicity cell. By taking the inhomogeneity scale ε to zero in the limit, we "homogenize" the microscopically inhomogeneous medium. For this reason the limit u_0 of the solution u_ε is called the homogenized or effective solution. The boundary value problems (1.1.3) and (1.1.5) are referred as the homogenized problems for (1.1.1) and (1.1.4), respectively. The constant matrix $\widehat{A}$, which describes the macroscopic characteristics of the inhomogeneous medium, is called the matrix of homogenized or effective coefficients.

There exists a vast literature on the homogenization theory of partial differential equations and systems. Some of the early work may be found in [93, 94, 40, 38, 39, 17, 18, 75, 99]. We refer the reader to the classical monographs [19, 111, 78], as well as to more recent books [28, 71, 25, 100] for further references on the subject. Much of the material in the existing books on homogenization mainly addresses the basic and qualitative questions in various settings,

- Does the solution u_ε of the boundary value problem converge as $\varepsilon \to 0$?
- If u_ε converges to u_0, does u_0 solve some limit boundary value problem?

In this monograph we shall be concerned with two fundamental issues in quantitative homogenization:

- The sharp convergence rates of u_ε to u_0.
- Regularity estimates of u_ε that are uniform in the small parameter ε.

As it turns out, these two issues are closely related. Results on convergence rates can be used to establish uniform regularity estimates; sharp regularity estimates lead to optimal results on convergence rates.

1.2 General presentation of the monograph

This book is designed as an introduction to quantitative homogenization. We will restrict ourself to the periodic setting and consider second-order linear elliptic systems in divergence form,

$$\mathcal{L}_\varepsilon = -\mathrm{div}\big(A(x/\varepsilon)\nabla\big), \tag{1.2.1}$$

in a bounded domain Ω in $\mathbb{R}^d$, where the coefficient matrix (tensor) $A = A(y)$ is assumed to be elliptic and periodic.

Qualitative homogenization theory

In Chapter 2 we present the qualitative homogenization theory for $\mathcal{L}_\varepsilon$, which has been well understood since 1970s. We start out with a review of basic facts on weak solutions for elliptic systems with bounded measurable coefficients, and use the method of (formal) asymptotic expansions to derive the formula for the homogenized (or effective) operator $\mathcal{L}_0$. We then prove the classical results on the homogenization of Dirichlet and Neumann boundary value problems for $\mathcal{L}_\varepsilon$.

Convergence rates

In Chapter 3 we address the issue of convergence rates for solutions and two-scale expansions. Various estimates in L^p and in H^1 are obtained without smoothness assumptions on the coefficient matrix A.

Interior and boundary regularity estimates

Chapters 4, 5, and 6 are devoted to the study of sharp regularity estimates, which are uniform in $\varepsilon > 0$, for solutions of $\mathcal{L}_\varepsilon(u_\varepsilon) = F$. The case of interior estimates is treated in Chapter 4. We use a compactness method to establish a Lipschitz estimate down to the microscopic scale ε under the ellipticity and periodicity assumptions. With additional smoothness assumptions on A, this, together with a simple blow-up argument, leads to full-scale Hölder and Lipschitz estimates. The compactness method, which originated from the study of the regularity theory in the calculus of variations and minimal surfaces, was introduced in the study of homogenization problems by M. Avellaneda and F. Lin [11]. In this chapter we also introduce a real-variable method for establishing L^p and $W^{1,p}$ estimates. This method, which originated in a paper by L. Caffarelli and I. Peral [24], may be regarded as a refined and dual version of the celebrated Calderón–Zygmund Lemma. As corollaries of the interior estimates, we obtain asymptotic expansions for the fundamental solution $\Gamma_\varepsilon(x, y)$ and its derivatives $\nabla_x\Gamma_\varepsilon(x, y)$, $\nabla_y\Gamma_\varepsilon(x, y)$ and $\nabla_x\nabla_y\Gamma_\varepsilon(x, y)$, as $\varepsilon \to 0$.

In Chapter 5 we study boundary regularity estimates for solutions of $\mathcal{L}_\varepsilon(u_\varepsilon) = F$ in Ω with the Dirichlet condition $u_\varepsilon = f$ on $\partial\Omega$. The boundary

Lipschitz estimate is proved by the compactness method mentioned above. A key step is to obtain the Lipschitz estimate for the so-called Dirichlet correctors. The real-variable method introduced in Chapter 4 is used to establish the boundary $W^{1,p}$ estimates. It effectively reduces to the problem to certain (weak) reverse Hölder inequalities.

In Chapter 6 we prove the boundary Hölder, Lipschitz, and $W^{1,p}$ estimates for solutions of $\mathcal{L}_\varepsilon(u_\varepsilon) = F$ in Ω with the Neumann condition $\frac{\partial u_\varepsilon}{\partial \nu_\varepsilon} = g$ in $\partial\Omega$. Here we introduce a general scheme, recently developed by S.N. Armstrong and C. Smart [7] in the study of stochastic homogenization, for establishing large-scale regularity estimates. Roughly speaking, the scheme states that if a function u is well approximated by $C^{1,\alpha}$ functions at every scale greater than ε, then u is Lipschitz continuous at every scale greater than ε.

The problem of convergence rates revisited

In Chapter 7 we revisit the problem of convergence rates. We establish an $O(\varepsilon)$ error estimate in H^1 for a two-scale expansion involving the Dirichlet correctors and use it to prove a convergence result, uniform in ε and k, for the Dirichlet eigenvalues $\lambda_{\varepsilon,k}$. We derive asymptotic expansions for the Green function $G_\varepsilon(x, y)$ and its derivatives, as $\varepsilon \to 0$. Analogous results are also obtained for the Neumann function $N_\varepsilon(x, y)$.

L^2 boundary value problems in Lipschitz domains

Chapter 8 is devoted to the study of L^2 boundary value problems for $\mathcal{L}_\varepsilon(u_\varepsilon) = 0$ in a bounded Lipschitz domain Ω. We establish optimal estimates in terms of nontangential maximal functions for Dirichlet problems with boundary data in $L^2(\partial\Omega)$ and in $H^1(\partial\Omega)$ as well as for the Neumann problem with boundary data in $L^2(\partial\Omega)$. This is achieved by the method of layer potentials – the classical method of integral equations. The results on the asymptotic expansions of the fundamental solution $\Gamma_\varepsilon(x, y)$ and its derivatives in Chapter 4 are used to establish the L^p boundedness of singular integrals on $\partial\Omega$, associated with the single- and double-layer potentials. The proof of Rellich estimates,

$$\left\| \frac{\partial u_\varepsilon}{\partial \nu_\varepsilon} \right\|_{L^2(\partial\Omega)} \approx \|\nabla_{\tan} u_\varepsilon\|_{L^2(\partial\Omega)}, \tag{1.2.2}$$

which are crucial in the use of the method of layer potentials in Lipschitz domains, is divided into two cases. In the small-scale case, where $\operatorname{diam}(\Omega) \le \varepsilon$, the estimates are obtained by using Rellich identities and a three-step approximation argument. The proof of (1.2.2) for the large-scale case, where $\operatorname{diam}(\Omega) > \varepsilon$, uses an error estimate in $H^1(\Omega)$ for a two-scale expansion obtained in Chapter 3. For the reader's convenience we also include a section in which we prove the estimate (1.2.2) and solve the L^2 Dirichlet, Neumann and regularity problems in Lipschitz domains for the simplest case $\mathcal{L} = -\Delta$.

1.3 Readership

This monograph is intended for advanced graduate students as well as for experienced researchers in the general areas of partial differential equations (PDEs) and analysis. We assume that the reader is familiar with the basic material usually covered in the first-year graduate courses on real analysis, functional analysis, and linear elliptic PDEs. If a result beyond the basic material is used, it will be stated explicitly with a reference for its proof. Since we will be working exclusively with second-order elliptic systems in divergence form, the books [26] by Ya-Zhe Chen and Lan-Cheng Wu and [50] by Mariano Giaquinta and Luca Martinazzi may be used as references for basic results on the regularity of weak solutions. We do not assume the reader has prior knowledge of the classical homogenization theory, which is presented in Chapter 2.

1.4 Notation

Throughout the monograph we will use C and c to denote positive constants that are independent of the parameter $\varepsilon > 0$. They may change from line to line and depend on A and/or Ω. We will use $\fint_E u$ to denote the L^1 average of a function u over a set E, i.e.,

$$\fint_E u = \frac{1}{|E|}\int_E u.$$

Let Ω be a domain in $\mathbb{R}^d$, $d \geq 2$. We use $L^p(\Omega;\mathbb{R}^m)$ to denote the L^p space of $\mathbb{R}^m$-valued functions in Ω. Similarly, $L^p(\partial\Omega;\mathbb{R}^m)$ denotes the space of $\mathbb{R}^m$-valued and L^p integrable functions (with respect to the surface measure) on $\partial\Omega$. Also, $C_0^\infty(\Omega;\mathbb{R}^m)$ is the set of $\mathbb{R}^m$-valued and C^∞ functions with compact support in Ω. Thus, if $m = 1$, $L^p(\Omega) = L^p(\Omega;\mathbb{R})$, $L^p(\partial\Omega) = L^p(\partial\Omega;\mathbb{R})$, and $C_0^\infty(\Omega) = C_0^\infty(\Omega;\mathbb{R})$.

The summation convention over repeated indices are summed will be used throughout this monograph. For example, if $A(x/\varepsilon) = \big(a_{ij}^{\alpha\beta}(x/\varepsilon)\big)$ with $1 \leq \alpha, \beta \leq m$ and $1 \leq i, j \leq d$, $u = (u^1, u^2, \dots, u^m)$ and $v = (v^1, v^2, \dots, v^m)$, then

$$A(x/\varepsilon)\nabla u \cdot \nabla v = a_{ij}^{\alpha\beta}(x/\varepsilon)\frac{\partial u^\beta}{\partial x_j}\frac{\partial v^\alpha}{\partial x_i}, \tag{1.4.1}$$

where $\nabla u = (\partial u^\alpha/\partial x_i)$ denotes the gradient of u. In the right-hand side (RHS) of (1.4.1), the repeated indices α, β are summed from 1 to m, and i, j from 1 to d.

We end this section with a list of symbols used frequently in the monograph.

Ω – a domain in $\mathbb{R}^d$

$B(x,r) = \{y \in \mathbb{R}^d : |y - x| < r\}$ – ball of radius r centered at x

$tB = B(x,tr)$ if $B = B(x,r)$

$$D_r = D(r, \psi) = \{(x', x_d) \in \mathbb{R}^d : \ |x'| < r \text{ and } \psi(x') < x_d < 10(M+1)\},$$
$$\Delta_r = \Delta(r, \psi) = \{(x', x_d) \in \mathbb{R}^d : \ |x'| < r \text{ and } x_d = \psi(x')\},$$

where ψ is a Lipschitz (or C^1, $C^{1,\alpha}$) function on $\mathbb{R}^{d-1}$, such that $\psi(0) = 0$ and $\|\nabla\psi\|_\infty \le M$.

$$(u)^*(x) = \sup\{|u(y)| : \ y \in \Omega \text{ and } |y - x| < C_0 \operatorname{dist}(y, \partial\Omega)\}$$
– nontangential maximal function of u

$$\mathcal{M}(f)(x) = \sup\left\{ \fint_B |f| : \ B \text{ is a ball containing } x \right\}$$
– Hardy–Littlewood maximal function in $\mathbb{R}^d$

$$\mathcal{M}_{\partial\Omega}(f)(x) = \sup\left\{ \fint_{B(x,r)\cap\partial\Omega} |f| : \ 0 < r < \operatorname{diam}(\partial\Omega) \right\}$$
– Hardy–Littlewood maximal function on $\partial\Omega$

$$\mathcal{L}_\varepsilon = -\operatorname{div}\big(A(x/\varepsilon)\nabla\big)$$
$$\mathcal{L}_0 = -\operatorname{div}\big(\widehat{A}\nabla\big)$$

$$A(y) = \big(a_{ij}^{\alpha\beta}(y)\big), \text{ with } 1 \le \alpha, \beta \le m \text{ and } 1 \le i, j \le d$$
$$\widehat{A} = \big(\widehat{a}_{ij}^{\alpha\beta}(y)\big) - \text{ matrix of homogenized coefficients}$$

$\Gamma_\varepsilon(x, y)$ – matrix of fundamental solutions for $\mathcal{L}_\varepsilon$ in $\mathbb{R}^d$
$G_\varepsilon(x, y)$ – matrix of Green functions for $\mathcal{L}_\varepsilon$ in Ω
$N_\varepsilon(x, y)$ – matrix of Neumann functions for $\mathcal{L}_\varepsilon$ in Ω

n – outward unit normal to $\partial\Omega$
$\nabla_{\tan} u$ – tangential gradient of u on $\partial\Omega$

$\dfrac{\partial u_\varepsilon}{\partial \nu_\varepsilon} = n \cdot A(x/\varepsilon)\nabla u_\varepsilon$ – conormal derivative of u_ε associated with $\mathcal{L}_\varepsilon$
$\dfrac{\partial u_0}{\partial \nu_0} = n \cdot \widehat{A}\nabla u_0$ – conormal derivative of u_0 associated with $\mathcal{L}_0$

Chapter 2

Second-Order Elliptic Systems with Periodic Coefficients

In this monograph we shall be concerned with a family of second-order linear elliptic operators in divergence form with rapidly oscillating periodic coefficients,

$$\mathcal{L}_\varepsilon = -\text{div}\big(A\,(x/\varepsilon)\,\nabla\big), \quad \varepsilon > 0, \tag{2.0.1}$$

in $\mathbb{R}^d$. The coefficient matrix (tensor) A in (2.0.1) is given by

$$A(y) = \big(a_{ij}^{\alpha\beta}(y)\big), \text{ with } 1 \le i, j \le d \text{ and } 1 \le \alpha, \beta \le m.$$

Thus, if $u = (u^1, u^2, \ldots, u^m)$,

$$\big(\mathcal{L}_\varepsilon(u)\big)^\alpha = -\frac{\partial}{\partial x_i}\left[a_{ij}^{\alpha\beta}\left(\frac{x}{\varepsilon}\right)\frac{\partial u^\beta}{\partial x_j}\right] \quad \text{for } \alpha = 1, 2, \ldots, m$$

(the repeated indices are summed). We will always assume that A is real, bounded measurable, and satisfies a certain ellipticity condition, to be specified later. We also assume that A is 1-periodic; i.e., for each $z \in \mathbb{Z}^d$,

$$A(y + z) = A(y) \quad \text{for a.e. } y \in \mathbb{R}^d. \tag{2.0.2}$$

Observe that by a linear transformation one may replace $\mathbb{Z}^d$ in (2.0.2) by any lattice in $\mathbb{R}^d$.

In this chapter we present the qualitative homogenization theory for $\mathcal{L}_\varepsilon$. We start out in Section 2.1 by going over some basic results on weak solutions of second-order elliptic systems in divergence form. In Section 2.2 we use the method of (formal) asymptotic expansions to derive the formula for the homogenized (or effective) operator $\mathcal{L}_0$ with *constant* coefficients. In Section 2.3 we prove some

Z. Shen, *Periodic Homogenization of Elliptic Systems*, Operator Theory: Advances and Applications 269, https://doi.org/10.1007/978-3-319-91214-1_2

classical theorems on the homogenization of boundary value problems for second-order elliptic systems. In particular, we will show that if $u_\varepsilon \in H_0^1(\Omega;\mathbb{R}^m)$ and $\mathcal{L}_\varepsilon(u_\varepsilon) = F$ in a bounded Lipschitz domain Ω, where $F \in H^{-1}(\Omega;\mathbb{R}^m)$, then $u_\varepsilon \to u_0$ strongly in $L^2(\Omega;\mathbb{R}^m)$ and weakly in $H_0^1(\Omega;\mathbb{R}^m)$, as $\varepsilon \to 0$. Moreover, the function $u_0 \in H_0^1(\Omega;\mathbb{R}^m)$ is a solution of $\mathcal{L}_0(u_0) = F$ in Ω. Section 2.4 is devoted to the qualitative homogenization of elliptic systems of linear elasticity.

2.1 Weak solutions

In this section we review basic results on weak solutions of second-order elliptic systems with bounded measurable coefficients. For the sake of convenience this will be done for the case of the operator $\mathcal{L}_\varepsilon$. However, the periodicity condition (2.0.2) is not used in the section.

For a domain Ω in $\mathbb{R}^d$ and $1 \le p \le \infty$, define

$$W^{1,p}(\Omega;\mathbb{R}^m) = \Big\{ u \in L^p(\Omega;\mathbb{R}^m) : \ \nabla u \in L^p(\Omega;\mathbb{R}^{m\times d}) \Big\},$$

where $u = (u^1, u^2, \dots, u^m)$ and $\nabla u = (\partial u^\alpha / \partial x_i)$ with $1 \le \alpha \le m$ and $1 \le i \le d$. Equipped with the norm

$$\|u\|_{W^{1,p}(\Omega)} := \Big\{ \|\nabla u\|^p_{L^p(\Omega)} + \|u\|^p_{L^p(\Omega)} \Big\}^{1/p}$$

for $1 \le p < \infty$, and $\|u\|_{W^{1,\infty}(\Omega)} := \|u\|_{L^\infty(\Omega)} + \|\nabla u\|_{L^\infty(\Omega)}$, $W^{1,p}(\Omega;\mathbb{R}^m)$ is a Banach space. We say $u \in W^{1,p}_{\text{loc}}(\Omega;\mathbb{R}^m)$ if $\varphi u \in W^{1,p}(\Omega;\mathbb{R}^m)$ for any $\varphi \in C_0^\infty(\Omega)$.

For $1 \le p < \infty$, let $W_0^{1,p}(\Omega;\mathbb{R}^m)$ denote the closure of the subspace $C_0^\infty(\Omega;\mathbb{R}^m)$ in $W^{1,p}(\Omega;\mathbb{R}^m)$ and $W^{-1,p'}(\Omega;\mathbb{R}^m)$ the dual of $W_0^{1,p}(\Omega;\mathbb{R}^m)$, where $p' = \frac{p}{p-1}$. If $p = 2$, we often use the notation: $H^1(\Omega;\mathbb{R}^m) = W^{1,2}(\Omega;\mathbb{R}^m)$, $H_0^1(\Omega;\mathbb{R}^m) = W_0^{1,2}(\Omega;\mathbb{R}^m)$, and $H^{-1}(\Omega;\mathbb{R}^m) = W^{-1,2}(\Omega;\mathbb{R}^m)$. Similarly, for $k \ge 1$, $H^k(\Omega;\mathbb{R}^m) = W^{k,2}(\Omega;\mathbb{R}^m)$, where $W^{k,p}(\Omega;\mathbb{R}^m)$ denotes the subspace of functions in $L^p(\Omega;\mathbb{R}^m)$ such that for all α with $|\alpha| \le k$, $D^\alpha u \in L^p(\Omega;\mathbb{R}^m)$.

Definition 2.1.1. Let $\mathcal{L}_\varepsilon = -\text{div}(A(x/\varepsilon)\nabla)$ with $A(y) = \big(a_{ij}^{\alpha\beta}(y)\big)$. For $F \in H^{-1}(\Omega;\mathbb{R}^m)$, we call $u_\varepsilon \in H^1_{\text{loc}}(\Omega;\mathbb{R}^m)$ a weak solution of $\mathcal{L}_\varepsilon(u_\varepsilon) = F$ in Ω, if

$$\int_\Omega A(x/\varepsilon)\nabla u_\varepsilon \cdot \nabla \varphi \, dx = \langle F, \varphi \rangle_{H^{-1}(\Omega)\times H_0^1(\Omega)} \tag{2.1.1}$$

for any test function $\varphi \in C_0^\infty(\Omega;\mathbb{R}^m)$.

To establish the existence of weak solutions for the Dirichlet problem, we introduce the following boundedness and ellipticity conditions: there exists a constant $\mu > 0$ such that

$$\|A\|_\infty \le \mu^{-1}, \tag{2.1.2}$$

$$\mu \int_{\mathbb{R}^d} |\nabla u|^2\, dx \le \int_{\mathbb{R}^d} A\nabla u \cdot \nabla u\, dx \quad \text{for any } u \in C_0^\infty(\mathbb{R}^d; \mathbb{R}^m), \tag{2.1.3}$$

where $\|\cdot\|_\infty$ denotes the L^∞ norm in $\mathbb{R}^d$. Observe that the condition (2.1.3), which is referred as V-ellipticity (coercivity), is invariant under translation and dilation. In particular, if $A = A(x)$ satisfies (2.1.2)–(2.1.3), so does $A^\varepsilon = A(x/\varepsilon)$ with the same constant μ.

Lemma 2.1.2. *The integral condition* (2.1.3) *implies the following algebraic condition,*

$$\mu|\xi|^2|\eta|^2 \le a_{ij}^{\alpha\beta}(y)\xi_i\xi_j\eta^\alpha\eta^\beta \tag{2.1.4}$$

for a.e. $y \in \mathbb{R}^d$, where $\xi = (\xi_1, \ldots, \xi_d) \in \mathbb{R}^d$ and $\eta = (\eta^1, \ldots, \eta^m) \in \mathbb{R}^m$.

Proof. Since A is real, it follows from (2.1.3) that

$$\mu \int_{\mathbb{R}^d} |\nabla u|^2\, dx \le \operatorname{Re} \int_{\mathbb{R}^d} A\nabla u \cdot \nabla \overline{u}\, dx \quad \text{for any } u \in C_0^\infty(\mathbb{R}^d; \mathbb{C}^m). \tag{2.1.5}$$

Fix $y \in \mathbb{R}^d$, $\xi \in \mathbb{R}^d$ and $\eta \in \mathbb{R}^m$. Let

$$u(x) = \varphi_\varepsilon(x) t^{-1} e^{it\xi\cdot x}\eta$$

in (2.1.5), where $t > 0$, $\varphi_\varepsilon(x) = \varepsilon^{-d/2}\varphi(\varepsilon^{-1}(x-y))$ and φ is a function in $C_0^\infty(\mathbb{R}^d)$ with $\int_{\mathbb{R}^d} |\varphi|^2\, dx = 1$. Using

$$\frac{\partial u^\beta}{\partial x_j} = i\varphi_\varepsilon e^{it\xi\cdot x}\xi_j\eta^\beta + \frac{\partial \varphi_\varepsilon}{\partial x_j} t^{-1} e^{it\xi\cdot x}\eta^\beta,$$

we see that as $t \to \infty$,

$$\int_{\mathbb{R}^d} |\nabla u|^2\, dx = |\xi|^2|\eta|^2 \int_{\mathbb{R}^d} |\varphi_\varepsilon|^2\, dx + O(t^{-1}),$$

$$\operatorname{Re} \int_{\mathbb{R}^d} A\nabla u \cdot \nabla\overline{u}\, dx = \int_{\mathbb{R}^d} a_{ij}^{\alpha\beta}(x)\xi_i\xi_j\eta^\alpha\eta^\beta|\varphi_\varepsilon|^2\, dx + O(t^{-1}).$$

In view of (2.1.5) this implies that

$$\mu|\xi|^2|\eta|^2 \int_{\mathbb{R}^d} |\varphi_\varepsilon|^2\, dx \le \int_{\mathbb{R}^d} a_{ij}^{\alpha\beta}(x)\xi_i\xi_j\eta^\alpha\eta^\beta|\varphi_\varepsilon|^2\, dx. \tag{2.1.6}$$

Since $\int_{\mathbb{R}^d} |\varphi_\varepsilon|^2\, dx = 1$ and $|\varphi_\varepsilon(x)|^2 = \varepsilon^{-d}|\varphi(\varepsilon^{-1}(x-y))|^2$ is a mollifier, the inequality (2.1.4) follows by letting $\varepsilon \to 0$ in (2.1.6). □

The ellipticity condition (2.1.4) is called the Legendre–Hadamard condition. It follows from Lemma 2.1.2 that in the scalar case ($m = 1$), conditions (2.1.3) and (2.1.4) are equivalent. By using the Plancherel Theorem one can also show the equivalency when $m \ge 2$ and A is constant. We leave this as an exercise for the reader.

Theorem 2.1.3. *Suppose A satisfies (2.1.2)–(2.1.3). Let $u_\varepsilon \in H^1_{\text{loc}}(\Omega; \mathbb{R}^m)$ be a weak solution of $\mathcal{L}_\varepsilon(u_\varepsilon) = F + \text{div}(G)$ in Ω, where $F \in L^2(\Omega; \mathbb{R}^m)$ and $G = (G_i^\alpha) \in L^2(\Omega; \mathbb{R}^{m\times d})$. Then, for any $\psi \in C_0^1(\Omega)$,*

$$\int_\Omega |\nabla u_\varepsilon|^2 |\psi|^2 \, dx \le C \left\{ \int_\Omega |u_\varepsilon|^2 |\nabla \psi|^2 \, dx + \int_\Omega |G|^2 |\psi|^2 \, dx + \int_\Omega |F||u_\varepsilon||\psi|^2 \, dx \right\}, \tag{2.1.7}$$

where C depends only on μ.[1]

Proof. Note that, by (2.1.1),

$$\int_\Omega A(x/\varepsilon)\nabla u_\varepsilon \cdot \nabla \varphi \, dx = \int_\Omega F^\alpha \varphi^\alpha \, dx - \int_\Omega G_i^\alpha \frac{\partial \varphi^\alpha}{\partial x_i} \, dx \tag{2.1.8}$$

for any $\varphi = (\varphi^\alpha) \in C_0^\infty(\Omega; \mathbb{R}^m)$. Since $u_\varepsilon \in H^1_{\text{loc}}(\Omega; \mathbb{R}^m)$, by a density argument, (2.1.8) continues to hold for $\varphi = \psi^2 u_\varepsilon$, where $\psi \in C_0^\infty(\Omega)$. Observe that

$$\begin{aligned} &A(x/\varepsilon)\nabla(\psi u_\varepsilon) \cdot \nabla(\psi u_\varepsilon) \\ &\quad = A(x/\varepsilon)\nabla u_\varepsilon \cdot \nabla(\psi^2 u_\varepsilon) + A(x/\varepsilon)(\nabla\psi) u_\varepsilon \cdot \nabla(\psi u_\varepsilon) \\ &\quad\quad - A(x/\varepsilon)\nabla(\psi u_\varepsilon) \cdot (\nabla \psi) u_\varepsilon + A(x/\varepsilon)(\nabla\psi) u_\varepsilon \cdot (\nabla\psi) u_\varepsilon, \end{aligned} \tag{2.1.9}$$

where $\psi \in C_0^1(\Omega)$. It follows that

$$\begin{aligned} &\int_\Omega A(x/\varepsilon)\nabla(u_\varepsilon \psi) \cdot \nabla(u_\varepsilon \psi) \, dx \\ &\quad = \int_\Omega A(x/\varepsilon)\nabla u_\varepsilon \cdot \nabla(u_\varepsilon \psi^2) \, dx + \int_\Omega A(x/\varepsilon)(\nabla\psi)u_\varepsilon \cdot \nabla(\psi u_\varepsilon) \, dx \\ &\quad\quad - \int_\Omega A(x/\varepsilon)\nabla(\psi u_\varepsilon) \cdot (\nabla\psi) u_\varepsilon \, dx + \int_\Omega A(x/\varepsilon) u_\varepsilon \nabla\psi \cdot u_\varepsilon \nabla\psi \, dx \\ &\quad = \int_\Omega F^\alpha u_\varepsilon^\alpha \psi^2 \, dx - \int_\Omega G_i^\alpha \frac{\partial}{\partial x_i}(u_\varepsilon^\alpha \psi^2) \, dx + \int_\Omega A(x/\varepsilon)(\nabla\psi)u_\varepsilon \cdot \nabla(\psi u_\varepsilon) \, dx \\ &\quad\quad - \int_\Omega A(x/\varepsilon)\nabla(\psi u_\varepsilon) \cdot (\nabla\psi) u_\varepsilon \, dx + \int_\Omega A(x/\varepsilon) u_\varepsilon \nabla\psi \cdot u_\varepsilon \nabla\psi \, dx, \end{aligned}$$

[1] The constants C and c in this monograph may also depend on d and m. However, this fact is irrelevant to our investigation and will be ignored.

where we have used (2.1.8) with $\varphi = u_\varepsilon \psi^2$ for the last step. Hence, by (2.1.2)–(2.1.3),

$$\begin{aligned}
\mu \int_\Omega |\nabla(u_\varepsilon \psi)|^2\, dx &\le \int_\Omega A(x/\varepsilon)\nabla(u_\varepsilon\psi)\cdot\nabla(u_\varepsilon\psi)\, dx \\
&\le \int_\Omega |F||u_\varepsilon||\psi|^2\, dx + C\int_\Omega |G\psi||\nabla(u_\varepsilon\psi)|\, dx + C\int_\Omega |G\psi||u_\varepsilon\nabla\psi|\, dx \\
&\quad + C\int_\Omega |\nabla(\psi u_\varepsilon)||u_\varepsilon||\nabla\psi|\, dx + C\int_\Omega |u_\varepsilon|^2|\nabla\psi|^2\, dx.
\end{aligned}$$

By applying the inequality

$$ab \le \delta a^2 + \frac{b^2}{4\delta}, \tag{2.1.10}$$

where $a, b \ge 0$ and $\delta > 0$, we obtain

$$\begin{aligned}
\mu \int_\Omega |\nabla(u_\varepsilon\psi)|^2\, dx \le \int_\Omega |F||u_\varepsilon||\psi|^2\, dx + C\delta\int_\Omega |\nabla(u_\varepsilon\psi)|^2\, dx \\
+ C\delta^{-1}\int_\Omega |G|^2|\psi|^2\, dx + C\delta^{-1}\int_\Omega |u_\varepsilon|^2|\nabla\psi|^2\, dx
\end{aligned}$$

for any $\delta > 0$. The inequality (2.1.7) now follows by choosing δ so that $C\delta = (1/2)\mu$. □

For a ball

$$B = B(x_0, r) = \{x \in \mathbb{R}^d : |x - x_0| < r\}$$

in $\mathbb{R}^d$, we will use tB to denote $B(x_0, tr)$, the ball with the same center and t times the radius as B. Let $u_\varepsilon \in H^1(2B; \mathbb{R}^m)$ be a weak solution of $\mathcal{L}_\varepsilon(u_\varepsilon) = F + \text{div}(G)$ in $2B$, where $F \in L^2(2B; \mathbb{R}^m)$ and $G = (G_i^\alpha) \in L^2(2B; \mathbb{R}^{m\times d})$. Then

$$\int_{sB} |\nabla u_\varepsilon|^2\, dx \le C\left\{\frac{1}{(t-s)^2 r^2}\int_{tB} |u_\varepsilon - E|^2\, dx + r^2\int_{tB} |F|^2\, dx + \int_{tB} |G|^2\, dx\right\} \tag{2.1.11}$$

for any $1 \le s < t \le 2$ and $E \in \mathbb{R}^m$, where C depends only on μ. The inequality (2.1.11) is called the (interior) Caccioppoli inequality. To get (2.1.11), one applies Theorem 2.1.3 to $u_\varepsilon - E$ and chooses $\psi \in C_0^1(tB)$ so that $0 \le \psi \le 1$, $\psi = 1$ on sB, and $|\nabla\psi| \le C(t-s)^{-1}r^{-1}$.

Theorem 2.1.4. *Suppose A satisfies (2.1.2)–(2.1.3). Let $u_\varepsilon \in H^1(2B; \mathbb{R}^m)$ be a weak solution of $\mathcal{L}_\varepsilon(u_\varepsilon) = 0$ in $2B$, where $B = B(x_0, r)$ for some $x_0 \in \mathbb{R}^d$ and $r > 0$. Then there exists $p > 2$, depending only on μ (and d, m), such that*

$$\left(\fint_B |\nabla u_\varepsilon|^p\, dx\right)^{1/p} \le C\left(\fint_{2B} |\nabla u_\varepsilon|^2\, dx\right)^{1/2}, \tag{2.1.12}$$

where C depends only on μ.

Proof. Suppose $\mathcal{L}_\varepsilon(u_\varepsilon)=0$ in $2B$. Let $E=\fint_{tB'} u_\varepsilon\, dx$, where $B'\subset B$. It follows by the Sobolev–Poincaré inequality that

$$\left(\fint_{tB'} |u_\varepsilon - E|^2\, dx\right)^{1/2} \le Cr'\left(\fint_{tB'} |\nabla u_\varepsilon|^q\, dx\right)^{1/q},$$

where r' is the radius of B', $1\le t\le 2$, $\frac{1}{q}=\frac{1}{2}+\frac{1}{d}$ for $d\ge 3$, and $1<q<2$ for $d=2$. In view of (2.1.11) we obtain

$$\left(\fint_{sB'} |\nabla u_\varepsilon|^2\, dx\right)^{1/2} \le \frac{C}{t-s}\left(\fint_{tB'} |\nabla u_\varepsilon|^q\, dx\right)^{1/q}, \tag{2.1.13}$$

for $1\le s<t\le 2$. This gives a reverse Hölder inequality (with increasing support). The fact that (2.1.13) for any $B'\subset B$ implies (2.1.12) follows from the so-called self-improving property of reverse Hölder inequalities. We refer the reader to [50, pp. 132–135] for a proof of this property. □

Let Ω be a bounded Lipschitz domain in $\mathbb{R}^d$. Roughly speaking, this means that $\partial\Omega$ is locally given by the graph of a Lipschitz continuous function (see Section 5.1 for the precise definition). We are interested in the Dirichlet boundary value problem

$$\begin{cases} \mathcal{L}_\varepsilon(u_\varepsilon) = F + \operatorname{div}(G) & \text{in } \Omega,\\ u_\varepsilon = f & \text{on } \partial\Omega, \end{cases} \tag{2.1.14}$$

and the Neumann boundary value problem

$$\begin{cases} \mathcal{L}_\varepsilon(u_\varepsilon) = F + \operatorname{div}(G) & \text{in } \Omega,\\ \dfrac{\partial u_\varepsilon}{\partial \nu_\varepsilon} = g - n\cdot G & \text{on } \partial\Omega, \end{cases} \tag{2.1.15}$$

with non-homogeneous boundary conditions, where the conormal derivative $\frac{\partial u_\varepsilon}{\partial \nu_\varepsilon}$ on $\partial\Omega$ is defined by

$$\left(\frac{\partial u_\varepsilon}{\partial \nu_\varepsilon}\right)^\alpha = n_i(x) a_{ij}^{\alpha\beta}(x/\varepsilon)\frac{\partial u_\varepsilon^\beta}{\partial x_j}, \tag{2.1.16}$$

and $n=(n_1,\dots,n_d)$ denotes the outward unit normal to $\partial\Omega$.

The space $H^{1/2}(\partial\Omega)$ may be defined as the subspace of $L^2(\partial\Omega)$ of functions f for which

$$\|f\|_{H^{1/2}(\partial\Omega)} := \left\{\int_{\partial\Omega} |f|^2\, d\sigma + \int_{\partial\Omega}\int_{\partial\Omega} \frac{|f(x)-f(y)|^2}{|x-y|^d}\, d\sigma(x) d\sigma(y)\right\}^{1/2} < \infty.$$

Theorem 2.1.5. *Assume A satisfies (2.1.2)–(2.1.3). Let Ω be a bounded Lipschitz domain in $\mathbb{R}^d$. Then, for any $f\in H^{1/2}(\partial\Omega;\mathbb{R}^m)$, $F\in L^2(\Omega;\mathbb{R}^m)$ and $G\in L^2(\Omega;\mathbb{R}^{m\times d})$, there exists a unique $u_\varepsilon\in H^1(\Omega;\mathbb{R}^m)$ such that $\mathcal{L}_\varepsilon(u_\varepsilon)=F+\operatorname{div}(G)$*

in Ω and $u_\varepsilon = f$ on $\partial\Omega$ in the sense of trace. Moreover, the solution satisfies the energy estimate

$$\|u_\varepsilon\|_{H^1(\Omega)} \le C\Big\{\|F\|_{L^2(\Omega)} + \|G\|_{L^2(\Omega)} + \|f\|_{H^{1/2}(\partial\Omega)}\Big\}, \tag{2.1.17}$$

where C depends only on μ and Ω.

Proof. In the case where $f = 0$ on $\partial\Omega$, this follows by applying the Lax–Milgram Theorem to the bilinear form

$$B[u,v] = \int_\Omega A(x/\varepsilon)\nabla u \cdot \nabla v\, dx \tag{2.1.18}$$

on the Hilbert space $H_0^1(\Omega;\mathbb{R}^m)$. In general, if $f \in H^{1/2}(\partial\Omega;\mathbb{R}^m)$, then f is the trace of a function w in $H^1(\Omega;\mathbb{R}^m)$ with

$$\|w\|_{H^1(\Omega)} \le C\,\|f\|_{H^{1/2}(\partial\Omega)}.$$

By considering $u_\varepsilon - w$, one can reduce the general case to the case where $f = 0$. □

We now consider the Neumann problem. Let $H^{-1/2}(\partial\Omega;\mathbb{R}^m)$ denote the dual of $H^{1/2}(\partial\Omega;\mathbb{R}^m)$.

Definition 2.1.6. We call $u_\varepsilon \in H^1(\Omega;\mathbb{R}^m)$ a weak solution of the Neumann problem (2.1.15) with data $F \in L^2(\Omega;\mathbb{R}^m)$, $G \in L^2(\Omega;\mathbb{R}^{m\times d})$ and $g \in H^{-1/2}(\partial\Omega;\mathbb{R}^m)$, if

$$\begin{aligned}&\int_\Omega A(x/\varepsilon)\nabla u_\varepsilon \cdot \nabla\varphi\, dx\\ &\quad= \int_\Omega F\cdot\varphi\, dx - \int_\Omega G\cdot\nabla\varphi\, dx + \langle g,\varphi\rangle_{H^{-1/2}(\partial\Omega)\times H^{1/2}(\partial\Omega)}\end{aligned} \tag{2.1.19}$$

for any $\varphi \in C^\infty(\mathbb{R}^d;\mathbb{R}^m)$.

If $m \ge 2$, the ellipticity condition (2.1.3) is not sufficient for the solvability of the Neumann problem. For this reason, we introduce the very strong ellipticity condition, also called the Legendre condition: there exists a constant $\mu > 0$ such that

$$\|A\|_\infty \le \mu^{-1} \quad\text{and}\quad \mu|\xi|^2 \le a_{ij}^{\alpha\beta}(y)\xi_i^\alpha\xi_j^\beta \tag{2.1.20}$$

for a.e. $y \in \mathbb{R}^d$, where $\xi = (\xi_i^\alpha) \in \mathbb{R}^{m\times d}$. It is easy to see that

$$(2.1.20) \Longrightarrow (2.1.2)\text{–}(2.1.3).$$

Theorem 2.1.7. *Let Ω be a bounded Lipschitz domain in $\mathbb{R}^d$ and let A satisfy (2.1.20). Assume that $F \in L^2(\Omega;\mathbb{R}^m)$, $G \in L^2(\Omega;\mathbb{R}^{m\times d})$ and $g \in H^{-1/2}(\partial\Omega;\mathbb{R}^m)$ satisfy the compatibility condition*

$$\int_\Omega F\cdot b\, dx + \langle g, b\rangle_{H^{-1/2}(\partial\Omega)\times H^{1/2}(\partial\Omega)} = 0 \tag{2.1.21}$$

for any $b \in \mathbb{R}^m$. *Then the Neumann problem* (2.1.15) *has a weak solution* u_ε, *unique up to a constant in* $\mathbb{R}^m$, *in* $H^1(\Omega; \mathbb{R}^m)$. *Moreover, this solution satisfies the energy estimate*

$$\|\nabla u_\varepsilon\|_{L^2(\Omega)} \le C\Big\{\|F\|_{L^2(\Omega)} + \|G\|_{L^2(\Omega)} + \|g\|_{H^{-1/2}(\partial\Omega)}\Big\}, \tag{2.1.22}$$

where C *depends only on* μ *and* Ω.

Proof. Using (2.1.20), one obtains

$$\mu \int_\Omega |\nabla u|^2 \, dx \le \int_\Omega A(x/\varepsilon)\nabla u \cdot \nabla u \, dx$$

for any $u \in H^1(\Omega; \mathbb{R}^m)$. The assertions of the theorem follow from the Lax–Milgram Theorem by considering the bilinear form (2.1.18) on the Hilbert space $H^1(\Omega; \mathbb{R}^m)/\mathbb{R}^m$. □

2.2 Two-scale asymptotic expansions and the homogenized operator

Let $\mathcal{L}_\varepsilon = -\text{div}(A(x/\varepsilon)\nabla)$ with matrix $A = A(y)$ satisfying (2.1.2)–(2.1.3). Assume that A is 1-periodic. In this section we use the method of formal two-scale asymptotic expansions to derive the formula for the homogenized (effective) operator for $\mathcal{L}_\varepsilon$.

Suppose that $\mathcal{L}(u_\varepsilon) = F$ in Ω. Let

$$Y = [0,1)^d \cong \mathbb{R}^d/\mathbb{Z}^d \tag{2.2.1}$$

be the elementary cell for the lattice $\mathbb{Z}^d$. In view of the assumptions on the coefficients of $\mathcal{L}_\varepsilon$, one seeks a solution u_ε in the form

$$u_\varepsilon(x) = u_0(x, x/\varepsilon) + \varepsilon u_1(x, x/\varepsilon) + \varepsilon^2 u_2(x, x/\varepsilon) + \cdots, \tag{2.2.2}$$

where the functions $u_j(x, y)$ are defined on $\Omega \times \mathbb{R}^d$ and 1-periodic in y, for any fixed $x \in \Omega$.

Note that if $\phi_\varepsilon(x) = \phi(x, y)$ with $y = x/\varepsilon$, then

$$\frac{\partial \phi_\varepsilon}{\partial x_j} = \frac{1}{\varepsilon}\frac{\partial \phi}{\partial y_j} + \frac{\partial \phi}{\partial x_j}.$$

It follows that

$$\begin{aligned}\mathcal{L}_\varepsilon\big(u_j(x, x/\varepsilon)\big) = \varepsilon^{-2} L^0\big(u_j(x,y)\big)(x, x/\varepsilon) + \varepsilon^{-1} L^1\big(u_j(x,y)\big)(x, x/\varepsilon)\\ + L^2\big(u_j(x,y)\big)(x, x/\varepsilon),\end{aligned} \tag{2.2.3}$$

where the operators L^0, L^1 and L^2 are defined by

$$\begin{aligned} L^0(\phi(x,y)) &= -\frac{\partial}{\partial y_i}\left\{a_{ij}^{\alpha\beta}(y)\frac{\partial \phi^\beta}{\partial y_j}\right\}, \\ L^1(\phi(x,y)) &= -\frac{\partial}{\partial x_i}\left\{a_{ij}^{\alpha\beta}(y)\frac{\partial \phi^\beta}{\partial y_j}\right\} - \frac{\partial}{\partial y_i}\left\{a_{ij}^{\alpha\beta}(y)\frac{\partial \phi^\beta}{\partial x_j}\right\}, \\ L^2(\phi(x,y)) &= -\frac{\partial}{\partial x_i}\left\{a_{ij}^{\alpha\beta}(y)\frac{\partial \phi^\beta}{\partial x_j}\right\}. \end{aligned} \tag{2.2.4}$$

From (2.2.2) and (2.2.3) we obtain, at least formally,

$$\begin{aligned} \mathcal{L}_\varepsilon(u_\varepsilon) = \varepsilon^{-2}L^0(u_0) &+ \varepsilon^{-1}\{L^1(u_0)+L^0(u_1)\} \\ &+ \{L^2(u_0)+L^1(u_1)+L^0(u_2)\} + \cdots . \end{aligned} \tag{2.2.5}$$

Since $\mathcal{L}_\varepsilon(u_\varepsilon)=F$, by identifying the powers of ε, it follows from (2.2.5) that

$$L^0(u_0)=0, \tag{2.2.6}$$

$$L^1(u_0)+L^0(u_1)=0, \tag{2.2.7}$$

$$L^2(u_0)+L^1(u_1)+L^0(u_2)=F. \tag{2.2.8}$$

Using the fact that $u_0(x,y)$ is 1-periodic in y, we deduce from (2.2.6) that

$$\int_Y A(y)\nabla_y u_0\cdot\nabla_y u_0\,dy=0.$$

Under the ellipticity condition (2.1.3) and periodicity condition (2.0.2), we will show that

$$\mu\int_Y |\nabla_y\phi|^2\,dy \le \int_Y A(y)\nabla_y\phi\cdot\nabla_y\phi\,dy \tag{2.2.9}$$

for any $\phi\in H^1_{\text{per}}(Y;\mathbb{R}^m)$. See Lemma 2.2.3. It follows that $\nabla_y u_0=0$. Thus $u_0(x,y)$ is independent of y; i.e.,

$$u_0(x,y)=u_0(x). \tag{2.2.10}$$

Here and henceforth, $H^k_{\text{per}}(Y;\mathbb{R}^m)$ denotes the closure in $H^k(Y;\mathbb{R}^m)$ of $C^\infty_{\text{per}}(Y;\mathbb{R}^m)$, the set of C^∞ 1-periodic $\mathbb{R}^m$-valued functions in $\mathbb{R}^d$.

To derive the equation for u_0, we first use (2.2.10) and (2.2.7) to obtain

$$\left(L^0(u_1)\right)^\alpha = -\left(L^1(u_0)\right)^\alpha = \frac{\partial}{\partial y_i}\left\{a_{ij}^{\alpha\beta}(y)\right\}\frac{\partial u_0^\beta}{\partial x_j}. \tag{2.2.11}$$

By the Lax–Milgram Theorem and (2.2.9) one can show that if h is 1-periodic and $h\in L^2_{\text{loc}}(\mathbb{R}^d;\mathbb{R}^{m\times d})$, the cell problem

$$\begin{cases} L^0(\phi) = \operatorname{div}(h) & \text{in } Y, \\ \phi\in H^1_{\text{per}}(Y;\mathbb{R}^m), \end{cases} \tag{2.2.12}$$

has a unique (up to constants) solution. In view of (2.2.11) we may write

$$u_1^\alpha(x,y) = \chi_j^{\alpha\beta}(y)\frac{\partial u_0^\beta}{\partial x_j}(x) + \widetilde{u}_1^\alpha(x), \tag{2.2.13}$$

where, for each $1 \le j \le d$ and $1 \le \beta \le m$, the function $\chi_j^\beta = (\chi_j^{1\beta}, \ldots, \chi_j^{m\beta}) \in H^1_{\text{per}}(Y;\mathbb{R}^m)$ is the unique solution of the following cell problem:

$$\begin{cases} L^0(\chi_j^\beta) = -L^0(P_j^\beta) & \text{in } Y, \\ \chi_j^\beta(y) \text{ is 1-periodic}, \\ \displaystyle\int_Y \chi_j^\beta \, dy = 0. \end{cases} \tag{2.2.14}$$

In (2.2.14) and henceforth, $P_j^\beta = P_j^\beta(y) = y_j e^\beta$, where $e^\beta = (0,\ldots,1,\ldots,0)$ with 1 in the βth position. Note that the αth component of $-L^0(P_j^\beta)$ is given by $\frac{\partial}{\partial y_i}\big(a_{ij}^{\alpha\beta}(y)\big)$.

We now use the equations (2.2.8) and (2.2.13) to obtain

$$\begin{aligned} \big(L^0(u_2)\big)^\alpha &= F^\alpha - \big(L^2(u_0)\big)^\alpha - \big(L^1(u_1)\big)^\alpha \\ &= F^\alpha(x) + a_{ij}^{\alpha\beta}(y)\frac{\partial^2 u_0^\beta}{\partial x_i \partial x_j} + a_{ij}^{\alpha\beta}(y)\frac{\partial^2 u_1^\beta}{\partial x_i \partial y_j} + \frac{\partial}{\partial y_i}\left\{a_{ij}^{\alpha\beta}(y)\frac{\partial u_1^\beta}{\partial x_j}\right\} \\ &= F^\alpha(x) + a_{ij}^{\alpha\beta}(y)\frac{\partial^2 u_0^\beta}{\partial x_i \partial x_j} + a_{ij}^{\alpha\beta}(y)\frac{\partial \chi_k^{\beta\gamma}}{\partial y_j}\cdot\frac{\partial^2 u_0^\gamma}{\partial x_i \partial x_k} + \frac{\partial}{\partial y_i}\left\{a_{ij}^{\alpha\beta}(y)\frac{\partial u_1^\beta}{\partial x_j}\right\}. \end{aligned}$$

It follows by integration in y over Y that for $1 \le \alpha \le m$,

$$-\fint_Y \left[a_{ij}^{\alpha\beta}(y) + a_{ik}^{\alpha\gamma}(y)\frac{\partial \chi_j^{\gamma\beta}}{\partial y_k}\right] dy \cdot \frac{\partial^2 u_0^\beta}{\partial x_i \partial x_j}(x) = F^\alpha(x) \tag{2.2.15}$$

in Ω.

Definition 2.2.1. Let $\widehat{A} = (\widehat{a}_{ij}^{\alpha\beta})$, where $1 \le i,j \le d$, $1 \le \alpha,\beta \le m$, and

$$\widehat{a}_{ij}^{\alpha\beta} = \fint_Y \left[a_{ij}^{\alpha\beta} + a_{ik}^{\alpha\gamma}\frac{\partial}{\partial y_k}\left(\chi_j^{\gamma\beta}\right)\right] dy, \tag{2.2.16}$$

and define

$$\mathcal{L}_0 = -\text{div}(\widehat{A}\nabla). \tag{2.2.17}$$

In summary we have formally deduced that the leading term u_0 in the expansion (2.2.2) depends only on x and that u_0 is a solution of $\mathcal{L}_0(u_0) = F$ in Ω. As we shall prove in the next section, the constant coefficient operator $\mathcal{L}_0$ is indeed the homogenized operator for $\mathcal{L}_\varepsilon$.

Correctors and effective coefficients

The constant matrix $\widehat{A}$ is called the matrix of effective or homogenized coefficients. Because of (2.2.13), we call the 1-periodic matrix

$$\chi(y) = \big(\chi_j^\beta(y)\big) = \big(\chi_j^{\alpha\beta}(y)\big),$$

with $1 \le j \le d$ and $1 \le \alpha, \beta \le m$, the matrix of (first-order) correctors for $\mathcal{L}_\varepsilon$. Define

$$a_{\text{per}}(\phi, \psi) = \fint_Y a_{ij}^{\alpha\beta}(y) \frac{\partial \phi^\beta}{\partial y_j} \cdot \frac{\partial \psi^\alpha}{\partial y_i}\, dy \tag{2.2.18}$$

for $\phi = (\phi^\alpha)$ and $\psi = (\psi^\alpha)$. In view of (2.2.14), the corrector χ_j^β is the unique function in $H^1_{\text{per}}(Y; \mathbb{R}^m)$ such that $\int_Y \chi_j^\beta\, dy = 0$ and

$$a_{\text{per}}\big(\chi_j^\beta, \psi\big) = -a_{\text{per}}\big(P_j^\beta, \psi\big) \qquad \text{for any } \psi \in H^1_{\text{per}}(Y; \mathbb{R}^m). \tag{2.2.19}$$

It follows from (2.2.9) and (2.2.19) with $\psi = \chi_j^\beta$ that

$$\|\chi_j^\beta\|_{H^1(Y)} \le C, \tag{2.2.20}$$

where C depends only on μ.

With the summation convention the first equation in (2.2.14) may be written as

$$\frac{\partial}{\partial y_i}\left[a_{ij}^{\alpha\beta} + a_{ik}^{\alpha\gamma} \frac{\partial}{\partial y_k}\left(\chi_j^{\gamma\beta}\right)\right] = 0 \quad \text{in } \mathbb{R}^d; \tag{2.2.21}$$

i.e., $\mathcal{L}_1(\chi_j^\beta + P_j^\beta) = 0$ in $\mathbb{R}^d$. It follows from the reverse Hölder estimate (2.1.12) and (2.2.20) that

$$\|\nabla \chi_j^\beta\|_{L^p(Y)} \le C_0 \qquad \text{for some } p > 2, \tag{2.2.22}$$

where p and C_0 depend only on μ. By the Sobolev embedding, this implies that χ_j^β are Hölder continuous if $d = 2$. By the classical De Giorgi–Nash theorem, χ_j^β is also Hölder continuous if $m = 1$ and $d \ge 3$. If $m \ge 2$ and $d \ge 3$, we can use the Sobolev embedding and (2.2.22) to obtain

$$\|\chi_j^\beta\|_{L^q(Y)} \le C \qquad \text{for some } q > \frac{2d}{d-2}. \tag{2.2.23}$$

We further note that by rescaling,

$$\mathcal{L}_\varepsilon\left\{P_j^\beta(x) + \varepsilon \chi_j^\beta(x/\varepsilon)\right\} = 0 \quad \text{in } \mathbb{R}^d \tag{2.2.24}$$

for any $\varepsilon > 0$.

We now proceed to prove the inequality (2.2.9), on which the existence of correctors (χ_j^β) depends. The proof uses the property of weak convergence for periodic functions.

Proposition 2.2.2. *Let $\{h_\ell\}$ be a sequence of 1-periodic functions. Assume that*

$$\|h_\ell\|_{L^2(Y)} \le C \quad \text{and} \quad \fint_Y h_\ell(y)dy \to c_0 \qquad \text{as } \ell \to \infty.$$

Let $\varepsilon_\ell \to 0$. Then $h_\ell(x/\varepsilon_\ell) \rightharpoonup c_0$ weakly in $L^2(\Omega)$ as $\ell \to \infty$, where Ω is a bounded domain in $\mathbb{R}^d$. In particular, if h is 1-periodic and $h \in L^2(Y)$, then

$$h(x/\varepsilon) \rightharpoonup \fint_Y h \quad \text{weakly in } L^2(\Omega), \text{ as } \varepsilon \to 0.$$

Proof. By considering the periodic function $h_\ell - \fint_Y h_\ell$, one may assume that $\int_Y h_\ell = 0$ and hence $c_0 = 0$. Let $u_\ell \in H^2_{\text{per}}(Y)$ be a 1-periodic function such that $\Delta u_\ell = h_\ell$ in Y. Let $g_\ell = \nabla u_\ell$. Then $h_\ell = \text{div}(g_\ell)$ and

$$\|g_\ell\|_{L^2(Y)} \le C\|h_\ell\|_{L^2(Y)} \le C.$$

Note that

$$h_\ell(x/\varepsilon_\ell) = \varepsilon_\ell \,\text{div}\big\{g_\ell(x/\varepsilon_\ell)\big\}.$$

It follows that, if $\varphi \in C_0^1(\Omega)$,

$$\int_\Omega h_\ell(x/\varepsilon_\ell)\varphi(x)\, dx = -\varepsilon_\ell \int_\Omega g_\ell(x/\varepsilon_\ell) \cdot \nabla\varphi(x)\, dx \to 0 \tag{2.2.25}$$

as $\varepsilon_\ell \to 0$. This is because, if $\Omega \subset B(0, R)$,

$$\begin{aligned}\int_\Omega |g_\ell(x/\varepsilon_\ell)|^2\, dx &\le \varepsilon_\ell^d \int_{B(0,R/\varepsilon_\ell)} |g_\ell(y)|^2\, dy \\ &\le C\, \|g_\ell\|^2_{L^2(Y)} \\ &\le C,\end{aligned}$$

where we have used the periodicity of g_ℓ for the second inequality, and where C depends on R. Similarly,

$$\|h_\ell(x/\varepsilon_\ell)\|_{L^2(\Omega)} \le C\, \|h_\ell\|_{L^2(Y)} \le C; \tag{2.2.26}$$

i.e., the sequence $\{h_\ell(x/\varepsilon_\ell)\}$ is bounded in $L^2(\Omega)$. In view of (2.2.25) we conclude that $h_\ell(x/\varepsilon_\ell) \rightharpoonup 0$ weakly in $L^2(\Omega)$. $\square$

Lemma 2.2.3. *Suppose that $A = A(y)$ is 1-periodic and satisfies (2.1.2)–(2.1.3). Then the inequality (2.2.9) holds for any $\phi \in H^1_{\text{per}}(Y; \mathbb{R}^m)$.*

Proof. Let $u_\varepsilon(x) = \varepsilon\eta(x)\phi(x/\varepsilon)$, where $\eta \in C_0^\infty(\mathbb{R}^d)$ with $\int_{\mathbb{R}^d} \eta^2\, dx = 1$ and ϕ is a 1-periodic function in $C^\infty(\mathbb{R}^d; \mathbb{R}^m)$. Since $A(x/\varepsilon)$ satisfies the condition (2.1.3), it follows that

$$\mu \int_{\mathbb{R}^d} |\nabla u_\varepsilon|^2\, dx \le \int_{\mathbb{R}^d} A(x/\varepsilon)\nabla u_\varepsilon \cdot \nabla u_\varepsilon\, dx. \tag{2.2.27}$$

We now take limits by letting $\varepsilon \to 0$ on both sides of (2.2.27). Using

$$\nabla u_\varepsilon(x) = \eta(x)\nabla\phi(x/\varepsilon) + \varepsilon\nabla\eta(x)\cdot\phi(x/\varepsilon)$$

and Proposition 2.2.2, we see that as $\varepsilon \to 0$,

$$\begin{aligned} \int_{\mathbb{R}^d} A(x/\varepsilon)\nabla u_\varepsilon \cdot \nabla u_\varepsilon\, dx &\to \fint_Y A\nabla\phi\cdot\nabla\phi\, dy \int_{\mathbb{R}^d} \eta^2\, dx, \\ \int_{\mathbb{R}^d} |\nabla u_\varepsilon|^2\, dx &\to \fint_Y |\nabla\phi|^2\, dy \int_{\mathbb{R}^d} \eta^2\, dx. \end{aligned}$$

This, together with (2.2.27), yields (2.2.9). □

The following lemma gives the ellipticity for the homogenized operator $\mathcal{L}_0$.

Lemma 2.2.4. *Suppose that $A = A(y)$ is 1-periodic and satisfies (2.1.2)–(2.1.3). Then*

$$\mu|\xi|^2|\eta|^2 \le \widehat{a}_{ij}^{\alpha\beta}\xi_i\xi_j\eta^\alpha\eta^\beta \quad \text{and} \quad |a_{ij}^{\alpha\beta}| \le \mu_1 \tag{2.2.28}$$

for any $\xi = (\xi_1, \dots, \xi_d) \in \mathbb{R}^d$ and $\eta = (\eta^1, \dots, \eta^m) \in \mathbb{R}^m$, where μ_1 depends only on μ (and d, m).

Proof. The second inequality in (2.2.28) follows readily from the estimate (2.2.20). To prove the first inequality, we will show that

$$\mu \int_{\mathbb{R}^d} |\nabla\phi|^2\, dx \le \int_{\mathbb{R}^d} \widehat{A}\nabla\phi\cdot\nabla\phi\, dx \quad \text{for any } \phi \in C_0^\infty(\mathbb{R}^d; \mathbb{R}^m). \tag{2.2.29}$$

As we pointed out earlier, since $\widehat{A}$ is constant, this is equivalent to the first inequality in (2.2.28).

To establish (2.2.29), we fix $\phi = (\phi^\alpha) \in C_0^\infty(\mathbb{R}^d; \mathbb{R}^m)$. Let

$$u_\varepsilon = \phi + \varepsilon\chi_j^\beta(x/\varepsilon)\frac{\partial\phi^\beta}{\partial x_j}$$

in (2.2.27) and then take the limits as $\varepsilon \to 0$. Using

$$\begin{aligned} \nabla u_\varepsilon &= \nabla\phi + \nabla\chi_j^\beta(x/\varepsilon)\frac{\partial\phi^\beta}{\partial x_j} + \varepsilon\chi_j^\beta(x/\varepsilon)\frac{\partial}{\partial x_j}\nabla\phi^\beta \\ &= \nabla(P_j^\beta + \chi_j^\beta)(x/\varepsilon)\frac{\partial\phi^\beta}{\partial x_j} + \varepsilon\chi_j^\beta(x/\varepsilon)\frac{\partial}{\partial x_j}\nabla\phi^\beta \end{aligned}$$

and Proposition 2.2.2, we see that as $\varepsilon \to 0$,

$$\begin{aligned}
&\int_{\mathbb{R}^d} A(x/\varepsilon)\nabla u_\varepsilon \cdot \nabla u_\varepsilon \, dx \\
&\qquad \to \fint_Y A\nabla\big(P_j^\beta + \chi_j^\beta\big) \cdot \nabla\big(P_i^\alpha + \chi_i^\alpha\big)\, dy \int_{\mathbb{R}^d} \frac{\partial \phi^\beta}{\partial x_j} \cdot \frac{\partial \phi^\alpha}{\partial x_i}\, dx \\
&\quad = \int_{\mathbb{R}^d} \widehat{A}\nabla\phi \cdot \nabla\phi \, dx.
\end{aligned}$$

Observe that $\nabla u_\varepsilon \rightharpoonup \nabla\phi$ weakly in $L^2(\mathbb{R}^d; \mathbb{R}^{m\times d})$. Since the norm in L^2 is lower semi-continuous with respect to the weak convergence, it follows that

$$\begin{aligned}
\mu \int_{\mathbb{R}^d} |\nabla\phi|^2 \, dx &\le \liminf_{\varepsilon\to 0} \mu \int_{\mathbb{R}^d} |\nabla u_\varepsilon|^2 \, dx \\
&\le \lim_{\varepsilon\to 0} \int_{\mathbb{R}^d} A(x/\varepsilon)\nabla u_\varepsilon \cdot \nabla u_\varepsilon \, dx \\
&= \int_{\mathbb{R}^d} \widehat{A}\nabla\phi \cdot \nabla\phi \, dx.
\end{aligned}$$

This completes the proof. □

We end this section with a useful observation on the homogenized matrix for the adjoint operator

$$\mathcal{L}_\varepsilon^* = -\text{div}\big(A^*(x/\varepsilon)\nabla\big).$$

Lemma 2.2.5. *Let $A^* = \big(a_{ij}^{*\alpha\beta}\big)$ denote the adjoint of A, where $a_{ij}^{*\alpha\beta} = a_{ji}^{\beta\alpha}$. Then $\widehat{A^*} = \big(\widehat{A}\big)^*$. In particular, if $A(y)$ is symmetric, i.e., $a_{ij}^{\alpha\beta}(y) = a_{ji}^{\beta\alpha}(y)$ for $1 \le i,j \le d$ and $1 \le \alpha,\beta \le m$, then so is $\widehat{A}$.*

Proof. Let $\chi^*(y) = \big(\chi_j^{*\beta}(y)\big) = \big(\chi_j^{*\alpha\beta}(y)\big)$ denote the matrix of correctors for $\mathcal{L}_\varepsilon^*$, i.e., $\chi_j^{*\beta}$ is the unique function in $H^1_{\text{per}}(Y;\mathbb{R}^m)$ such that $\int_Y \chi_j^{*\beta} = 0$ and

$$a_{\text{per}}^*(\chi_j^{*\beta}, \psi) = -a_{\text{per}}^*(P_j^\beta, \psi) \qquad \text{for any } \psi \in H^1_{\text{per}}(Y;\mathbb{R}^m), \tag{2.2.30}$$

where $a_{\text{per}}^*(\phi,\psi) = a_{\text{per}}(\psi,\phi)$. Observe that, by (2.2.19) and (2.2.30),

$$\begin{aligned}
\widehat{a}_{ij}^{\alpha\beta} &= a_{\text{per}}\big(P_j^\beta + \chi_j^\beta, P_i^\alpha\big) = a_{\text{per}}\big(P_j^\beta + \chi_j^\beta, P_i^\alpha + \chi_i^{*\alpha}\big) \\
&= a_{\text{per}}^*\big(P_i^\alpha + \chi_i^{*\alpha}, P_j^\beta + \chi_j^\beta\big) = a_{\text{per}}^*\big(P_i^\alpha + \chi_i^{*\alpha}, P_j^\beta\big) \\
&= a_{\text{per}}^*\big(P_i^\alpha + \chi_i^{*\alpha}, P_j^\beta + \chi_j^{*\beta}\big) \\
&= \widehat{a}_{ji}^{*\beta\alpha},
\end{aligned} \tag{2.2.31}$$

for $1 \le \alpha,\beta \le m$ and $1 \le i,j \le d$. This shows that $\big(\widehat{A}\big)^* = \widehat{A^*}$. □

2.3 Homogenization of elliptic systems

We start with a Div-Curl Lemma. Recall that for $u = (u^1, u^2, \dots, u^d)$,

$$\operatorname{div}(u) = \sum_{i=1}^d \frac{\partial u^i}{\partial x_i} \quad \text{and} \quad \operatorname{curl}(u) = \left(\frac{\partial u^i}{\partial x_j} - \frac{\partial u^j}{\partial x_i}\right)_{1\le i,j\le d}.$$

Theorem 2.3.1. *Let $\{u_\ell\}$ and $\{v_\ell\}$ be two bounded sequences in $L^2(\Omega;\mathbb{R}^d)$. Suppose that*

1. *$u_\ell \rightharpoonup u$ and $v_\ell \rightharpoonup v$ weakly in $L^2(\Omega;\mathbb{R}^d)$;*
2. *$\operatorname{curl}(u_\ell) = 0$ in Ω and $\operatorname{div}(v_\ell) \to f$ strongly in $H^{-1}(\Omega)$.*

Then

$$\int_\Omega (u_\ell \cdot v_\ell)\,\varphi\,dx \to \int_\Omega (u\cdot v)\,\varphi\,dx,$$

as $\ell\to\infty$, for any scalar function $\varphi \in C_0^1(\Omega)$.

Proof. By considering

$$u_\ell \cdot v_\ell = (u_\ell - u)\cdot(v_\ell - v) - u\cdot v + u_\ell \cdot v + u\cdot v_\ell,$$

we may assume that $u_\ell \rightharpoonup 0$, $v_\ell \rightharpoonup 0$ weakly in $L^2(\Omega;\mathbb{R}^d)$ and that $\operatorname{div}(v_\ell)\to 0$ strongly in $H^{-1}(\Omega)$. Using a partition of unity we may also assume that $\varphi \in C_0^1(B)$ for some ball $B\subset\Omega$.

Since $\operatorname{curl}(u_\ell) = 0$ in Ω, there exists $U_\ell \in H^1(B)$ such that $u_\ell = \nabla U_\ell$ in B and $\int_B U_\ell\,dx = 0$. It follows that

$$\begin{aligned}\int_B (u_\ell\cdot v_\ell)\varphi\,dx &= \int_B (\nabla U_\ell \cdot v_\ell)\varphi\,dx\\ &= -\langle \operatorname{div}(v_\ell), U_\ell\varphi\rangle_{H^{-1}(B)\times H_0^1(B)} - \int_B U_\ell (v_\ell\cdot\nabla\varphi)\,dx.\end{aligned}$$

Hence,

$$\begin{aligned}\left|\int_B (u_\ell\cdot v_\ell)\varphi\,dx\right| &\le \|\operatorname{div}(v_\ell)\|_{H^{-1}(B)}\|U_\ell\varphi\|_{H_0^1(B)}\\ &\quad + \|U_\ell\|_{L^2(B)}\|v_\ell\cdot\nabla\varphi\|_{L^2(B)}.\end{aligned}\tag{2.3.1}$$

We will show that both terms in the RHS of (2.3.1) converge to zero.

By the Poincaré inequality,

$$\|U_\ell\|_{L^2(B)} \le C\,\|u_\ell\|_{L^2(B)} \le C.$$

Thus,

$$\|\operatorname{div}(v_\ell)\|_{H^{-1}(B)}\|U_\ell\varphi\|_{H_0^1(B)} \to 0.$$

Since $\|U_\ell\|_{L^2(B)} \le C$, $\nabla U_\ell = u_\ell \rightharpoonup 0$ weakly in $L^2(B; \mathbb{R}^d)$, and $\int_B U_\ell = 0$, we see that if $\{U_{\ell_k}\}$ is a subsequence of $\{U_\ell\}$ and converges weakly in $L^2(B)$, then it must converge weakly to zero. This implies that the full sequence $U_\ell \rightharpoonup 0$ weakly in $L^2(B)$. It follows that $U_\ell \rightharpoonup 0$ weakly in $H^1(B)$ and therefore $U_\ell \to 0$ strongly in $L^2(B)$. Consequently,

$$\|U_\ell\|_{L^2(B)} \|v_\ell \cdot \nabla \varphi\|_{L^2(B)} \le C\, \|U_\ell\|_{L^2(B)} \to 0,$$

as $\ell \to \infty$. This completes the proof. □

The next theorem shows that the sequence of operators $\{\mathcal{L}^\ell_{\varepsilon_\ell}\}$ is H-compact in the sense of H-convergence [100].

Theorem 2.3.2. *Let $\{A_\ell(y)\}$ be a sequence of 1-periodic matrices satisfying* (2.1.2)–(2.1.3) *with the same constant μ. Let $F_\ell \in H^{-1}(\Omega; \mathbb{R}^m)$. Suppose that*

$$\mathcal{L}^\ell_{\varepsilon_\ell}(u_\ell) = F_\ell \quad \text{in } \Omega, \tag{2.3.2}$$

where $\varepsilon_\ell \to 0$, $u_\ell \in H^1(\Omega; \mathbb{R}^m)$, and

$$\mathcal{L}^\ell_{\varepsilon_\ell} = -\text{div}\big(A_\ell(x/\varepsilon_\ell)\nabla\big).$$

We further assume that

$$\begin{cases} F_\ell \to F \text{ in } H^{-1}(\Omega; \mathbb{R}^m), \\ u_\ell \rightharpoonup u \text{ weakly in } H^1(\Omega; \mathbb{R}^m), \\ \widehat{A_\ell} \to A^0, \end{cases} \tag{2.3.3}$$

where $\widehat{A_\ell}$ denotes the matrix of effective coefficients for A_ℓ. Then

$$A_\ell(x/\varepsilon_\ell)\nabla u_\ell \rightharpoonup A^0 \nabla u \quad \text{weakly in } L^2(\Omega; \mathbb{R}^{m\times d}), \tag{2.3.4}$$

A^0 is a constant matrix satisfying the ellipticity condition (2.2.28), *and u is a weak solution of*

$$-\text{div}(A^0 \nabla u) = F \quad \text{in } \Omega. \tag{2.3.5}$$

Proof. We first note that since $\widehat{A_\ell} \to A^0$ and $\widehat{A_\ell}$ satisfies (2.2.28), so does A^0. Also, (2.3.5) follows directly from (2.3.2) and (2.3.4). To see (2.3.4), we let $\{u_{\ell'}\}$ be a subsequence such that

$$A_{\ell'}(x/\varepsilon_{\ell'})\nabla u_{\ell'} \rightharpoonup H \quad \text{weakly in } L^2(\Omega; \mathbb{R}^{m\times d})$$

for some $H \in L^2(\Omega; \mathbb{R}^{m\times d})$ and show that $H = A^0 \nabla u$. This will imply that the full sequence $A_\ell(x/\varepsilon_\ell)\nabla u_\ell$ converges weakly to $A^0 \nabla u$ in $L^2(\Omega; \mathbb{R}^{m\times d})$.

Without loss of generality we assume that

$$A_\ell(x/\varepsilon_\ell)\nabla u_\ell \rightharpoonup H \quad \text{weakly in } L^2(\Omega; \mathbb{R}^{m\times d}) \tag{2.3.6}$$

for some $H = (H_i^\alpha) \in L^2(\Omega; \mathbb{R}^{m\times d})$. Let $\chi_\ell^*(y) = \big(\chi_{k,\ell}^{*\beta}(y)\big)$ denote the correctors associated with the matrix A_ℓ^*, the adjoint of A_ℓ. Fix $1 \le k \le d$, $1 \le \gamma \le m$ and consider the identity

$$\begin{aligned}\int_\Omega A_\ell(x/\varepsilon_\ell)\nabla u_\ell \cdot \nabla\Big(P_k^\gamma + \varepsilon_\ell \chi_{k,\ell}^{*\gamma}(x/\varepsilon_\ell)\Big)\cdot \psi\, dx \\ = \int_\Omega \nabla u_\ell \cdot A_\ell^*(x/\varepsilon_\ell)\nabla\Big(P_k^\gamma + \varepsilon_\ell \chi_{k,\ell}^{*\gamma}(x/\varepsilon_\ell)\Big)\cdot \psi\, dx,\end{aligned} \tag{2.3.7}$$

where $\psi \in C_0^1(\Omega)$. By Proposition 2.2.2,

$$\begin{aligned}\nabla\Big(P_k^\gamma + \varepsilon_\ell \chi_{k,\ell}^{*\gamma}(x/\varepsilon_\ell)\Big) &= \nabla P_k^\gamma + \nabla \chi_{k,\ell}^{*\gamma}(x/\varepsilon_\ell)\\ &\rightharpoonup \nabla P_k^\gamma\end{aligned} \tag{2.3.8}$$

weakly in $L^2(\Omega)$, where we have used the fact that $\fint_Y \nabla \chi_{k,\ell}^{*\gamma}\, dy = 0$.

Since $\mathcal{L}_{\varepsilon_\ell}^\ell(u_{\varepsilon_\ell}) = F_\ell$ in Ω, in view of (2.3.6) and (2.3.8), it follows by Theorem 2.3.1 that the LHS of (2.3.7) converges to

$$\int_\Omega (H\cdot \nabla P_k^\gamma)\,\psi\, dx = \int_\Omega H_k^\gamma \psi\, dx.$$

Similarly, note that $\nabla u_\ell \rightharpoonup \nabla u$ and

$$\begin{aligned}A_\ell^*(x/\varepsilon_\ell)\nabla\Big(P_k^\gamma + \varepsilon_\ell\chi_{k,\ell}^{*\gamma}(x/\varepsilon_\ell)\Big) &\rightharpoonup \lim_{\ell\to\infty} \fint_Y A_\ell^*\Big(\nabla P_k^\gamma + \nabla\chi_{k,\ell}^{*\gamma}\Big)\, dy\\ &= \lim_{\ell\to\infty} \widehat{A_\ell^*}\,\nabla P_k^\gamma\\ &= (A^0)^*\,\nabla P_k^\gamma\end{aligned}$$

weakly in $L^2(\Omega)$, where we have used Proposition 2.2.2 as well as Lemma 2.2.5. Since

$$\mathcal{L}_{\varepsilon_\ell}^{\ell *}\{P_k^\gamma + \varepsilon_\ell \chi_k^{*\gamma}(x/\varepsilon_\ell)\} = 0 \quad \text{in } \mathbb{R}^d,$$

we may use Theorem 2.3.1 again to conclude that the RHS of (2.3.7) converges to

$$\int_\Omega \big(\nabla u \cdot (A^0)^*\nabla P_k^\gamma\big)\,\psi\, dx.$$

As a result, since $\psi \in C_0^1(\Omega)$ is arbitrary, it follows that

$$H_k^\gamma = \nabla u\cdot (A^0)^*\nabla P_k^\gamma = A^0\nabla u\cdot \nabla P_k^\gamma \quad \text{in } \Omega. \tag{2.3.9}$$

This shows that $H = A^0\nabla u$ and completes the proof. □

We now use Theorem 2.3.2 to establish the qualitative homogenization of the Dirichlet and Neumann problems for $\mathcal{L}_\varepsilon$. The proof only uses a special case of

Theorem 2.3.2, where $A_\ell = A$ is fixed. The general case is essential in a compactness argument we will use in Chapters 4 and 5 for regularity estimates that are uniform in $\varepsilon > 0$.

Homogenization of the Dirichlet Problem (2.1.14).

Assume that A satisfies the boundedness and ellipticity conditions (2.1.2)–(2.1.3) and is 1-periodic. Let $F \in L^2(\Omega; \mathbb{R}^m)$, $G \in L^2(\Omega; \mathbb{R}^{m\times d})$ and $f \in H^{1/2}(\partial\Omega; \mathbb{R}^m)$. By Theorem 2.1.5, there exists a unique $u_\varepsilon \in H^1(\Omega; \mathbb{R}^m)$ such that

$$\mathcal{L}_\varepsilon(u_\varepsilon) = F + \operatorname{div}(G) \quad \text{in } \Omega \quad \text{and} \quad u_\varepsilon = f \quad \text{on } \partial\Omega$$

(the boundary data are taken in the sense of trace). Furthermore, the solution u_ε satisfies

$$\|u_\varepsilon\|_{H^1(\Omega)} \le C\Big\{\|F\|_{L^2(\Omega)} + \|G\|_{L^2(\Omega)} + \|f\|_{H^{1/2}(\partial\Omega)}\Big\},$$

where C depends only on μ and Ω.

Let $\{u_{\varepsilon'}\}$ be a subsequence of $\{u_\varepsilon\}$ such that as $\varepsilon' \to 0$,

$$u_{\varepsilon'} \rightharpoonup u \text{ weakly in } H^1(\Omega; \mathbb{R}^m)$$

for some $u \in H^1(\Omega; \mathbb{R}^m)$. It readily follows from Theorem 2.3.2 that

$$A(x/\varepsilon')\nabla u_{\varepsilon'} \rightharpoonup \widehat{A}\nabla u$$

and $\mathcal{L}_0(u) = F + \operatorname{div}(G)$ in Ω.

Since $f \in H^{1/2}(\partial\Omega; \mathbb{R}^m)$, there exists $\Phi \in H^1(\Omega; \mathbb{R}^m)$ such that $\Phi = f$ on $\partial\Omega$. Since

$$u_{\varepsilon'} - \Phi \rightharpoonup u - \Phi \quad \text{weakly in } H^1(\Omega; \mathbb{R}^m)$$

and $u_{\varepsilon'} - \Phi \in H_0^1(\Omega; \mathbb{R}^m)$, we see that $u - \Phi \in H_0^1(\Omega; \mathbb{R}^m)$. Hence, $u = f$ on $\partial\Omega$. Consequently, u is the unique weak solution to the Dirichlet problem,

$$\mathcal{L}_0(u_0) = F + \operatorname{div}(G) \quad \text{in } \Omega \quad \text{and} \quad u_0 = f \quad \text{on } \partial\Omega.$$

Since $\{u_\varepsilon\}$ is bounded in $H^1(\Omega; \mathbb{R}^m)$ and thus any sequence $\{u_{\varepsilon_\ell}\}$ with $\varepsilon_\ell \to 0$ contains a subsequence that converges weakly in $H^1(\Omega; \mathbb{R}^m)$, we conclude that as $\varepsilon \to 0$,

$$\begin{cases} A(x/\varepsilon)\nabla u_\varepsilon \rightharpoonup \widehat{A}\nabla u_0 & \text{weakly in } L^2(\Omega; \mathbb{R}^{m\times d}), \\ u_\varepsilon \rightharpoonup u_0 & \text{weakly in } H^1(\Omega; \mathbb{R}^m). \end{cases} \tag{2.3.10}$$

By the compactness of the embedding $H^1(\Omega; \mathbb{R}^m) \hookrightarrow L^2(\Omega; \mathbb{R}^m)$, we also obtain

$$u_\varepsilon \to u_0 \quad \text{strongly in } L^2(\Omega; \mathbb{R}^m). \tag{2.3.11}$$

Homogenization of the Neumann Problem (2.1.15).

Assume that A satisfies the Legendre ellipticity condition (2.1.20) and is 1-periodic. To establish the homogenization theorem for the Neumann problem, we first show that the homogenized matrix $\widehat{A}$ also satisfies the Legendre condition. This ensures that the corresponding Neumann problem for $\mathcal{L}_0$ is well posed.

Lemma 2.3.3. *Suppose that $A = A(y)$ is 1-periodic and satisfies the Legendre condition (2.1.20). Then $\widehat{A}$ also satisfies the Legendre condition. In fact,*

$$\mu|\xi|^2 \le \widehat{a}_{ij}^{\alpha\beta}\xi_i^\alpha \xi_j^\beta \quad \text{and} \quad |\widehat{a}_{ij}^{\alpha\beta}| \le \mu_1 \tag{2.3.12}$$

for any $\xi = (\xi_i^\alpha) \in \mathbb{R}^{m\times d}$, where $\mu_1 > 0$ depends only on μ (as well as d, m).

Proof. The proof for the second inequality in (2.3.12) is the same as in the proof of Lemma 2.2.4. To see the first, we fix $\xi = (\xi_i^\alpha) \in \mathbb{R}^{m\times d}$ and let $\phi = \xi_i^\alpha P_i^\alpha$, $\psi = \xi_i^\alpha \chi_i^\alpha$. Observe that, by (2.1.20),

$$\begin{aligned}
\widehat{a}_{ij}^{\alpha\beta}\xi_i^\alpha \xi_j^\beta &= a_{\text{per}}(\phi+\psi, \phi+\psi)\\
&\ge \mu \fint_Y |\nabla\phi + \nabla\psi|^2\, dy\\
&= \mu \fint_Y |\nabla \phi|^2\, dy + \mu \fint_Y |\nabla\psi|^2\, dy,
\end{aligned}$$

where we have also used the fact that $\int_Y \nabla \chi_i^\alpha\, dy = 0$. It follows that

$$\begin{aligned}
\widehat{a}_{ij}^{\alpha\beta}\xi_i^\alpha \xi_j^\beta &\ge \mu \fint_Y |\nabla\phi|^2\, dy\\
&= \mu|\xi|^2.
\end{aligned}$$

This finishes the proof. □

Let $F \in L^2(\Omega;\mathbb{R}^m)$, $G \in L^2(\Omega;\mathbb{R}^{m\times d})$, and $g \in H^{-1/2}(\partial\Omega;\mathbb{R}^m)$, the dual of $H^{1/2}(\partial\Omega;\mathbb{R}^m)$. Assume that F, G and g satisfy the compatibility condition (2.1.21). By Theorem 2.1.7, the Neumann problem (2.1.15) has a unique (up to a constant in $\mathbb{R}^m$) solution. Furthermore, if $\int_\Omega u_\varepsilon\, dx = 0$, by (2.1.22) and the Poincaré inequality,

$$\|u_\varepsilon\|_{H^1(\Omega)} \le C\Big\{\|F\|_{L^2(\Omega)} + \|G\|_{L^2(\Omega)} + \|g\|_{H^{-1/2}(\partial\Omega)}\Big\},$$

where C depends only on μ and Ω. Let $\{u_{\varepsilon'}\}$ be a subsequence of $\{u_\varepsilon\}$ such that $u_{\varepsilon'} \rightharpoonup u_0$ weakly in $H^1(\Omega;\mathbb{R}^m)$ for some $u_0 \in H^1(\Omega;\mathbb{R}^m)$. It follows by Theorem 2.3.2 that

$$A(x/\varepsilon')\nabla u_{\varepsilon'} \rightharpoonup \widehat{A}\nabla u_0 \quad \text{weakly in } L^2(\Omega;\mathbb{R}^{m\times d}).$$

By taking limits in (2.1.19) we see that u_0 is a weak solution to the Neumann problem:

$$\mathcal{L}_0(u_0) = F + \operatorname{div}(G) \quad \text{in } \Omega \quad \text{and} \quad \frac{\partial u_0}{\partial \nu_0} = g - n \cdot G \quad \text{on } \partial\Omega, \tag{2.3.13}$$

and that $\int_\Omega u_0 \, dx = 0$, where

$$\left(\frac{\partial u_0}{\partial \nu_0}\right)^\alpha = n_i \widehat{a}_{ij}^{\alpha\beta} \frac{\partial u_0^\beta}{\partial x_j} \tag{2.3.14}$$

is the conormal derivative associated with the operator $\mathcal{L}_0$. Since such u_0 is unique, we conclude that as $\varepsilon \to 0$, $u_\varepsilon \to u_0$ weakly in $H^1(\Omega; \mathbb{R}^m)$ and thus strongly in $L^2(\Omega; \mathbb{R}^m)$. As in the case of the Dirichlet problem, we also obtain

$$A(x/\varepsilon)\nabla u_\varepsilon \rightharpoonup \widehat{A}\nabla u_0 \text{ weakly in } L^2(\Omega; \mathbb{R}^{m\times d}).$$

2.4 Elliptic systems of linear elasticity

In this section we consider the elliptic system of elasticity with the operator $\mathcal{L}_\varepsilon = -\operatorname{div}\big(A(x/\varepsilon)\nabla\big)$. We assume that the coefficient matrix $A(y) = \big(a_{ij}^{\alpha\beta}(y)\big)$, with $1 \le i, j, \alpha, \beta \le d$, is 1-periodic and satisfies the elasticity condition, denoted by $A \in E(\kappa_1, \kappa_2)$,

$$\begin{aligned} &a_{ij}^{\alpha\beta}(y) = a_{ji}^{\beta\alpha}(y) = a_{\alpha j}^{i\beta}(y), \\ &\kappa_1|\xi|^2 \le a_{ij}^{\alpha\beta}(y)\xi_i^\alpha \xi_j^\beta \le \kappa_2|\xi|^2 \end{aligned} \tag{2.4.1}$$

for a.e. $y \in \mathbb{R}^d$ and for any *symmetric* matrix $\xi = (\xi_i^\alpha) \in \mathbb{R}^{d\times d}$, where κ_1, κ_2 are positive constants.

Lemma 2.4.1. *Let Ω be a bounded domain in $\mathbb{R}^d$. Then*

$$\sqrt{2}\,\|\nabla u\|_{L^2(\Omega)} \le \|\nabla u + (\nabla u)^T\|_{L^2(\Omega)} \tag{2.4.2}$$

for any $u \in H_0^1(\Omega; \mathbb{R}^d)$, where $(\nabla u)^T$ denotes the transpose of ∇u.

Proof. By a density argument, to prove (2.4.2), which is called the first Korn inequality, we may assume that $u \in C_0^\infty(\Omega; \mathbb{R}^d)$. This allows us to use integration by parts to obtain

$$\begin{aligned} \int_\Omega |\nabla u + (\nabla u)^T|^2 \, dx &= \int_\Omega \left(\frac{\partial u^\alpha}{\partial x_i} + \frac{\partial u^i}{\partial x_\alpha}\right)\left(\frac{\partial u^\alpha}{\partial x_i} + \frac{\partial u^i}{\partial x_\alpha}\right) dx \\ &= 2\int_\Omega |\nabla u|^2 \, dx + 2\int_\Omega \frac{\partial u^\alpha}{\partial x_i}\frac{\partial u^i}{\partial x_\alpha} \, dx \\ &= 2\int_\Omega |\nabla u|^2 \, dx - 2\int_\Omega u^\alpha \frac{\partial}{\partial x_\alpha}\big(\operatorname{div}(u)\big) \, dx \end{aligned}$$

$$= 2\int_\Omega |\nabla u|^2\,dx + 2\int_\Omega |\mathrm{div}(u)|^2\,dx$$
$$\geq 2\int_\Omega |\nabla u|^2\,dx,$$

from which the inequality (2.4.2) follows. □

Lemma 2.4.2. *Suppose* $A = \left(a_{ij}^{\alpha\beta}\right) \in E(\kappa_1, \kappa_2)$. *Then*

$$\frac{\kappa_1}{4}|\xi + \xi^T|^2 \leq a_{ij}^{\alpha\beta}\xi_i^\alpha \xi_j^\beta \leq \frac{\kappa_2}{4}|\xi + \xi^T|^2 \tag{2.4.3}$$

for any $\xi = (\xi_i^\alpha) \in \mathbb{R}^{d\times d}$. *Moreover,*

$$|a_{ij}^{\alpha\beta}\xi_i^\alpha \zeta_j^\beta| \leq \frac{\kappa_2}{4}|\xi + \xi^T||\zeta + \zeta^T| \tag{2.4.4}$$

for any $\xi = (\xi_i^\alpha), \zeta = (\zeta_i^\alpha) \in \mathbb{R}^{d\times d}$.

Proof. Note that by the symmetry conditions in (2.4.1),

$$a_{ij}^{\alpha\beta} = a_{ji}^{\beta\alpha} = a_{\alpha j}^{i\beta} = a_{j\alpha}^{\beta i} = a_{\beta\alpha}^{ji} = a_{\alpha\beta}^{ij} = a_{i\beta}^{\alpha j}. \tag{2.4.5}$$

It follows that for any $\xi = (\xi_i^\alpha) \in \mathbb{R}^{d\times d}$,

$$a_{ij}^{\alpha\beta}\xi_i^\alpha \xi_j^\beta = \frac{1}{4}a_{ij}^{\alpha\beta}(\xi_i^\alpha + \xi_\alpha^i)(\xi_j^\beta + \xi_\beta^j),$$

from which (2.4.3) follows readily by using (2.4.1). To see (2.4.4), one considers the quadratic function

$$Q(t) = a_{ij}^{\alpha\beta}(\xi_i^\alpha + t\zeta_i^\alpha)(\xi_j^\beta + t\zeta_j^\beta).$$

Note that by (2.4.3), $Q(t) \geq 0$ for any $t \in \mathbb{R}$. It follows that its discriminant is less than or equal to zero. This, together with the fact $A^* = A$, leads to

$$|a_{ij}^{\alpha\beta}\xi_i^\alpha \zeta_j^\beta| \leq |a_{ij}^{\alpha\beta}\xi_i^\alpha \xi_j^\beta|^{1/2}|a_{ij}^{\alpha\beta}\zeta_i^\alpha \zeta_j^\beta|^{1/2}$$
$$\leq \frac{\kappa_2}{4}|\xi + \xi^T||\zeta + \zeta^T|,$$

where we have used (2.4.3) for the last step. □

It follows from Lemmas 2.4.1 and 2.4.2 that

$$\int_{\mathbb{R}^d} A\nabla u \cdot \nabla u\,dx \geq \frac{\kappa_1}{4}\int_{\mathbb{R}^d} |\nabla u + (\nabla u)^T|^2\,dx$$
$$\geq \frac{\kappa_1}{2}\int_{\mathbb{R}^d} |\nabla u|^2\,dx,$$

for any $u \in C_0^1(\mathbb{R}^d; \mathbb{R}^d)$. This, together with (2.4.4), shows that the elasticity condition (2.4.1) implies (2.1.2)–(2.1.3) for some $\mu > 0$ depending only on κ_1 and κ_2. Consequently, all results proved in the previous sections under the conditions (2.1.2)–(2.1.3) hold for the elasticity system. In particular, the matrix of homogenized coefficients may be defined and satisfies the ellipticity condition (2.2.28). However, a stronger result can be proved.

Theorem 2.4.3. *Suppose that* $A = \big(a_{ij}^{\alpha\beta}\big) \in E(\kappa_1, \kappa_2)$ *and is* 1*-periodic. Let* $\widehat{A} = \big(\widehat{a}_{ij}^{\alpha\beta}\big)$ *be its matrix of effective coefficients. Then* $\widehat{A} \in E(\kappa_1, \kappa_2)$.

Proof. Let the bilinear form $a_{\text{per}}(\cdot,\cdot)$ be defined by (2.2.18). Observe that

$$\begin{aligned} \widehat{a}_{ij}^{\alpha\beta} &= a_{\text{per}}\big(P_j^\beta + \chi_j^\beta, P_i^\alpha\big) \\ &= a_{\text{per}}\big(P_j^\beta + \chi_j^\beta, P_i^\alpha \pm \chi_i^\alpha\big), \end{aligned} \tag{2.4.6}$$

where we have used (2.2.19) and $\chi_i^\alpha \in H_{\text{per}}^1(Y;\mathbb{R}^d)$ for the second equality. Since $a_{ij}^{\alpha\beta} = a_{ji}^{\beta\alpha}$, we have $a_{\text{per}}(\phi,\psi) = a_{\text{per}}(\psi,\phi)$. It follows that $\widehat{a}_{ij}^{\alpha\beta} = \widehat{a}_{ji}^{\beta\alpha}$. Also, using $a_{ij}^{\alpha\beta} = a_{\alpha j}^{i\beta}$ and (2.2.16), we obtain $\widehat{a}_{ij}^{\alpha\beta} = \widehat{a}_{\alpha j}^{i\beta}$.

Let $\xi = (\xi_i^\alpha) \in \mathbb{R}^{d\times d}$ be a symmetric matrix. Let $\phi = \xi_j^\beta P_j^\beta$ and $\psi = \xi_j^\beta \chi_j^\beta$. It follows from (2.4.6) and (2.4.3) that

$$\begin{aligned} \widehat{a}_{ij}^{\alpha\beta}\xi_i^\alpha\xi_j^\beta &= a_{\text{per}}\big(\phi+\psi, \phi+\psi\big) \\ &\geq \frac{\kappa_1}{4}\fint_Y |\nabla\phi + \nabla\psi + (\nabla\phi)^T + (\nabla\psi)^T|^2\, dy \\ &= \frac{\kappa_1}{4}\fint_Y |\nabla\phi + (\nabla\phi)^T|^2\, dy + \frac{\kappa_1}{4}\fint_Y |\nabla\psi + (\nabla\psi)^T|^2\, dy, \end{aligned}$$

where for the last step we have used the observation $\fint_Y \nabla\chi_j^\beta\, dy = 0$. Since $\nabla\phi = \xi = \xi^T$, this implies that

$$\begin{aligned} \widehat{a}_{ij}^{\alpha\beta}\xi_i^\alpha\xi_j^\beta &\geq \frac{\kappa_1}{4}\fint_Y |\nabla\phi + (\nabla\phi)^T|^2\, dy \\ &= \frac{\kappa_1}{4}|\xi + \xi^T|^2 \\ &= \kappa_1|\xi|^2. \end{aligned} \tag{2.4.7}$$

Also, note that by (2.4.6),

$$\begin{aligned} \widehat{a}_{ij}^{\alpha\beta}\xi_i^\alpha\xi_j^\beta &= a_{\text{per}}\big(\phi+\psi, \phi-\psi\big) \\ &= a_{\text{per}}(\phi,\phi) - a_{\text{per}}(\psi,\psi) \\ &\leq a_{\text{per}}(\phi,\phi) \\ &\leq \kappa_2|\xi|^2, \end{aligned}$$

where we have used the fact $a_{\text{per}}(\phi,\psi) = a_{\text{per}}(\psi,\phi)$ and $a_{\text{per}}(\psi,\psi) \geq 0$. ∎

As we pointed out earlier, the results for the Dirichlet problem in Section 2.3 hold for the elasticity operator. Additional work is needed for the Neumann problem (2.1.15), as the elasticity condition (2.4.1) does not imply the Legendre ellipticity condition.

Let

$$\mathcal{R} = \Big\{ \phi = Bx + b : \ B \in \mathbb{R}^{d\times d} \text{ is skew-symmetric and } b \in \mathbb{R}^d \Big\} \tag{2.4.8}$$

denote the space of rigid displacements, with

$$\dim(\mathcal{R}) = \frac{d(d+1)}{2}.$$

Using the symmetry condition $a_{ij}^{\alpha\beta} = a_{\alpha j}^{i\beta}$, we see that $A(x/\varepsilon)\nabla u \cdot \nabla \phi = 0$ for any $\phi \in \mathcal{R}$. Consequently, the existence of solutions of (2.1.15) implies that

$$\int_\Omega F \cdot \phi \, dx - \int_\Omega G \cdot \nabla \phi \, dx + \langle g, \phi \rangle_{H^{-1/2}(\partial\Omega)\times H^{1/2}(\partial\Omega)} = 0 \tag{2.4.9}$$

for any $\phi \in \mathcal{R}$.

Theorem 2.4.4. *Let Ω be a bounded Lipschitz domain in $\mathbb{R}^d$ and $A \in E(\kappa_1, \kappa_2)$. Assume that $F \in L^2(\Omega; \mathbb{R}^d)$, $G \in L^2(\Omega; \mathbb{R}^{d\times d})$ and $g \in H^{-1/2}(\partial\Omega; \mathbb{R}^d)$ satisfy the compatibility condition* (2.4.9). *Then the Neumann problem* (2.1.15) *has a weak solution u_ε, unique up to an element of $\mathcal{R}$, in $H^1(\Omega; \mathbb{R}^d)$. Moreover, the solution satisfies the energy estimate*

$$\|\nabla u_\varepsilon\|_{L^2(\Omega)} \le C \left\{ \|F\|_{L^2(\Omega)} + \|G\|_{L^2(\Omega)} + \|g\|_{H^{-1/2}(\partial\Omega)} \right\}, \tag{2.4.10}$$

where C depends only on κ_1, κ_2 and Ω.

Proof. This again follows from the Lax–Milgram Theorem by considering the bilinear form (2.1.18) on the Hilbert space $H^1(\Omega; \mathbb{R}^d)/\mathcal{R}$. To prove that $B[u, v]$ is coercive, one applies the second Korn inequality

$$\int_\Omega |\nabla u|^2 \, dx \le C \int_\Omega |\nabla u + (\nabla u)^T|^2 \, dx \tag{2.4.11}$$

for any $u \in H^1(\Omega; \mathbb{R}^d)$ with the property that $u \perp \mathcal{R}$ in $H^1(\Omega; \mathbb{R}^d)$ or in $L^2(\Omega; \mathbb{R}^d)$. We refer the reader to [78, Chapter I] for a proof of (2.4.11). □

Theorem 2.4.5. *Assume that A is 1-periodic and satisfies the elasticity condition* (2.4.1). *Let $u_\varepsilon \in H^1(\Omega; \mathbb{R}^d)$ be the weak solution to the Neumann problem* (2.1.15) *with $\int_\Omega u_\varepsilon \cdot \phi \, dx = 0$ for any $\phi \in \mathcal{R}$, given by Theorem* 2.4.4. *Then*

$$\begin{cases} u_\varepsilon \rightharpoonup u_0 & \text{weakly in } H^1(\Omega; \mathbb{R}^d), \\ A(x/\varepsilon)\nabla u_\varepsilon \rightharpoonup \widehat{A}\nabla u_0 & \text{weakly in } L^2(\Omega; \mathbb{R}^{d\times d}), \end{cases} \tag{2.4.12}$$

where u_0 is the unique weak solution to the Neumann problem

$$\mathcal{L}_0(u_0) = F + \operatorname{div}(G) \quad \text{in } \Omega \quad \text{and} \quad \frac{\partial u_0}{\partial \nu_0} = g - n \cdot G \quad \text{on } \partial\Omega, \tag{2.4.13}$$

with $\int_\Omega u_0 \cdot \phi \, dx = 0$ for any $\phi \in \mathcal{R}$.

Proof. The proof is similar to that for the Neumann problem under the Legendre ellipticity condition. We point out that Theorem 2.4.3 is needed for the existence and uniqueness of the Neumann problem (2.4.13) for $\mathcal{L}_0$. □

We end this section with some observations on elliptic systems of linear elasticity.

Let $A(y) = \big(a_{ij}^{\alpha\beta}(y)\big) \in E(\kappa_1, \kappa_2)$. Define

$$\widetilde{a}_{ij}^{\alpha\beta}(y) = a_{ij}^{\alpha\beta}(y) + \mu \delta_{i\alpha}\delta_{j\beta} - \mu \delta_{i\beta}\delta_{j\alpha}, \tag{2.4.14}$$

where $0 < \mu \le \kappa_1/2$. The following proposition shows that $\widetilde{A} = (\widetilde{a}_{ij}^{\alpha\beta})$ is symmetric and satisfies the Legendre ellipticity condition.

Proposition 2.4.6. *Let $A \in E(\kappa_1, \kappa_2)$ and $\widetilde{a}_{ij}^{\alpha\beta}$ be defined by (2.4.14). Then $\widetilde{a}_{ij}^{\alpha\beta} = \widetilde{a}_{ji}^{\beta\alpha}$, and*

$$\widetilde{a}_{ij}^{\alpha\beta} \xi_i^\alpha \xi_j^\beta \ge \mu |\xi|^2 \tag{2.4.15}$$

for any $\xi = (\xi_i^\alpha) \in \mathbb{R}^{d\times d}$.

Proof. The symmetry property is obvious. To see (2.4.15), we let $\xi = (\xi_i^\alpha) \in \mathbb{R}^{d\times d}$ and recall that by (2.4.3),

$$a_{ij}^{\alpha\beta} \xi_i^\alpha \xi_j^\beta \ge \frac{\kappa_1}{4} |\xi + \xi^T|^2.$$

It follows that

$$\begin{aligned}
\widetilde{a}_{ij}^{\alpha\beta} \xi_i^\alpha \xi_j^\beta &\ge \frac{\kappa_1}{4} |\xi + \xi^T|^2 + \mu|\xi|^2 - \mu \xi_i^j \xi_j^i \\
&= \frac{\kappa_1}{2} \big(|\xi|^2 + \xi_i^j \xi_j^i\big) + \mu|\xi|^2 - \mu \xi_i^j \xi_j^i \\
&= \mu|\xi|^2 + \frac{1}{2}\Big(\frac{\kappa_1}{2} - \mu\Big) |\xi + \xi^T|^2 \\
&\ge \mu|\xi|^2,
\end{aligned}$$

where we have used the assumption $\mu \le \kappa_1/2$ for the last step. □

Proposition 2.4.7. *Let $\widetilde{A}(y) = (\widetilde{a}_{ij}^{\alpha\beta})$ be defined by (2.4.14). Then*

$$\int_\Omega A(x/\varepsilon)\nabla u \cdot \nabla \varphi \, dx = \int_\Omega \widetilde{A}(x/\varepsilon)\nabla u \cdot \nabla \varphi \, dx \tag{2.4.16}$$

for any $u \in H^1_{\text{loc}}(\Omega; \mathbb{R}^d)$ and $\varphi \in C_0^\infty(\Omega; \mathbb{R}^d)$.

Proof. Let $u \in C^\infty(\Omega;\mathbb{R}^d)$ and $\varphi \in C_0^\infty(\Omega;\mathbb{R}^d)$. Note that

$$\begin{aligned}
&\int_\Omega \left(\widetilde{A}(x/\varepsilon) - A(x/\varepsilon)\right) \nabla u \cdot \nabla \varphi \, dx \\
&= \mu \int_\Omega \left(\delta_{i\alpha}\delta_{j\beta} - \delta_{i\beta}\delta_{j\alpha}\right) \frac{\partial u^\beta}{\partial x_j} \cdot \frac{\partial \varphi^\alpha}{\partial x_i} \, dx \\
&= \mu \int_\Omega \operatorname{div}(u) \cdot \operatorname{div}(\varphi) \, dx - \mu \int_\Omega \frac{\partial u^\beta}{\partial x_\alpha} \cdot \frac{\partial \varphi^\alpha}{\partial x_\beta} \, dx \\
&= 0,
\end{aligned} \tag{2.4.17}$$

where we have used integration by parts for the last step. By a density argument one may deduce that (2.4.17) continues to hold for any $u \in H^1_{\text{loc}}(\Omega;\mathbb{R}^d)$ and $\varphi \in C_0^\infty(\Omega;\mathbb{R}^d)$. □

Let $\widetilde{\mathcal{L}}_\varepsilon = -\operatorname{div}(\widetilde{A}(x/\varepsilon)\nabla)$. It follows from Proposition 2.4.7 that if $u_\varepsilon \in H^1_{\text{loc}}(\Omega;\mathbb{R}^d)$, then

$$\mathcal{L}_\varepsilon(u_\varepsilon) = F \quad \text{in } \Omega \quad \text{if and only if} \quad \widetilde{\mathcal{L}}_\varepsilon(u_\varepsilon) = F \quad \text{in } \Omega, \tag{2.4.18}$$

where $F \in \left(C_0^\infty(\Omega;\mathbb{R}^d)\right)'$ is a distribution. In view of Proposition 2.4.6 this allows us to treat the system of linear elasticity as a special case of elliptic systems satisfying the Legendre condition and the symmetry condition. Indeed, the approach works well for interior regularity estimates as well as for boundary estimates with the Dirichlet condition. However, we point out that since the Neumann boundary condition depends on the coefficient matrix, the re-writing of the system of elasticity changes the Neumann problem. Indeed, if $\frac{\partial u_\varepsilon}{\partial \widetilde{\nu}_\varepsilon}$ denotes the conormal derivative associated with $\widetilde{\mathcal{L}}_\varepsilon$, then

$$\left(\frac{\partial u_\varepsilon}{\partial \widetilde{\nu}_\varepsilon}\right)^\alpha = \left(\frac{\partial u_\varepsilon}{\partial \nu_\varepsilon}\right)^\alpha + \mu n_\alpha \operatorname{div}(u_\varepsilon) - \mu n_\beta \frac{\partial u_\varepsilon^\beta}{\partial x_\alpha}.$$

2.5 Notes

The material in Section 2.1, which may be found in [26, 50], is standard for second-order linear elliptic systems in divergence form with bounded measurable coefficients.

The formal asymptotic expansions as well as other results presented in Section 2.2 may be found in [19].

Much of the material in Sections 2.3 and 2.4 is more or less well known and may be found in [19, 111, 78, 25].

Chapter 3

Convergence Rates, Part I

Let $\mathcal{L}_\varepsilon = -\text{div}(A(x/\varepsilon)\nabla)$ for $\varepsilon > 0$, where $A(y) = \big(a_{ij}^{\alpha\beta}(y)\big)$ is 1-periodic and satisfies a certain ellipticity condition. Let $\mathcal{L}_0 = -\text{div}(\widehat{A}\nabla)$, where $\widehat{A} = \big(\widehat{a}_{ij}^{\alpha\beta}\big)$ denotes the matrix of effective coefficients, given by (2.2.16). For $F \in L^2(\Omega; \mathbb{R}^m)$ and $\varepsilon \geq 0$, consider the Dirichlet problem

$$\begin{cases} \mathcal{L}_\varepsilon(u_\varepsilon) = F & \text{in } \Omega, \\ u_\varepsilon = f & \text{on } \partial\Omega, \end{cases} \tag{3.0.1}$$

and the Neumann problem

$$\begin{cases} \mathcal{L}_\varepsilon(u_\varepsilon) = F & \text{in } \Omega, \\ \dfrac{\partial u_\varepsilon}{\partial \nu_\varepsilon} = g & \text{on } \partial\Omega, \\ u_\varepsilon \perp \mathbb{R}^m \text{ in } L^2(\Omega; \mathbb{R}^m), \end{cases} \tag{3.0.2}$$

where $f \in H^1(\partial\Omega; \mathbb{R}^m)$ and $g \in L^2(\partial\Omega; \mathbb{R}^m)$. We have shown in Section 2.3 that as $\varepsilon \to 0$, u_ε converges to u_0 weakly in $H^1(\Omega; \mathbb{R}^m)$ and strongly in $L^2(\Omega; \mathbb{R}^m)$. In this chapter we investigate the problem of convergence rates in H^1 and L^2.

In Section 3.1 we introduce the flux correctors and study the properties of an ε-smoothing operator S_ε. Sections 3.2 and 3.3 are devoted to error estimates of two-scale expansions in H^1 for the Dirichlet and Neumann problems in a Lipschitz domain Ω, respectively. We will show that

$$\|u_\varepsilon - u_0 - \varepsilon\chi(x/\varepsilon)\eta_\varepsilon S_\varepsilon^2(\nabla u_0)\|_{H^1(\Omega)} \leq C\sqrt{\varepsilon}\, \|u_0\|_{H^2(\Omega)}, \tag{3.0.3}$$

where η_ε is a cut-off function satisfying (3.2.1). Moreover, if A satisfies the sym-

Z. Shen, *Periodic Homogenization of Elliptic Systems*, Operator Theory: Advances and Applications 269, https://doi.org/10.1007/978-3-319-91214-1_3

metry condition $A^* = A$, we obtain

$$\|u_\varepsilon - u_0 - \varepsilon\chi(x/\varepsilon)\eta_\varepsilon S_\varepsilon^2(\nabla u_0)\|_{H^1(\Omega)} \le \begin{cases} C\sqrt{\varepsilon}\Big\{\|F\|_{L^q(\Omega)} + \|f\|_{H^1(\partial\Omega)}\Big\} & \text{for problem (3.0.1)},\\ C\sqrt{\varepsilon}\Big\{\|F\|_{L^q(\Omega)} + \|g\|_{L^2(\partial\Omega)}\Big\} & \text{for problem (3.0.2)}, \end{cases} \tag{3.0.4}$$

for $0 < \varepsilon < 1$, where $q = \frac{2d}{d+1}$. The constants C in (3.0.3)–(3.0.4) depend only on the ellipticity constant μ and Ω.

The $O(\sqrt{\varepsilon})$ rates in $H^1(\Omega)$ given by (3.0.3) and (3.0.4) are more or less sharp. Note that the error estimates (3.0.3)–(3.0.4) also imply the $O(\sqrt{\varepsilon})$ rate for $u_\varepsilon - u_0$ in $L^2(\Omega)$. However, this is not sharp. In fact, it will be proved in Sections 3.4 and 3.5 that if Ω is a bounded Lipschitz domain and $A^* = A$, the scale-invariant estimate

$$\|u_\varepsilon - u_0\|_{L^p(\Omega)} \le C\varepsilon\, \|u_0\|_{W^{2,q}(\Omega)}, \tag{3.0.5}$$

holds for $p = \frac{2d}{d-1}$ and $q = p' = \frac{2d}{d+1}$, where C depends only on μ and Ω. Without the symmetry condition, it is shown that

$$\|u_\varepsilon - u_0\|_{L^2(\Omega)} \le C\varepsilon\, \|u_0\|_{H^2(\Omega)}, \tag{3.0.6}$$

under the assumption that Ω is a bounded $C^{1,1,}$ domain. In Section 3.6 we address the problem of convergence rates for elliptic systems of linear elasticity.

No smoothness condition will be imposed on the coefficient matrix A in this chapter. Further results on convergence rates may be found in Chapter 7 under additional smoothness conditions on A.

3.1 Flux correctors and ε-smoothing

Throughout this section we assume that $A = A(y)$ is 1-periodic and satisfies the boundedness and V-ellipticity conditions (2.1.2)–(2.1.3). For $1 \le i, j \le d$ and $1 \le \alpha, \beta \le m$, let

$$b_{ij}^{\alpha\beta}(y) = a_{ij}^{\alpha\beta}(y) + a_{ik}^{\alpha\gamma}(y)\frac{\partial}{\partial y_k}\Big(\chi_j^{\gamma\beta}(y)\Big) - \widehat{a}_{ij}^{\alpha\beta}, \tag{3.1.1}$$

where the repeated index k is summed from 1 to d and γ from 1 to m. Observe that the matrix $B(y) = \big(b_{ij}^{\alpha\beta}(y)\big)$ is 1-periodic and that $\|B\|_{L^p(Y)} \le C_0$ for some $p > 2$ and $C_0 > 0$ depending on μ. Moreover, it follows from the definitions of χ_j^β and $\widehat{a}_{ij}^{\alpha\beta}$ in Section 2.2 that

$$\frac{\partial}{\partial y_i}\Big(b_{ij}^{\alpha\beta}\Big) = 0 \quad \text{and} \quad \int_Y b_{ij}^{\alpha\beta}(y)\, dy = 0. \tag{3.1.2}$$

Proposition 3.1.1. *There exist $\phi_{kij}^{\alpha\beta} \in H^1_{\text{per}}(Y)$, where $1 \le i, j, k \le d$ and $1 \le \alpha, \beta \le m$, such that*

$$b_{ij}^{\alpha\beta} = \frac{\partial}{\partial y_k}\left(\phi_{kij}^{\alpha\beta}\right) \qquad \text{and} \qquad \phi_{kij}^{\alpha\beta} = -\phi_{ikj}^{\alpha\beta}. \tag{3.1.3}$$

Moreover, if $\chi = (\chi_j^\beta)$ is Hölder continuous, then $\phi_{kij}^{\alpha\beta} \in L^\infty(Y)$.

Proof. Since $\int_Y b_{ij}^{\alpha\beta}\, dy = 0$, there exists $f_{ij}^{\alpha\beta} \in H^2_{\text{per}}(Y)$ such that $\int_Y f_{ij}^{\alpha\beta}\, dy = 0$ and

$$\Delta f_{ij}^{\alpha\beta} = b_{ij}^{\alpha\beta} \quad \text{in } Y. \tag{3.1.4}$$

Moreover,

$$\|f_{ij}^{\alpha\beta}\|_{H^2(Y)} \le C\, \|b_{ij}^{\alpha\beta}\|_{L^2(Y)} \le C.$$

Define

$$\phi_{kij}^{\alpha\beta}(y) = \frac{\partial}{\partial y_k}\left(f_{ij}^{\alpha\beta}\right) - \frac{\partial}{\partial y_i}\left(f_{kj}^{\alpha\beta}\right).$$

Clearly, $\phi_{kij}^{\alpha\beta} \in H^1_{\text{per}}(Y)$ and $\phi_{kij}^{\alpha\beta} = -\phi_{ikj}^{\alpha\beta}$. Since $\frac{\partial}{\partial y_i}\{b_{ij}^{\alpha\beta}\} = 0$, we deduce from (3.1.4) that $\frac{\partial}{\partial y_i}\{f_{ij}^{\alpha\beta}\}$ is a 1-periodic harmonic function and thus is constant. Hence,

$$\begin{aligned}
\frac{\partial}{\partial y_k}\left\{\phi_{kij}^{\alpha\beta}\right\} &= \Delta\left\{f_{ij}^{\alpha\beta}\right\} - \frac{\partial^2}{\partial y_k \partial y_i}\left\{f_{kj}^{\alpha\beta}\right\} \\
&= \Delta\left\{f_{ij}^{\alpha\beta}\right\} \\
&= b_{ij}^{\alpha\beta}.
\end{aligned}$$

Suppose that the corrector χ is Hölder continuous. Recall that

$$\mathcal{L}_1\big(\chi_j^\beta + P_j^\beta\big) = 0 \quad \text{in } \mathbb{R}^d.$$

By Caccioppoli's inequality,

$$\int_{B(y,r)} |\nabla \chi|^2\, dx \le \frac{C}{r^2} \int_{B(y,2r)} |\chi(x) - \chi(y)|^2\, dx + C\, r^d.$$

This implies that $|\nabla \chi|$ is in the Morrey space $L^{2,\rho}(Y)$ for some $\rho > d-2$; i.e.,

$$\int_{B(y,r)} |\nabla \chi|^2\, dx \le C\, r^\rho \qquad \text{for } y \in Y \text{ and } 0 < r < 1.$$

Consequently, $b_{ij}^{\alpha\beta} \in L^{2,\rho}(Y)$ for some $\rho > d-2$ and

$$\sup_{x \in Y} \int_Y \frac{|b_{ij}^{\alpha\beta}(y)|}{|x-y|^{d-1}}\, dy \le C.$$

In view of (3.1.4), by considering $\Delta(f_{ij}^{\alpha\beta}\varphi)$, where φ is a cut-off function, and using a potential representation for the Laplace equation, one deduces that

$$\begin{aligned}\|\nabla f_{ij}^{\alpha\beta}\|_{L^\infty(Y)} &\le C\,\|f_{ij}^{\alpha\beta}\|_{L^2(Y)} + C\sup_{x\in Y}\int_Y \frac{|b_{ij}^{\alpha\beta}(y)|}{|x-y|^{d-1}}\,dy\\ &\le C.\end{aligned}$$

It follows that $\phi_{kij}^{\alpha\beta}\in L^\infty(Y)$. □

Remark 3.1.2. Recall that if $d=2$ or $m=1$, the function χ is Hölder continuous. As a result, we obtain $\|\phi_{kij}^{\alpha\beta}\|_\infty\le C$. In the case where $d\ge 3$ and $m\ge 2$, we have $b_{ij}^{\alpha\beta}\in L^p(Y)$ for some $p>2$. It follows that $\nabla^2 f_{ij}^{\alpha\beta}\in L^p(Y)$ for some $p>2$. By the Sobolev embedding this implies that $\phi_{kij}^{\alpha\beta}\in L^q(Y)$ for some $q>\frac{2d}{d-2}$. We mention that if the coefficient matrix A is in VMO($\mathbb{R}^d$) (see (4.0.1) for the definition), then $\chi(y)$ is Hölder continuous, and consequently $\phi=\left(\phi_{kij}^{\alpha\beta}\right)$ is bounded.

Remark 3.1.3. A key property of $\phi=\left(\phi_{kij}^{\alpha\beta}\right)$, which follows from (3.1.3), is the identity:

$$b_{ij}^{\alpha\beta}(x/\varepsilon)\,\frac{\partial\psi^\alpha}{\partial x_i}=\varepsilon\,\frac{\partial}{\partial x_k}\left\{\phi_{kij}^{\alpha\beta}\,(x/\varepsilon)\,\frac{\partial\psi^\alpha}{\partial x_i}\right\} \tag{3.1.5}$$

for any $\psi=(\psi^\alpha)\in H^2(\Omega;\mathbb{R}^m)$.

Let $u_\varepsilon\in H^1(\Omega;\mathbb{R}^m)$, $u_0\in H^2(\Omega;\mathbb{R}^m)$, and

$$w_\varepsilon=u_\varepsilon-u_0-\varepsilon\,\chi(x/\varepsilon)\nabla u_0.$$

A direct computation shows that

$$\begin{aligned}&A(x/\varepsilon)\nabla u_\varepsilon-\widehat{A}\nabla u_0-B(x/\varepsilon)\nabla u_0\\ &\qquad=A(x/\varepsilon)\nabla w_\varepsilon+\varepsilon\,A(x/\varepsilon)\chi(x/\varepsilon)\nabla^2 u_0,\end{aligned} \tag{3.1.6}$$

where $B(y)=\left(b_{ij}^{\alpha\beta}(y)\right)$ is defined by (3.1.1). It follows that

$$\begin{aligned}&\|A(x/\varepsilon)\nabla u_\varepsilon-\widehat{A}\nabla u_0-B(x/\varepsilon)\nabla u_0\|_{L^2(\Omega)}\\ &\quad\le C\,\|\nabla w_\varepsilon\|_{L^2(\Omega)}+C\,\varepsilon\,\|\chi(x/\varepsilon)\nabla^2 u_0\|_{L^2(\Omega)}\\ &\quad\le C\,\|\nabla u_\varepsilon-\nabla u_0-\nabla\chi(x/\varepsilon)\nabla u_0\|_{L^2(\Omega)}+C\varepsilon\,\|\chi(x/\varepsilon)\nabla^2 u_0\|_{L^2(\Omega)},\end{aligned} \tag{3.1.7}$$

where C depends only on μ. This indicates that the 1-periodic matrix-valued function $B(y)$ plays the same role for the flux $A(x/\varepsilon)\nabla u_\varepsilon$ as $\nabla\chi(y)$ does for ∇u_ε. Since $b_{ij}^{\alpha\beta}=\frac{\partial}{\partial y_k}\left(\phi_{kij}^{\alpha\beta}\right)$, the 1-periodic function $\phi=\left(\phi_{kij}^{\alpha\beta}\right)$ is called the flux corrector.

To deal with the fact that the correctors χ and ϕ may be unbounded (if $d\ge 3$ and $m\ge 2$), we introduce an ε-smoothing operator S_ε.

Definition 3.1.4. Fix $\rho \in C_0^\infty(B(0,1/2))$ such that $\rho \geq 0$ and $\int_{\mathbb{R}^d} \rho\, dx = 1$. For $\varepsilon > 0$, define

$$S_\varepsilon(f)(x) = \rho_\varepsilon * f(x) = \int_{\mathbb{R}^d} f(x-y)\rho_\varepsilon(y)\, dy, \tag{3.1.8}$$

where $\rho_\varepsilon(y) = \varepsilon^{-d}\rho(y/\varepsilon)$.

The following two propositions contain the most useful properties of S_ε for us.

Proposition 3.1.5. *Let $f \in L^p_{\text{loc}}(\mathbb{R}^d)$ for some $1 \leq p < \infty$. Then for any $g \in L^p_{\text{loc}}(\mathbb{R}^d)$,*

$$\|g(x/\varepsilon)\, S_\varepsilon(f)\|_{L^p(\mathcal{O})} \leq C \sup_{x\in\mathbb{R}^d} \left(\fint_{B(x,1/2)} |g|^p \right)^{1/p} \|f\|_{L^p(\mathcal{O}(\varepsilon/2))}, \tag{3.1.9}$$

where $\mathcal{O} \subset \mathbb{R}^d$ is open,

$$\mathcal{O}(t) = \big\{ x \in \mathbb{R}^d : \operatorname{dist}(x, \mathcal{O}) < t \big\},$$

and C depends only on p.

Proof. By Hölder's inequality,

$$|S_\varepsilon(f)(x)|^p \leq \int_{\mathbb{R}^d} |f(y)|^p \rho_\varepsilon(x-y)\, dy,$$

where we also used the fact that $\int_{\mathbb{R}^d} \rho_\varepsilon\, dx = 1$. This, together with Fubini's Theorem, gives (3.1.9) for the case $\mathcal{O} = \mathbb{R}^d$. The general case follows from the observation that $S_\varepsilon(f)(x) = S_\varepsilon(f\chi_{\mathcal{O}(\varepsilon/2)})(x)$ for any $x \in \mathcal{O}$. $\square$

It follows from (3.1.9) that if g is 1-periodic and belongs to $L^p(Y)$, then

$$\|g(x/\varepsilon) S_\varepsilon(f)\|_{L^p(\mathcal{O})} \leq C\, \|g\|_{L^p(Y)} \|f\|_{L^p(\mathcal{O}(\varepsilon/2))}, \tag{3.1.10}$$

where C depends only on p. Since

$$\int_{\mathbb{R}^d} |\nabla \rho_\varepsilon|\, dx \leq C\varepsilon^{-1},$$

a similar argument gives

$$\|g(x/\varepsilon) \nabla S_\varepsilon(f)\|_{L^p(\mathcal{O})} \leq C\varepsilon^{-1} \|g\|_{L^p(Y)} \|f\|_{L^p(\mathcal{O}(\varepsilon/2))}. \tag{3.1.11}$$

Proposition 3.1.6. *Let $f \in W^{1,p}(\mathbb{R}^d)$ for some $1 \leq p \leq \infty$. Then*

$$\|S_\varepsilon(f) - f\|_{L^p(\mathbb{R}^d)} \leq \varepsilon\, \|\nabla f\|_{L^p(\mathbb{R}^d)}. \tag{3.1.12}$$

Moreover, if $q = \frac{2d}{d+1}$,

$$\begin{aligned} \|S_\varepsilon(f)\|_{L^2(\mathbb{R}^d)} &\leq C\varepsilon^{-1/2} \|f\|_{L^q(\mathbb{R}^d)}, \\ \|S_\varepsilon(f) - f\|_{L^2(\mathbb{R}^d)} &\leq C\varepsilon^{1/2} \|\nabla f\|_{L^q(\mathbb{R}^d)}, \end{aligned} \tag{3.1.13}$$

where C depends only on d.

Proof. Using the relation

$$f(x+y)-f(x)=\int_0^1 \nabla f(x+ty)\cdot y\,dt$$

and Minkowski's inequality for integrals, we see that

$$\|f(\cdot+y)-f(\cdot)\|_{L^p(\mathbb{R}^d)}\le |y|\|\nabla f\|_{L^p(\mathbb{R}^d)}$$

for any $y\in\mathbb{R}^d$. By Minkowski's inequality again,

$$\begin{aligned}
\|S_\varepsilon(f)-f\|_{L^p(\mathbb{R}^d)} &\le \int_{\mathbb{R}^d}\|f(\cdot-\varepsilon y)-f(\cdot)\|_{L^p(\mathbb{R}^d)}\,\rho(y)\,dy\\
&\le \int_{\mathbb{R}^d}\varepsilon|y|\rho(y)\,dy\,\|\nabla f\|_{L^p(\mathbb{R}^d)}\\
&\le \varepsilon\,\|\nabla f\|_{L^p(\mathbb{R}^d)},
\end{aligned}$$

which gives (3.1.12).

Next, we note that the Fourier transform of $S_\varepsilon(f)$ is given by $\widehat{\rho}(\varepsilon\xi)\widehat{f}(\xi)$. Let $q=\frac{2d}{d+1}$. By Plancherel's theorem and Hölder's inequality,

$$\begin{aligned}
\int_{\mathbb{R}^d}|S_\varepsilon(f)|^2\,dx &= \int_{\mathbb{R}^d}|\widehat{\rho}(\varepsilon\xi)|^2|\widehat{f}(\xi)|^2\,d\xi\\
&\le \left(\int_{\mathbb{R}^d}|\widehat{\rho}(\varepsilon\xi)|^{2d}\,d\xi\right)^{1/d}\|\widehat{f}\|^2_{L^{q'}(\mathbb{R}^d)}\\
&\le C\varepsilon^{-1}\|f\|^2_{L^q(\mathbb{R}^d)},
\end{aligned}$$

where in the last step we have used the Hausdorff–Young inequality

$$\|\widehat{f}\|_{L^{q'}(\mathbb{R}^d)}\le \|f\|_{L^q(\mathbb{R}^d)}.$$

This gives the first inequality in (3.1.13). Similarly, using $\widehat{\rho}(0)=\int_{\mathbb{R}^d}\rho=1$, we obtain

$$\begin{aligned}
\|S_\varepsilon(f)-f\|_{L^2(\mathbb{R}^d)} &= \|(\widehat{\rho}(\varepsilon\xi)-1)\widehat{f}\|_{L^2(\mathbb{R}^d)}\\
&\le C\left(\int_{\mathbb{R}^d}|\widehat{\rho}(\varepsilon\xi)-\widehat{\rho}(0)|^{2d}|\xi|^{-2d}\,d\xi\right)^{1/(2d)}\|\widehat{\nabla f}\|_{L^{q'}(\mathbb{R}^d)}\\
&= C\varepsilon^{1/2}\|(\widehat{\rho}(\xi)-\widehat{\rho}(0))|\xi|^{-1}\|_{L^{2d}(\mathbb{R}^d)}\|\widehat{\nabla f}\|_{L^{q'}(\mathbb{R}^d)}\\
&\le C\varepsilon^{1/2}\|\nabla f\|_{L^q(\mathbb{R}^d)},
\end{aligned}$$

where we have also used that $|\widehat{\rho}(\xi)-\widehat{\rho}(0)|\le C|\xi|$ for $|\xi|\le 1$ in the last step. □

Next we establish some estimates for integrals on boundary layers. Let

$$\Omega_t=\big\{x\in\Omega:\ \operatorname{dist}(x,\partial\Omega)<t\big\}, \tag{3.1.14}$$

where $t>0$.

Proposition 3.1.7. *Let Ω be a bounded Lipschitz domain in $\mathbb{R}^d$ and $q = \frac{2d}{d+1}$. Then for any $f \in W^{1,q}(\Omega)$,*

$$\|f\|_{L^2(\Omega_t)} \le Ct^{1/2}\|f\|_{W^{1,q}(\Omega)} \quad \text{and} \quad \|f\|_{L^2(\partial\Omega)} \le C\,\|f\|_{W^{1,q}(\Omega)}, \tag{3.1.15}$$

where C depends only on Ω.

Proof. Let $x = (x', x_d) \in \mathbb{R}^d$. Using the fundamental theorem of calculus, it is not hard to show that

$$\begin{aligned} &\int_{|x'|<r} |f(x', s)|^2\, dx' \\ &\quad \le \frac{C}{r}\int_{|x'|<r}\int_0^r |f(x', x_d)|^2\, dx_d dx' + C\int_{|x'|<r}\int_0^r |f|\,|\nabla f|\, dx_d dx' \end{aligned}$$

for any $s \in (0, r)$. It follows that

$$\begin{aligned} &\int_0^t\int_{|x'|<r} |f(x', s)|^2\, dx' ds \\ &\quad \le \frac{Ct}{r}\int_{|x'|<r}\int_0^r |f(x', x_d)|^2\, dx_d dx' + Ct\int_{|x'|<r}\int_0^r |f|\,|\nabla f|\, dx_d dx' \end{aligned}$$

for any $t \in (0, r)$. A change of variables yields

$$\begin{aligned} &\int_{|x'|<r}\int_{\psi(x')}^{\psi(x')+t} |f(x', s)|^2\, ds dx' \\ &\le \frac{Ct}{r}\int_{|x'|<r}\int_{\psi(x')}^{\psi(x')+r} |f|^2\, dx_d dx' + Ct\int_{|x'|<r}\int_{\psi(x')}^{\psi(x')+r} |f|\,|\nabla f|\, dx_d dx', \end{aligned}$$

where ψ is a Lipschitz function in $\mathbb{R}^{d-1}$ and C depends on $\|\nabla\psi\|_\infty$. By covering $\partial\Omega$ with coordinate patches, we obtain

$$\begin{aligned} \int_{\Omega_t} |f|^2\, dx &\le Ct\int_\Omega |f|^2\, dx + Ct\int_\Omega |f|\,|\nabla f|\, dx \\ &\le Ct\,\|f\|^2_{L^2(\Omega)} + Ct\,\|f\|_{L^{q'}(\Omega)}\|\nabla f\|_{L^q(\Omega)}, \end{aligned}$$

where $q = \frac{2d}{d+1}$. This, together with the Sobolev inequality

$$\|f\|_{L^{q'}(\Omega)} \le C\,\|f\|_{W^{1,q}(\Omega)},$$

gives the first inequality in (3.1.15). To obtain the second one, we note that by a similar argument

$$\int_{\partial\Omega} |f|^2\, d\sigma \le C\int_\Omega |f|^2\, dx + C\int_\Omega |f|\,|\nabla f|\, dx,$$

which is bounded by $C\|f\|^2_{W^{1,q}(\Omega)}$. □

Proposition 3.1.8. *Let Ω be a bounded Lipschitz domain in $\mathbb{R}^d$ and $q = \frac{2d}{d+1}$. Let $g \in L^2_{\text{loc}}(\mathbb{R}^d)$ be a 1-periodic function. Then, for any $f \in W^{1,q}(\Omega)$,*

$$\int_{\Omega_{2t}\setminus\Omega_t} |g(x/\varepsilon)|^2 |S_\varepsilon(f)|^2\, dx \le Ct\, \|g\|^2_{L^2(Y)} \|f\|^2_{W^{1,q}(\Omega)}, \tag{3.1.16}$$

where $t \ge \varepsilon$ and C depends only on Ω.

Proof. We may assume that t is small. Let $\mathcal{O} = \Omega_{2t} \setminus \Omega_t$. It follows from (3.1.10) that

$$\begin{aligned}\int_{\Omega_{2t}\setminus\Omega_t} |g(x/\varepsilon)|^2 |S_\varepsilon(f)|^2\, dx &\le C\, \|g\|^2_{L^2(Y)} \|f\|^2_{L^2(\mathcal{O}(\varepsilon/2))}\\ &\le Ct\, \|g\|^2_{L^2(Y)} \|f\|^2_{W^{1,q}(\Omega)},\end{aligned}$$

where we have used (3.1.15) for the last inequality. □

We end this section with the fractional integral estimates.

Proposition 3.1.9. *For $0 < \alpha < d$, define*

$$T_\alpha(f)(x) = \int_{\mathbb{R}^d} \frac{f(y)}{|x-y|^{d-\alpha}}\, dy.$$

Then, if $1 < p < \frac{d}{\alpha}$ and $\frac{1}{q} = \frac{1}{p} - \frac{\alpha}{d}$,

$$\|T_\alpha(f)\|_{L^q(\mathbb{R}^d)} \le C\, \|f\|_{L^p(\mathbb{R}^d)}, \tag{3.1.17}$$

where C depends only on α and p.

Proof. The inequality (3.1.17) is referred as the Hardy–Littlewood–Sobolev inequality. We refer the reader to [50, pp. 162–163] for its proof. □

3.2 Convergence rates in H^1 for the Dirichlet problem

Throughout this section we assume that A is 1-periodic and satisfies (2.1.2)–(2.1.3). The symmetry condition $A^* = A$ will be imposed for some sharp results in Lipschitz domains.

Fix a cut-off function $\eta_\varepsilon \in C_0^\infty(\Omega)$ such that

$$\begin{cases} 0 \le \eta_\varepsilon \le 1, & |\nabla \eta_\varepsilon| \le C/\varepsilon,\\ \eta_\varepsilon(x) = 1 & \text{if } \operatorname{dist}(x, \partial\Omega) \ge 4\varepsilon,\\ \eta_\varepsilon(x) = 0 & \text{if } \operatorname{dist}(x, \partial\Omega) \le 3\varepsilon. \end{cases} \tag{3.2.1}$$

Let $S_\varepsilon^2 = S_\varepsilon \circ S_\varepsilon$ and define

$$w_\varepsilon = u_\varepsilon - u_0 - \varepsilon \chi(x/\varepsilon)\eta_\varepsilon S^2_\varepsilon(\nabla u_0), \tag{3.2.2}$$

where $u_\varepsilon \in H^1(\Omega; \mathbb{R}^m)$ is the weak solution of the Dirichlet problem (3.0.1) and u_0 the homogenized solution.

Lemma 3.2.1. *Let Ω be a bounded Lipschitz domain in $\mathbb{R}^d$ and Ω_t be defined by* (3.1.14). *Then, for any $\psi \in H_0^1(\Omega; \mathbb{R}^m)$,*

$$\begin{aligned} &\Big|\int_\Omega A(x/\varepsilon)\nabla w_\varepsilon \cdot \nabla\psi\, dx\Big| \\ &\le C\, \|\nabla\psi\|_{L^2(\Omega)} \Big\{ \varepsilon \|S_\varepsilon(\nabla^2 u_0)\|_{L^2(\Omega\setminus\Omega_{3\varepsilon})} + \|\nabla u_0 - S_\varepsilon(\nabla u_0)\|_{L^2(\Omega\setminus\Omega_{2\varepsilon})}\Big\} \\ &\qquad + C\, \|\nabla\psi\|_{L^2(\Omega_{4\varepsilon})} \|\nabla u_0\|_{L^2(\Omega_{5\varepsilon})}, \end{aligned} \tag{3.2.3}$$

where w_ε is given by (3.2.2) *and C depends only on μ and Ω.*

Proof. Since $w_\varepsilon \in H^1(\Omega; \mathbb{R}^m)$, by a density argument, it suffices to prove (3.2.3) for any $\psi \in C_0^\infty(\Omega; \mathbb{R}^m)$. Note that

$$\begin{aligned} A(x/\varepsilon)\nabla w_\varepsilon &= A(x/\varepsilon)\nabla u_\varepsilon - A(x/\varepsilon)\nabla u_0 - A(x/\varepsilon)\nabla\chi(x/\varepsilon)\eta_\varepsilon S_\varepsilon^2(\nabla u_0) \\ &\quad - \varepsilon A(x/\varepsilon)\chi(x/\varepsilon)\nabla\big(\eta_\varepsilon S_\varepsilon^2(\nabla u_0)\big) \\ &= A(x/\varepsilon)\nabla u_\varepsilon - \widehat{A}\nabla u_0 + \big(\widehat{A} - A(x/\varepsilon)\big)\big[\nabla u_0 - \eta_\varepsilon S_\varepsilon^2(\nabla u_0)\big] \\ &\quad - B(x/\varepsilon)\eta_\varepsilon S_\varepsilon^2(\nabla u_0) - \varepsilon A(x/\varepsilon)\chi(x/\varepsilon)\nabla\big(\eta_\varepsilon S_\varepsilon^2(\nabla u_0)\big), \end{aligned}$$

where we have used the fact

$$B(y) = A(y) + A(y)\nabla\chi(y) - \widehat{A}. \tag{3.2.4}$$

Since

$$\int_\Omega A(x/\varepsilon)\nabla u_\varepsilon \cdot \nabla\psi\, dx = \int_\Omega \widehat{A}\nabla u_0 \cdot \nabla\psi\, dx \tag{3.2.5}$$

for any $\psi \in C_0^\infty(\Omega; \mathbb{R}^m)$, we obtain

$$\begin{aligned} &\Big|\int_\Omega A(x/\varepsilon)\nabla w_\varepsilon \cdot \nabla\psi\, dx\Big| \\ &\le C\int_\Omega (1-\eta_\varepsilon)|\nabla u_0||\nabla\psi|\, dx + C\int_\Omega \eta_\varepsilon|\nabla u_0 - S_\varepsilon^2(\nabla u_0)|\,|\nabla\psi|\, dx \\ &\quad + \Big|\int_\Omega \eta_\varepsilon B(x/\varepsilon) S_\varepsilon^2(\nabla u_0)\cdot\nabla\psi\, dx\Big| \\ &\quad + C\varepsilon\int_\Omega |\chi(x/\varepsilon)\nabla\big(\eta_\varepsilon S_\varepsilon^2(\nabla u_0)\big)||\nabla\psi|\, dx. \end{aligned} \tag{3.2.6}$$

Since $\eta_\varepsilon = 1$ in $\Omega \setminus \Omega_{4\varepsilon}$, by the Cauchy inequality, the first term in the RHS of (3.2.6) is bounded by

$$C\, \|\nabla u_0\|_{L^2(\Omega_{4\varepsilon})} \|\nabla\psi\|_{L^2(\Omega_{4\varepsilon})}.$$

Using that $\eta_\varepsilon = 0$ in $\Omega_{3\varepsilon}$ and

$$\begin{aligned} &\|\nabla u_0 - S_\varepsilon^2(\nabla u_0)\|_{L^2(\Omega\setminus\Omega_{3\varepsilon})} \\ &\quad \le \|\nabla u_0 - S_\varepsilon(\nabla u_0)\|_{L^2(\Omega\setminus\Omega_{3\varepsilon})} + \|S_\varepsilon(\nabla u_0) - S_\varepsilon^2(\nabla u_0)\|_{L^2(\Omega\setminus\Omega_{3\varepsilon})} \\ &\quad \le C\, \|\nabla u_0 - S_\varepsilon(\nabla u_0)\|_{L^2(\Omega\setminus\Omega_{2\varepsilon})}, \end{aligned}$$

we can bound the second term by

$$C\,\|\nabla u_0 - S_\varepsilon(\nabla u_0)\|_{L^2(\Omega\setminus\Omega_{2\varepsilon})}\|\nabla\psi\|_{L^2(\Omega)}.$$

Also, by the Cauchy inequality and (3.1.10), the fourth term in the RHS of (3.2.6) is dominated by

$$C\,\|\nabla u_0\|_{L^2(\Omega_{5\varepsilon})}\|\nabla\psi\|_{L^2(\Omega_{4\varepsilon})} + C\varepsilon\,\|S_\varepsilon(\nabla^2 u_0)\|_{L^2(\Omega\setminus\Omega_{2\varepsilon})}\|\nabla\psi\|_{L^2(\Omega)}.$$

Finally, to handle the third term in the RHS of (3.2.6), we use the identity (3.1.5) to obtain

$$\begin{aligned}\eta_\varepsilon B(x/\varepsilon)S_\varepsilon^2(\nabla u_0)\cdot\nabla\psi &= b_{ij}^{\alpha\beta}(x/\varepsilon)S_\varepsilon^2\left(\frac{\partial u_0^\beta}{\partial x_j}\right)\frac{\partial\psi^\alpha}{\partial x_i}\eta_\varepsilon\\ &= \varepsilon\frac{\partial}{\partial x_k}\left(\phi_{kij}^{\alpha\beta}(x/\varepsilon)\frac{\partial\psi^\alpha}{\partial x_i}\right)S_\varepsilon^2\left(\frac{\partial u_0^\beta}{\partial x_j}\right)\eta_\varepsilon.\end{aligned} \tag{3.2.7}$$

It follows from (3.2.7), via integration by parts, that

$$\begin{aligned}&\left|\int_\Omega \eta_\varepsilon B(x/\varepsilon)S_\varepsilon^2(\nabla u_0)\cdot\nabla\psi\,dx\right|\\ &\quad\le C\varepsilon\int_\Omega \eta_\varepsilon|\phi(x/\varepsilon)||\nabla\psi||S_\varepsilon^2(\nabla^2 u_0)|\,dx\\ &\qquad + C\varepsilon\int_\Omega |\nabla\eta_\varepsilon||\phi(x/\varepsilon)||\nabla\psi||S_\varepsilon^2(\nabla u_0)|\,dx\\ &\quad\le C\varepsilon\,\|\nabla\psi\|_{L^2(\Omega)}\|S_\varepsilon(\nabla^2 u_0)\|_{L^2(\Omega\setminus\Omega_{2\varepsilon})} + C\,\|\nabla\psi\|_{L^2(\Omega_{4\varepsilon})}\|\nabla u_0\|_{L^2(\Omega_{5\varepsilon})},\end{aligned}$$

where we have used the Cauchy inequality and (3.1.10) for the last step. This completes the proof. □

The next theorem gives the $O(\sqrt{\varepsilon})$ convergence rate in H^1. We will use $W^{k,p}(\Omega;\mathbb{R}^m)$ to denote the Sobolev space of functions in $L^p(\Omega;\mathbb{R}^m)$ whose weak derivatives up to the kth order are in $L^p(\Omega)$. As usual, $H^k(\Omega;\mathbb{R}^m) = W^{k,2}(\Omega;\mathbb{R}^m)$.

Theorem 3.2.2. *Assume that A is 1-periodic and satisfies (2.1.2)–(2.1.3). Let Ω be a bounded Lipschitz domain in $\mathbb{R}^d$. Let w_ε be given by (3.2.2). Then for $0<\varepsilon<1$,*

$$\|\nabla w_\varepsilon\|_{L^2(\Omega)} \le C\Big\{\varepsilon\|\nabla^2 u_0\|_{L^2(\Omega\setminus\Omega_\varepsilon)} + \|\nabla u_0\|_{L^2(\Omega_{5\varepsilon})}\Big\}. \tag{3.2.8}$$

Consequently,

$$\|w_\varepsilon\|_{H_0^1(\Omega)} \le C\sqrt{\varepsilon}\,\|u_0\|_{H^2(\Omega)}. \tag{3.2.9}$$

The constant C depends only on μ and Ω.

Proof. Since $w_\varepsilon \in H_0^1(\Omega; \mathbb{R}^m)$, we may take $\psi = w_\varepsilon$ in (3.2.3). This, together with the ellipticity condition (2.1.3), gives

$$\|\nabla w_\varepsilon\|_{L^2(\Omega)} \le C\varepsilon \, \|\nabla^2 u_0\|_{L^2(\Omega\setminus\Omega_{2\varepsilon})} + C \, \|\nabla u_0 - S_\varepsilon(\nabla u_0)\|_{L^2(\Omega\setminus\Omega_{2\varepsilon})} + C \, \|\nabla u_0\|_{L^2(\Omega_{5\varepsilon})}. \tag{3.2.10}$$

Choose $\widetilde{\eta}_\varepsilon \in C_0^\infty(\Omega)$ such that $0 \le \widetilde{\eta}_\varepsilon \le 1$, $\widetilde{\eta}_\varepsilon = 0$ in Ω_ε, $\widetilde{\eta}_\varepsilon = 1$ in $\Omega \setminus \Omega_{3\varepsilon/2}$, and $|\nabla \widetilde{\eta}| \le C\varepsilon^{-1}$. It follows that

$$\begin{aligned} \|\nabla u_0 - S_\varepsilon(\nabla u_0)\|_{L^2(\Omega\setminus\Omega_{2\varepsilon})} &\le \|\widetilde{\eta}_\varepsilon(\nabla u_0) - S_\varepsilon(\widetilde{\eta}_\varepsilon \nabla u_0)\|_{L^2(\mathbb{R}^d)} \\ &\le C\varepsilon \, \|\nabla(\widetilde{\eta}_\varepsilon \nabla u_0)\|_{L^2(\mathbb{R}^d)} \\ &\le C\Big\{\varepsilon \|\nabla^2 u_0\|_{L^2(\Omega\setminus\Omega_\varepsilon)} + \|\nabla u_0\|_{L^2(\Omega_{2\varepsilon})}\Big\}, \end{aligned} \tag{3.2.11}$$

where for the second inequality we have used (3.1.12). The estimate (3.2.8) now follows from (3.2.10) and (3.2.11). Note that, by (3.1.15),

$$\|\nabla u_0\|_{L^2(\Omega_{5\varepsilon})} \le C\sqrt{\varepsilon}\, \|u_0\|_{H^2(\Omega)}. \tag{3.2.12}$$

The inequality (3.2.9) follows readily from (3.2.8) and (3.2.12). □

Under the additional symmetry condition $A^* = A$, i.e., $a_{ij}^{\alpha\beta} = a_{ji}^{\beta\alpha}$, an improved estimate can be obtained using sharp regularity estimates for $\mathcal{L}_0$.

Theorem 3.2.3. *Suppose that A is 1-periodic and satisfies (2.1.2)–(2.1.3). Also assume that $A^* = A$. Let Ω be a bounded Lipschitz domain in $\mathbb{R}^d$. Let w_ε be given by (3.2.2). Then, for $0 < \varepsilon < 1$,*

$$\|w_\varepsilon\|_{H_0^1(\Omega)} \le C\sqrt{\varepsilon}\Big\{\|F\|_{L^q(\Omega)} + \|f\|_{H^1(\partial\Omega)}\Big\}, \tag{3.2.13}$$

where $q = \frac{2d}{d+1}$ and C depends only on μ and Ω.

Definition 3.2.4. For a continuous function u in a bounded Lipschitz domain Ω, the nontangential maximal function of u is defined by

$$(u)^*(x) = \sup\Big\{|u(y)| : \ y \in \Omega \text{ and } |y - x| < C_0 \operatorname{dist}(y, \partial\Omega)\Big\} \tag{3.2.14}$$

for $x \in \partial\Omega$, where $C_0 > 1$ is a sufficiently large constant depending on Ω. See Section 8.1 for more details on $(u)^*$.

The proof of Theorem 3.2.3 relies on the following regularity result for solutions of $\mathcal{L}_0(u) = 0$ in Ω.

Lemma 3.2.5. *Assume that A satisfies the same conditions as in Theorem 3.2.3. Let Ω be a bounded Lipschitz domain in $\mathbb{R}^d$. Let $u \in H^1(\Omega; \mathbb{R}^m)$ be a weak solution to the Dirichlet problem: $\mathcal{L}_0(u) = 0$ in Ω and $u = f$ on $\partial\Omega$, where $f \in H^1(\partial\Omega; \mathbb{R}^m)$. Then*

$$\|(\nabla u)^*\|_{L^2(\partial\Omega)} \le C \, \|f\|_{H^1(\partial\Omega)}, \tag{3.2.15}$$

where C depends only on μ and Ω.

Proof. By Lemmas 2.2.4 and 2.2.5, $\widehat{A}$ satisfies the Legendre–Hadamard ellipticity condition (2.2.28) and is symmetric. As a result, the estimate (3.2.15) follows from [42, 43]. We refer the reader to Chapter 8 for estimates in terms of nontangential maximal functions in Lipschitz domains. In particular, the estimate (3.2.15) is proved for solutions of $\mathcal{L}_\varepsilon(u_\varepsilon) = 0$ in Ω under the conditions (2.1.20) and (2.0.2) as well as the Hölder continuity condition on A. □

Proof of Theorem 3.2.3. We start by taking $\psi = w_\varepsilon \in H_0^1(\Omega; \mathbb{R}^m)$ in (3.2.3). By the ellipticity condition (2.1.3), this gives

$$\begin{aligned} \|\nabla w_\varepsilon\|_{L^2(\Omega)} \le C\varepsilon \, \|S_\varepsilon(\nabla^2 u_0)\|_{L^2(\Omega\setminus\Omega_{3\varepsilon})} + C \, \|\nabla u_0\|_{L^2(\Omega_{5\varepsilon})} \\ + C \, \|\nabla u_0 - S_\varepsilon(\nabla u_0)\|_{L^2(\Omega\setminus\Omega_{2\varepsilon})}. \end{aligned} \tag{3.2.16}$$

To bound the RHS of (3.2.16), we write $u_0 = v_0 + \phi$, where

$$v_0(x) = \int_\Omega \Gamma_0(x-y) F(y)\, dy \tag{3.2.17}$$

and $\Gamma_0(x)$ denotes the matrix of fundamental solutions for the homogenized operator $\mathcal{L}_0$ in $\mathbb{R}^d$, with pole at the origin. It follows from the standard Calderón–Zygmund estimates for singular integrals (see Section 4.2 and [95, Chapter 2]) that

$$\|\nabla^2 v_0\|_{L^r(\mathbb{R}^d)} \le C \, \|F\|_{L^r(\Omega)}, \tag{3.2.18}$$

for any $1 < r < \infty$. Since $|\nabla \Gamma_0(x)| \le C|x|^{1-d}$, by the fractional integral estimates in Proposition 3.1.9,

$$\|\nabla v_0\|_{L^p(\mathbb{R}^d)} \le C \, \|F\|_{L^q(\Omega)}, \tag{3.2.19}$$

where $p = \frac{2d}{d-1}$ and $q = p' = \frac{2d}{d+1}$. These estimates, together with (3.1.13) and (3.1.15), yield that

$$\begin{aligned} \varepsilon \|S_\varepsilon(\nabla^2 v_0)\|_{L^2(\Omega\setminus\Omega_{3\varepsilon})} &\le C\varepsilon^{1/2} \|\nabla^2 v_0\|_{L^q(\mathbb{R}^d)} \\ &\le C\varepsilon^{1/2} \|F\|_{L^q(\Omega)}, \end{aligned}$$

and that

$$\begin{aligned} \|\nabla v_0\|_{L^2(\Omega_{5\varepsilon})} &\le C\varepsilon^{1/2} \|\nabla v_0\|_{W^{1,q}(\Omega)} \\ &\le C\varepsilon^{1/2} \|F\|_{L^q(\Omega)}. \end{aligned}$$

Also, note that by (3.1.13),

$$\begin{aligned} \|\nabla v_0 - S_\varepsilon(\nabla v_0)\|_{L^2(\Omega\setminus\Omega_{2\varepsilon})} &\le C\varepsilon^{1/2} \|\nabla^2 v_0\|_{L^q(\mathbb{R}^d)} \\ &\le C\varepsilon^{1/2} \|F\|_{L^q(\Omega)}. \end{aligned}$$

In summary, we have proved that

$$\begin{aligned} \varepsilon \|S_\varepsilon(\nabla^2 v_0)\|_{L^2(\Omega\setminus\Omega_{3\varepsilon})} + \|\nabla v_0\|_{L^2(\Omega_{5\varepsilon})} \\ + \|\nabla v_0 - S_\varepsilon(\nabla v_0)\|_{L^2(\Omega\setminus\Omega_{2\varepsilon})} \le C\varepsilon^{1/2} \|F\|_{L^q(\Omega)}. \end{aligned} \tag{3.2.20}$$

It remains to bound the LHS of (3.2.20), with v_0 replaced by ϕ. To this end we first note that $\mathcal{L}_0(\phi) = 0$ in Ω and $\phi = f - v_0$ on $\partial\Omega$. This allows us to apply Lemma 3.2.5. Since

$$\|v_0\|_{H^1(\partial\Omega)} \le C\, \|v_0\|_{W^{2,q}(\Omega)} \le C\, \|F\|_{L^q(\Omega)},$$

where we have used (3.1.15) for the first inequality, we obtain

$$\begin{aligned} \|(\nabla\phi)^*\|_{L^2(\partial\Omega)} &\le C\Big\{\|f\|_{H^1(\partial\Omega)} + \|v_0\|_{H^1(\partial\Omega)}\Big\} \\ &\le C\Big\{\|f\|_{H^1(\partial\Omega)} + \|F\|_{L^q(\Omega)}\Big\}. \end{aligned} \tag{3.2.21}$$

It follows that

$$\begin{aligned} \|\nabla\phi\|_{L^2(\Omega_{5\varepsilon})} &\le C\varepsilon^{1/2}\|(\nabla\phi)^*\|_{L^2(\partial\Omega)} \\ &\le C\varepsilon^{1/2}\Big\{\|f\|_{H^1(\partial\Omega)} + \|F\|_{L^q(\Omega)}\Big\}. \end{aligned}$$

Next, we use the interior estimate for $\mathcal{L}_0$,

$$|\nabla^2\phi(x)|^2 \le \frac{C}{r^{d+2}} \int_{B(x,r)} |\nabla\phi(y)|^2\, dy,$$

where $r = \text{dist}(x, \partial\Omega)/8$, and Fubini's Theorem to obtain

$$\begin{aligned} \int_{\Omega\setminus\Omega_\varepsilon} |\nabla^2\phi(x)|^2\, dx &\le C \int_{\Omega\setminus\Omega_{\varepsilon/2}} |\nabla\phi(x)|^2 \big[\text{dist}(x, \partial\Omega)\big]^{-2}\, dx \\ &\le C\varepsilon^{-1} \int_{\partial\Omega} |(\nabla\phi)^*|^2\, d\sigma. \end{aligned}$$

Hence,

$$\begin{aligned} \varepsilon\|S_\varepsilon(\nabla^2\phi)\|_{L^2(\Omega\setminus\Omega_{3\varepsilon})} &\le C\varepsilon\, \|\nabla^2\phi\|_{L^2(\Omega\setminus\Omega_{2\varepsilon})} \\ &\le C\varepsilon^{1/2}\|(\nabla\phi)^*\|_{L^2(\partial\Omega)} \\ &\le C\varepsilon^{1/2}\Big\{\|f\|_{H^1(\partial\Omega)} + \|F\|_{L^q(\Omega)}\Big\}. \end{aligned}$$

Finally, we observe that as in (3.2.11),

$$\begin{aligned} \|\nabla\phi - S_\varepsilon(\nabla\phi)\|_{L^2(\Omega\setminus\Omega_{2\varepsilon})} &\le C\Big\{\varepsilon\, \|\nabla^2\phi\|_{L^2(\Omega\setminus\Omega_\varepsilon)} + \|\nabla\phi\|_{L^2(\Omega_{2\varepsilon})}\Big\} \\ &\le C\varepsilon^{1/2}\Big\{\|f\|_{H^1(\partial\Omega)} + \|F\|_{L^q(\Omega)}\Big\}, \end{aligned} \tag{3.2.22}$$

where for the second inequality we have used (3.1.12). As a result, we have proved that

$$\begin{aligned} &\varepsilon\, \|S_\varepsilon(\nabla^2\phi)\|_{L^2(\Omega\setminus\Omega_{3\varepsilon})} + \|\nabla\phi\|_{L^2(\Omega_{5\varepsilon})} + \|\nabla\phi - S_\varepsilon(\nabla\phi)\|_{L^2(\Omega\setminus\Omega_{2\varepsilon})} \\ &\qquad \le C\varepsilon^{1/2}\Big\{\|f\|_{H^1(\partial\Omega)} + \|F\|_{L^q(\Omega)}\Big\}. \end{aligned}$$

This, together with (3.2.16) and (3.2.20), gives (3.2.13). □

Remark 3.2.6. The proof of Theorem 3.2.3 only uses the symmetry of $\widehat{A}$. Thus we may drop the assumption $A^* = A$ in the scalar case $m = 1$, as $\mathcal{L}_0(u) = (1/2)(\mathcal{L}_0 + \mathcal{L}_0^*)(u)$. Also, since

$$\|F\|_{L^q(\Omega)} = \|\mathcal{L}_0(u_0)\|_{L^q(\Omega)} \leq C\,\|\nabla^2 u_0\|_{L^q(\Omega)}$$

and $\|f\|_{H^1(\partial\Omega)} \leq C\,\|u_0\|_{W^{2,q}(\Omega)}$, it follows from (3.2.13) that

$$\|w_\varepsilon\|_{H_0^1(\Omega)} \leq C\sqrt{\varepsilon}\,\|u_0\|_{W^{2,q}(\Omega)}, \tag{3.2.23}$$

where C depends only on μ and Ω.

We now consider the two-scale expansions without the ε-smoothing.

Theorem 3.2.7. *Assume that A is 1-periodic and satisfies (2.1.2)–(2.1.3). Let Ω be a bounded Lipschitz domain in $\mathbb{R}^d$. Let $u_\varepsilon \in H^1(\Omega;\mathbb{R}^m)$ be the weak solution of the Dirichlet problem (3.0.1) and u_0 the homogenized solution. Then, if $u_0 \in W^{2,d}(\Omega;\mathbb{R}^m)$,*

$$\|u_\varepsilon - u_0 - \varepsilon\chi(x/\varepsilon)\nabla u_0\|_{H^1(\Omega)} \leq C\sqrt{\varepsilon}\,\|u_0\|_{W^{2,d}(\Omega)}, \tag{3.2.24}$$

where $0 < \varepsilon < 1$ and C depends only on μ and Ω. Furthermore, if the corrector χ is bounded and $u_0 \in H^2(\Omega;\mathbb{R}^m)$, then

$$\|u_\varepsilon - u_0 - \varepsilon\chi(x/\varepsilon)\nabla u_0\|_{H^1(\Omega)} \leq C\sqrt{\varepsilon}\,\|u_0\|_{H^2(\Omega)}, \tag{3.2.25}$$

where $0 < \varepsilon < 1$ and C depends only on μ, $\|\chi\|_\infty$ and Ω.

To prove Theorem 3.2.7, we need to control the L^2 norm of $\nabla\chi(x/\varepsilon)\psi$.

Lemma 3.2.8. *Let $\chi = \left(\chi_j^{\alpha\beta}\right)$ be the matrix of correctors defined by (2.2.14). Then*

$$\|\nabla\chi(x/\varepsilon)\psi\|_{L^2(\Omega)} \leq C\Big\{\varepsilon\,\|\nabla\psi\|_{L^d(\Omega)} + \|\psi\|_{L^d(\Omega)}\Big\} \tag{3.2.26}$$

for any $\psi \in W^{1,d}(\Omega)$, where C depends only on μ and Ω. Moreover, if χ is bounded, then

$$\|\nabla\chi(x/\varepsilon)\psi\|_{L^2(\Omega)} \leq C(1 + \|\chi\|_\infty)\Big\{\varepsilon\,\|\nabla\psi\|_{L^2(\Omega)} + \|\psi\|_{L^2(\Omega)}\Big\} \tag{3.2.27}$$

for any $\psi \in H^1(\Omega)$, where C depends only on μ and Ω.

Proof. Let

$$u_\varepsilon = \varepsilon\chi_j^\beta(x/\varepsilon) + P_j^\beta(x - x_0).$$

Since $\mathcal{L}_\varepsilon(u_\varepsilon) = 0$ in $\mathbb{R}^d$, it follows by Theorem 2.1.3 that

$$\int_{\mathbb{R}^d} |\nabla u_\varepsilon|^2|\varphi|^2\,dx \leq C\int_{\mathbb{R}^d} |u_\varepsilon|^2|\nabla\varphi|^2\,dx$$

for any $\varphi \in C_0^1(\mathbb{R}^d)$. Thus, if $\varphi \in C_0^1(B(x_0, 2\varepsilon))$, then

$$\int_{B(x_0,2\varepsilon)} |\nabla\chi(x/\varepsilon)|^2|\varphi|^2\, dx \le C\int_{B(x_0,2\varepsilon)} |\varphi|^2 + C\varepsilon^2 \int_{B(x_0,2\varepsilon)} |\nabla\varphi|^2\, dx \\ + C\varepsilon^2 \int_{B(x_0,2\varepsilon)} |\chi(x/\varepsilon)|^2|\nabla\varphi|^2\, dx.$$

Let $\varphi = \psi\widetilde{\eta}_\varepsilon$, where $\psi \in C_0^1(\mathbb{R}^d)$ and $\widetilde{\eta}_\varepsilon$ is a cut-off function in $C_0^1(B(x_0, 2\varepsilon))$ with the properties that $0 \le \widetilde{\eta}_\varepsilon \le 1$, $\widetilde{\eta}_\varepsilon = 1$ on $B(x_0, \varepsilon)$ and $|\nabla\widetilde{\eta}_\varepsilon| \le C/\varepsilon$. We obtain

$$\int_{B(x_0,\varepsilon)} |\nabla\chi(x/\varepsilon)|^2|\psi|^2\, dx \le C\int_{B(x_0,2\varepsilon)} (1 + |\chi(x/\varepsilon)|^2)|\psi|^2\, dx \\ + C\varepsilon^2 \int_{B(x_0,2\varepsilon)} (1 + |\chi(x/\varepsilon)|^2)|\nabla\psi|^2\, dx.$$

By integrating the inequality above in x_0 over $\mathbb{R}^d$, we see that

$$\begin{aligned} &\int_{\mathbb{R}^d} |\nabla\chi(x/\varepsilon)|^2|\psi|^2\, dx \\ &\le C\int_{\mathbb{R}^d} (1 + |\chi(x/\varepsilon)|^2)|\psi|^2\, dx + C\varepsilon^2 \int_{\mathbb{R}^d} (1 + |\chi(x/\varepsilon)|^2)|\nabla\psi|^2\, dx \end{aligned} \tag{3.2.28}$$

for any $\psi \in C_0^1(\mathbb{R}^d)$.

If χ is bounded, it follows from (3.2.28) that

$$\|\nabla\chi(x/\varepsilon)\psi\|_{L^2(\Omega)} \le C(1 + \|\chi\|_\infty)\Big\{\|\psi\|_{L^2(\mathbb{R}^d)} + \varepsilon\, \|\nabla\psi\|_{L^2(\mathbb{R}^d)}\Big\}$$

for any $\psi \in C_0^1(\mathbb{R}^d)$. By a limiting argument, the inequality holds for any $\psi \in H^1(\mathbb{R}^d)$. This gives (3.2.27), using the fact that for any $\psi \in H^1(\Omega)$, one may extend it to a function $\widetilde{\psi}$ in $H^1(\mathbb{R}^d)$ so that $\|\widetilde{\psi}\|_{L^2(\mathbb{R}^d)} \le C\, \|\psi\|_{L^2(\Omega)}$ and $\|\widetilde{\psi}\|_{H^1(\mathbb{R}^d)} \le C\, \|\psi\|_{H^1(\Omega)}$ [95, Chpater VI].

Finally, recall that if $d \ge 3$ and $m \ge 2$, $|\chi| \in L^q(Y)$ for $q = \frac{2d}{d-2}$. For any $\psi \in W^{1,d}(\Omega)$, we may extend it to a function $\widetilde{\psi}$ with compact support in $W^{1,d}(\mathbb{R}^d)$ so that

$$\|\widetilde{\psi}\|_{L^d(\mathbb{R}^d)} \le C\, \|\psi\|_{L^d(\Omega)}, \quad \|\widetilde{\psi}\|_{W^{1,d}(\mathbb{R}^d)} \le C\, \|\psi\|_{W^{1,d}(\Omega)},$$

and $\widetilde{\psi}(x) = 0$ if $\text{dist}(x, \Omega) \ge 1$. In view of (3.2.28) we obtain

$$\begin{aligned} \|\nabla\chi(x/\varepsilon)\psi\|_{L^2(\Omega)} &\le C\Big\{\|\widetilde{\psi}\|_{L^d(\mathbb{R}^d)} + \varepsilon\, \|\nabla\widetilde{\psi}\|_{L^d(\mathbb{R}^d)}\Big\} \\ &\le C\Big\{\|\psi\|_{L^d(\Omega)} + \varepsilon\, \|\nabla\psi\|_{L^d(\Omega)}\Big\}, \end{aligned}$$

using Hölder's inequality and the fact that

$$\|\chi(x/\varepsilon)\|_{L^q(B(x_0,R))} \le CR^{d/q}\|\chi\|_{L^q(Y)}.$$

This completes the proof. □

Proof of Theorem 3.2.7. Suppose χ is bounded. To prove (3.2.25), in view of (3.2.9), it suffices to show that

$$\|\varepsilon\chi(x/\varepsilon)\nabla u_0 - \varepsilon\chi(x/\varepsilon)\eta_\varepsilon S_\varepsilon^2(\nabla u_0)\|_{H^1(\Omega)} \le C\sqrt{\varepsilon}\,\|u_0\|_{H^2(\Omega)}. \tag{3.2.29}$$

To this end, we note that the LHS of (3.2.29) is bounded by

$$\begin{aligned}
&C\varepsilon\,\|\chi(x/\varepsilon)\big(\nabla u_0 - \eta_\varepsilon S_\varepsilon^2(\nabla u_0)\big)\|_{L^2(\Omega)}\\
&\qquad + C\,\|\nabla\chi(x/\varepsilon)\big(\nabla u_0 - \eta_\varepsilon S_\varepsilon^2(\nabla u_0)\big)\|_{L^2(\Omega)}\\
&\qquad + C\varepsilon\,\|\chi(x/\varepsilon)\nabla\big(\nabla u_0 - \eta_\varepsilon S_\varepsilon^2(\nabla u_0)\big)\|_{L^2(\Omega)}\\
&\quad \le C\varepsilon\,\|\nabla(\nabla u_0 - \eta_\varepsilon S_\varepsilon^2(\nabla u_0))\|_{L^2(\Omega)} + C\,\|\nabla u_0 - \eta_\varepsilon S_\varepsilon^2(\nabla u_0)\|_{L^2(\Omega)},
\end{aligned}$$

where we have used (3.2.27). Note that

$$\begin{aligned}
\varepsilon\,\|\nabla(\nabla u_0 - \eta_\varepsilon S_\varepsilon^2(\nabla u_0))\|_{L^2(\Omega)} &\le \varepsilon\,\|\nabla^2 u_0\|_{L^2(\Omega)} + \varepsilon\,\|\nabla\big(\eta_\varepsilon S_\varepsilon^2(\nabla u_0)\big)\|_{L^2(\Omega)}\\
&\le C\varepsilon\,\|u_0\|_{H^2(\Omega)} + C\,\|\nabla u_0\|_{L^2(\Omega_{5\varepsilon})}\\
&\le C\sqrt{\varepsilon}\,\|u_0\|_{H^2(\Omega)},
\end{aligned}$$

and

$$\begin{aligned}
\|\nabla u_0 - \eta_\varepsilon S_\varepsilon^2(\nabla u_0)\|_{L^2(\Omega)} &\le C\,\|\nabla u_0\|_{L^2(\Omega_{5\varepsilon})} + \|\nabla u_0 - S_\varepsilon^2(\nabla u_0)\|_{L^2(\Omega\setminus\Omega_{4\varepsilon})}\\
&\le C\sqrt{\varepsilon}\,\|u_0\|_{H^2(\Omega)},
\end{aligned}$$

where we also used (3.2.22) for the last inequality.

To prove (3.2.24) under the assumption $u_0 \in W^{2,d}(\Omega;\mathbb{R}^m)$, we extend u_0 to a function $\widetilde{u}_0$ in $W^{2,d}(\mathbb{R}^d;\mathbb{R}^m)$ so that $\|\widetilde{u}_0\|_{W^{2,d}(\mathbb{R}^d)} \le C\,\|u_0\|_{W^{2,d}(\Omega)}$. Observe that

$$\begin{aligned}
&\|\varepsilon\chi(x/\varepsilon)\nabla u_0 - \varepsilon\chi(x/\varepsilon)S_\varepsilon^2(\nabla\widetilde{u}_0)\|_{H^1(\Omega)}\\
&\quad\le C\varepsilon\,\|\chi(x/\varepsilon)\big(\nabla u_0 - S_\varepsilon^2(\nabla\widetilde{u}_0)\big)\|_{L^2(\Omega)}\\
&\qquad + C\,\|\nabla\chi(x/\varepsilon)\big(\nabla u_0 - S_\varepsilon^2(\nabla\widetilde{u}_0)\big)\|_{L^2(\Omega)}\\
&\qquad + C\varepsilon\,\|\chi(x/\varepsilon)\big(\nabla^2 u_0 - S_\varepsilon^2(\nabla^2\widetilde{u}_0)\big)\|_{L^2(\Omega)}\\
&\quad\le C\varepsilon\,\|\widetilde{u}_0\|_{W^{2,d}(\mathbb{R}^d)} + C\,\|\nabla\widetilde{u}_0 - S_\varepsilon^2(\nabla\widetilde{u}_0)\|_{L^d(\mathbb{R}^d)}\\
&\quad\le C\varepsilon\,\|u_0\|_{W^{2,d}(\Omega)},
\end{aligned}$$

where we have used Hölder's inequality and (3.2.26) for the second inequality and (3.1.12) for the last. Also, it follows from (3.1.16) that

$$\begin{aligned}
&\|\varepsilon\chi(x/\varepsilon)S_\varepsilon^2(\nabla\widetilde{u}_0) - \varepsilon\chi(x/\varepsilon)\eta_\varepsilon S_\varepsilon^2(\nabla u_0)\|_{H^1(\Omega)}\\
&\quad = \varepsilon\,\|\chi(x/\varepsilon)S_\varepsilon^2(\nabla\widetilde{u}_0)(1-\eta_\varepsilon)\|_{H^1(\Omega)}\\
&\quad \le C\sqrt{\varepsilon}\,\|\widetilde{u}_0\|_{H^2(\mathcal{O})} \le C\sqrt{\varepsilon}\,\|\widetilde{u}_0\|_{W^{2,d}(\mathbb{R}^d)}\\
&\quad \le C\sqrt{\varepsilon}\,\|u_0\|_{W^{2,d}(\Omega)},
\end{aligned}$$

where $\mathcal{O} = \{x \in \mathbb{R}^d : \operatorname{dist}(x, \Omega) < 1\}$. Thus, we have proved that

$$\|\varepsilon\chi(x/\varepsilon)\nabla u_0 - \varepsilon\chi(x/\varepsilon)\eta_\varepsilon S_\varepsilon^2(\nabla u_0)\|_{H^1(\Omega)} \le C\sqrt{\varepsilon}\,\|u_0\|_{W^{2,d}(\Omega)}.$$

This, together with (3.2.23), yields the estimate (3.2.24). □

Remark 3.2.9. For $\varepsilon \ge 0$, let $u_\varepsilon \in H^1(\Omega; \mathbb{R}^m)$ be the weak solution to the Dirichlet problem

$$\mathcal{L}_\varepsilon(u_\varepsilon) = F_\varepsilon \quad \text{in } \Omega \quad \text{and} \quad u_\varepsilon = f_\varepsilon \quad \text{on } \partial\Omega, \tag{3.2.30}$$

where $F_\varepsilon \in H^{-1}(\Omega; \mathbb{R}^m)$ and $f_\varepsilon \in H^{1/2}(\Omega; \mathbb{R}^m)$. Then

$$\begin{aligned} &\|u_\varepsilon - u_0 - \varepsilon\chi(x/\varepsilon)\nabla u_0\|_{H^1(\Omega)} \\ &\le \begin{cases} C\Big\{\sqrt{\varepsilon}\,\|u_0\|_{W^{2,d}(\Omega)} + \|F_\varepsilon - F_0\|_{H^{-1}(\Omega)} + \|f_\varepsilon - f_0\|_{H^{1/2}(\partial\Omega)}\Big\}, \\ C\Big\{\sqrt{\varepsilon}\,\|u_0\|_{H^2(\Omega)} + \|F_\varepsilon - F_0\|_{H^{-1}(\Omega)} + \|f_\varepsilon - f_0\|_{H^{1/2}(\partial\Omega)}\Big\} \\ \text{if } \chi \text{ is bounded.} \end{cases} \end{aligned}$$

To see this one applies Theorem 3.2.7 to the weak solution of $\mathcal{L}_\varepsilon(v_\varepsilon) = F_0$ in Ω with $v_\varepsilon = f_0$ on $\partial\Omega$ and uses Theorem 2.1.5 to estimate $\|u_\varepsilon - v_\varepsilon\|_{H^1(\Omega)}$.

3.3 Convergence rates in H^1 for the Neumann problem

In this section we extend the results in Section 3.2 to solutions of the Neumann problem (3.0.2). Throughout this section we assume that A satisfies the Legendre ellipticity condition (2.1.20).

Lemma 3.3.1. *Let u_ε be the solution of* (3.0.2) *with $\int_\Omega u_\varepsilon\,dx = 0$, and u_0 the homogenized solution. Let w_ε be defined as in* (3.2.2). *Then the inequality* (3.2.3) *holds for any $\psi \in H^1(\Omega; \mathbb{R}^m)$.*

Proof. Since $w_\varepsilon \in H^1(\Omega; \mathbb{R}^m)$, it suffices to prove (3.2.3) for $\psi \in C^\infty(\mathbb{R}^d; \mathbb{R}^m)$. Since

$$\int_\Omega A(x/\varepsilon)\nabla u_\varepsilon \cdot \nabla\psi\,dx = \int_\Omega \widehat{A}\nabla u_0 \cdot \nabla\psi\,dx$$

for any $\psi \in C^\infty(\mathbb{R}^d; \mathbb{R}^m)$, we see that the inequality (3.2.6) continues to hold for any $\psi \in C^\infty(\mathbb{R}^d; \mathbb{R}^m)$. The rest of the proof is exactly the same as that of Lemma 3.2.1. □

The following theorem gives the $O(\sqrt{\varepsilon})$ convergence rate for the Neumann problem.

Theorem 3.3.2. *Assume that A is 1-periodic and satisfies* (2.1.20). *Let Ω be a bounded Lipschitz domain in $\mathbb{R}^d$. Let w_ε be the same as in Lemma* 3.3.1. *Then*

$$\|w_\varepsilon\|_{H^1(\Omega)} \le C\Big\{\varepsilon\,\|\nabla^2 u_0\|_{L^2(\Omega\setminus\Omega_\varepsilon)} + \varepsilon\,\|\nabla u_0\|_{L^2(\Omega)} + \|\nabla u_0\|_{L^2(\Omega_{5\varepsilon})}\Big\} \tag{3.3.1}$$

for $0<\varepsilon<1$. Consequently, if $u_0\in H^2(\Omega;\mathbb{R}^m)$,

$$\|w_\varepsilon\|_{H^1(\Omega)} \le C\sqrt{\varepsilon}\, \|u_0\|_{H^2(\Omega)}. \tag{3.3.2}$$

The constant C depends only on μ and Ω.

Proof. Since $w_\varepsilon\in H^1(\Omega;\mathbb{R}^m)$, by Lemma 3.3.1, we may take $\psi=w_\varepsilon$ in (3.2.3). This, together with the ellipticity condition (2.1.20), gives

$$\begin{aligned}
&\|\nabla w_\varepsilon\|_{L^2(\Omega)}\\
&\le C\Big\{\varepsilon\, \|\nabla^2 u_0\|_{L^2(\Omega\setminus\Omega_{2\varepsilon})} + \|\nabla u_0 - S_\varepsilon(\nabla u_0)\|_{L^2(\Omega\setminus\Omega_{2\varepsilon})} + \|\nabla u_0\|_{L^2(\Omega_{5\varepsilon})}\Big\}\\
&\le C\Big\{\varepsilon\, \|\nabla^2 u_0\|_{L^2(\Omega\setminus\Omega_{\varepsilon})} + \|\nabla u_0\|_{L^2(\Omega_{5\varepsilon})}\Big\},
\end{aligned} \tag{3.3.3}$$

where the second inequality follows by the same argument as in the proof of Theorem 3.2.2. By the Poincaré inequality, we obtain

$$\begin{aligned}
\|w_\varepsilon\|_{H^1(\Omega)} &\le C\, \|\nabla w_\varepsilon\|_{L^2(\Omega)} + C\Big|\int_\Omega \varepsilon\chi(x/\varepsilon)\eta_\varepsilon S_\varepsilon^2(\nabla u_0)\, dx\Big|\\
&\le C\, \|\nabla w_\varepsilon\|_{L^2(\Omega)} + C\varepsilon\, \|\nabla u_0\|_{L^2(\Omega)},
\end{aligned}$$

where we have used the fact that $\int_\Omega u_\varepsilon\, dx = \int_\Omega u_0\, dx = 0$. Estimates (3.3.1) and (3.3.2) now follow from (3.3.3). ☐

As in the case of the Dirichlet problem, the estimates in Theorem 3.3.2 can be improved under the additional symmetry condition.

Lemma 3.3.3. *Assume that $(\widehat{A})^* = \widehat{A}$. Let $u\in H^1(\Omega;\mathbb{R}^m)$ be a weak solution to the Neumann problem: $\mathcal{L}_0(u)=0$ in Ω and $\frac{\partial u}{\partial \nu_0}=g$ on $\partial\Omega$, where $g\in L^2(\partial\Omega;\mathbb{R}^m)$ and $\int_{\partial\Omega} g\, d\sigma=0$. Then*

$$\|(\nabla u)^*\|_{L^2(\partial\Omega)} \le C\, \|g\|_{L^2(\partial\Omega)}, \tag{3.3.4}$$

where C depends only on μ and Ω.

Proof. This was proved in [42, 36]. See Chapter 8. ☐

Theorem 3.3.4. *Suppose that A is 1-periodic and satisfies (2.1.20). Further, assume that $A^*=A$. Let Ω be a bounded Lipschitz domain in $\mathbb{R}^d$. Let w_ε be the same as in Lemma 3.3.1. Then*

$$\|w_\varepsilon\|_{H^1(\Omega)} \le C\sqrt{\varepsilon}\Big\{\|F\|_{L^q(\Omega)} + \|g\|_{L^2(\partial\Omega)}\Big\}, \tag{3.3.5}$$

where $q=\frac{2d}{d+1}$ and C depends only on μ and Ω.

Proof. It follows from Lemma 3.3.1 by letting $\psi = w_\varepsilon \in H^1(\Omega;\mathbb{R}^m)$ and using (2.1.20) that

$$\begin{aligned} \|\nabla w_\varepsilon\|_{L^2(\Omega)} \le C\varepsilon \,\|S_\varepsilon(\nabla^2 u_0)\|_{L^2(\Omega\setminus\Omega_{2\varepsilon})} \\ + C\,\|\nabla u_0 - S_\varepsilon(\nabla u_0)\|_{L^2(\Omega\setminus\Omega_{2\varepsilon})} + C\,\|\nabla u_0\|_{L^2(\Omega_{5\varepsilon})}. \end{aligned} \tag{3.3.6}$$

To prove (3.3.5) under the assumption $A^* = A$, it suffices to bound the RHS of (3.3.6) by the RHS of (3.3.5). We proceed as in the proof of Theorem 3.2.3 by writing $u_0 = v_0 + \phi$, where v_0 is given by (3.2.17). The terms involving v_0 are handled exactly in the same manner as before. Similarly, to control the terms involving ϕ, it suffices to bound the $L^2(\partial\Omega)$ norm of $(\nabla\phi)^*$. To do this, we note that

$$\mathcal{L}_0(\phi) = 0 \quad \text{in } \Omega \quad \text{and} \quad \frac{\partial \phi}{\partial \nu_0} = g - \frac{\partial v_0}{\partial \nu_0} \quad \text{on } \partial\Omega.$$

By Lemma 3.3.3,

$$\|(\nabla\phi)^*\|_{L^2(\partial\Omega)} \le C\,\|g\|_{L^2(\partial\Omega)} + C\,\|\nabla v_0\|_{L^2(\partial\Omega)}.$$

Note that

$$\|\nabla v_0\|_{L^2(\partial\Omega)} \le C\,\|v_0\|_{W^{2,q}(\Omega)} \le C\,\|F\|_{L^q(\Omega)}.$$

Thus,

$$\|(\nabla\phi)^*\|_{L^2(\partial\Omega)} \le C\Big\{\|g\|_{L^2(\partial\Omega)} + \|F\|_{L^q(\Omega)}\Big\},$$

which completes the proof. □

Theorem 3.3.5. *Assume that A is 1-periodic and satisfies (2.1.20). Let Ω be a bounded Lipschitz domain in $\mathbb{R}^d$. Let $u_\varepsilon \in H^1(\Omega;\mathbb{R}^m)$ be the weak solution of the Neumann problem (3.0.2) with $\int_\Omega u_\varepsilon\, dx = 0$. Let u_0 be the homogenized solution. Then, if $u_0 \in W^{2,d}(\Omega;\mathbb{R}^m)$ and $0 < \varepsilon < 1$,*

$$\|u_\varepsilon - u_0 - \varepsilon\chi(x/\varepsilon)\nabla u_0\|_{H^1(\Omega)} \le C\sqrt{\varepsilon}\,\|u_0\|_{W^{2,d}(\Omega)}, \tag{3.3.7}$$

where C depends only on μ and Ω. Furthermore, if the corrector χ is bounded and $u_0 \in H^2(\Omega;\mathbb{R}^m)$, then

$$\|u_\varepsilon - u_0 - \varepsilon\chi(x/\varepsilon)\nabla u_0\|_{H^1(\Omega)} \le C\sqrt{\varepsilon}\,\|u_0\|_{H^2(\Omega)} \tag{3.3.8}$$

for any $0 < \varepsilon < 1$, where C depends only on μ, $\|\chi\|_\infty$ and Ω.

Proof. The proof is exactly the same as that of Theorem 3.2.7, where the Dirichlet boundary condition is never used. □

Remark 3.3.6. For $\varepsilon \ge 0$, let $u_\varepsilon \in H^1(\Omega;\mathbb{R}^m)$ be the weak solution to the Neumann problem

$$\mathcal{L}_\varepsilon(u_\varepsilon) = F_\varepsilon \quad \text{in } \Omega \quad \text{and} \quad \frac{\partial u_\varepsilon}{\partial \nu_\varepsilon} = g_\varepsilon, \quad \text{on } \partial\Omega, \tag{3.3.9}$$

with $\int_\Omega u_\varepsilon \, dx = 0$, where $F_\varepsilon \in L^2(\Omega; \mathbb{R}^m)$, $g_\varepsilon \in L^2(\partial\Omega; \mathbb{R}^m)$ and $\int_{\partial\Omega} g_\varepsilon \, d\sigma = 0$. Then

$$
\begin{aligned}
&\|u_\varepsilon - u_0 - \varepsilon\chi(x/\varepsilon)\nabla u_0\|_{H^1(\Omega)} \\
&\le \begin{cases} C\left\{\sqrt{\varepsilon}\, \|u_0\|_{W^{2,d}(\Omega)} + \|F_\varepsilon - F_0\|_{L^2(\Omega)} + \|g_\varepsilon - g_0\|_{H^{-1/2}(\partial\Omega)}\right\}, \\ C\left\{\sqrt{\varepsilon}\, \|u_0\|_{H^2(\Omega)} + \|F_\varepsilon - F_0\|_{L^2(\Omega)} + \|g_\varepsilon - g_0\|_{H^{-1/2}(\partial\Omega)}\right\} \\ \text{if } \chi \text{ is bounded.} \end{cases}
\end{aligned}
$$

To see this, we use the inequality

$$
\begin{aligned}
&\|u_\varepsilon - u_0 - \varepsilon\chi(x/\varepsilon)\nabla u_0\|_{H^1(\Omega)} \\
&\qquad \le \|v_\varepsilon - u_0 - \varepsilon\chi(x/\varepsilon)\nabla u_0\|_{H^1(\Omega)} + \|u_\varepsilon - v_\varepsilon\|_{H^1(\Omega)},
\end{aligned}
$$

where v_ε is the solution of

$$
\mathcal{L}_\varepsilon(v_\varepsilon) = F_0 \quad \text{in } \Omega \quad \text{and} \quad \frac{\partial v_\varepsilon}{\partial \nu_\varepsilon} = g_0,
$$

such that $\int_\Omega v_\varepsilon \, dx = 0$. By Theorem 3.3.5, $\|v_\varepsilon - u_0 - \varepsilon\chi(x/\varepsilon)\nabla u_0\|_{H^1(\Omega)}$ is bounded by $C\sqrt{\varepsilon}\, \|u_0\|_{W^{2,d}(\Omega)}$, and by $C\sqrt{\varepsilon}\, \|u_0\|_{H^2(\Omega)}$ if χ is bounded. Since $\mathcal{L}_\varepsilon(u_\varepsilon - v_\varepsilon) = F_\varepsilon - F_0$ in Ω and $\frac{\partial}{\partial \nu_\varepsilon}(u_\varepsilon - v_\varepsilon) = g_\varepsilon - g_0$ on $\partial\Omega$, Theorem 2.1.7 shows that

$$
\|u_\varepsilon - v_\varepsilon\|_{H^1(\Omega)} \le C\left\{\|F_\varepsilon - F_0\|_{L^2(\Omega)} + \|g_\varepsilon - g_0\|_{H^{-1/2}(\partial\Omega)}\right\}.
$$

3.4 Convergence rates in L^p for the Dirichlet problem

In this section we establish a sharp $O(\varepsilon)$ convergence rate in L^p with $p = \frac{2d}{d-1}$ for the Dirichlet problem (3.0.1), under the assumptions that A satisfies (2.1.2)–(2.1.3) and $A^* = A$. Without the symmetry condition we obtain an $O(\varepsilon)$ convergence rate in L^2.

Lemma 3.4.1. *Let Ω be a bounded Lipschitz domain in $\mathbb{R}^d$. Let w_ε be given by (3.2.2), where u_ε is the weak solution of (3.0.1) and u_0 the homogenized solution. Assume that $u_0 \in W^{2,q}(\Omega; \mathbb{R}^m)$ for $q = \frac{2d}{d+1}$. Then, for any $\psi \in C_0^\infty(\Omega; \mathbb{R}^m)$,*

$$
\begin{aligned}
&\left| \int_{\mathbb{R}^d} A(x/\varepsilon)\nabla w_\varepsilon \cdot \nabla\psi \, dx \right| \\
&\qquad \le C\, \|u_0\|_{W^{2,q}(\Omega)} \left\{ \varepsilon\, \|\nabla\psi\|_{L^p(\Omega)} + \sqrt{\varepsilon}\, \|\nabla\psi\|_{L^2(\Omega_{4\varepsilon})} \right\},
\end{aligned} \tag{3.4.1}
$$

where $p = q' = \frac{2d}{d-1}$ and C depends only on μ and Ω.

Proof. An inspection of the proof of Lemma 3.2.1 shows that

$$\begin{aligned}\Big|\int_\Omega A(x/\varepsilon)\nabla w_\varepsilon\cdot\nabla\psi\,dx\Big| &\le C\,\|\nabla\psi\|_{L^p(\Omega)}\Big\{\varepsilon\,\|S_\varepsilon(\nabla^2 u_0)\|_{L^q(\Omega\setminus\Omega_{3\varepsilon})}+\|\nabla u_0-S_\varepsilon(\nabla u_0)\|_{L^q(\Omega\setminus\Omega_{2\varepsilon})}\Big\}\\&\quad+C\,\|\nabla\psi\|_{L^2(\Omega_{4\varepsilon})}\|\nabla u_0\|_{L^2(\Omega_{5\varepsilon})}.\end{aligned} \tag{3.4.2}$$

Note that $\|\nabla u_0\|_{L^2(\Omega_{5\varepsilon})}\le C\sqrt{\varepsilon}\,\|u_0\|_{W^{2,q}(\Omega)}$ and

$$\|S_\varepsilon(\nabla^2 u_0)\|_{L^q(\Omega\setminus\Omega_{3\varepsilon})}\le C\,\|\nabla^2 u_0\|_{L^q(\Omega)}.$$

Since $u_0\in W^{2,q}(\Omega;\mathbb{R}^m)$, there exists $\widetilde{u}_0\in W^{2,q}(\mathbb{R}^d;\mathbb{R}^m)$ such that $\widetilde{u}_0=u_0$ in Ω and

$$\|\widetilde{u}_0\|_{W^{2,q}(\mathbb{R}^d)}\le C\,\|u_0\|_{W^{2,q}(\Omega)}.$$

It follows that

$$\begin{aligned}\|\nabla u_0-S_\varepsilon(\nabla u_0)\|_{L^q(\Omega\setminus\Omega_{3\varepsilon})}&\le\|\nabla\widetilde{u}_0-S_\varepsilon(\nabla\widetilde{u}_0)\|_{L^q(\mathbb{R}^d)}\\&\le C\varepsilon\,\|\nabla^2\widetilde{u}_0\|_{L^q(\mathbb{R}^d)}\\&\le C\varepsilon\,\|u_0\|_{W^{2,q}(\Omega)},\end{aligned} \tag{3.4.3}$$

where we have used (3.1.12) for the second inequality. In conjunction with (3.4.2), this establishes (3.4.1). □

Lemma 3.4.2. *Assume that $\big(\widehat{A}\big)^*=\widehat{A}$. Let Ω be a bounded Lipschitz domain in $\mathbb{R}^d$. Let $u\in H_0^1(\Omega;\mathbb{R}^m)$ be a weak solution to the Dirichlet problem: $\mathcal{L}_0(u)=F$ in Ω and $u=0$ on $\partial\Omega$, where $F\in C_0^\infty(\Omega;\mathbb{R}^m)$. Then*

$$\|\nabla u\|_{L^2(\Omega_t)}\le Ct^{1/2}\|F\|_{L^q(\Omega)}, \tag{3.4.4}$$

for $0<t<\operatorname{diam}(\Omega)$, and

$$\|\nabla u\|_{L^p(\Omega)}\le C\,\|F\|_{L^q(\Omega)}, \tag{3.4.5}$$

where $p=\frac{2d}{d-1}$, $q=p'=\frac{2d}{d+1}$, and C depends only on μ and Ω.

Proof. Write $u=\phi+v_0$, where v_0 is defined by (3.2.17). The proof of (3.4.4) is essentially contained in that of Theorem 3.2.3. We leave it as an exercise for the reader. To see (3.4.5), we recall that, by the fractional integral estimates, $\|\nabla v_0\|_{L^p(\Omega)}\le C\|F\|_{L^q(\Omega)}$. To estimate $\nabla\phi$, we use the inequality

$$\left(\int_\Omega|\psi|^p\,dx\right)^{1/p}\le C\left(\int_{\partial\Omega}|(\psi)^*|^2\,d\sigma\right)^{1/2}, \tag{3.4.6}$$

where $p = \frac{2d}{d-1}$, ψ is a continuous function in Ω and $(\psi)^*$ denotes the nontangential maximal function of ψ. This gives

$$\begin{aligned} \|\nabla \phi\|_{L^p(\Omega)} &\le C \,\|(\nabla \phi)^*\|_{L^2(\partial\Omega)} \le C\, \|\nabla v_0\|_{L^2(\partial\Omega)} \\ &\le C\, \|v_0\|_{W^{2,q}(\Omega)} \le C\, \|F\|_{L^q(\Omega)}, \end{aligned}$$

where we have used Lemma 3.2.5 for the second inequality and (3.1.15) for the third.

Finally, to prove (3.4.6), we observe that for any $x \in \Omega$ and $\hat{x} \in \partial\Omega$ with $|x - \hat{x}| = \text{dist}(x, \partial\Omega) = r$,

$$|\psi(x)| \le (\psi)^*(y),$$

if $y \in \partial\Omega$ and $|y - \hat{x}| \le cr$. It follows that

$$\begin{aligned} |\psi(x)| &\le \frac{C}{r^{d-1}} \int_{B(\hat{x}, cr)\cap\partial\Omega} |(\psi)^*|\, d\sigma \\ &\le C \int_{\partial\Omega} \frac{(\psi)^*(y)}{|x-y|^{d-1}}\, d\sigma(y). \end{aligned} \tag{3.4.7}$$

Let $f \in C_0^1(\Omega)$ and

$$g(y) = \int_\Omega \frac{|f(x)|}{|x-y|^{d-1}} dx.$$

Note that, by (3.4.7),

$$\begin{aligned} \Big| \int_\Omega \psi(x) f(x)\, dx \Big| &\le C \int_\Omega \int_{\partial\Omega} \frac{(\psi)^*(y)|f(x)|}{|x-y|^{d-1}}\, d\sigma(y) dx \\ &= C \int_{\partial\Omega} (\psi)^* g\, d\sigma \\ &\le C\, \|(\psi)^*\|_{L^2(\partial\Omega)} \|g\|_{L^2(\partial\Omega)} \\ &\le C\, \|(\psi)^*\|_{L^2(\partial\Omega)} \|g\|_{W^{1,q}(\Omega)} \\ &\le C\, \|(\psi)^*\|_{L^2(\partial\Omega)} \|f\|_{L^q(\Omega)}, \end{aligned}$$

where $q = \frac{2d}{d+1}$ and we have used (3.1.15) for the third inequality and singular integral estimates for the fourth. The inequality (3.4.6) follows by a duality argument. □

The next theorem gives a sharp $O(\varepsilon)$ convergence rate in L^p with $p = \frac{2d}{d-1}$.

Theorem 3.4.3. *Assume that A is 1-periodic and satisfies (2.1.2)–(2.1.3). Also assume that $A^* = A$. Let Ω be a bounded Lipschitz domain in $\mathbb{R}^d$. Let u_ε ($\varepsilon \ge 0$) be the weak solution of (3.0.1). Assume that $u_0 \in W^{2,q}(\Omega; \mathbb{R}^d)$ for $q = \frac{2d}{d+1}$. Then*

$$\|u_\varepsilon - u_0\|_{L^p(\Omega)} \le C\varepsilon\, \|u_0\|_{W^{2,q}(\Omega)}, \tag{3.4.8}$$

where $p = q' = \frac{2d}{d-1}$ and C depends only on μ and Ω.

Proof. Let w_ε be given by (3.2.2). For any $G \in C_0^\infty(\Omega; \mathbb{R}^m)$, let $v_\varepsilon \in H_0^1(\Omega; \mathbb{R}^m)$ ($\varepsilon \geq 0$) be the weak solution to the Dirichlet problem,

$$\mathcal{L}_\varepsilon^*(v_\varepsilon) = G \quad \text{in } \Omega \quad \text{and} \quad v_\varepsilon = 0 \quad \text{on } \partial\Omega. \tag{3.4.9}$$

Define

$$r_\varepsilon = v_\varepsilon - v_0 - \varepsilon\chi(x/\varepsilon)\eta_\varepsilon S_\varepsilon^2(\nabla v_0),$$

where η_ε is a cut-off function satisfying (3.2.1). Observe that

$$\begin{aligned}
&\Big|\int_\Omega w_\varepsilon \cdot G\, dx\Big| \\
&= \Big|\int_\Omega A(x/\varepsilon)\nabla w_\varepsilon \cdot \nabla v_\varepsilon\, dx\Big| \\
&\leq \Big|\int_\Omega A(x/\varepsilon)\nabla w_\varepsilon \cdot \nabla r_\varepsilon\, dx\Big| + \Big|\int_\Omega A(x/\varepsilon)\nabla w_\varepsilon \cdot \nabla v_0\, dx\Big| \\
&\quad + \Big|\int_\Omega A(x/\varepsilon)\nabla w_\varepsilon \cdot \nabla\big(\varepsilon\chi(x/\varepsilon)\eta_\varepsilon S_\varepsilon^2(\nabla v_0)\big)\, dx\Big| \\
&= I_1 + I_2 + I_3.
\end{aligned} \tag{3.4.10}$$

To estimate I_1, we note that by (3.2.23) and (3.2.13),

$$\|\nabla w_\varepsilon\|_{L^2(\Omega)} \leq C\sqrt{\varepsilon}\, \|u_0\|_{W^{2,q}(\Omega)} \quad \text{and} \quad \|\nabla r_\varepsilon\|_{L^2(\Omega)} \leq C\sqrt{\varepsilon}\, \|G\|_{L^q(\Omega)},$$

where we have used the assumption $A^* = A$. By the Cauchy inequality, this gives

$$I_1 \leq C\varepsilon\, \|u_0\|_{W^{2,q}(\Omega)} \|G\|_{L^q(\Omega)}. \tag{3.4.11}$$

Next, to bound I_2, we apply Lemma 3.4.1 to obtain

$$\begin{aligned}
I_2 &\leq C\, \|u_0\|_{W^{2,q}(\Omega)} \Big\{\varepsilon\, \|\nabla v_0\|_{L^p(\Omega)} + \sqrt{\varepsilon}\, \|\nabla v_0\|_{L^2(\Omega_{4\varepsilon})}\Big\} \\
&\leq C\varepsilon\, \|u_0\|_{W^{2,q}(\Omega)} \|G\|_{L^q(\Omega)},
\end{aligned} \tag{3.4.12}$$

where we used Lemma 3.4.2 for the last step.

To estimate I_3, we let

$$\varphi_\varepsilon = \varepsilon\chi(x/\varepsilon)\eta_\varepsilon S_\varepsilon^2(\nabla v_0).$$

Using (3.1.9) as well as the observation

$$\|\nabla S_\varepsilon^2(f)\|_{L^p(\mathbb{R}^d)} \leq C\varepsilon^{-1} \|f\|_{L^p(\mathbb{R}^d)},$$

it is not hard to show that

$$\varepsilon\, \|\nabla\varphi_\varepsilon\|_{L^p(\Omega)} + \sqrt{\varepsilon}\, \|\nabla\varphi_\varepsilon\|_{L^2(\Omega_{4\varepsilon})} \leq C\varepsilon\, \|\nabla v_0\|_{L^p(\Omega)} + C\sqrt{\varepsilon}\, \|\nabla v_0\|_{L^2(\Omega_{5\varepsilon})}.$$

As in the case of I_2, by Lemma 3.4.1, this implies that

$$\begin{aligned} I_3 &\le C\,\|u_0\|_{W^{2,q}(\Omega)}\Big\{\varepsilon\,\|\nabla\varphi_\varepsilon\|_{L^p(\Omega)}+\sqrt{\varepsilon}\,\|\nabla\varphi_\varepsilon\|_{L^2(\Omega_{4\varepsilon})}\Big\}\\ &\le C\varepsilon\,\|u_0\|_{W^{2,q}(\Omega)}\|G\|_{L^q(\Omega)}. \end{aligned} \tag{3.4.13}$$

Thus, in view of (3.4.10)–(3.4.13), we have proved that

$$\Big|\int_\Omega w_\varepsilon\cdot G\,dx\Big| \le C\varepsilon\,\|u_0\|_{W^{2,q}(\Omega)}\|G\|_{L^q(\Omega)}, \tag{3.4.14}$$

where C depends only on μ and Ω. By duality, this implies that

$$\|w_\varepsilon\|_{L^p(\Omega)} \le C\varepsilon\,\|u_0\|_{W^{2,q}(\Omega)}.$$

It follows that

$$\begin{aligned} \|u_\varepsilon-u_0\|_{L^p(\Omega)} &\le \|w_\varepsilon\|_{L^p(\Omega)}+\|\varepsilon\chi(x/\varepsilon)\eta_\varepsilon S_\varepsilon^2(\nabla u_0)\|_{L^p(\Omega)}\\ &\le C\varepsilon\,\|u_0\|_{W^{2,q}(\Omega)}. \end{aligned}$$

□

Remark 3.4.4. Since $p=\frac{2d}{d-1}$ and $q=\frac{2d}{d+1}$, we have

$$\frac{1}{q}-\frac{1}{p}=\frac{1}{d}.$$

It follows that the estimate (3.4.8) is scale-invariant. Consequently, the constant C in the estimate can be made to be independent of $\operatorname{diam}(\Omega)$, if $\operatorname{diam}(\Omega)$ is large. Indeed, suppose that $\mathcal{L}_\varepsilon(u_\varepsilon)=\mathcal{L}_0(u_0)$ in Ω_R and $u_\varepsilon=u_0$ on $\partial\Omega_R$, where $R>1$ and $\Omega_R=\{x\in\mathbb{R}^d : x/R\in\Omega\}$. Let $v_\varepsilon(x)=u_\varepsilon(Rx)$ and $v_0(x)=u_0(Rx)$ for $x\in\Omega$. Then

$$\mathcal{L}_{\frac{\varepsilon}{R}}(v_\varepsilon)=\mathcal{L}_0(v_0) \quad \text{in } \Omega \quad \text{and} \quad v_\varepsilon=v_0 \quad \text{on } \partial\Omega.$$

It follows that

$$\|v_\varepsilon-v_0\|_{L^p(\Omega)} \le C\left(\frac{\varepsilon}{R}\right)\|v_0\|_{W^{2,q}(\Omega)}.$$

Using $\frac{1}{q}-\frac{1}{p}=\frac{1}{d}$, by a change of variables, we see that

$$\|u_\varepsilon-u_0\|_{L^p(\Omega_R)} \le C\varepsilon\,\|u_0\|_{W^{2,q}(\Omega_R)},$$

where C depends only on μ and Ω.

The next theorem gives the $O(\varepsilon)$ convergence rate in L^2 without the symmetry condition, assuming that Ω is a bounded $C^{1,1}$ domain (see Section 8.1 for the definition of $C^{k,\alpha}$ domains). The smoothness assumption on Ω ensures the H^2 estimate for $\mathcal{L}_0$.

Theorem 3.4.5. *Assume that A is 1-periodic and satisfies (2.1.2)–(2.1.3). Let Ω be a bounded $C^{1,1}$ domain in $\mathbb{R}^d$. Let $u_\varepsilon \in H^1(\Omega;\mathbb{R}^m)$ ($\varepsilon \geq 0$) be the weak solution of problem (3.0.1). Assume that $u_0 \in H^2(\Omega;\mathbb{R}^m)$. Then*

$$\|u_\varepsilon - u_0\|_{L^2(\Omega)} \leq C\varepsilon \, \|u_0\|_{H^2(\Omega)}, \tag{3.4.15}$$

where C depends only on μ and Ω.

Proof. The proof is similar to that of Theorem 3.4.3. By Theorem 3.2.2, the estimates

$$\|w_\varepsilon\|_{H_0^1(\Omega)} \leq C\sqrt{\varepsilon}\,\|u_0\|_{H^2(\Omega)} \quad \text{and} \quad \|r_\varepsilon\|_{H_0^1(\Omega)} \leq C\sqrt{\varepsilon}\,\|v_0\|_{H^2(\Omega)}$$

hold without the symmetry condition. It follows from the proof of Theorem 3.4.3 that

$$\left| \int_\Omega w_\varepsilon \cdot G \, dx \right| \leq C\varepsilon \, \|u_0\|_{H^2(\Omega)} \|v_0\|_{H^2(\Omega)}. \tag{3.4.16}$$

Recall that $v_0 \in H_0^1(\Omega;\mathbb{R}^m)$ is a solution of $\mathcal{L}_0(v_0) = G$ in Ω. Since Ω is $C^{1,1}$ and $\mathcal{L}_0$ is a second-order elliptic operator with constant coefficients, it is known that $v_0 \in H^2(\Omega;\mathbb{R}^m)$ and $\|v_0\|_{H^2(\Omega)} \leq C\,\|G\|_{L^2(\Omega)}$ (see [26, Chapter 8] or [76, Chapter 4]). This, together with (3.4.16), gives

$$\left| \int_\Omega w_\varepsilon \cdot G \, dx \right| \leq C\varepsilon \, \|u_0\|_{H^2(\Omega)} \|G\|_{L^2(\Omega)}.$$

By duality, we obtain $\|w_\varepsilon\|_{L^2(\Omega)} \leq C\varepsilon\,\|u_0\|_{H^2(\Omega)}$. Thus

$$\begin{aligned}\|u_\varepsilon - u_0\|_{L^2(\Omega)} &\leq C\varepsilon\,\|u_0\|_{H^2(\Omega)} + \|\varepsilon\chi(x/\varepsilon)\eta_\varepsilon S_\varepsilon^2(\nabla u_0)\|_{L^2(\Omega)} \\ &\leq C\varepsilon\,\|u_0\|_{H^2(\Omega)},\end{aligned}$$

which completes the proof. □

3.5 Convergence rates in L^p for the Neumann problem

In this section we establish the $O(\varepsilon)$ convergence rate in L^2 for the Neumann problem (3.0.2) in a bounded $C^{1,1}$ domain Ω, under the assumptions that A is 1-periodic and satisfies the Legendre condition (2.1.20). With the additional symmetry condition, the $O(\varepsilon)$ rate is obtained in L^p, with $p = \frac{2d}{d-1}$, in a bounded Lipschitz domain.

Lemma 3.5.1. *Suppose that A is 1-periodic and satisfies (2.1.20). Let Ω be a bounded Lipschitz domain. Let w_ε be given by (3.2.2), where u_ε is the weak solution of problem (3.0.2) and u_0 the homogenized solution. Assume that $u_0 \in W^{2,q}(\Omega;\mathbb{R}^m)$ for $q = \frac{2d}{d+1}$. Then the inequality (3.4.1) holds for any $\psi \in C_0^\infty(\mathbb{R}^d;\mathbb{R}^m)$.*

Proof. The proof is exactly the same as that for Lemma 3.4.1. □

The symmetry condition is needed for the next lemma.

Lemma 3.5.2. *Suppose that A is 1-periodic and satisfies (2.1.20). Further, assume that $\widehat{A^*} = \widehat{A}$. Let Ω be a bounded Lipschitz domain. Let $u \in H^1(\Omega; \mathbb{R}^m)$ be a weak solution to the Neumann problem: $\mathcal{L}_0(u) = G$ in Ω and $\frac{\partial u}{\partial \nu_0} = 0$ on $\partial\Omega$, where $G \in C^\infty(\mathbb{R}^d; \mathbb{R}^m)$ and $\int_\Omega G\, dx = 0$. Then*

$$\|\nabla u\|_{L^2(\Omega_t)} \le C t^{1/2} \|G\|_{L^q(\Omega)}, \tag{3.5.1}$$

for $0 < t < \text{diam}(\Omega)$, and

$$\|\nabla u\|_{L^p(\Omega)} \le C\, \|G\|_{L^q(\Omega)}, \tag{3.5.2}$$

where $p = \frac{2d}{d-1}$, $q = p' = \frac{2d}{d+1}$, and C depends only on μ and Ω.

Proof. The proof is similar to that of Lemma 3.4.2. We leave it to the reader as an exercise. □

Theorem 3.5.3. *Suppose that A and Ω satisfy the same conditions as in Lemma 3.5.2. Let $u_\varepsilon \in H^1(\Omega; \mathbb{R}^m)$ ($\varepsilon \ge 0$) be the weak solution of the Neumann problem (3.0.2). Then, if $u_0 \in W^{2,q}(\Omega; \mathbb{R}^m)$,*

$$\|u_\varepsilon - u_0\|_{L^p(\Omega)} \le C\varepsilon\, \|u_0\|_{W^{2,q}(\Omega)}, \tag{3.5.3}$$

where $q = p' = \frac{2d}{d+1}$ and C depends only on μ and Ω.

Proof. The proof of (3.5.3) is similar to that in the case of the Dirichlet condition, using Theorem 3.3.4, Lemmas 3.5.1 and 3.5.2, and a duality argument.

Fix $G \in C^\infty(\mathbb{R}^d; \mathbb{R}^m)$ with $\int_\Omega G\, dx = 0$. Let $v_\varepsilon \in H^1(\Omega; \mathbb{R}^d)$ ($\varepsilon \ge 0$) be the weak solution to the Neumann problem,

$$\mathcal{L}_\varepsilon(v_\varepsilon) = G \quad \text{in } \Omega \quad \text{and} \quad \frac{\partial v_\varepsilon}{\partial \nu_\varepsilon} = 0 \quad \text{on } \partial\Omega, \tag{3.5.4}$$

with $\int_\Omega v_\varepsilon\, dx = 0$. As in the case of the Dirichlet problem, we can show that

$$\left| \int_\Omega w_\varepsilon \cdot G\, dx \right| \le C\varepsilon\, \|u_0\|_{W^{2,q}(\Omega)} \|G\|_{L^q(\Omega)},$$

where $w_\varepsilon = u_\varepsilon - u_0 - \varepsilon\chi(x/\varepsilon)\eta_\varepsilon S_\varepsilon^2(\nabla u_0)$. Since

$$\|\varepsilon\chi(x/\varepsilon)\eta_\varepsilon S_\varepsilon^2(\nabla u_0)\|_{L^p(\Omega)} \le C\varepsilon\, \|u_0\|_{W^{2,q}(\Omega)},$$

we then obtain

$$\left| \int_\Omega (u_\varepsilon - u_0) \cdot G\, dx \right| \le C\varepsilon\, \|u_0\|_{W^{2,q}(\Omega)} \|G\|_{L^q(\Omega)}.$$

Since $\int_\Omega (u_\varepsilon - u_0)\, dx = 0$, by duality, this gives the estimate (3.5.3). □

Without the symmetry condition, as in the case of the Dirichlet problem, an $O(\varepsilon)$ convergence rate is obtained in L^2 in a $C^{1,1}$ domain.

Theorem 3.5.4. *Assume that A is 1-periodic and satisfies (2.1.20). Let Ω be a bounded $C^{1,1}$ domain in $\mathbb{R}^d$. Let $u_\varepsilon \in H^1(\Omega; \mathbb{R}^m)$ ($\varepsilon \geq 0$) be the weak solution of problem (3.0.2). Assume that $u_0 \in H^2(\Omega; \mathbb{R}^m)$. Then*

$$\|u_\varepsilon - u_0\|_{L^2(\Omega)} \leq C\varepsilon \, \|u_0\|_{H^2(\Omega)}, \tag{3.5.5}$$

where C depends only on μ and Ω.

Proof. The proof is similar to that of Theorem 3.4.5. The $C^{1,1}$ assumption on Ω is used to ensure the H^2 estimate $\|\nabla^2 v\|_{L^2(\Omega)} \leq C\|G\|_{L^2(\Omega)}$ for solutions to the elliptic system $\mathcal{L}_0(v) = G$ in Ω with the Neumann condition $\frac{\partial v}{\partial \nu_0} = 0$ on $\partial\Omega$ [76]. The details are left to the reader. □

3.6 Convergence rates for elliptic systems of elasticity

In this section we study the convergence rates for elliptic systems of linear elasticity. Since the elasticity condition (2.4.1) implies (2.1.2)–(2.1.3) and the symmetry condition $A^* = A$, results obtained in Sections 3.2 and 3.4 for the Dirichlet problem hold for the system of elasticity.

Theorem 3.6.1. *Assume that $A \in E(\kappa_1, \kappa_2)$ and is 1-periodic. Let Ω be a bounded Lipschitz domain in $\mathbb{R}^d$. Let w_ε be defined by (3.2.2), where u_ε is the weak solution to the Dirichlet problem: $\mathcal{L}_\varepsilon(u_\varepsilon) = F$ in Ω and $u_\varepsilon = f$ on $\partial\Omega$. Then, for $0 < \varepsilon < 1$,*

$$\|w_\varepsilon\|_{H^1(\Omega)} \leq C\sqrt{\varepsilon}\left\{\|F\|_{L^q(\Omega)} + \|f\|_{H^1(\partial\Omega)}\right\}, \tag{3.6.1}$$

and

$$\|u_\varepsilon - u_0\|_{L^p(\Omega)} \leq C\varepsilon \, \|u_0\|_{W^{2,q}(\Omega)}, \tag{3.6.2}$$

where $q = \frac{2d}{d+1}$, $p = \frac{2d}{d-1}$, and C depends only on κ_1, κ_2, and Ω.

In the following we extend the results in Sections 3.3 and 3.5 to the Neumann problem

$$\begin{cases} \mathcal{L}_\varepsilon(u_\varepsilon) = F & \text{in } \Omega, \\ \dfrac{\partial u_\varepsilon}{\partial \nu_\varepsilon} = g & \text{on } \partial\Omega, \\ u_\varepsilon \perp \mathcal{R} \text{ in } L^2(\Omega; \mathbb{R}^d), \end{cases} \tag{3.6.3}$$

where $F \in L^2(\Omega; \mathbb{R}^d)$ and $g \in L^2(\partial\Omega; \mathbb{R}^d)$ satisfy the compatibility condition (2.4.9), and $\mathcal{R}$ denotes the space of rigid displacements, given by (2.4.8).

Lemma 3.6.2. *Let u_ε be the solution of (3.6.3) and u_0 the homogenized solution. Let w_ε be defined as in (3.2.2). Then the inequality (3.2.3) holds for any $\psi \in H^1(\Omega; \mathbb{R}^d)$.*

Proof. The proof is similar to that of Lemma 3.3.1. □

Let

$$L^p_{\mathcal{R}}(\partial\Omega) = \Big\{ g \in L^p(\partial\Omega; \mathbb{R}^d) : \int_{\partial\Omega} g \cdot \phi \, d\sigma = 0 \text{ for any } \phi \in \mathcal{R} \Big\}. \tag{3.6.4}$$

Lemma 3.6.3. *Let $u \in H^1(\Omega; \mathbb{R}^d)$ be a weak solution to the Neumann problem: $\mathcal{L}_0(u) = 0$ in Ω and $\frac{\partial u}{\partial \nu_0} = g$ on $\partial\Omega$, where $g \in L^2_{\mathcal{R}}(\partial\Omega)$. Suppose that $u \perp \mathcal{R}$ in $L^2(\Omega; \mathbb{R}^d)$. Then*

$$\|(\nabla u)^*\|_{L^2(\partial\Omega)} \le C \, \|g\|_{L^2(\partial\Omega)}, \tag{3.6.5}$$

where C depends only on κ_1, κ_2, and Ω.

Proof. This was proved in [36, 105]. Also see [47]. □

Theorem 3.6.4. *Let Ω be a bounded Lipschitz domain in $\mathbb{R}^d$. Let w_ε be the same as in Lemma 3.6.2. Then, for $0 < \varepsilon < 1$,*

$$\|w_\varepsilon\|_{H^1(\Omega)} \le C\sqrt{\varepsilon} \Big\{ \|F\|_{L^q(\Omega)} + \|g\|_{L^2(\partial\Omega)} \Big\}, \tag{3.6.6}$$

where $q = \frac{2d}{d+1}$ and C depends only on κ_1, κ_2, and Ω.

Proof. It follows from Lemma 3.6.2 that

$$\begin{aligned} &\|\nabla w_\varepsilon + (\nabla w_\varepsilon)^T\|^2_{L^2(\Omega)} \\ &\quad \le C\varepsilon \, \|\nabla w_\varepsilon\|_{L^2(\Omega)} \|S_\varepsilon(\nabla^2 u_0)\|_{L^2(\Omega\setminus\Omega_{2\varepsilon})} \\ &\qquad + C \, \|\nabla w_\varepsilon\|_{L^2(\Omega)} \Big\{ \|\nabla u_0 - S_\varepsilon(\nabla u_0)\|_{L^2(\Omega\setminus\Omega_{2\varepsilon})} + \|\nabla u_0\|_{L^2(\Omega_{5\varepsilon})} \Big\}. \end{aligned}$$

We then apply the second Korn inequality,

$$\|u\|_{H^1(\Omega)} \le C \, \|\nabla u + (\nabla u)^T\|_{L^2(\Omega)} + C \sum_{k=1}^{\ell} \Big| \int_\Omega u \cdot \phi_k \, dx \Big|, \tag{3.6.7}$$

where $\ell = \frac{d(d+1)}{2}$ and $\{\phi_k : k = 1, \dots, \ell\}$ forms an orthonormal basis for $\mathcal{R}$ as a subspace of $L^2(\Omega; \mathbb{R}^d)$. This leads to

$$\begin{aligned} \|w_\varepsilon\|_{H^1(\Omega)} &\le C\Big\{ \varepsilon \, \|S_\varepsilon(\nabla^2 u_0)\|_{L^2(\Omega\setminus\Omega_{2\varepsilon})} + \|\nabla u_0 - S_\varepsilon(\nabla u_0)\|_{L^2(\Omega\setminus\Omega_{2\varepsilon})} \Big\} \\ &\qquad + C \, \|\nabla u_0\|_{L^2(\Omega_{5\varepsilon})} + C \sum_{k=1}^{\ell} \Big| \int_\Omega \varepsilon \chi(x/\varepsilon) \eta_\varepsilon S_\varepsilon(\nabla u_0) \cdot \phi_k \, dx \Big| \\ &\le C\Big\{ \varepsilon \, \|S_\varepsilon(\nabla^2 u_0)\|_{L^2(\Omega\setminus\Omega_{2\varepsilon})} + \|\nabla u_0 - S_\varepsilon(\nabla u_0)\|_{L^2(\Omega\setminus\Omega_{2\varepsilon})} \\ &\qquad + \|\nabla u_0\|_{L^2(\Omega_{5\varepsilon})} + \varepsilon \, \|\nabla u_0\|_{L^2(\Omega)} \Big\}, \end{aligned}$$

where we have used the assumptions that $u_\varepsilon, u_0 \perp \mathcal{R}$ in $L^2(\Omega; \mathbb{R}^d)$.

Next, we proceed as in the proof of Theorem 3.2.3 by writing $u_0 = v_0 + \phi$, where v_0 is given by (3.2.17). The terms involving v_0 are handled exactly in the same manner as before. Similarly, to control the terms involving ϕ, it suffices to bound the $L^2(\partial\Omega)$ norm of $(\nabla\phi)^*$. To do this, we note that $\mathcal{L}_0(\phi) = 0$ in Ω and $\frac{\partial \phi}{\partial \nu_0} = g - \frac{\partial v_0}{\partial \nu_0}$ on $\partial\Omega$. It follows by Lemma 3.6.3 that

$$\|(\nabla\phi)^*\|_{L^2(\partial\Omega)} \le C\,\|g\|_{L^2(\partial\Omega)} + C\,\|\nabla v_0\|_{L^2(\partial\Omega)} + C\sum_{k=1}^{\ell}\left|\int_\Omega \phi\cdot\phi_k\, dx\right|.$$

Observe that

$$\|\nabla v_0\|_{L^2(\partial\Omega)} \le C\,\|v_0\|_{W^{2,q}(\Omega)} \le C\,\|F\|_{L^q(\Omega)},$$

and that

$$\left|\int_\Omega \phi\cdot\phi_k\, dx\right| = \left|\int_\Omega v_0\cdot\phi_k\, dx\right| \le C\,\|F\|_{L^q(\Omega)},$$

where we have used the fact that $u_0 \perp \mathcal{R}$ in $L^2(\Omega;\mathbb{R}^d)$. Thus,

$$\|(\nabla\phi)^*\|_{L^2(\partial\Omega)} \le C\Big\{\|g\|_{L^2(\partial\Omega)} + \|F\|_{L^q(\Omega)}\Big\}.$$

This completes the proof. □

Theorem 3.6.5. *Let Ω be a bounded Lipschitz domain in $\mathbb{R}^d$. Let u_ε be the weak solution in $H^1(\Omega;\mathbb{R}^d)$ of the Neumann problem (3.6.3). Let u_0 be the homogenized solution. Then, if $u_0 \in W^{2,d}(\Omega;\mathbb{R}^d)$,*

$$\|u_\varepsilon - u_0 - \varepsilon\chi(x/\varepsilon)\nabla u_0\|_{H^1(\Omega)} \le C\sqrt{\varepsilon}\,\|u_0\|_{W^{2,d}(\Omega)} \tag{3.6.8}$$

for any $0 < \varepsilon < 1$, where C depends only on κ_1, κ_2, and Ω. Furthermore, if the corrector χ is bounded and $u_0 \in H^2(\Omega;\mathbb{R}^d)$, then

$$\|u_\varepsilon - u_0 - \varepsilon\chi(x/\varepsilon)\nabla u_0\|_{H^1(\Omega)} \le C\sqrt{\varepsilon}\,\|u_0\|_{H^2(\Omega)} \tag{3.6.9}$$

for any $0 < \varepsilon < 1$, where C depends only on κ_1, κ_2, $\|\chi\|_\infty$ and Ω.

Proof. The proof is exactly the same as that of Theorem 3.2.7, where the Dirichlet boundary condition is never used. □

We now move to the sharp convergence rate in L^p with $p = \frac{2d}{d-1}$ for (3.6.3).

Lemma 3.6.6. *Let w_ε be given by (3.2.2), where u_ε is the weak solution of (3.6.3) and u_0 the homogenized solution. Assume that $u_0 \in W^{2,q}(\Omega;\mathbb{R}^d)$ for $q = \frac{2d}{d+1}$. Then the inequality (3.4.1) holds for any $\psi \in C_0^\infty(\mathbb{R}^d;\mathbb{R}^d)$.*

Proof. The proof is exactly the same as that for Lemma 3.4.1. □

Lemma 3.6.7. *Let $u \in H^1(\Omega; \mathbb{R}^d)$ be a weak solution to the Neumann problem: $\mathcal{L}_0(u) = G$ in Ω and $\frac{\partial u}{\partial \nu_0} = 0$ on $\partial\Omega$, where $G \in C^\infty(\mathbb{R}^d; \mathbb{R}^d)$ and $G \perp \mathcal{R}$ in $L^2(\Omega; \mathbb{R}^d)$. Assume that $u \perp \mathcal{R}$ in $L^2(\Omega; \mathbb{R}^d)$. Then*

$$\|\nabla u\|_{L^2(\Omega_t)} \le C t^{1/2} \|G\|_{L^q(\Omega)}, \tag{3.6.10}$$

$$\|\nabla u\|_{L^p(\Omega)} \le C \|G\|_{L^q(\Omega)}, \tag{3.6.11}$$

where $p = \frac{2d}{d-1}$, $q = p' = \frac{2d}{d+1}$, and C depends only on κ_1, κ_2, and Ω.

Proof. With Lemma 3.6.3 at our disposal, the proof is similar to that of Lemma 3.4.2. □

Theorem 3.6.8. *Let Ω be a bounded Lipschitz domain in $\mathbb{R}^d$. Let u_ε ($\varepsilon \ge 0$) be the weak solution in $H^1(\Omega; \mathbb{R}^d)$ of the Neumann problem (3.6.3). Then, if $u_0 \in W^{2,q}(\Omega; \mathbb{R}^d)$,*

$$\|u_\varepsilon - u_0\|_{L^p(\Omega)} \le C\varepsilon \|u_0\|_{W^{2,q}(\Omega)}, \tag{3.6.12}$$

where $q = p' = \frac{2d}{d+1}$ and C depends only on κ_1, κ_2, and Ω.

Proof. The proof of (3.6.12) is similar to that in the case of the Dirichlet boundary condition, using Theorem 3.6.4, 3.6.6, 3.6.7, and a duality argument.

Fix $G \in C^\infty(\mathbb{R}^d; \mathbb{R}^d)$ such that $G \perp \mathcal{R}$ in $L^2(\Omega; \mathbb{R}^d)$. Let $v_\varepsilon \in H^1(\Omega; \mathbb{R}^d)$ ($\varepsilon \ge 0$) be the weak solution to the Neumann problem,

$$\mathcal{L}_\varepsilon(v_\varepsilon) = G \quad \text{in } \Omega \quad \text{and} \quad \frac{\partial v_\varepsilon}{\partial \nu_\varepsilon} = 0 \quad \text{on } \partial\Omega, \tag{3.6.13}$$

with the property that $v_\varepsilon \perp \mathcal{R}$ in $L^2(\Omega; \mathbb{R}^d)$. As in the case of the Dirichlet problem, we have

$$\left| \int_\Omega w_\varepsilon \cdot G \, dx \right| \le C\varepsilon \|u_0\|_{W^{2,q}(\Omega)} \|G\|_{L^q(\Omega)},$$

where $w_\varepsilon = u_\varepsilon - u_0 - \varepsilon \chi(x/\varepsilon) \eta_\varepsilon S_\varepsilon^2(\nabla u_0)$. Using

$$\|\varepsilon \chi(x/\varepsilon) \eta_\varepsilon S_\varepsilon^2(\nabla u_\varepsilon)\|_{L^p(\Omega)} \le C\varepsilon \|u_0\|_{W^{2,q}(\Omega)},$$

we then obtain

$$\left| \int_\Omega (u_\varepsilon - u_0) \cdot G \, dx \right| \le C\varepsilon \|u_0\|_{W^{2,q}(\Omega)} \|G\|_{L^q(\Omega)}.$$

Since $u_\varepsilon, u_0 \perp \mathcal{R}$ in $L^2(\Omega; \mathbb{R}^d)$, by duality, this gives the estimate (3.6.12). □

3.7 Notes

There is an extensive literature on the problem of convergence rates in periodic homogenization. Early results, proved under smoothness conditions on the correctors $\chi = (\chi_j^\beta)$, may be found in books [19, 111, 78]. The flux correctors, defined by (3.1.3), were already used in [111].

In [53, 54, 109, 110, 112, 81, 79], various ε-smoothing techniques were introduced to treat the case where χ and $\nabla\chi$ may be unbounded. In particular, error estimates similar to (3.0.3) were proved in [112] for solutions of scalar second-order elliptic equations with the Dirichlet or Neumann boundary conditions. Further extensions were made in [80, 98], where the error estimate (3.0.3) in H^1 was established for a broader class of elliptic systems.

The error estimate (3.0.4) in H^1 for two-scale expansions in Lipschitz domains was proved in [90]. It follows from (3.0.4) that if $\mathcal{L}_\varepsilon(u_\varepsilon) = 0$ in a bounded Lipschitz domain Ω, then

$$\begin{cases} \|\nabla u_\varepsilon\|_{L^2(\Omega_\varepsilon)} \le C\sqrt{\varepsilon}\, \|u_\varepsilon\|_{H^1(\partial\Omega)}, \\ \|\nabla u_\varepsilon\|_{L^2(\Omega_\varepsilon)} \le C\sqrt{\varepsilon}\, \left\| \dfrac{\partial u_\varepsilon}{\partial \nu_\varepsilon} \right\|_{L^2(\partial\Omega)}, \end{cases} \tag{3.7.1}$$

where C depends only on μ and Ω. The inequalities in (3.7.1) should be regarded as large-scale Rellich estimates. Such estimates play an essential role in the study of L^2 boundary value problems for $\mathcal{L}_\varepsilon$ in Lipschitz domains. See Chapter 8 and [67, 68, 90, 47] for details.

The sharp convergence rate (3.0.6) in L^2 was obtained in [53, 54, 97, 98]. The duality method was used first in [97]. Also see related work in [63, 56, 107, 57, 91, 77] and references therein.

The scale-invariant estimate (3.0.5) is new. A weaker estimate,

$$\|u_\varepsilon - u_0\|_{L^p(\Omega)} \le C\varepsilon\, \|u_0\|_{H^2(\Omega)},$$

where $p = \frac{2d}{d-1}$, was proved in [90].

Chapter 4

Interior Estimates

In this chapter we establish interior Hölder ($C^{0,\alpha}$) estimates, $W^{1,p}$ estimates, and Lipschitz ($C^{0,1}$) estimates, that are uniform in $\varepsilon > 0$, for solutions of $\mathcal{L}_\varepsilon(u_\varepsilon) = F$, where $\mathcal{L}_\varepsilon = -\text{div}(A(x/\varepsilon)\nabla)$. As a result, we obtain uniform size estimates of $\Gamma_\varepsilon(x,y)$, $\nabla_x\Gamma_\varepsilon(x,y)$, $\nabla_y\Gamma_\varepsilon(x,y)$, and $\nabla_x\nabla_y\Gamma_\varepsilon(x,y)$, where $\Gamma_\varepsilon(x,y)$ denotes the matrix of fundamental solutions for $\mathcal{L}_\varepsilon$ in $\mathbb{R}^d$. This in turn allows us to derive asymptotic expansions, as $\varepsilon \to 0$, of $\Gamma_\varepsilon(x,y)$, $\nabla_x\Gamma(x,y)$, $\nabla_y\Gamma_\varepsilon(x,y)$, and $\nabla_x\nabla_y\Gamma_\varepsilon(x,y)$. Note that if $u_\varepsilon(x) = P_j^\beta(x) + \varepsilon\chi_j^\beta(x/\varepsilon)$, then $\nabla u_\varepsilon = \nabla P_j^\beta + \nabla\chi_j^\beta(x/\varepsilon)$ and $\mathcal{L}_\varepsilon(u_\varepsilon) = 0$ in $\mathbb{R}^d$. Thus no uniform regularity beyond Lipschitz estimates should be expected (unless $\text{div}(A) = 0$, which would imply $\chi_j^\beta = 0$).

In Section 4.1 we use a compactness method to establish a Lipschitz estimate down to the microscopic scale ε under the ellipticity and periodicity conditions. This, together with a simple blow-up argument, leads to the full Lipschitz estimate under an additional smoothness condition. The compactness method, which originated from the study of the regularity theory in the calculus of variations and minimal surfaces, was introduced to the study of homogenization problems by M. Avellaneda and F. Lin [11]. It also will be used in Chapter 5 to establish uniform boundary estimates for solutions of $\mathcal{L}_\varepsilon(u_\varepsilon) = F$ with Dirichlet conditions.

In Section 4.2 we present a real-variable method, which originated in a paper by L. Caffarelli and I. Peral [24]. The method may be regarded as a dual and refined version of the celebrated Calderón–Zygmund Lemma. It is used in Section 4.3 to study the $W^{1,p}$ estimates for the elliptic system $\mathcal{L}_\varepsilon(u_\varepsilon) = \text{div}(G)$, and reduces effectively the problem to certain reverse Hölder inequalities for local solutions of $\mathcal{L}_\varepsilon(u_\varepsilon) = 0$. In Section 4.4 we investigate the asymptotic behavior of fundamental solutions and their derivatives.

Throughout this chapter we will assume that the coefficient matrix $A = (a_{ij}^{\alpha\beta})$, with $1 \le \alpha, \beta \le m$ and $1 \le i, j \le d$, is 1-periodic and satisfies the boundedness and V-ellipticity conditions (2.1.2)–(2.1.3). This in particular includes the case of elasticity operators, where $m = d$ and $A \in E(\kappa_1, \kappa_2)$. We will also need

Z. Shen, *Periodic Homogenization of Elliptic Systems*, Operator Theory: Advances and Applications 269, https://doi.org/10.1007/978-3-319-91214-1_4

to impose some smoothness condition to ensure regularity estimates at the microscopic scale. We say $A \in \mathrm{VMO}(\mathbb{R}^d)$ if

$$\lim_{r\to 0} \sup_{x\in\mathbb{R}^d} \fint_{B(x,r)} \left| A - \fint_{B(x,r)} A \right| = 0. \tag{4.0.1}$$

Observe that $A \in \mathrm{VMO}(\mathbb{R}^d)$ if A is uniformly continuous in $\mathbb{R}^d$. The uniform Hölder and $W^{1,p}$ estimates will be proved under the condition (4.0.1). A stronger smoothness condition,

$$|A(x) - A(y)| \le \tau |x-y|^\lambda \quad \text{for any } x, y \in \mathbb{R}^d, \tag{4.0.2}$$

where $\tau \ge 0$ and $\lambda \in (0,1]$, will be imposed for the uniform Lipschitz estimate.

A very important feature of the family of operators $\{\mathcal{L}_\varepsilon, \varepsilon > 0\}$ is the following rescaling property:

$$\begin{aligned} &\text{if } \mathcal{L}_\varepsilon(u_\varepsilon) = F \text{ and } v(x) = u_\varepsilon(rx), \\ &\text{then } \mathcal{L}_{\frac{\varepsilon}{r}}(v) = G, \text{ where } G(x) = r^2 F(rx). \end{aligned} \tag{4.0.3}$$

It plays an essential role in the compactness method as well as in numerous other rescaling arguments in this monograph. The translation property is also important:

$$\begin{aligned} &\text{if } -\operatorname{div}\big(A(x/\varepsilon)\nabla u_\varepsilon\big) = F \text{ and } v_\varepsilon(x) = u_\varepsilon(x - x_0), \\ &\text{then } -\operatorname{div}\big(\widetilde{A}(x/\varepsilon)\nabla v_\varepsilon\big) = \widetilde{F}, \\ &\text{where } \widetilde{A}(y) = A(y - \varepsilon^{-1}x_0) \text{ and } \widetilde{F}(x) = F(x - x_0). \end{aligned} \tag{4.0.4}$$

Observe that the matrix $\widetilde{A}$ is 1-periodic and satisfies the same ellipticity condition as A. It also satisfies the same smoothness condition that we will impose on A, uniformly in $\varepsilon > 0$ and $x_0 \in \mathbb{R}^d$.

4.1 Interior Lipschitz estimates

In this section we prove the interior Lipschitz estimate, using a compactness method. As a corollary, we also establish a Liouville property for entire solutions of $\mathcal{L}_1(u) = 0$ in $\mathbb{R}^d$.

Theorem 4.1.1. *Assume that the matrix A satisfies (2.1.2)–(2.1.3) and is 1-periodic. Let $u_\varepsilon \in H^1(B;\mathbb{R}^m)$ be a weak solution of $\mathcal{L}_\varepsilon(u_\varepsilon) = F$ in B, where $B = B(x_0,R)$ for some $x_0 \in \mathbb{R}^d$ and $R > 0$, and $F \in L^p(B;\mathbb{R}^m)$ for some $p > d$. Suppose that $0 < \varepsilon < R$. Then*

$$\left(\fint_{B(x_0,\varepsilon)} |\nabla u_\varepsilon|^2 \right)^{1/2} \le C_p \left\{ \left(\fint_{B(x_0,R)} |\nabla u_\varepsilon|^2 \right)^{1/2} + R \left(\fint_{B(x_0,R)} |F|^p \right)^{1/p} \right\}, \tag{4.1.1}$$

where C_p depends only on μ and p.

Estimate (4.1.1) should be regarded as a Lipschitz estimate down to the microscopic scale ε. In fact, under some smoothness condition on A, the full-scale Lipschitz estimate follows readily from Theorem 4.1.1 by a blow-up argument.

Theorem 4.1.2 (Interior Lipschitz estimate). *Suppose that A satisfies (2.1.2)–(2.1.3) and is 1-periodic. Also assume that A satisfies the smoothness condition (4.0.2). Let $u_\varepsilon \in H^1(B;\mathbb{R}^m)$ be a weak solution to $\mathcal{L}_\varepsilon(u_\varepsilon) = F$ in B for some ball $B = B(x_0, R)$, where $F \in L^p(B;\mathbb{R}^m)$ for some $p > d$. Then*

$$|\nabla u_\varepsilon(x_0)| \le C_p \left\{ \left(\fint_B |\nabla u_\varepsilon|^2 \right)^{1/2} + R \left(\fint_B |F|^p \right)^{1/p} \right\}, \tag{4.1.2}$$

where C_p depends only on p, μ and (λ, τ).

Proof. We give the proof of Theorem 4.1.2, assuming Theorem 4.1.1. By translation and dilation we may assume that $x_0 = 0$ and $R = 1$.

The boundedness and ellipticity conditions (2.1.2)–(2.1.3), together with the Hölder continuity assumption (4.0.2), allow us to use the following local regularity result: if $-\operatorname{div}(A(x)\nabla u) = F$ in $B(0,1)$, where $F \in L^p(B(0,1);\mathbb{R}^m)$ for some $p > d$, then

$$|\nabla u(0)| \le C_p \left\{ \left(\fint_{B(0,1)} |\nabla u|^2 \right)^{1/2} + \left(\fint_{B(0,1)} |F|^p \right)^{1/p} \right\}, \tag{4.1.3}$$

where C_p depends only on μ, p, λ and τ (see, e.g., [50]). We may also assume that $0 < \varepsilon < (1/2)$, as the case $\varepsilon \ge (1/2)$ follows directly from (4.1.3) and the observation that the coefficient matrix $A(x/\varepsilon)$ is uniformly Hölder continuous for $\varepsilon \ge (1/2)$.

To handle the case $0 < \varepsilon < (1/2)$, we use a blow-up argument and the estimate (4.1.1). Let $w(x) = \varepsilon^{-1} u_\varepsilon(\varepsilon x)$. Since $\mathcal{L}_1(w) = \varepsilon F(\varepsilon x)$ in $B(0,1)$, it follows again from (4.1.3) that

$$\begin{aligned} |\nabla w(0)| &\le C \left\{ \left(\fint_{B(0,1)} |\nabla w|^2 \right)^{1/2} + \left(\fint_{B(0,1)} |\varepsilon F(\varepsilon x)|^p \, dx \right)^{1/p} \right\} \\ &\le C \left\{ \left(\fint_{B(0,\varepsilon)} |\nabla u_\varepsilon|^2 \right)^{1/2} + \varepsilon^{1-\frac{d}{p}} \left(\fint_{B(0,1)} |F|^p \right)^{1/p} \right\}, \end{aligned}$$

where C depends only on p, μ, λ and τ. This, together with (4.1.1) and the fact that $\nabla w(0) = \nabla u_\varepsilon(0)$, gives the estimate (4.1.2). □

In the rest of this section we will assume that A satisfies (2.1.2)–(2.1.3) and is 1-periodic. No smoothness condition on A is needed. The proof uses only the interior $C^{1,\alpha}$ estimate for elliptic systems with constant coefficients satisfying the Legendre–Hadamard condition [50].

Recall that $P_j^\beta(x) = x_j e^\beta$ and $e^\beta = (0, \dots, 1, \dots, 0)$ with 1 in the βth position.

Lemma 4.1.3 (One-step improvement). *Let $0 < \sigma < \rho < 1$ and $\rho = 1 - \frac{d}{p}$. There exist constants $\varepsilon_0 \in (0, 1/2)$ and $\theta \in (0, 1/4)$, depending only on μ, σ and ρ, such that*

$$\left(\fint_{B(0,\theta)} \left| u_\varepsilon(x) - \fint_{B(0,\theta)} u_\varepsilon - \left(P_j^\beta(x) + \varepsilon \chi_j^\beta(x/\varepsilon) \right) \fint_{B(0,\theta)} \frac{\partial u_\varepsilon^\beta}{\partial x_j} \right|^2 dx \right)^{1/2}$$
$$\le \theta^{1+\sigma} \max \left\{ \left(\fint_{B(0,1)} |u_\varepsilon|^2 \right)^{1/2}, \left(\fint_{B(0,1)} |F|^p \right)^{1/p} \right\}, \tag{4.1.4}$$

whenever $0 < \varepsilon < \varepsilon_0$, and $u_\varepsilon \in H^1(B(0,1); \mathbb{R}^m)$ is a weak solution of

$$\mathcal{L}_\varepsilon(u_\varepsilon) = F \quad \text{in } B(0,1). \tag{4.1.5}$$

Proof. Estimate (4.1.4) is proved by contradiction, using Theorem 2.3.2 and the following observation: for any $\theta \in (0, 1/4)$,

$$\begin{aligned}
&\sup_{|x| \le \theta} \left| u(x) - \fint_{B(0,\theta)} u - x_j \fint_{B(0,\theta)} \frac{\partial u}{\partial x_j} \right| \\
&\quad \le C\, \theta^{1+\rho} \|\nabla u\|_{C^{0,\rho}(B(0,\theta))} \\
&\quad \le C\, \theta^{1+\rho} \|\nabla u\|_{C^{0,\rho}(B(0,1/4))} \\
&\quad \le C_0\, \theta^{1+\rho} \left\{ \left(\fint_{B(0,1/2)} |u|^2 \right)^{1/2} + \left(\fint_{B(0,1/2)} |F|^p \right)^{1/p} \right\},
\end{aligned} \tag{4.1.6}$$

where u is a solution of $-\text{div}(A^0 \nabla u) = F$ in $B(0, 1/2)$ and A^0 is a constant matrix satisfying the Legendre–Hadamard condition (2.2.28). We mention that the last inequality in (4.1.6) is a standard $C^{1,\rho}$ estimate for second-order elliptic systems with constant coefficients, and that the constant C_0 depends only on μ and ρ [50].

Since $\sigma < \rho$, we may choose $\theta \in (0, 1/4)$ so small that $2^{d+1} C_0 \theta^{1+\rho} < \theta^{1+\sigma}$. We claim that the estimate (4.1.4) holds for this θ and some $\varepsilon_0 \in (0, 1/2)$, which depends only on μ, σ and ρ.

Suppose this is not the case. Then there exist sequences $\{\varepsilon_\ell\} \subset (0, 1/2)$, $\{A_\ell\}$ satisfying (2.1.2)–(2.1.3) and (2.0.2), $\{F_\ell\} \subset L^p(B(0,1); \mathbb{R}^m)$, and $\{u_\ell\} \subset H^1(B(0,1); \mathbb{R}^m)$, such that $\varepsilon_\ell \to 0$,

$$-\text{div}\big(A_\ell(x/\varepsilon_\ell) \nabla u_\ell\big) = F_\ell \quad \text{in } B(0,1),$$

$$\left(\fint_{B(0,1)} |u_\ell|^2 \right)^{1/2} \le 1, \quad \left(\fint_{B(0,1)} |F_\ell|^p \right)^{1/p} \le 1, \tag{4.1.7}$$

and

$$\left(\fint_{B(0,\theta)} \left| u_\ell(x) - \fint_{B(0,\theta)} u_\ell - \left(P_j^\beta(x) + \varepsilon_\ell \chi_{\ell,j}^\beta(x/\varepsilon_\ell) \right) \fint_{B(0,\theta)} \frac{\partial u_\ell^\beta}{\partial x_j} \right|^2 dx \right)^{1/2} > \theta^{1+\sigma}, \tag{4.1.8}$$

where $\chi_{\ell,j}^\beta$ denote the correctors associated with the 1-periodic matrix A_ℓ. Observe that by (4.1.7) and Caccioppoli's inequality, the sequence $\{u_\ell\}$ is bounded in $H^1(B(0,1/2);\mathbb{R}^m)$. By passing to subsequences, we may assume that

$$\begin{cases} u_\ell \rightharpoonup u \text{ weakly in } L^2(B(0,1);\mathbb{R}^m), \\ u_\ell \rightharpoonup u \text{ weakly in } H^1(B(0,1/2);\mathbb{R}^m), \\ F_\ell \rightharpoonup F \text{ weakly in } L^p(B(0,1);\mathbb{R}^m), \\ \widehat{A_\ell} \to A^0, \end{cases}$$

where $\widehat{A_\ell}$ denotes the homogenized matrix for A_ℓ. Since $p > d$, $F_\ell \rightharpoonup F$ weakly in $L^p(B(0,1);\mathbb{R}^m)$ implies that $F_\ell \to F$ strongly in $H^{-1}(B(0,1);\mathbb{R}^m)$. It then follows by Theorem 2.3.2 that $u \in H^1(B(0,1/2);\mathbb{R}^m)$ is a solution of $-\text{div}(A^0\nabla u) = F$ in $B(0,1/2)$.

We now let $\ell \to \infty$ in (4.1.7) and (4.1.8). This leads to

$$\left(\fint_{B(0,1)} |u|^2 \right)^{1/2} \le 1, \quad \left(\fint_{B(0,1)} |F|^p \right)^{1/p} \le 1, \tag{4.1.9}$$

and

$$\left(\fint_{B(0,\theta)} \left| u(x) - \fint_{B(0,\theta)} u - x_j \fint_{B(0,\theta)} \frac{\partial u}{\partial x_j} \right|^2 dx \right)^{1/2} \ge \theta^{1+\sigma}, \tag{4.1.10}$$

where we have used the fact that $u_\ell \to u$ strongly in $L^2(B(0,1/2);\mathbb{R}^m)$. Here we also have used the fact that the sequence $\{\chi_{\ell,j}^\beta\}$ is bounded in $L^2(Y;\mathbb{R}^m)$. Finally, we note that by (4.1.10), (4.1.6) and (4.1.9),

$$\begin{aligned} \theta^{1+\sigma} &\le C_0 \theta^{1+\rho} \left\{ \left(\fint_{B(0,1/2)} |u|^2 \right)^{1/2} + \left(\fint_{B(0,1/2)} |F|^p \right)^{1/p} \right\} \\ &< 2^{d+1} C_0 \theta^{1+\rho}, \end{aligned}$$

which is in contradiction with the choice of θ. This completes the proof. □

Remark 4.1.4. Since

$$\inf_{\alpha\in\mathbb{R}^d} \fint_E \left| f-\alpha\right|^2 = \fint_E \left| f - \fint_E f\right|^2$$

for any $f \in L^2(E;\mathbb{R}^m)$, we may replace $\fint_{B(0,\theta)} u_\varepsilon$ in (4.1.4) by the average

$$\fint_{B(0,\theta)} \left[u_\varepsilon - \left(P_j^\beta(x) + \varepsilon\, \chi_j^\beta(x/\varepsilon)\right) \fint_{B(0,\theta)} \frac{\partial u_\varepsilon^\beta}{\partial x_j}\right] dx.$$

The observation will be used in the proof of the next lemma.

Lemma 4.1.5 (Iteration). *Let $0<\sigma<\rho<1$ and $\rho = 1-\frac{d}{p}$. Let (ε_0, θ) be the constants given by Lemma 4.1.3. Suppose that $0<\varepsilon<\theta^{k-1}\varepsilon_0$ for some $k\ge 1$ and u_ε is a solution of $\mathcal{L}_\varepsilon(u_\varepsilon)=F$ in $B(0,1)$. Then there exist constants $E(\varepsilon,\ell) = \left(E_j^\beta(\varepsilon,\ell)\right) \in \mathbb{R}^{m\times d}$ for $1\le \ell\le k$, such that if*

$$v_\varepsilon = u_\varepsilon - \left(P_j^\beta + \varepsilon \chi_j^\beta(x/\varepsilon)\right) E_j^\beta(\varepsilon,\ell),$$

then

$$\begin{aligned} &\left(\fint_{B(0,\theta^\ell)} \left| v_\varepsilon - \fint_{B(0,\theta^\ell)} v_\varepsilon\right|^2\right)^{1/2} \\ &\qquad \le \theta^{\ell(1+\sigma)} \max\left\{ \left(\fint_{B(0,1)} |u_\varepsilon|^2\right)^{1/2}, \left(\fint_{B(0,1)} |F|^p\right)^{1/p}\right\}. \end{aligned} \tag{4.1.11}$$

Moreover, the constants $E(\varepsilon,\ell)$ satisfy

$$\begin{aligned} |E(\varepsilon,\ell)| &\le C \max\left\{ \left(\fint_{B(0,1)} |u_\varepsilon|^2\right)^{1/2}, \left(\fint_{B(0,1)} |F|^p\right)^{1/p}\right\}, \\ |E(\varepsilon,\ell+1)-E(\varepsilon,\ell)| &\le C\, \theta^{\ell\sigma} \max\left\{ \left(\fint_{B(0,1)} |u_\varepsilon|^2\right)^{1/2}, \left(\fint_{B(0,1)} |F|^p\right)^{1/p}\right\}, \end{aligned} \tag{4.1.12}$$

where C depends only on μ, σ and ρ.

Proof. We prove (4.1.11)–(4.1.12) by induction on ℓ. The case $\ell=1$ follows readily from Lemma 4.1.3 and Remark 4.1.4, with

$$E_j^\beta(\varepsilon,1) = \fint_{B(0,\theta)} \frac{\partial u_\varepsilon^\beta}{\partial x_j}$$

(set $E(\varepsilon, 0) = 0$). Suppose now that the desired constants $E(\varepsilon, \ell)$ exist for all integers up to some ℓ, where $1 \le \ell \le k-1$. To construct $E(\varepsilon, \ell+1)$, consider the function

$$
\begin{aligned}
w(x) = u_\varepsilon(\theta^\ell x) &- \left\{P_j^\beta(\theta^\ell x) + \varepsilon \chi_j^\beta(\theta^\ell x/\varepsilon)\right\} E_j^\beta(\varepsilon, \ell) \\
&- \fint_{B(0,\theta^\ell)} \left[u_\varepsilon - \left(P_j^\beta + \varepsilon \chi_j^\beta(y/\varepsilon)\right) E_j^\beta(\varepsilon, \ell)\right] dy.
\end{aligned}
$$

By the rescaling property (4.0.3) and the equation (2.2.24) for correctors,

$$
\mathcal{L}_{\frac{\varepsilon}{\theta^\ell}}(w) = F_\ell \quad \text{in } B(0,1),
$$

where $F_\ell(x) = \theta^{2\ell} F(\theta^\ell x)$. Since $\varepsilon \theta^{-\ell} \le \varepsilon \theta^{-k+1} \le \varepsilon_0$, it follows from Lemma 4.1.3 and Remark 4.1.4 that

$$
\begin{aligned}
&\Bigg(\fint_{B(0,\theta)} \Bigg| w - \left\{P_j^\beta + \varepsilon \theta^{-\ell} \chi_j^\beta(\theta^\ell x/\varepsilon)\right\} \fint_{B(0,\theta)} \frac{\partial w^\beta}{\partial x_j} \\
&\qquad - \fint_{B(0,\theta)} \left[w - \left(P_j^\beta + \theta^{-\ell} \varepsilon \chi_j^\beta(\theta^\ell y/\varepsilon)\right) \fint_{B(0,\theta)} \frac{\partial w^\beta}{\partial x_j}\right] dy \Bigg|^2 dx \Bigg)^{1/2} \\
&\le \theta^{1+\sigma} \max\left\{ \left(\fint_{B(0,1)} |w|^2\right)^{1/2}, \left(\fint_{B(0,1)} |F_\ell|^p\right)^{1/p} \right\}.
\end{aligned}
\tag{4.1.13}
$$

Observe that by the induction assumption,

$$
\left(\fint_{B(0,1)} |w|^2\right)^{1/2} \le \theta^{\ell(1+\sigma)} \max\left\{ \left(\fint_{B(0,1)} |u_\varepsilon|^2\right)^{1/2}, \left(\fint_{B(0,1)} |F|^p\right)^{1/p} \right\}.
\tag{4.1.14}
$$

Also, since $0 < \rho = 1 - \frac{d}{p}$,

$$
\left(\fint_{B(0,1)} |F_\ell|^p\right)^{1/p} \le \theta^{\ell(1+\rho)} \left(\fint_{B(0,1)} |F|^p\right)^{1/p}.
\tag{4.1.15}
$$

Hence, the RHS of (4.1.13) is bounded by

$$
\theta^{(\ell+1)(1+\sigma)} \max\left\{ \left(\fint_{B(0,1)} |u_\varepsilon|^2\right)^{1/2}, \left(\fint_{B(0,1)} |F|^p\right)^{1/p} \right\}.
$$

Finally, note that the LHS of (4.1.13) may be written as

$$
\begin{aligned}
&\Bigg(\fint_{B(0,\theta^{\ell+1})} \Bigg| u_\varepsilon - \left\{P_j^\beta + \varepsilon \chi_j^\beta(x/\varepsilon)\right\} E_j^\beta(\varepsilon, \ell+1) \\
&\qquad - \fint_{B(0,\theta^{\ell+1})} \left[u_\varepsilon - \left(P_j^\beta + \varepsilon \chi_j^\beta(y/\varepsilon)\right) E_j^\beta(\varepsilon, \ell+1)\right] dy \Bigg|^2 dx \Bigg)^{1/2}
\end{aligned}
$$

with

$$E_j^\beta(\varepsilon, \ell+1) = E_j^\beta(\varepsilon, \ell) + \theta^{-\ell} \fint_{B(0,\theta)} \frac{\partial w}{\partial x_j}.$$

By Caccioppoli's inequality,

$$\begin{aligned}
|E(\varepsilon, \ell+1) - E(\varepsilon, \ell)| &\le \theta^{-\ell} \left(\fint_{B(0,\theta)} |\nabla w|^2 \right)^{1/2} \\
&\le C\theta^{-\ell} \max \left\{ \left(\fint_{B(0,1)} |w|^2 \right)^{1/2}, \left(\fint_{B(0,1)} |F_\ell|^p \right)^{1/2} \right\} \\
&\le C\, \theta^{\ell\sigma} \max \left\{ \left(\fint_{B(0,1)} |u_\varepsilon|^2 \right)^{1/2}, \left(\fint_{B(0,1)} |F|^p \right)^{1/p} \right\},
\end{aligned}$$

where we have used estimates (4.1.14) and (4.1.15) for the last inequality. Thus we have established the second inequality in (4.1.12), from which the first follows by summation. This completes the induction argument and thus the proof. □

We now give the proof of Theorem 4.1.1.

Proof of Theorem 4.1.1. By translation and dilation we may assume that $x_0 = 0$ and $R = 1$. We may also assume that $0 < \varepsilon < \varepsilon_0 \theta$, where ε_0, θ are the constants given by Lemma 4.1.3. The case $\varepsilon_0\theta \le \varepsilon < 1$ is trivial.

Now suppose that $0 < \varepsilon < \varepsilon_0\theta$. Choose $k \ge 2$ so that $\varepsilon_0\theta^k \le \varepsilon < \varepsilon_0\theta^{k-1}$. It follows by Lemma 4.1.5 that

$$\begin{aligned}
&\left(\fint_{B(0,\theta^{k-1})} \left| u_\varepsilon - \fint_{B(0,\theta^{k-1})} u_\varepsilon \right|^2 \right)^{1/2} \\
&\qquad \le C\,\theta^k \left\{ \left(\fint_{B(0,1)} |u_\varepsilon|^2 \right)^{1/2} + \left(\fint_{B(0,1)} |F|^p \right)^{1/p} \right\}.
\end{aligned}$$

This, together with Caccioppoli's inequality, gives

$$\begin{aligned}
\left(\fint_{B(0,\varepsilon)} |\nabla u_\varepsilon|^2 \right)^{1/2} &\le C \left(\fint_{B(0,\varepsilon_0\theta^{k-1})} |\nabla u_\varepsilon|^2 \right)^{1/2} \\
&\le C \left\{ \theta^{-k} \left(\fint_{B(0,\theta^{k-1})} \left| u_\varepsilon - \fint_{B(0,\theta^{k-1})} u_\varepsilon \right|^2 \right)^{1/2} + \theta^k \left(\fint_{B(0,\theta^{k-1})} |F|^2 \right)^{1/2} \right\} \\
&\le C \left\{ \left(\fint_{B(0,1)} |u_\varepsilon|^2 \right)^{1/2} + \left(\fint_{B(0,1)} |F|^p \right)^{1/p} \right\}.
\end{aligned}$$

By replacing u_ε with $u_\varepsilon - \fint_{B(0,1)} u_\varepsilon$, we obtain

$$\left(\fint_{B(0,\varepsilon)} |\nabla u_\varepsilon|^2\right)^{1/2} \le C\left\{\left(\fint_{B(0,1)} \left|u_\varepsilon - \fint_{B(0,1)} u_\varepsilon\right|^2\right)^{1/2} + \left(\fint_{B(0,1)} |F|^p\right)^{1/p}\right\},$$

from which the estimate (4.1.1) follows by the Poincaré inequality. □

A Liouville property

We end this section by proving a Liouville property for entire solutions of elliptic systems with periodic coefficients. This is done by using the $C^{1,\sigma}$ estimates on mesoscopic scales given by Lemma 4.1.5. No smoothness condition is imposed on the coefficient matrix A.

Theorem 4.1.6. *Suppose that A satisfies (2.1.2)–(2.1.3) and is 1-periodic. Let $u \in H^1_{\text{loc}}(\mathbb{R}^d; \mathbb{R}^m)$ be a weak solution of $\mathcal{L}_1(u) = 0$ in $\mathbb{R}^d$. Assume that there exist constants $C_u > 0$ and $\sigma \in (0,1)$ such that*

$$\left(\fint_{B(0,R)} |u|^2\right)^{1/2} \le C_u\, R^{1+\sigma} \tag{4.1.16}$$

for all $R > 1$. Then

$$u(x) = H + \big(P_j^\beta(x) + \chi_j^\beta(x)\big)E_j^\beta \qquad \text{in } \mathbb{R}^d \tag{4.1.17}$$

for some constants $H \in \mathbb{R}^m$ and $E = (E_j^\beta) \in \mathbb{R}^{m\times d}$.

Proof. We begin by choosing σ_1 and ρ such that $\sigma < \sigma_1 < \rho < 1$. Let ε_0, θ be the positive constants given by Lemma 4.1.3 for $0 < \sigma_1 < \rho < 1$. Suppose that $k \ge 1$ and $\theta^{k+1} < \varepsilon_0$. Let $v(x) = u(Rx)$, where $R = \theta^{-k-\ell}$ for some $\ell \ge 1$. Note that $\mathcal{L}_\varepsilon(v) = 0$ in $B(0,1)$, where $\varepsilon = R^{-1}$. Since $\varepsilon = \theta^{k+\ell} < \varepsilon_0 \theta^{\ell-1}$, in view of Lemma 4.1.5, there exist constants $H(\varepsilon,\ell) \in \mathbb{R}^d$ and $E(\varepsilon,\ell) = \big(E_j^\beta(\varepsilon,\ell)\big) \in \mathbb{R}^{d\times d}$, such that

$$\left(\fint_{B(0,\theta^\ell)} \left|v(x) - \Big(P_j^\beta(x) + \varepsilon\chi_j^\beta(x/\varepsilon)\Big)\, E_j^\beta(\varepsilon,\ell) - H(\varepsilon,\ell)\right|^2 dx\right)^{1/2}$$
$$\le \theta^{\ell(\sigma_1+1)}\left(\fint_{B(0,1)} |v|^2\right)^{1/2}.$$

By a change of variables this gives

$$\left(\fint_{B(0,\theta^\ell R)} \left|u(x) - \varepsilon\Big(P_j^\beta(x) + \chi_j^\beta(x)\Big)\, E_j^\beta(\varepsilon,\ell) - H(\varepsilon,\ell)\right|^2 dx\right)^{1/2}$$
$$\le \theta^{\ell(\sigma_1+1)}\left(\fint_{B(0,R)} |u|^2\right)^{1/2}.$$

Since $R = \theta^{-k-\ell}$, it follows that

$$\inf_{\substack{E=(E_j^\beta)\in\mathbb{R}^{d\times m}\\ H\in\mathbb{R}^m}} \left(\fint_{B(0,\theta^{-k})} \left| u - \left(P_j^\beta + \chi_j^\beta\right) E_j^\beta - H \right|^2 \right)^{1/2} \le C_u\, \theta^{\ell(\sigma_1+1)}\, \theta^{(-k-\ell)(\sigma+1)}, \tag{4.1.18}$$

where we have used assumption (4.1.16). Since $\sigma_1 > \sigma$ and $\theta \in (0,1)$, we may let $\ell \to \infty$ in (4.1.18) to conclude that for each k with $\theta^{k+1} < \varepsilon_0$, there exist constants $H^k \in \mathbb{R}^m$ and $E^k = (E_j^{k\beta}) \in \mathbb{R}^{m\times d}$ such that

$$u(x) = H^k + \left(P_j^\beta(x) + \chi_j^\beta(x)\right) E_j^{k\beta} \qquad \text{in } B(0,\theta^{-k}).$$

Finally, note that $\nabla u = \left(\nabla P_j^\beta + \nabla \chi_j^\beta\right) E_j^{k\beta}$ in Y. Since $\int_Y \nabla \chi_j^\beta = 0$, we obtain

$$\int_Y (\nabla u)\, dx = \int_Y \nabla P_j^\beta \cdot E_j^{k\beta}\, dx.$$

This implies that $E_j^{k_1\beta} = E_j^{k_2\beta}$ for any k_1, k_2 large. As a consequence, we also obtain $H^{k_1} = H^{k_2}$ for any k_1, k_2 large. Hence, (4.1.17) holds in $\mathbb{R}^d$ for some $H \in \mathbb{R}^m$ and $E = (E_j^\beta) \in \mathbb{R}^{m\times d}$. □

Remark 4.1.7. Let u be a weak solution of $\mathcal{L}_1(u) = 0$ in $\mathbb{R}^d$. Suppose that there exist $C_u > 0$ and $\sigma \in (0,1)$ such that

$$\left(\fint_{B(0,R)} |u|^2 \right)^{1/2} \le C_u\, R^\sigma \tag{4.1.19}$$

for any $R > 1$. By Theorem 4.1.6, the solution u is of form (4.1.17). This, together with (4.1.19), implies that u is constant.

4.2 A real-variable method

In this section we introduce a real-variable method for L^p estimates. The method, which will be used to establish $W^{1,p}$ estimates for $\mathcal{L}_\varepsilon$, may be regarded as a refined and dual version of the celebrated Calderón–Zygmund Lemma.

Definition 4.2.1. For $f \in L^1_{\text{loc}}(\mathbb{R}^d)$, the Hardy–Littlewood maximal function $\mathcal{M}(f)$ is defined by

$$\mathcal{M}(f)(x) = \sup\left\{ \fint_B |f| : B \text{ is a ball containing } x \right\}. \tag{4.2.1}$$

It is well known that the operator $\mathcal{M}$ is bounded on $L^p(\mathbb{R}^d)$ for $1 < p \le \infty$, and is of weak type $(1,1)$:

$$\big|\{x \in \mathbb{R}^d : \mathcal{M}(f)(x) > t\}\big| \le \frac{C}{t} \int_{\mathbb{R}^d} |f|\, dx \qquad \text{for any } t > 0, \tag{4.2.2}$$

where C depends only on d (see, e.g., [95, Chapter 1] for a proof). For a fixed ball B in $\mathbb{R}^d$, the localized Hardy–Littlewood maximal function $\mathcal{M}_B(f)$ is defined by

$$\mathcal{M}_B(f)(x) = \sup \left\{ \fint_{B'} |f| : \ x \in B' \text{ and } B' \subset B \right\}. \tag{4.2.3}$$

Since $\mathcal{M}_B(f)(x) \le \mathcal{M}(f\chi_B)(x)$ for any $x \in B$, it follows that $\mathcal{M}_B$ is bounded on $L^p(B)$ for $1 < p \le \infty$, and is of weak type $(1,1)$.

In the proof of Theorem 4.2.3 we will perform a Calderón–Zygmund decomposition. It will be convenient to work with (open) cubes Q in $\mathbb{R}^d$ with sides parallel to the coordinate hyperplanes. By tQ we denote the cube that has the same center and t times the side length as Q. We say Q' is a dyadic subcube of Q if Q' may be obtained from Q by repeatedly bisecting the sides of Q. If Q' is obtained from Q by bisecting each side of Q once, we will call Q the dyadic parent of Q'.

Lemma 4.2.2. *Let Q be a cube in $\mathbb{R}^d$. Suppose that $E \subset Q$ is open and $|E| < 2^{-d}|Q|$. Then there exists a sequence of disjoint dyadic subcubes $\{Q_k\}$ of Q such that*

1. $Q_k \subset E$;
2. *the dyadic parent of Q_k in Q is not contained in E;*
3. $|E \setminus \bigcup_k Q_k| = 0$.

Proof. This is a dyadic version of the Calderón–Zygmund decomposition. To prove the lemma, one simply collects all dyadic subcubes Q' of Q with the property that $Q' \subset E$ and its dyadic parent is not contained in E; i.e., Q' is maximal among all dyadic subcubes of Q that are contained in E. Note that since E is open in Q, the set $E \setminus \bigcup_k Q_k$ is contained in the union Z of boundaries of all dyadic subcubes of Q. Since Z has measure zero, one obtains $|E \setminus \bigcup_k Q_k| = 0$. □

For the proof of the next theorem, we recall the formula

$$\int_E |f|^p\, dx = p \int_0^\infty t^{p-1} \big|\{x \in E : |f(x)| > t\}\big|\, dt, \tag{4.2.4}$$

where $E \subset \mathbb{R}^d$ and $p > 0$.

Theorem 4.2.3. *Let B_0 be a ball in $\mathbb{R}^d$ and $F \in L^2(4B_0)$. Let $q > 2$ and $f \in L^p(4B_0)$ for some $2 < p < q$. Suppose that for each ball $B \subset 2B_0$ with $|B| \le c_1|B_0|$,* *there exist two measurable functions F_B and R_B on $2B$, such that $|F| \le |F_B| + |R_B|$ on $2B$, and*

$$\begin{aligned} \left(\fint_{2B} |R_B|^q \right)^{1/q} &\le N_1 \left\{ \left(\fint_{4B} |F|^2 \right)^{1/2} + \sup_{4B_0 \supset B' \supset B} \left(\fint_{B'} |f|^2 \right)^{1/2} \right\}, \\ \left(\fint_{2B} |F_B|^2 \right)^{1/2} &\le N_2 \sup_{4B_0 \supset B' \supset B} \left(\fint_{B'} |f|^2 \right)^{1/2} + \eta \left(\fint_{4B} |F|^2 \right)^{1/2}, \end{aligned} \tag{4.2.5}$$

where $N_1, N_2 > 1$, $0 < c_1 < 1$, and $\eta \ge 0$. Then there exists $\eta_0 > 0$, depending only on p, q, c_1, N_1, N_2, with the property that if $0 \le \eta < \eta_0$, then $F \in L^p(B_0)$ and

$$\left(\fint_{B_0} |F|^p \right)^{1/p} \le C \left\{ \left(\fint_{4B_0} |F|^2 \right)^{1/2} + \left(\fint_{4B_0} |f|^p \right)^{1/p} \right\}, \tag{4.2.6}$$

where C depends only on N_1, N_2, c_1, p and q.

Proof. Let Q_0 be a cube such that $2Q_0 \subset 2B_0$ and $|Q_0| \approx |B_0|$. We will show that

$$\left(\fint_{Q_0} |F|^p \right)^{1/p} \le C \left\{ \left(\fint_{4B_0} |F|^2 \right)^{1/2} + \left(\fint_{4B_0} |f|^p \right)^{1/p} \right\}, \tag{4.2.7}$$

where C depends only on N_1, N_2, c_1, p, q, and $|Q_0|/|B_0|$. Estimate (4.2.6) follows from (4.2.7) by covering B_0 with a finite number of non-overlapping $\overline{Q_0}$ of the same size such that $2Q_0 \subset 2B_0$.

To prove (4.2.7), let

$$E(t) = \left\{ x \in Q_0 : \mathcal{M}_{4B_0}(|F|^2)(x) > t \right\} \qquad \text{for } t > 0.$$

We claim that if $0 \le \eta < \eta_0$ and $\eta_0 = \eta_0(p, q, c_1, N_1, N_2)$ is sufficiently small, then one can choose three constants $\delta, \gamma \in (0,1)$, and $C_0 > 0$, depending only on N_1, N_2, c_1, p and q, such that

$$|E(\alpha t)| \le \delta |E(t)| + \left| \left\{ x \in Q_0 : \mathcal{M}_{4B_0}(|f|^2)(x) > \gamma t \right\} \right| \tag{4.2.8}$$

for all $t > t_0$, where

$$\alpha = (2\delta)^{-2/p} \qquad \text{and} \qquad t_0 = C_0 \fint_{4B_0} |F|^2. \tag{4.2.9}$$

Assume the claim (4.2.8) for a moment. We multiply both sides of (4.2.8) by $t^{\frac{p}{2}-1}$ and then integrate the resulting inequality in t over the interval (t_0, T). This leads to

$$\int_{t_0}^{T} t^{\frac{p}{2}-1} |E(\alpha t)| \, dt \le \delta \int_{t_0}^{T} t^{\frac{p}{2}-1} |E(t)| \, dt + C_\gamma \int_{4B_0} |f|^p \, dx, \tag{4.2.10}$$

where we have used (4.2.4) as well as the boundedness of $\mathcal{M}_{4B_0}$ on $L^{p/2}(4B_0)$. By a change of variables in the LHS of (4.2.10), we may deduce that for any $T > 0$,

$$\alpha^{-\frac{p}{2}}(1-\delta\alpha^{\frac{p}{2}}) \int_0^T t^{\frac{p}{2}-1}|E(t)|\, dt \leq C|Q_0|t_0^{\frac{p}{2}} + C_\gamma \int_{4B_0} |f|^p\, dx. \tag{4.2.11}$$

Note that $\delta\alpha^{p/2} = (1/2)$. By letting $T \to \infty$ in (4.2.11) and using (4.2.4) we obtain

$$\int_{Q_0} |F|^p\, dx \leq C|Q_0|t_0^{\frac{p}{2}} + C\int_{4B_0} |f|^p\, dx, \tag{4.2.12}$$

which, in view of (4.2.9), gives (4.2.7).

It remains to prove (4.2.8). To this end we first note that by the weak (1, 1) estimate for $\mathcal{M}_{4B_0}$,

$$|E(t)| \leq \frac{C_d}{t}\int_{4B_0} |F|^2\, dx,$$

where C_d depends only on d. It follows that $|E(t)| < \delta|Q_0|$ for any $t > t_0$, if we choose

$$C_0 = 2\delta^{-1}C_d|4B_0|/|Q_0|$$

in (4.2.9) with $\delta \in (0, 1)$ to be determined. We now fix $t > t_0$. Since $E(t)$ is open in Q_0, by Lemma 4.2.2,

$$\bigcup_k Q_k \subset E(t) \quad \text{and} \quad |E(t) \setminus \bigcup_k Q_k| = 0,$$

where $\{Q_k\}$ are (disjoint) maximal dyadic subcubes of Q_0 contained in $E(t)$. By choosing δ sufficiently small, we may assume that $|Q_k| < c_1|Q_0|$. We will show that if $0 \leq \eta < \eta_0$ and η_0 is sufficiently small, then one can choose $\delta, \gamma \in (0, 1)$ so small that

$$|E(\alpha t) \cap Q_k| \leq \delta|Q_k|, \tag{4.2.13}$$

whenever

$$\Big\{x \in Q_k : \mathcal{M}_{4B_0}(|f|^2)(x) \leq \gamma t\Big\} \neq \emptyset. \tag{4.2.14}$$

It is not hard to see that (4.2.8) follows from (4.2.13) by summation. Indeed,

$$\begin{aligned}
|E(\alpha t)| &= |E(\alpha t) \cap E(t)| \\
&= |E(\alpha t) \cap \bigcup_k Q_k| \\
&\leq \sum_{k'} |E(\alpha t) \cap Q_{k'}| + \Big|\Big\{x \in Q_0 : \mathcal{M}_{4B_0}(|f|^2)(x) > \gamma t\Big\}\Big| \\
&\leq \delta \sum_{k'} |Q_{k'}| + \Big|\Big\{x \in Q_0 : \mathcal{M}_{4B_0}(|f|^2)(x) > \gamma t\Big\}\Big| \\
&\leq \delta|E(t)| + \Big|\Big\{x \in Q_0 : \mathcal{M}_{4B_0}(|f|^2)(x) > \gamma t\Big\}\Big|,
\end{aligned}$$

where the summation is taken over only those $Q_{k'}$ for which (4.2.14) hold.

Finally, to see (4.2.13), we fix Q_k that satisfies the condition (4.2.14). Observe that

$$\mathcal{M}_{4B_0}(|F|^2)(x) \leq \max\left\{\mathcal{M}_{2B_k}(|F|^2)(x), C_d\, t\right\} \tag{4.2.15}$$

for any $x \in Q_k$, where B_k is the ball that has the same center and diameter as Q_k. This is because Q_k is maximal and so its dyadic parent is not contained in $E(t)$, which in turn implies that

$$\fint_{B'} |F|^2 \leq C_d\, t \tag{4.2.16}$$

for any ball $B' \subset 4B_0$ such that $B' \cap B_k \neq \emptyset$ and $\operatorname{diam}(B') \geq \operatorname{diam}(B_k)$. Clearly we may assume $\alpha > C_d$ by choosing δ small. In view of (4.2.15) this implies that

$$\begin{aligned}
|E(\alpha t) \cap Q_k| &\leq \left|\{x \in Q_k : \mathcal{M}_{2B_k}(|F|^2)(x) > \alpha t\}\right| \\
&\leq \left|\left\{x \in Q_k : \mathcal{M}_{2B_k}(|F_{B_k}|^2)(x) > \frac{\alpha t}{4}\right\}\right| \\
&\quad + \left|\left\{x \in Q_k : \mathcal{M}_{2B_k}(|R_{B_k}|^2)(x) > \frac{\alpha t}{4}\right\}\right| \\
&\leq \frac{C_d}{\alpha t}\int_{2B_k} |F_{B_k}|^2\, dx + \frac{C_{d,q}}{(\alpha t)^{\frac{q}{2}}}\int_{2B_k} |R_{B_k}|^q\, dx,
\end{aligned} \tag{4.2.17}$$

where we have used the assumption

$$|F| \leq |F_{B_k}| + |R_{B_k}| \quad \text{on } 2B_k$$

as well as the weak $(1,1)$ and weak $(q/2, q/2)$ bounds of $\mathcal{M}_{2B_k}$.

By the second inequality in the assumption (4.2.5), we have

$$\begin{aligned}
\fint_{2B_k} |F_{2B_k}|^2\, dx &\leq 2N_2^2 \sup_{4B_0 \supset B' \supset B_k} \fint_{B'} |f|^2 + 2\eta^2 \fint_{2B_k} |F|^2 \\
&\leq 2N_2^2 \cdot \gamma t + 2\eta^2 C_d\, t,
\end{aligned} \tag{4.2.18}$$

where the last inequality follows from (4.2.14) and (4.2.16). Similarly, we use the first inequality in (4.2.5) and (4.2.16) to obtain

$$\begin{aligned}
\fint_{2B_k} |R_{B_k}|^q\, dx &\leq N_1^q \left\{\left(\fint_{4B_k} |F|^2\right)^{1/2} + (\gamma t)^{1/2}\right\}^q \\
&\leq N_1^q C_{d,q} t^{q/2}.
\end{aligned} \tag{4.2.19}$$

We now use (4.2.18) and (4.2.19) to bound the right side of (4.2.17). This yields

$$\begin{aligned}
&|E(\alpha t) \cap Q_k| \\
&\leq |Q_k|\left\{C_d \cdot N_2^2 \cdot \gamma \cdot \alpha^{-1} + \eta^2 \cdot C_d \cdot \alpha^{-1} + C_{d,q} \cdot N_1^q \cdot \alpha^{-q/2}\right\} \\
&\leq \delta |Q_k| \left\{C_d \cdot N_2^2 \cdot \gamma \cdot \delta^{\frac{2}{p}-1} + C_d \cdot \eta_0^2 \cdot \delta^{\frac{2}{p}-1} + C_{d,q} \cdot N_1^q \cdot \delta^{\frac{q}{p}-1}\right\},
\end{aligned} \tag{4.2.20}$$

where we have used the fact $\alpha = (2\delta)^{-\frac{2}{p}}$. Note that since $p < q$, it is possible to choose $\delta \in (0,1)$ so small that

$$C_{d,q} N_1^q \delta^{\frac{q}{p}-1} \le (1/4).$$

After δ is chosen, we take $\gamma \in (0,1)$ and $\eta_0 \in (0,1)$ so small that

$$C_d \cdot N_2^2 \cdot \gamma \cdot \delta^{\frac{2}{p}-1} + C_d \cdot \eta_0^2 \cdot \delta^{\frac{2}{p}-1} \le (1/4).$$

This gives

$$\begin{aligned}|E(\alpha t) \cap Q_k| &\le (1/2)\delta |Q_k| \\ &< \delta |Q_k|\end{aligned}$$

and finishes the proof. □

Remark 4.2.4. The fact that L^2 is a Hilbert space plays no role in the proof of Theorem 4.2.3. Consequently, one can replace the L^2 average in the assumption (4.2.5) by the L^{p_0} average for some $1 \le p_0 < q$, and obtain

$$\left(\fint_{B_0} |F|^p \right)^{1/p} \le C \left\{ \left(\fint_{4B_0} |F|^{p_0} \right)^{1/p_0} + \left(\fint_{4B_0} |f|^p \right)^{1/p} \right\},$$

for $p_0 < p < q$, in the place of (4.2.6).

An operator T is called sublinear if there exists a constant K such that

$$|T(f+g)| \le K\{|T(f)| + |T(g)|\}. \tag{4.2.21}$$

Theorem 4.2.5. *Let T be a bounded sublinear operator on $L^2(\mathbb{R}^d)$ with*

$$\|T\|_{L^2 \to L^2} \le C_0.$$

Let $q > 2$. Suppose that

$$\left(\fint_B |T(g)|^q \right)^{1/q} \le N \left\{ \left(\fint_{2B} |T(g)|^2 \right)^{1/2} + \sup_{B' \supset B} \left(\fint_{B'} |g|^2 \right)^{1/2} \right\} \tag{4.2.22}$$

for any ball B in $\mathbb{R}^d$ and for any $g \in C_0^\infty(\mathbb{R}^d)$ with $\operatorname{supp}(g) \subset \mathbb{R}^d \setminus 4B$. Then for any $f \in C_0^\infty(\mathbb{R}^d)$,

$$\|T(f)\|_{L^p(\mathbb{R}^d)} \le C_p \, \|f\|_{L^p(\mathbb{R}^d)}, \tag{4.2.23}$$

where $2 < p < q$ and C_p depends at most on p, q, C_0, N, and K in (4.2.21).

Proof. Let $f \in C_0^\infty(\mathbb{R}^d)$ and $F = T(f)$. Suppose that $\operatorname{supp}(f) \subset B(0,\rho)$ for some $\rho > 1$. Let $B_0 = B(0,R)$, where $R > 100\rho$. For each ball $B \subset 2B_0$ with $|B| \le (100)^{-1}|B_0|$, we define

$$F_B = KT(f\varphi_B) \quad \text{and} \quad R_B = KT(f(1-\varphi_B)),$$

where $\varphi_B \in C_0^\infty(9B)$ such that $0 \le \varphi_B \le 1$ and $\varphi_B = 1$ in $8B$. Clearly, by (4.2.21), $|F| \le |F_B| + |R_B|$ in $\mathbb{R}^d$. By the L^2 boundedness of T, we have

$$\left(\fint_{2B} |F_B|^2 \right)^{1/2} \le C \left(\fint_{9B} |f|^2 \right)^{1/2}.$$

In view of the assumption (4.2.22) we obtain

$$\begin{aligned}
\left(\fint_{2B} |R_B|^q \right)^{1/q} &\le C \left(\fint_{4B} |R_B|^2 \right)^{1/2} + C \sup_{4B_0 \supset B' \supset B} \left(\fint_{B'} |f|^2 \right)^{1/2} \\
&\le C \left(\fint_{4B} |T(f)|^2 \right)^{1/2} + C \left(\fint_{4B} |T(f\varphi_B)|^2 \right)^{1/2} \\
&\qquad + C \sup_{4B_0 \supset B' \supset B} \left(\fint_{B'} |f|^2 \right)^{1/2} \\
&\le C \left(\fint_{4B} |F|^2 \right)^{1/2} + C \sup_{4B_0 \supset B' \supset B} \left(\fint_{B'} |f|^2 \right)^{1/2},
\end{aligned}$$

where we have used the L^2 boundedness of T in the last inequality. It now follows from Theorem 4.2.3 (with $\eta = 0$) that

$$\begin{aligned}
&\left(\int_{B_0} |T(f)|^p \, dx \right)^{1/p} \\
&\le C |B_0|^{\frac{1}{p} - \frac{1}{2}} \left(\int_{4B_0} |T(f)|^2 \, dx \right)^{1/2} + C \left(\int_{4B_0} |f|^p \, dx \right)^{1/p}
\end{aligned} \tag{4.2.24}$$

for $2 < p < q$. By letting $R \to \infty$ in (4.2.24) and using the fact that $T(f) \in L^2(\mathbb{R}^d)$, we obtain the estimate (4.2.23). □

One may use Theorem 4.2.5 to show that the Calderón–Zygmund operators are bounded in $L^p(\mathbb{R}^d)$ for any $1 < p < \infty$. Indeed, suppose that T is a Calderón–Zygmund operator. This means that

1. T is a bounded linear operator on $L^2(\mathbb{R}^d)$;
2. T is associated with a Calderón–Zygmund kernel $K(x, y)$ in the sense that

$$T(f)(x) = \int_{\mathbb{R}^d} K(x, y) f(y) \, dy \qquad \text{for } x \notin \text{supp}(f),$$

   if f is a bounded measurable function with compact support.

A measurable function $K(x, y)$ in $\mathbb{R}^d \times \mathbb{R}^d$ is called a Calderón–Zygmund kernel, if there exist $C > 0$ and $\delta \in (0, 1]$ such that

$$|K(x, y)| \le C|x - y|^{-d}$$

for any $x, y \in \mathbb{R}^d$, $x \ne y$, and

$$|K(x, y + h) - K(x, y)| + |K(x + h, y) - K(x, y)| \le \frac{C|h|^\delta}{|x - y|^{d+\delta}}$$

for any $x, y, h \in \mathbb{R}^d$ and $|h| < (1/2)|x-y|$ (see, e.g., [96]). Suppose $\text{supp}(f) \subset \mathbb{R}^d \setminus 4B$. Then for any $x, z \in B$,

$$\begin{aligned}|T(f)(x) - T(f)(z)| &\leq \int_{\mathbb{R}^d} |K(x,y) - K(z,y)|\,|f(y)|\,dy \\ &\leq C|x-z|^\delta \int_{\mathbb{R}^d\setminus 4B} \frac{|f(y)|\,dy}{|z-y|^{d+\delta}} \\ &\leq Cr^\delta \sum_{j=2}^\infty \frac{1}{(2^j r)^{d+\delta}} \int_{2^{j+1}B\setminus 2^j B} |f(y)|\,dy \\ &\leq C \sup_{B'\supset B} \fint_{B'} |f|,\end{aligned}$$

where r is the radius of B. This implies that

$$\|T(f)\|_{L^\infty(B)} \leq \fint_B |T(f)| + C \sup_{B'\supset B} \fint_{B'} |f|.$$

It follows that T satisfies the condition (4.2.22) for any $q > 2$. Since T is bounded on $L^2(\mathbb{R}^d)$, by Theorem 4.2.5, we may conclude that T is bounded on $L^p(\mathbb{R}^d)$ for any $2 < p < \infty$. The boundedness of T on $L^p(\mathbb{R}^d)$ for $1 < p < 2$ follows by duality.

The next two results extend Theorems 4.2.3 and 4.2.5 to bounded Lipschitz domains. Regarding the Lipschitz character of a Lipschitz domain we refer the reader to Section 8.1.

Theorem 4.2.6. *Let $q > 2$ and Ω be a bounded Lipschitz domain. Let $F \in L^2(\Omega)$ and $f \in L^p(\Omega)$ for some $2 < p < q$. Suppose that for each ball B with the property that $|B| \leq c_0|\Omega|$ and either $4B \subset \Omega$ or B is centered on $\partial\Omega$, there exist two measurable functions F_B and R_B on $\Omega \cap 2B$, such that $|F| \leq |F_B| + |R_B|$ on $\Omega \cap 2B$, and*

$$\begin{aligned}\left(\fint_{\Omega\cap 2B} |R_B|^q\right)^{1/q} &\leq N_1 \left\{ \left(\fint_{\Omega\cap 4B} |F|^2\right)^{1/2} + \sup_{4B_0\supset B'\supset B} \left(\fint_{\Omega\cap B'} |f|^2\right)^{1/2} \right\}, \\ \left(\fint_{\Omega\cap 2B} |F_B|^2\right)^{1/2} &\leq N_2 \sup_{4B_0\supset B'\supset B} \left(\fint_{\Omega\cap B'} |f|^2\right)^{1/2} + \eta \left(\fint_{\Omega\cap 4B} |F|^2\right)^{1/2},\end{aligned} \tag{4.2.25}$$

where $N_1, N_2 > 1$ and $0 < c_0 < 1$. Then there exists $\eta_0 > 0$, depending only on N_1, N_2, c_0, p, q, and the Lipschitz character of Ω, with the property that if $0 \leq \eta < \eta_0$, then $F \in L^p(\Omega)$ and

$$\left(\fint_\Omega |F|^p\right)^{1/p} \leq C \left\{ \left(\fint_\Omega |F|^2\right)^{1/2} + \left(\fint_\Omega |f|^p\right)^{1/p} \right\}, \tag{4.2.26}$$

where C depends at most on N_1, N_2, c_0, p, q, and the Lipschitz character of Ω.

Proof. We will deduce the theorem from Theorem 4.2.3. Choose a ball B_0 so that $\Omega \subset B_0$ and $\text{diam}(B_0) = \text{diam}(\Omega)$. Consider the functions $\widetilde{f} = f\chi_\Omega$ and $\widetilde{F} = F\chi_\Omega$ in $4B_0$. Let $B = B(x_0, r)$ be a ball such that $B \subset 2B_0$ and $|B| \le (100)^{-1} c_0 |\Omega|$. If $4B \subset \Omega$, we simply choose $\widetilde{F}_B = F_B$ and $\widetilde{R}_B = R_B$. If $2B \cap \Omega = \emptyset$, we let $\widetilde{F}_B = \widetilde{R}_B = 0$. Note that in the remaining case, we have

$$\Omega \cap 4B \neq \emptyset \quad \text{and} \quad 4B \cap (\mathbb{R}^d \setminus \overline{\Omega}) \neq \emptyset.$$

This implies that there exists $y_0 \in \partial\Omega$ such that $4B \subset B(y_0, 8r)$. Let $B_1 = B(y_0, 8r)$ and define $\widetilde{F}_B$, $\widetilde{R}_B$ by

$$\widetilde{F}_B = F_{B_1}\chi_\Omega \quad \text{and} \quad \widetilde{R}_B = R_{B_1}\chi_\Omega.$$

It is not hard to verify that the functions $\widetilde{F}_B$ and $\widetilde{R}_B$ in all three cases satisfy the conditions in Theorem 4.2.3. As a result the inequality (4.2.26) follows from (4.2.6). □

Theorem 4.2.7. *Let $q > 2$ and Ω be a bounded Lipschitz domain in $\mathbb{R}^d$. Let T be a bounded sublinear operator on $L^2(\Omega)$ with $\|T\|_{L^2 \to L^2} \le C_0$. Suppose that for any ball B in $\mathbb{R}^d$ with the property that $|B| \le c_0|\Omega|$ and either $2B \subset \Omega$ or B is centered on $\partial\Omega$,*

$$\begin{aligned} &\left(\fint_{\Omega \cap B} |T(g)|^q \right)^{1/q} \\ &\le N \left\{ \left(\fint_{\Omega \cap 2B} |T(g)|^2 \right)^{1/2} + \sup_{\Omega \supset B' \supset B} \left(\fint_{\Omega \cap B'} |g|^2 \right)^{1/2} \right\} \end{aligned} \tag{4.2.27}$$

where $g \in L^2(\Omega)$ and $g = 0$ in $\Omega \cap 4B$. Then, for any $f \in L^p(\Omega)$,

$$\|T(f)\|_{L^p(\Omega)} \le C_p \|f\|_{L^p(\Omega)}, \tag{4.2.28}$$

where $2 < p < q$ and C_p depends at most on p, q, c_0, C_0, N, K in (4.2.21), and the Lipschitz character of Ω.

Theorem 4.2.7 follows from Theorem 4.2.6 in the same manner as Theorem 4.2.5 follows from Theorem 4.2.3. We leave the details to the reader.

4.3 Interior $W^{1,p}$ estimates

In this section we establish interior $W^{1,p}$ estimates for solutions of $\mathcal{L}_\varepsilon(u_\varepsilon) = F$ under the assumptions that A satisfies the boundedness and ellipticity conditions (2.1.2)–(2.1.3), is 1-periodic, and belongs to $\text{VMO}(\mathbb{R}^d)$. In order to quantify the smoothness, we impose the following condition:

$$\sup_{x \in \mathbb{R}^d} \fint_{B(x,t)} \left| A - \fint_{B(x,t)} A \right| \le \omega(t) \qquad \text{for } 0 < t \le 1, \tag{4.3.1}$$

where ω is a nondecreasing continuous function on $[0, 1]$ with $\omega(0) = 0$.

Theorem 4.3.1 (Interior $W^{1,p}$ estimate). *Suppose that the matrix A satisfies (2.1.2)–(2.1.3) and is 1-periodic. Also assume that A satisfies the* VMO *condition* (4.3.1). *Let $H \in L^p(2B;\mathbb{R}^m)$ and $G \in L^p(2B;\mathbb{R}^{m\times d})$ for some $2 < p < \infty$ and ball $B = B(x_0, r)$. Suppose that $u_\varepsilon \in H^1(2B;\mathbb{R}^m)$ and $\mathcal{L}_\varepsilon(u_\varepsilon) = H + \text{div}(G)$ in $2B$. Then $|\nabla u_\varepsilon| \in L^p(B)$ and*

$$\left(\fint_B |\nabla u_\varepsilon|^p\right)^{1/p} \le C_p \left\{ \left(\fint_{2B} |\nabla u_\varepsilon|^2\right)^{1/2} + r\left(\fint_{2B} |H|^p\right)^{1/p} + \left(\fint_{2B} |G|^p\right)^{1/p} \right\}, \tag{4.3.2}$$

where C_p depends only on μ, p, and the function $\omega(t)$ in (4.3.1).

Theorem 4.3.1 is proved by using the real-variable argument given in the previous section. Roughly speaking, the argument reduces the $W^{1,p}$ estimate to a (weak) reverse Hölder inequality for some exponent $q > p$.

Lemma 4.3.2. *Suppose that A satisfies* (2.1.2)–(2.1.3) *and* (4.3.1). *Let*

$$u \in H^1(2B;\mathbb{R}^m)$$

be a weak solution of $\mathcal{L}_1(u) = 0$ in $2B$ for some $B = B(x_0, r)$ with $0 < r \le 1$. Then $|\nabla u| \in L^p(B)$ for any $2 < p < \infty$, and

$$\left(\fint_B |\nabla u|^p\right)^{1/p} \le C_p \left(\fint_{2B} |\nabla u|^2\right)^{1/2}, \tag{4.3.3}$$

where C_p depends only on μ, p, and $\omega(t)$.

Proof. This is a local $W^{1,p}$ estimate for second-order elliptic systems in divergence form with VMO coefficients. We prove it by using Theorem 4.2.3 with $F = |\nabla u|$ and $f = 0$. Let $B' \subset B$ with $|B'| \le (1/8)^d |B|$. Let $v \in H^1(2B';\mathbb{R}^m)$ be the weak solution to the Dirichlet problem:

$$\text{div}\big(A^0 \nabla v\big) = 0 \quad \text{in } 3B' \quad \text{and} \quad v = u \quad \text{on } \partial(3B'), \tag{4.3.4}$$

where

$$A^0 = \fint_{3B'} A$$

is a constant matrix. Recall that (2.1.2)–(2.1.3) implies (2.1.4). It follows that A^0 satisfies the Legendre–Hadamard ellipticity condition and thus (2.1.2)–(2.1.3). Consequently, the Dirichlet problem (4.3.4) is well posed, and the solution satisfies

$$\int_{3B'} |\nabla v|^2 \, dx \le C \int_{3B'} |\nabla u|^2 \, dx, \tag{4.3.5}$$

where C depends only on μ.

Let

$$F_{B'} = |\nabla(u - v)| \quad \text{and} \quad R_{B'} = |\nabla v|.$$

We will show that $F_{B'}$ and $R_{B'}$ satisfy the conditions in Theorem 4.2.3. Clearly, $F = |\nabla u| \le F_{B'} + R_{B'}$ on $2B'$. By the interior Lipschitz estimate for elliptic systems with constant coefficients,

$$\begin{aligned}\max_{2B'} R_{B'} = \max_{2B'} |\nabla v| &\le C \left(\fint_{3B'} |\nabla v|^2 \right)^{1/2} \\ &\le C \left(\fint_{3B'} |\nabla u|^2 \right)^{1/2} \\ &\le C \left(\fint_{4B'} |F|^2 \right)^{1/2},\end{aligned}$$

where C depends only on μ. Since $\mathcal{L}_1(u) = 0$ in $4B' \subset 2B$, the reverse Hölder inequality (2.1.12) yields

$$\left(\fint_{3B'} |\nabla u|^q \right)^{1/q} \le C \left(\fint_{4B'} |\nabla u|^2 \right)^{1/2},$$

where $C > 0$ and $q > 2$ depend only on μ. Note that $u - v \in H_0^1(3B'; \mathbb{R}^m)$ and

$$\operatorname{div}\big(A^0 \nabla (u - v)\big) = \operatorname{div}\big((A^0 - A)\nabla u\big) \quad \text{in } 3B'.$$

It follows that

$$\begin{aligned}\mu \left(\fint_{3B'} |\nabla (u - v)|^2 \right)^{1/2} &\le \left(\fint_{3B'} |(A - A^0)\nabla u|^2 \right)^{1/2} \\ &\le \left(\fint_{3B'} |A - A^0|^{2p_0'} \right)^{1/2p_0'} \left(\fint_{3B'} |\nabla u|^{2p_0} \right)^{1/2p_0} \\ &\le \eta(r) \left(\fint_{4B'} |\nabla u|^2 \right)^{1/2},\end{aligned}$$

where $p_0 = (q/2) > 1$ and

$$\eta(r) = \sup_{\substack{x \in \mathbb{R}^d \\ 0 < s < r}} \left(\fint_{B(x,s)} \left| A - \fint_{B(x,s)} A \right|^{2p_0'} \right)^{1/2p_0'}.$$

Hence,

$$\left(\fint_{3B'} |F_{B'}|^2 \right)^{1/2} \le C\, \eta(r) \left(\fint_{4B'} |F|^2 \right)^{1/2}.$$

Finally, we observe that by the John–Nirenberg inequality (see, e.g., [96, 50]), if $A \in \mathrm{VMO}(\mathbb{R}^d)$, we have $\eta(r) \to 0$ as $r \to 0$. Moreover, given any $\eta_0 \in (0, 1)$, there exists $r_0 > 0$, depending only on μ and the function $\omega(t)$ in (4.3.1), such that $0 \le C\eta(r) < \eta_0$ if $0 < r < r_0$. As a result, if $0 < r < r_0$, the functions $F_{B'}$ and $R_{B'}$ satisfy the conditions in Theorem 4.2.3. Consequently, the estimate (4.3.3) holds for $0 < r < r_0$. By a simple covering argument it also holds for any $0 < r \le 1$. □

Lemma 4.3.3. *Suppose that A satisfies* (2.1.2)–(2.1.3) *and* (4.3.1). *Also assume A is 1-periodic. Let $u_\varepsilon \in H^1(2B;\mathbb{R}^m)$ be a weak solution of $\mathcal{L}_\varepsilon(u_\varepsilon) = 0$ in $2B$ for some $B = B(x_0, r)$. Then $|\nabla u_\varepsilon| \in L^p(B)$ for any $2 < p < \infty$, and*

$$\left(\fint_B |\nabla u_\varepsilon|^p \right)^{1/p} \le C_p \left(\fint_{2B} |\nabla u_\varepsilon|^2 \right)^{1/2}, \tag{4.3.6}$$

where C_p depends only on μ, p, and $\omega(t)$.

Proof. By translation and dilation we may assume $B = B(0,1)$. The case $\varepsilon \ge (1/4)$ follows readily from Lemma 4.3.2, as the coefficient matrix $A(x/\varepsilon)$ satisfies (4.3.1) uniformly in $\varepsilon \ge (1/2)$. Suppose now that $0 < \varepsilon < (1/4)$, $p > 2$, and $\mathcal{L}_\varepsilon(u_\varepsilon) = 0$ in $B(0,2)$. By a simple blow-up argument we may deduce from Lemma 4.3.2 that

$$\left(\fint_{B(y,\varepsilon)} |\nabla u_\varepsilon|^p \right)^{1/p} \le C \left(\fint_{B(y,2\varepsilon)} |\nabla u_\varepsilon|^2 \right)^{1/2}$$

for any $y \in B(0,1)$. This, together with Theorem 4.1.1, gives

$$\left(\fint_{B(y,\varepsilon)} |\nabla u_\varepsilon|^p \right)^{1/p} \le C \left(\fint_{B(y,1)} |\nabla u_\varepsilon|^2 \right)^{1/2}.$$

It follows that

$$\int_{B(y,\varepsilon)} |\nabla u_\varepsilon|^p \, dx \le C\varepsilon^d \, \|\nabla u_\varepsilon\|^p_{L^2(B(0,2))},$$

which yields estimate (4.3.6) by an integration in y over $B(0,1)$. □

Remark 4.3.4. Suppose that A satisfies the same conditions as in Lemma 4.3.3. Let $\mathcal{L}_\varepsilon(u_\varepsilon) = 0$ in $B = B(x_0, r)$. Then, for any $0 < t < r$ and $\sigma \in (0,1)$,

$$\left(\fint_{B(x_0,t)} |\nabla u_\varepsilon|^2 \right)^{1/2} \le C_\sigma \left(\frac{t}{r} \right)^{\sigma - 1} \left(\fint_{B(x_0,r)} |\nabla u_\varepsilon|^2 \right)^{1/2}, \tag{4.3.7}$$

where C_σ depends only on σ, μ and $\omega(t)$ in (4.3.1). This follows from Lemma 4.3.3 and Hölder's inequality.

Proof of Theorem 4.3.1. Suppose that $\mathcal{L}_\varepsilon(u_\varepsilon) = H + \text{div}(G)$ in $2B_0$, where $B = B(x_0, r_0)$. By dilation we may assume that $r_0 = 1$. We shall apply Theorem 4.2.3 with $q = p + 1$, $\eta = 0$,

$$F = |\nabla u_\varepsilon| \quad \text{and} \quad f = |H| + |G|.$$

For each ball B' such that $4B' \subset 2B_0$, we write $u_\varepsilon = v_\varepsilon + w_\varepsilon$ in $2B'$, where $v_\varepsilon \in H_0^1(4B'; \mathbb{R}^m)$ is the weak solution to $\mathcal{L}_\varepsilon(v_\varepsilon) = H + \text{div}(G)$ in $4B'$ and $v_\varepsilon = 0$ on $\partial(4B')$. Let

$$F_{B'} = |\nabla v_\varepsilon| \quad \text{and} \quad R_{B'} = |\nabla w_\varepsilon|.$$

Clearly, $|F| \le F_{B'} + R_{B'}$ in $2B'$. It is also easy to see that by Theorem 2.1.5,

$$\begin{aligned} \left(\fint_{4B'} |F_{B'}|^2 \right)^{1/2} &\le C \left(\fint_{4B'} (r'|H| + |G|)^2 \right)^{1/2} \\ &\le C \left(\fint_{4B'} |f|^2 \right)^{1/2}, \end{aligned}$$

where r' is the radius of B'. To verify the remaining condition in (4.2.5), we note that $w_\varepsilon \in H^1(4B'; \mathbb{R}^m)$ and $\mathcal{L}_\varepsilon(w_\varepsilon) = 0$ in $4B'$. It then follows from Lemma 4.3.3 that

$$\begin{aligned} \left(\fint_{2B'} |\nabla w_\varepsilon|^q \right)^{1/q} &\le C \left(\fint_{4B'} |\nabla w_\varepsilon|^2 \right)^{1/2} \\ &\le C \left(\fint_{4B'} |\nabla u_\varepsilon|^2 \right)^{1/2} + C \left(\fint_{4B'} |\nabla v_\varepsilon|^2 \right)^{1/2} \\ &\le C \left(\fint_{4B'} |F|^2 \right)^{1/2} + C \left(\fint_{4B'} |f|^2 \right)^{1/2}. \end{aligned}$$

Using Theorem 4.2.3 we obtain

$$\left(\fint_{B} |\nabla u_\varepsilon|^p \right)^{1/p} \le C \left(\fint_{4B} |\nabla u_\varepsilon|^2 \right)^{1/2} + C \left(\fint_{4B} |f|^p \right)^{1/p} \tag{4.3.8}$$

for any ball B such that $4B \subset 2B_0$. By a simple covering argument this implies

$$\begin{aligned} &\left(\fint_{B_0} |\nabla u_\varepsilon|^p \right)^{1/p} \\ &\le C \left(\fint_{2B_0} |\nabla u_\varepsilon|^2 \right)^{1/2} + C \left(\fint_{2B_0} |f|^p \right)^{1/p} \\ &\le C \left(\fint_{2B_0} |\nabla u_\varepsilon|^2 \right)^{1/2} + C \left(\fint_{2B_0} |H|^p \right)^{1/p} + C \left(\fint_{2B_0} |G|^p \right)^{1/p}, \end{aligned} \tag{4.3.9}$$

where C depends only on μ, p and $\omega(t)$ in (4.3.1). □

Consider the homogeneous Sobolev space

$$\dot{W}^{1,2}(\mathbb{R}^d, \mathbb{R}^m) = \Big\{ u \in L^2_{\text{loc}}(\mathbb{R}^d; \mathbb{R}^m) : \ \nabla u \in L^2(\mathbb{R}^d; \mathbb{R}^{m \times d}) \Big\}.$$

Elements of $\dot{W}^{1,2}(\mathbb{R}^d; \mathbb{R}^m)$ are equivalence classes of functions under the relation that $u \sim v$ if $u - v$ is constant. It follows from the boundedness and ellipticity condition (2.1.2)–(2.1.3) and the Lax–Milgram Theorem that for any $f = (f_i^\alpha) \in L^2(\mathbb{R}^d; \mathbb{R}^{m\times d})$, there exists a unique $u_\varepsilon \in \dot{W}^{1,2}(\mathbb{R}^d; \mathbb{R}^m)$ such that $\mathcal{L}_\varepsilon(u_\varepsilon) = \operatorname{div}(f)$ in $\mathbb{R}^d$. Moreover, the solution satisfies the estimate

$$\|\nabla u_\varepsilon\|_{L^2(\mathbb{R}^d)} \le C \|f\|_{L^2(\mathbb{R}^d)},$$

where C depends only on μ. The following theorem gives the $W^{1,p}$ estimate in $\mathbb{R}^d$.

Theorem 4.3.5. *Suppose that A satisfies (2.1.2)–(2.1.3) and is 1-periodic. Also assume that A satisfies the VMO condition (4.3.1). Let $f \in C_0^1(\mathbb{R}^d, \mathbb{R}^{m\times d})$ and $1 < p < \infty$. Then the unique solution in $\dot{W}^{1,2}(\mathbb{R}^d; \mathbb{R}^d)$ to $\mathcal{L}_\varepsilon(u_\varepsilon) = \operatorname{div}(f)$ in $\mathbb{R}^d$ satisfies the estimate*

$$\|\nabla u_\varepsilon\|_{L^p(\mathbb{R}^d)} \le C_p \|f\|_{L^p(\mathbb{R}^d)}, \tag{4.3.10}$$

where C_p depends only on μ, p, and the function $\omega(t)$.

Proof. We first consider the case $p > 2$. Let $B = B(0, r)$. It follows from (4.3.2) that

$$\begin{aligned} \|\nabla u_\varepsilon\|_{L^p(B)} &\le C |B|^{\frac{1}{p}-\frac{1}{2}} \|\nabla u_\varepsilon\|_{L^2(2B)} + C \|f\|_{L^p(2B)} \\ &\le C |B|^{\frac{1}{p}-\frac{1}{2}} \|\nabla u_\varepsilon\|_{L^2(\mathbb{R}^d)} + C \|f\|_{L^p(\mathbb{R}^d)}. \end{aligned}$$

By letting $r \to \infty$, this gives the estimate (4.3.10).

The case $1 < p < 2$ follows by a duality argument. Indeed, suppose $\mathcal{L}_\varepsilon(u_\varepsilon) = \operatorname{div}(f)$ in $\mathbb{R}^d$ and $\mathcal{L}_\varepsilon^*(v_\varepsilon) = \operatorname{div}(g)$ in $\mathbb{R}^d$, where $u_\varepsilon, v_\varepsilon \in \dot{W}^{1,2}(\mathbb{R}^d; \mathbb{R}^m)$ and $f, g \in C_0^1(\mathbb{R}^d; \mathbb{R}^{m\times d})$. Then

$$\int_{\mathbb{R}^d} f \cdot \nabla v_\varepsilon \, dx = -\int_{\mathbb{R}^d} A(x/\varepsilon) \nabla u_\varepsilon \cdot \nabla v_\varepsilon \, dx = \int_{\mathbb{R}^d} g \cdot \nabla u_\varepsilon \, dx.$$

Since $\|\nabla v_\varepsilon\|_{L^q(\mathbb{R}^d)} \le C \|g\|_{L^q(\mathbb{R}^d)}$ for $q = p' > 2$, we obtain

$$\left| \int_{\mathbb{R}^d} g \cdot \nabla u_\varepsilon \, dx \right| \le C \|f\|_{L^p(\mathbb{R}^d)} \|g\|_{L^q(\mathbb{R}^d)}.$$

By duality, this yields $\|\nabla u_\varepsilon\|_{L^p(\mathbb{R}^d)} \le C \|f\|_{L^p(\mathbb{R}^d)}$. ☐

Remark 4.3.6. Without the periodicity and VMO condition on A, the estimate (4.3.10) holds if

$$\left| \frac{1}{p} - \frac{1}{2} \right| < \delta,$$

where $\delta > 0$ depends only on μ. This result, which is due to N. Meyers [72], follows readily from the proof of Theorem 4.3.5, using the reverse Hölder inequality (2.1.12).

The next theorem gives the interior Hölder estimates.

Theorem 4.3.7 (Interior Hölder estimate). *Suppose that A satisfies* (2.1.2)–(2.1.3) *and* (4.3.1). *Also assume that A is 1-periodic. Let $u_\varepsilon \in H^1(2B;\mathbb{R}^m)$ and $\mathcal{L}_\varepsilon(u_\varepsilon) = F + \text{div}(G)$ in $2B$ for some ball $B = B(x_0, r)$. Then for $p > d$,*

$$\begin{aligned} &|u_\varepsilon(x) - u_\varepsilon(y)| \\ &\le C_p \left(\frac{|x-y|}{r}\right)^{1-\frac{d}{p}} \left\{ \left(\fint_{2B} |u_\varepsilon|^2\right)^{1/2} + r^2 \left(\fint_{2B} |F|^p\right)^{1/p} + r \left(\fint_{2B} |G|^p\right)^{1/p} \right\} \end{aligned} \tag{4.3.11}$$

for any $x, y \in B$, where C_p depends only on μ, p, and $\omega(t)$.

Proof. This follows from Theorem 4.3.1 by the Sobolev embedding,

$$|u_\varepsilon(x) - u_\varepsilon(y)| \le C_p\, r \left(\frac{|x-y|}{r}\right)^{1-\frac{d}{p}} \left(\fint_B |\nabla u_\varepsilon|^p\right)^{1/p},$$

for any $x, y \in B$, where $p > d$. □

Remark 4.3.8. It follows from (4.3.11) that if $\mathcal{L}_\varepsilon(u_\varepsilon) = \text{div}(f)$ in $2B$, then

$$\|u_\varepsilon\|_{L^\infty(B)} \le C_p \left\{ \left(\fint_{2B} |u_\varepsilon|^2\right)^{1/2} + r \left(\fint_{2B} |f|^p\right)^{1/p} \right\} \tag{4.3.12}$$

for $p > d$.

Remark 4.3.9. The VMO assumption (4.3.1) on the coefficient matrix A is used only in Lemma 4.3.2 to establish local $W^{1,p}$ estimates for solutions of $\mathcal{L}_1(u) = 0$. Consequently, the smoothness condition in Theorems 4.3.1 and 4.3.7 may be weakened if one is able to weaken the condition in Lemma 4.3.2. We refer the reader to [69, 41] and references therein for local $W^{1,p}$ estimates for elliptic and parabolic operators with partially VMO coefficients.

4.4 Asymptotic expansions of fundamental solutions

Let $d \ge 3$. A fundamental solution in $\mathbb{R}^d$ can be constructed for any scalar elliptic operator in divergence form with real, bounded measurable coefficients (see [55]). In fact, the construction in [55] can be extended to any system of second-order elliptic operators in divergence form with complex, bounded measurable coefficients, provided that solutions of the system and its adjoint satisfy the De Giorgi–Nash type local Hölder continuity estimates; i.e., there exist $\sigma \in (0,1)$ and $H_0 > 0$ such that

$$\left(\fint_{B(x,t)} |\nabla u|^2\right)^{1/2} \le H_0 \left(\frac{t}{r}\right)^{\sigma-1} \left(\fint_{B(x,r)} |\nabla u|^2\right)^{1/2} \tag{4.4.1}$$

for $0 < t < r < \infty$, whenever $\mathcal{L}(u) = 0$ or $\mathcal{L}^*(u) = 0$ in $B(x,r)$ (see [59]). As a result, in view of Remark 4.3.4, if A is 1-periodic and satisfies (2.1.2)–(2.1.3) and

(4.3.1), one can construct an $m \times m$ matrix $\Gamma_\varepsilon(x,y) = \big(\Gamma_\varepsilon^{\alpha\beta}(x,y)\big)$ such that for each $y \in \mathbb{R}^d$, $\nabla_x \Gamma_\varepsilon(x,y)$ is locally integrable and

$$\phi^\gamma(y) = \int_{\mathbb{R}^d} a_{ij}^{\alpha\beta}(x/\varepsilon) \frac{\partial}{\partial x_j} \Big\{ \Gamma_\varepsilon^{\beta\gamma}(x,y) \Big\} \frac{\partial \phi^\alpha}{\partial x_i}\, dx \tag{4.4.2}$$

for $\phi = (\phi^\alpha) \in C_0^1(\mathbb{R}^d; \mathbb{R}^m)$. Moreover, the matrix $\Gamma_\varepsilon(x,y)$ satisfies the size estimate

$$|\Gamma_\varepsilon(x,y)| \le C\, |x-y|^{2-d} \tag{4.4.3}$$

for any $x, y \in \mathbb{R}^d$ and $x \neq y$, where the constant C depends only on μ and the function $\omega(t)$. Such matrix, which is unique, is called the matrix of fundamental solutions for $\mathcal{L}_\varepsilon$.

If $m = 1$, the periodicity condition and VMO condition are not needed; the constant C in (4.4.3) depends only on μ. This is also the case for $d = 2$, where the estimate (4.4.3) is replaced by $\|\Gamma_\varepsilon(\cdot, y)\|_{\text{BMO}(\mathbb{R}^2)} \le C$ and

$$\Big| \Gamma_\varepsilon(x,y) - \fint_{B(x,1)} \Gamma_\varepsilon(x,z)\, dz \Big| \le C \Big\{ 1 + \big| \log |x-y| \big| \Big\} \tag{4.4.4}$$

for any $x, y \in \mathbb{R}^2$ and $x \neq y$. It is also known that

$$|\Gamma_\varepsilon(x,y) - \Gamma_\varepsilon(x,z)| \le \frac{C|y-z|^\sigma}{|x-y|^\sigma} \tag{4.4.5}$$

for $x, y \in \mathbb{R}^2$ and $|y-z| < (1/2)|x-y|$, where σ depends only on μ. See [101].

It can be shown that the matrix of fundamental solutions

$$\Gamma_\varepsilon^*(x,y) = \big(\Gamma_\varepsilon^{*\alpha\beta}(x,y) \big)$$

for $\mathcal{L}_\varepsilon^*$ is given by $\big(\Gamma_\varepsilon(y,x)\big)^T$, the matrix transpose of $\Gamma_\varepsilon(y,x)$; i.e.,

$$\Gamma_\varepsilon^{*\alpha\beta}(x,y) = \Gamma_\varepsilon^{\beta\alpha}(y,x). \tag{4.4.6}$$

By uniqueness and the rescaling property of $\mathcal{L}_\varepsilon$,

$$\Gamma_\varepsilon(x,y) = \varepsilon^{2-d}\Gamma_1(\varepsilon^{-1}x, \varepsilon^{-1}y). \tag{4.4.7}$$

Theorem 4.4.1. *Suppose that A satisfies (2.1.2)–(2.1.3) and is 1-periodic. Also assume that A satisfies the Hölder continuity condition (4.0.2). Then, for any $x, y \in \mathbb{R}^d$ and $x \neq y$,*

$$|\nabla_x \Gamma_\varepsilon(x,y)| + |\nabla_y \Gamma_\varepsilon(x,y)| \le C\, |x-y|^{1-d}, \tag{4.4.8}$$

and

$$|\nabla_y \nabla_x \Gamma_\varepsilon(x,y)| \le C\, |x-y|^{-d}, \tag{4.4.9}$$

where C depends only on μ, λ, and τ.

Proof. We first consider the case $d \ge 3$. Fix $x_0, y_0 \in \mathbb{R}^d$ and $r = |x_0 - y_0| > 0$. Let $u_\varepsilon(x) = \Gamma_\varepsilon^\beta(x, y_0)$, where $\Gamma_\varepsilon^\beta(x,y_0) = (\Gamma_\varepsilon^{1\beta}(x,y_0), \dots, \Gamma_\varepsilon^{m\beta}(x,y_0))$. Note that

$\mathcal{L}_\varepsilon(u_\varepsilon) = 0$ in $B(x_0, r/2)$ and by (4.4.3),

$$\|u_\varepsilon\|_{L^\infty(B(x_0,r/2))} \le C r^{2-d}.$$

It follows from the Lipschitz estimates in Theorem 4.1.2 that

$$\|\nabla u_\varepsilon\|_{L^\infty(B(x_0,r/4))} \le C r^{1-d}.$$

This gives $|\nabla_x \Gamma_\varepsilon(x_0, y_0)| \le C\,|x_0 - y_0|^{1-d}$. In view of (4.4.6), the same argument also yields the estimate $|\nabla_y \Gamma_\varepsilon(x, y)| \le C\,|x - y|^{1-d}$. To see (4.4.9), we apply the Lipschitz estimate to

$$v_\varepsilon(x) = \Gamma_\varepsilon(x, y_0) - \Gamma_\varepsilon(x, y_1),$$

where $y_1 \in B(y_0, r/4)$, and use the estimate (4.4.8). It follows that

$$\begin{aligned}
|\nabla_x \Gamma_\varepsilon(x_0, y_0) - \nabla_x \Gamma_\varepsilon(x_0, y_1)| &\le \frac{C}{r} \max_{x \in B(x_0, r/4)} |\Gamma_\varepsilon(x, y_0) - \Gamma_\varepsilon(x, y_1)|\\
&\le \frac{C|y_0 - y_1|}{r^d},
\end{aligned}$$

which yields (4.4.9).

Finally, we note that if $d = 2$, (4.4.8) is a consequence of the interior Lipschitz estimate and the fact that $\|\Gamma_\varepsilon(\cdot, y)\|_{\text{BMO}(\mathbb{R}^2)} \le C$. The estimate (4.4.9) follows from (4.4.8) by the same argument as in the case $d \ge 3$. □

In the remaining of this section we study the asymptotic behavior, as $\varepsilon \to 0$, of $\Gamma_\varepsilon(x, y)$, $\nabla_x \Gamma_\varepsilon(x, y)$, $\nabla_y \Gamma_\varepsilon(x, y)$, and $\nabla_x \nabla_y \Gamma_\varepsilon(x, y)$. Let $\Gamma_0(x, y)$ denote the matrix of fundamental solutions for the homogenized operator $\mathcal{L}_0$. Since $\mathcal{L}_0$ is a second-order elliptic operator with constant coefficients, we have $\Gamma_0(x, y) = \widetilde{\Gamma}_0(x - y)$ and

$$|\nabla^k \widetilde{\Gamma}_0(x)| \le C_k |x|^{2-d-k} \quad \text{for any integer } k \ge 1.$$

Theorem 4.4.2. *Suppose that A satisfies* (2.1.2)–(2.1.3) *and is* 1*-periodic. If $m \ge 2$, we also assume $A \in \text{VMO}(\mathbb{R}^d)$. Then, if $d \ge 3$,*

$$|\Gamma_\varepsilon(x, y) - \Gamma_0(x, y)| \le C\varepsilon\,|x - y|^{1-d} \tag{4.4.10}$$

for any $x, y \in \mathbb{R}^d$ and $x \ne y$.

The proof of Theorem 4.4.2 is based on a local L^∞ estimate for $u_\varepsilon - u_0$.

Lemma 4.4.3. *Assume that A satisfies the same conditions as in Theorem* 4.4.2. *Let $u_\varepsilon \in H^1(2B; \mathbb{R}^m)$ and $u_0 \in C^2(2B; \mathbb{R}^m)$ for some ball $B = B(x_0, r)$. Suppose that $\mathcal{L}_\varepsilon(u_\varepsilon) = \mathcal{L}_0(u_0)$ in $2B$. Then*

$$\begin{aligned}
\|u_\varepsilon - u_0\|_{L^\infty(B)} \le C \left\{ \fint_{2B} |u_\varepsilon - u_0|^2 \right\}^{1/2} &+ C\varepsilon\,\|\nabla u_0\|_{L^\infty(2B)}\\
&+ C\varepsilon\, r\,\|\nabla^2 u_0\|_{L^\infty(2B)},
\end{aligned} \tag{4.4.11}$$

where C depends only on μ and $\omega(t)$ in (4.3.1) *(if $m \ge 2$).*

Proof. By translation and dilation we may assume that $x_0 = 0$ and $r = 1$. Consider

$$w_\varepsilon = u_\varepsilon - u_0 - \varepsilon \chi_j^\beta(x/\varepsilon) \frac{\partial u_0^\beta}{\partial x_j} \tag{4.4.12}$$

in $2B$. Since $\mathcal{L}_\varepsilon(u_\varepsilon) = \mathcal{L}_0(u_0)$, we see that

$$\begin{aligned} \mathcal{L}_\varepsilon(w_\varepsilon) &= \operatorname{div}\big((A(x/\varepsilon) - \widehat{A})\nabla u_0\big) + \operatorname{div}\Big(A(x/\varepsilon)\nabla\big(\varepsilon\chi_j^\beta(x/\varepsilon)\frac{\partial u_0^\beta}{\partial x_j}\big)\Big) \\ &= \operatorname{div}\big(B(x/\varepsilon)\nabla u_0\big) + \varepsilon \operatorname{div}\Big(A(x/\varepsilon)\chi_j^\beta(x/\varepsilon)\frac{\partial}{\partial x_j}\nabla u_0^\beta\Big), \end{aligned} \tag{4.4.13}$$

where $B(y) = A(y) + A(y)\nabla\chi(y) - \widehat{A}$. By (3.1.5) we obtain

$$\big(\mathcal{L}_\varepsilon(w_\varepsilon)\big)^\alpha = -\varepsilon \frac{\partial}{\partial x_i}\left\{\Big[\phi_{jik}^{\alpha\gamma}(x/\varepsilon) - a_{ij}^{\alpha\beta}(x/\varepsilon)\chi_k^{\beta\gamma}(x/\varepsilon)\Big]\frac{\partial^2 u_0^\gamma}{\partial x_j \partial x_k}\right\}. \tag{4.4.14}$$

In view of the L^∞ estimate (4.3.12), this implies that

$$\|w_\varepsilon\|_{L^\infty(B)} \le C\left\{\left(\fint_{2B}|w_\varepsilon|^2\right)^{1/2} + \varepsilon\,\|\nabla^2 u_0\|_{L^\infty(2B)}\right\}, \tag{4.4.15}$$

from which the estimate (4.4.11) follows easily. We point out that under the smoothness assumption on A in Theorem 4.4.2, the correctors $\chi_j^{\alpha\beta}$ as well as the flux correctors $\phi_{jik}^{\alpha\beta}$ are bounded. □

Proof of Theorem 4.4.2. Fix $x_0, y_0 \in \mathbb{R}^d$ and let $r = |x_0 - y_0|/4$. It suffices to consider the case $0 < \varepsilon < r$, since the estimate for the case $\varepsilon \ge r$ is trivial and follows directly from the size estimate (4.4.3). Furthermore, by (4.4.7), we may assume that $r = 1$.

For $f \in C_0^1(B(y_0, 1); \mathbb{R}^m)$, define

$$u_\varepsilon(x) = \int_{\mathbb{R}^d} \Gamma_\varepsilon(x,y) f(y)\, dy \quad \text{and} \quad u_0(x) = \int_{\mathbb{R}^d} \Gamma_0(x,y) f(y)\, dy.$$

By the Calderón–Zygmund estimates for singular integrals we have

$$\|\nabla^2 u_0\|_{L^p(\mathbb{R}^d)} \le C_p\,\|f\|_{L^p(\mathbb{R}^d)} \quad \text{for any } 1 < p < \infty$$

(see [95, Chapter II]). Also, the fractional integral estimates give

$$\|\nabla u_0\|_{L^q(\mathbb{R}^d)} \le C\,\|f\|_{L^p(\mathbb{R}^d)}$$

for

$$1 < p < q < \infty \quad \text{and} \quad \frac{1}{q} = \frac{1}{p} - \frac{1}{d}$$

(see [95, Chapter V]). Let w_ε be defined by (4.4.12). In view of (4.4.14) and the fact that $w_\varepsilon(x) = O(|x|^{2-d})$ as $|x| \to \infty$, we obtain $|\nabla w_\varepsilon| \in L^2(\mathbb{R}^d)$ and

$$\|\nabla w_\varepsilon\|_{L^2(\mathbb{R}^d)} \le C\varepsilon \|\nabla^2 u_0\|_{L^2(\mathbb{R}^d)} \le C\varepsilon \|f\|_{L^2(\mathbb{R}^d)}.$$

By Hölder's inequality and the Sobolev embedding this implies that

$$\|w_\varepsilon\|_{L^2(B(x_0,1))} \le C \|w_\varepsilon\|_{L^{2^*}(\mathbb{R}^d)} \le C\varepsilon \|f\|_{L^2(\mathbb{R}^d)},$$

where $2^* = \frac{2d}{d-2}$. Hence,

$$\begin{aligned} \|u_\varepsilon - u_0\|_{L^2(B(x_0,1))} &\le C\varepsilon \|f\|_{L^2(\mathbb{R}^d)} + C\varepsilon \|\nabla u_0\|_{L^2(B(x_0,1))} \\ &\le C\varepsilon \|f\|_{L^2(B(y_0,1))}. \end{aligned} \tag{4.4.16}$$

Note that $\mathcal{L}_\varepsilon(u_\varepsilon) = \mathcal{L}_0(u_0) = 0$ in $B(x_0, 3)$. We may apply Lemma 4.4.3 to obtain

$$|u_\varepsilon(x_0) - u_0(x_0)| \le C\varepsilon \|f\|_{L^2(B(y_0,1))}, \tag{4.4.17}$$

where we have used (4.4.16) and the observation that

$$\|\nabla u_0\|_{L^\infty(B(x_0,1))} \le C \|\nabla u_0\|_{L^2(B(x_0,2))} \le C \|f\|_{L^2(B(y_0,1))}.$$

By duality, the estimate (4.4.17) yields that

$$\|\Gamma_\varepsilon(x_0, \cdot) - \Gamma_0(x_0, \cdot)\|_{L^2(B(y_0,1))} \le C\varepsilon.$$

Finally, since $\mathcal{L}_\varepsilon^*\big(\Gamma_\varepsilon(x_0,\cdot)\big) = \mathcal{L}_0^*\big(\Gamma_0(x_0,\cdot)\big)$ in $B(y_0, 3)$, we may invoke Lemma 4.4.3 again to conclude that

$$\begin{aligned} |\Gamma_\varepsilon(x_0, y_0) - \Gamma_0(x_0, y_0)| &\le C \|\Gamma_\varepsilon(x_0, \cdot) - \Gamma_0(x_0, \cdot)\|_{L^2(B(y_0,1))} \\ &\qquad + C\varepsilon \|\nabla_y \Gamma_0(x_0, \cdot)\|_{L^\infty(B(y_0,1))} \\ &\qquad + C\varepsilon \|\nabla_y^2 \Gamma_0(x_0, \cdot)\|_{L^\infty(B(y_0,1))} \\ &\le C\varepsilon. \end{aligned}$$

This completes the proof. □

Remark 4.4.4. Let $d = 2$. Suppose that A is 1-periodic and satisfies (2.1.2)–(2.1.3). Then

$$\|\Gamma_\varepsilon(x, \cdot) - \Gamma_0(x, \cdot) - E\|_{L^\infty(B(y,r))} \le \frac{C\varepsilon}{|x-y|}, \tag{4.4.18}$$

for any $x, y \in \mathbb{R}^2$ and $x \ne y$, where $r = (1/4)|x - y|$ and E denotes the L^1 average of $\Gamma_\varepsilon(x, \cdot) - \Gamma_0(x, \cdot)$ over $B(x, r)$. This follows from the proof of Theorem 4.4.2, with a few modifications.

As in the case $d \ge 3$, we fix $x_0, y_0 \in \mathbb{R}^2$ and let $r = (1/4)|x_0 - y_0|$. We may assume that $r = 1$ and $0 < \varepsilon < 1$. Let $f \in C_0^1(B(y_0, 1); \mathbb{R}^m)$ with $\int_{B(y_0,1)} f\, dx = 0$.

Define u_ε and u_0 as in the proof of Theorem 4.4.2. Using (4.4.5) and the assumption that $\int_{B(y_0,1)} f\, dx = 0$, we can show that

$$u_\varepsilon(y) = O(|y|^{-\sigma}) \quad \text{as } |y| \to \infty.$$

By Caccioppoli's inequality this implies that $|\nabla w_\varepsilon| \in L^2(\mathbb{R}^2)$. Moreover, by Remark 4.3.6, there exists some $q < 2$ such that

$$\|\nabla w_\varepsilon\|_{L^q(\mathbb{R}^2)} \le C\varepsilon\, \|\nabla^2 u_0\|_{L^q(\mathbb{R}^2)} \le C\varepsilon\, \|f\|_{L^q(\mathbb{R}^2)}.$$

It follows by Hölder's inequality and the Sobolev inequality that

$$\begin{aligned}\|w_\varepsilon\|_{L^2(B(x_0,1))} &\le C\, \|w_\varepsilon\|_{L^p(\mathbb{R}^2)} \le C\, \|\nabla w_\varepsilon\|_{L^q(\mathbb{R}^2)}\\ &\le C\varepsilon\, \|f\|_{L^2(B(y_0,1))},\end{aligned}$$

where $\frac{1}{p} = \frac{1}{q} - \frac{1}{2}$. As in the proof of Theorem 4.4.2, this leads to

$$|u_\varepsilon(x_0) - u_0(x_0)| \le C\varepsilon\, \|f\|_{L^2(B(y_0,1))}.$$

By duality, we obtain

$$\|\Gamma_\varepsilon(x_0,\cdot) - \Gamma_0(x_0,\cdot) - E\|_{L^2(B(y_0,1))} \le C\varepsilon,$$

where E denotes the L^1 average of $\Gamma_\varepsilon(x_0,y) - \Gamma_0(x_0,y)$ over $B(y_0,1)$. The desired estimate now follows by Lemma 4.4.3.

The next theorem gives an asymptotic expansion for $\nabla_x \Gamma_\varepsilon(x,y)$.

Theorem 4.4.5. *Suppose that A is 1-periodic and satisfies* (2.1.2)–(2.1.3). *Also assume that A satisfies the Hölder continuity condition* (4.0.2). *Then*

$$\big|\nabla_x \Gamma_\varepsilon(x,y) - \nabla_x \Gamma_0(x,y) - \nabla\chi(x/\varepsilon)\nabla_x\Gamma_0(x,y)\big| \le \frac{C\varepsilon \ln[\varepsilon^{-1}|x-y|+2]}{|x-y|^d}.$$

More precisely,

$$\begin{aligned}\Big|\frac{\partial}{\partial x_i}\Big\{\Gamma_\varepsilon^{\alpha\beta}(x,y)\Big\} - \frac{\partial}{\partial x_i}\Big\{\Gamma_0^{\alpha\beta}(x,y)\Big\} - \frac{\partial}{\partial x_i}\Big\{\chi_j^{\alpha\gamma}\Big\}(x/\varepsilon)\cdot\frac{\partial}{\partial x_j}\Big\{\Gamma_0^{\gamma\beta}(x,y)\Big\}\Big|&\\ \le \frac{C\varepsilon\ln[\varepsilon^{-1}|x-y|+2]}{|x-y|^d}&\end{aligned} \tag{4.4.19}$$

for any $x, y \in \mathbb{R}^d$ and $x \neq y$, where C depends only on μ, λ, and τ.

The proof of Theorem 4.4.5 relies on the following Lipschitz estimate.

Lemma 4.4.6. *Assume that A satisfies the same conditions as in Theorem 4.4.5. Suppose that $u_\varepsilon \in H^1(4B;\mathbb{R}^m)$, $u_0 \in C^{2,\eta}(4B;\mathbb{R}^m)$, and $\mathcal{L}_\varepsilon(u_\varepsilon) = \mathcal{L}_0(u_0)$ in $4B$ for some ball $B = B(x_0, r)$. Then, if $0 < \varepsilon < r$,*

$$\begin{aligned}
&\left\| \frac{\partial u_\varepsilon^\alpha}{\partial x_i} - \frac{\partial u_0^\alpha}{\partial x_i} - \frac{\partial}{\partial x_i}\left\{\chi_j^{\alpha\beta}\right\}\left(\frac{x}{\varepsilon}\right) \cdot \frac{\partial u_0^\beta}{\partial x_j} \right\|_{L^\infty(B)} \\
&\le C r^{-1} \left\{ \fint_{4B} |u_\varepsilon - u_0|^2 \right\}^{1/2} + C\varepsilon r^{-1} \|\nabla u_0\|_{L^\infty(4B)} \\
&\quad + C\varepsilon \ln[\varepsilon^{-1} r + 2] \|\nabla^2 u_0\|_{L^\infty(4B)} + C\varepsilon^{1+\lambda} \|\nabla^2 u_0\|_{C^{0,\lambda}(4B)},
\end{aligned} \tag{4.4.20}$$

where C depends only on μ, λ, and τ.

Proof. Let $w_\varepsilon = u_\varepsilon - u_0 - \varepsilon \chi_j^\beta(x/\varepsilon) \frac{\partial u_0^\beta}{\partial x_j}$. It suffices to show that $\|\nabla w_\varepsilon\|_{L^\infty(B)}$ is bounded by the RHS of (4.4.20). Choose $\varphi \in C_0^\infty(3B)$ such that $0 \le \varphi \le 1$, $\varphi = 1$ in $2B$, and $|\nabla \varphi| \le C r^{-1}$, $|\nabla^2 \varphi| \le C r^{-2}$. Note that

$$\big(\mathcal{L}_\varepsilon(w_\varepsilon \varphi)\big)^\alpha = \big(\mathcal{L}_\varepsilon(w_\varepsilon)\big)^\alpha \varphi - a_{ij}^{\alpha\beta}(x/\varepsilon) \frac{\partial w_\varepsilon^\beta}{\partial x_j} \frac{\partial \varphi}{\partial x_i} - \frac{\partial}{\partial x_i}\left\{ a_{ij}^{\alpha\beta}(x/\varepsilon) w_\varepsilon^\beta \frac{\partial \varphi}{\partial x_j} \right\}.$$

This, together with (4.4.14), implies that

$$\begin{aligned}
w_\varepsilon^\alpha(x) &= -\varepsilon \int_{3B} \frac{\partial}{\partial y_i}\left\{\Gamma_\varepsilon^{\alpha\beta}(x,y)\right\} f_i^\beta(y) \varphi(y)\, dy - \varepsilon \int_{3B} \Gamma_\varepsilon^{\alpha\beta}(x,y) f_i^\beta(y) \frac{\partial \varphi}{\partial y_i}\, dy \\
&- \int_{3B} \Gamma_\varepsilon^{\alpha\beta}(x,y) a_{ij}^{\beta\gamma}(y/\varepsilon) \frac{\partial w_\varepsilon^\gamma}{\partial y_j} \frac{\partial \varphi}{\partial y_i}\, dy + \int_{3B} \frac{\partial}{\partial y_i}\left\{\Gamma_\varepsilon^{\alpha\beta}(x,y)\right\} a_{ij}^{\beta\gamma}(y/\varepsilon) w_\varepsilon^\gamma \frac{\partial \varphi}{\partial y_j}\, dy \\
&= I_1 + I_2 + I_3 + I_4
\end{aligned} \tag{4.4.21}$$

for any $x \in B$, where $f = (f_i^\alpha)$ and

$$f_i^\alpha(x) = \left[-\phi_{jik}^{\alpha\gamma}(x/\varepsilon) + a_{ij}^{\alpha\beta}(x/\varepsilon) \chi_k^{\beta\gamma}(x/\varepsilon) \right] \frac{\partial^2 u_0^\gamma}{\partial x_j \partial x_k}. \tag{4.4.22}$$

Using (4.4.8) and (4.4.9), we see that for $x \in B$,

$$\begin{aligned}
&|\nabla_x \{I_2 + I_3 + I_4\}| \\
&\le \frac{C\varepsilon}{r^d} \int_{3B} |f| + \frac{C}{r^d} \int_{3B} |\nabla w_\varepsilon| + \frac{C}{r^{d+1}} \int_{3B} |w_\varepsilon| \\
&\le \frac{C}{r} \left\{ \fint_{4B} |w_\varepsilon|^2 \right\}^{1/2} + C\varepsilon \left\{ \fint_{4B} |\nabla^2 u_0|^2 \right\}^{1/2} \\
&\le \frac{C}{r} \left\{ \fint_{4B} |u_\varepsilon - u_0|^2 \right\}^{1/2} + \frac{C\varepsilon}{r} \|\nabla u_0\|_{L^\infty(4B)} + C\varepsilon \|\nabla^2 u_0\|_{L^\infty(4B)},
\end{aligned} \tag{4.4.23}$$

where we have used Caccioppoli's inequality for the second inequality.

Finally, to estimate ∇I_1, we write

$$
\begin{aligned}
I_1 = &-\varepsilon \int_{3B} \frac{\partial}{\partial y_i}\Big\{\Gamma_\varepsilon^{\alpha\beta}(x,y)\varphi(y)\Big\}\Big\{f_i^\beta(y) - f_i^\beta(x)\Big\}\, dy \\
&+\varepsilon \int_{3B} \Gamma_\varepsilon^{\alpha\beta}(x,y)\frac{\partial \varphi}{\partial y_i} f_i^\beta(y)\, dy.
\end{aligned}
$$

It follows that

$$
\begin{aligned}
|\nabla_x I_1| &\le C\varepsilon \int_{3B} \frac{|f(y)-f(x)|}{|x-y|^d}\, dy + \frac{C\varepsilon}{r^d}\int_{3B} |f(y)|\, dy \\
&\le C\varepsilon \int_{B(x,\varepsilon)} \frac{|f(y)-f(x)|}{|x-y|^d}\, dy + C\varepsilon \ln[\varepsilon^{-1}r+2]\|f\|_{L^\infty(3B)}.
\end{aligned}
\tag{4.4.24}
$$

Observe that

$$
\|f\|_{C^{0,\lambda}(4B)} \le C\varepsilon^{-\lambda}\|\nabla^2 u_0\|_{L^\infty(4B)} + C\,\|\nabla^2 u_0\|_{C^{0,\lambda}(4B)}.
$$

In view of (4.4.24), we obtain

$$
\begin{aligned}
|\nabla_x I_1| \le C\varepsilon\,\|\nabla^2 u_0\|_{L^\infty(4B)} &+ C\varepsilon^{1+\lambda}\|\nabla^2 u_0\|_{C^{0,\lambda}(4B)} \\
&+ C\varepsilon \ln[\varepsilon^{-1}r+2]\|\nabla^2 u_0\|_{L^\infty(4B)}.
\end{aligned}
$$

This, together with (4.4.23), gives the estimate (4.4.20). □

Proof of Theorem 4.4.5. Fix $x_0, y_0 \in \mathbb{R}^d$ and let $r = |x_0 - y_0|/8$. We may assume that $\varepsilon < r$, since the estimate for the case $\varepsilon \ge r$ is trivial and follows directly from (4.4.8).

Let $u_\varepsilon(x) = \Gamma_\varepsilon(x, y_0)$ and $u_0 = \Gamma_0(x, y_0)$. Then $\mathcal{L}_\varepsilon(u_\varepsilon) = \mathcal{L}_0(u_0) = 0$ in $B(x_0, 4r)$. Note that, by Theorem 4.4.2,

$$
\|u_\varepsilon - u_0\|_{L^\infty(4B)} \le C\varepsilon\, r^{1-d}
$$

for $d \ge 3$. If $d = 2$, we may deduce from Remark 4.4.4 that

$$
\|u_\varepsilon - u_0 - E\|_{L^\infty(4B)} \le C\varepsilon\, r^{-1},
$$

where E denotes the L^1 average of $\Gamma_\varepsilon(\cdot, y_0) - \Gamma_0(\cdot, y_0)$ over $B(x_0, r)$. Also, since $\mathcal{L}_0$ is a second-order elliptic operator with constant coefficients,

$$
\begin{aligned}
&\|\nabla u_0\|_{L^\infty(4B)} \le Cr^{1-d}, \quad \|\nabla^2 u_0\|_{L^\infty(4B)} \le Cr^{-d}, \\
&\text{and} \quad \|\nabla^2 u_0\|_{C^{0,\lambda}(4B)} \le Cr^{-d-\lambda}.
\end{aligned}
$$

It then follows from Lemma 4.4.6 that

$$
\Big\| \frac{\partial u_\varepsilon}{\partial x_i} - \frac{\partial u_0}{\partial x_i} - \frac{\partial \chi_j^\beta}{\partial x_i}(x/\varepsilon)\frac{\partial u_0^\beta}{\partial x_j}\Big\|_{L^\infty(B)} \le C\varepsilon r^{-d}\ln[\varepsilon^{-1}r+2].
$$

This finishes the proof. □

Using (4.4.6), we may deduce from Theorem 4.4.5 that

$$\begin{aligned}\Big|\frac{\partial}{\partial y_i}\Big\{\Gamma_\varepsilon^{\alpha\beta}(x,y)\Big\}-\frac{\partial}{\partial y_i}\Big\{\Gamma_0^{\alpha\beta}(x,y)\Big\}-\frac{\partial}{\partial y_i}\Big\{\chi_j^{*\beta\gamma}\Big\}(y/\varepsilon)\cdot\frac{\partial}{\partial y_j}\Big\{\Gamma_0^{\alpha\gamma}(x,y)\Big\}\Big|\\ \le\frac{C\varepsilon\ln[\varepsilon^{-1}|x-y|+2]}{|x-y|^d},\end{aligned}\tag{4.4.25}$$

for any $x, y \in \mathbb{R}^d$ and $x \neq y$, where $\chi^* = (\chi_j^{*\beta\gamma})$ denotes the matrix of correctors for $\mathcal{L}_\varepsilon^*$.

The following theorem gives an asymptotic expansion of $\nabla_x\nabla_y\Gamma_\varepsilon(x,y)$.

Theorem 4.4.7. *Suppose that A satisfies the same conditions as in Theorem 4.4.5. Then*

$$\begin{aligned}&\Big|\frac{\partial^2}{\partial x_i\partial y_j}\Big\{\Gamma_\varepsilon^{\alpha\beta}(x,y)\Big\}\\ &\quad-\frac{\partial}{\partial x_i}\Big\{\delta^{\alpha\gamma}x_k+\varepsilon\chi_k^{\alpha\gamma}(x/\varepsilon)\Big\}\frac{\partial^2}{\partial x_k\partial y_\ell}\Big\{\Gamma_0^{\gamma\sigma}(x,y)\Big\}\frac{\partial}{\partial y_j}\Big\{\delta^{\beta\sigma}y_\ell+\varepsilon\chi_\ell^{*\beta\sigma}(y/\varepsilon)\Big\}\Big|\\ &\le\frac{C\varepsilon\ln[\varepsilon^{-1}|x-y|+2]}{|x-y|^{d+1}}\end{aligned}\tag{4.4.26}$$

for any $x, y \in \mathbb{R}^d$ and $x \neq y$, where C depends only on μ, λ, and τ.

Proof. Fix $x_0, y_0 \in \mathbb{R}^d$ and let $r = |x_0 - y_0|/8$. Again, it suffices to consider the case $0 < \varepsilon < r$. For each $1 \le j \le d$ and $1 \le \beta \le m$, let

$$\begin{cases}u_\varepsilon^\alpha(x)=\dfrac{\partial}{\partial y_j}\Big\{\Gamma_\varepsilon^{\alpha\beta}\Big\}(x,y_0),\\ u_0^\alpha(x)=\dfrac{\partial}{\partial y_\ell}\Big\{\Gamma_0^{\alpha\sigma}\Big\}(x,y_0)\cdot\Big\{\delta^{\beta\sigma}\delta_{j\ell}+\dfrac{\partial}{\partial y_j}(\chi_\ell^{*\beta\sigma})(y_0/\varepsilon)\Big\}\end{cases}$$

in $4B = B(x_0, 4r)$. In view of (4.4.25) we have

$$\|u_\varepsilon-u_0\|_{L^\infty(B(x_0,r))}\le C\varepsilon\, r^{-d}\ln[\varepsilon^{-1}+2].$$

Also note that $\|\nabla u_0\|_{L^\infty(4B)} \le Cr^{-d}$,

$$\|\nabla^2 u_0\|_{L^\infty(4B)}\le Cr^{-d-1}\quad\text{and}\quad\|\nabla^2 u_0\|_{C^{0,\lambda}(4B)}\le Cr^{-d-1-\lambda}.$$

By Lemma 4.4.6 we obtain

$$\Big\|\frac{\partial u_\varepsilon^\alpha}{\partial x_i}-\Big\{\delta^{\alpha\gamma}\delta_{ik}+\frac{\partial}{\partial x_i}\Big\{\chi_k^{\alpha\gamma}\Big\}(x/\varepsilon)\Big\}\frac{\partial u_0^\gamma}{\partial x_k}\Big\|_{L^\infty(B)}\le\frac{C\varepsilon\ln[\varepsilon^{-1}r+2]}{r^{d+1}},$$

which gives the desired estimate. □

4.5 Notes

The compactness method used Section 4.1 was introduced to the study of homogenization problems by M. Avellaneda and F. Lin [11, 13]. The uniform interior Hölder and Lipschitz estimates as well as boundary estimates with the Dirichlet condition were proved in [11]. In [15] a general Liouville property for $\mathcal{L}_1$ was proved under some additional smoothness condition on A.

Theorems 4.2.3 and 4.2.5 are taken from [87, 85] by Z. Shen. The real-variable argument in Section 4.2 is motivated by a paper [24] of L. Caffarelli and I. Peral. Also see related work for L^p estimates in [106, 22, 9, 23, 21, 8, 44].

The matrices of fundamental solutions were studied in [16] and used to prove the $W^{1,p}$ estimates in Theorem 4.3.5 as well as a weak $(1,1)$ estimate. Also see [102] for L^p bounds of Riesz transforms for second-order elliptic operators with periodic coefficients. Early work on asymptotic behavior of fundamental solutions for scalar equations with periodic coefficients may be found in [84, 70]. Asymptotic results in Section 4.4 are stronger than those obtained in [16]. The approach was developed in [66], where the asymptotic behaviors of Green and Neumann functions in bounded domains were studied (see Chapter 7).

Related work on uniform interior estimates in periodic homogenization may be found in [58, 45, 108, 82, 77] and references therein.

Chapter 5

Regularity for the Dirichlet Problem

This chapter is devoted to the study of uniform boundary regularity estimates for the Dirichlet problem

$$\begin{cases} \mathcal{L}_\varepsilon(u_\varepsilon) = F & \text{in } \Omega, \\ u_\varepsilon = g & \text{on } \partial\Omega, \end{cases} \tag{5.0.1}$$

where $\mathcal{L}_\varepsilon = -\text{div}(A(x/\varepsilon)\nabla)$. Assuming that the coefficient matrix $A = A(y)$ is elliptic, periodic, and belongs to $\text{VMO}(\mathbb{R}^d)$, we establish uniform boundary Hölder and $W^{1,p}$ estimates in C^1 domains Ω. We also prove uniform boundary Lipschitz estimates in $C^{1,\alpha}$ domains under the assumption that A is elliptic, periodic, and Hölder continuous. As in the previous chapter for interior estimates, boundary Hölder and Lipschitz estimates are proved by a compactness method. The boundary $W^{1,p}$ estimates are obtained by combining the boundary Hölder estimates with the interior $W^{1,p}$ estimates, via the real-variable method introduced in Section 4.2.

We point out that the boundary $W^{1,p}$ estimates may fail in Lipschitz domains for p large ($p > 3$ for $d \geq 3$ and $p > 4$ for $d = 2$), even for Laplace's equation [61]. Also, the $C^{1,\alpha}$ assumption on the domain Ω for the boundary Lipschitz estimate is more or less sharp. In general, one should not expect Lipschitz estimates in C^1 domains, even for harmonic functions. The compactness method we use for the boundary Lipschitz estimate is similar to that in Section 4.1 for the interior Lipschitz estimate. To control the influence of the boundary layers, a Dirichlet corrector is introduced. The crucial step in this compactness scheme is to show that the corrector is uniformly Lipschitz continuous.

Throughout this chapter we will assume that $A(y) = (a_{ij}^{\alpha\beta}(y))$, with $1 \leq i, j \leq d$ and $1 \leq \alpha, \beta \leq m$, is 1-periodic and satisfies the boundedness and V-ellipticity conditions (2.1.2)–(2.1.3). As in the case of interior estimates studied in the last chapter, all results in this chapter hold for elliptic systems of linear elasticity.

Z. Shen, *Periodic Homogenization of Elliptic Systems*, Operator Theory: Advances and Applications 269, https://doi.org/10.1007/978-3-319-91214-1_5

5.1 Boundary localization in the periodic setting

In this section we make a few observations on changes of coordinate systems by translations and rotations and their effects on the operator $\mathcal{L}_\varepsilon$.

Let $\psi : \mathbb{R}^{d-1} \to \mathbb{R}$ be a Lipschitz (continuous) function, i.e., there exists a nonnegative constant M such that

$$|\psi(x') - \psi(y')| \le M|x' - y'| \qquad \text{for all } x', y' \in \mathbb{R}^{d-1}. \tag{5.1.1}$$

It is known that any Lipschitz function is almost everywhere (a.e.) differentiable in the ordinary sense that for a.e. $x' \in \mathbb{R}^{d-1}$, there exists a vector $\nabla\psi(x') \in \mathbb{R}^{d-1}$ such that

$$\lim_{|y'|\to 0} \frac{|\psi(x'+y') - \psi(x') - \langle \nabla\psi(x'), y'\rangle|}{|y'|} = 0.$$

Furthermore, one has $\|\nabla\psi\|_\infty \le M$. This is a classical result due to Denjoy, Rademacher and Stepanov.

Definition 5.1.1. Let Ω be a bounded domain in $\mathbb{R}^d$. We say Ω is Lipschitz (resp. C^1) if there exist $r_0 > 0$, $M_0 > 0$, and $\{z_k : k = 1, 2, \ldots, N_0\} \subset \partial\Omega$ such that

$$\partial\Omega \subset \bigcup_k B(z_k, r_0),$$

and for each k, there exist a Lipschitz (resp. C^1) function ψ_k in $\mathbb{R}^{d-1}$ and a coordinate system, obtained from the standard Euclidean system through translation and rotation, so that $z_k = (0, 0)$ and

$$\begin{aligned} &B(z_k, C_0 r_0) \cap \Omega \\ &\quad = B(z_k, C_0 r_0) \cap \Big\{(x', x_d) \in \mathbb{R}^d : \ x' \in \mathbb{R}^{d-1} \text{ and } x_d > \psi_k(x')\Big\}, \end{aligned} \tag{5.1.2}$$

where $C_0 = 10\sqrt{d}(M_0 + 1)$ and ψ_k satisfies

$$\psi_k(0) = 0 \quad \text{and} \quad \|\nabla\psi_k\|_\infty \le M_0. \tag{5.1.3}$$

We will call Ω a $C^{1,\alpha}$ domain for some $\alpha \in (0, 1]$, if each $\psi = \psi_k$ satisfies

$$\psi(0) = 0, \quad \|\nabla\psi\|_\infty \le M_0 \quad \text{and} \quad \|\nabla\psi\|_{C^{0,\alpha}(\mathbb{R}^{d-1})} \le M_0. \tag{5.1.4}$$

Here we have used the notation,

$$\|u\|_{C^{0,\alpha}(E)} = \sup\left\{\frac{|u(x) - u(y)|}{|x - y|^\alpha} : \ x, y \in E \text{ and } x \ne y\right\}. \tag{5.1.5}$$

Suppose that $\mathcal{L}_\varepsilon(u_\varepsilon) = F$ in Ω, where Ω is a bounded Lipschitz domain. Recall that if $v_\varepsilon(x) = u_\varepsilon(x + x_0)$ for some $x_0 \in \mathbb{R}^d$, then $-\text{div}\big(B(x/\varepsilon)\nabla v_\varepsilon\big) = G$

in $\widetilde{\Omega}$, where $B(y) = A(y + (x_0/\varepsilon))$, $G(x) = F(x + x_0)$, and $\widetilde{\Omega} = \{x : x + x_0 \in \Omega\}$. It should be easy to verify that all of our assumptions on A and Ω are translation invariant.

To handle the rotation we invoke a theorem, whose proof may be found in [83], on the approximation of real orthogonal matrices by orthogonal matrices with rational entries.

Theorem 5.1.2. *Let $O = (O_{ij})$ be a $d \times d$ orthogonal matrix. For any $\delta > 0$, there exists a $d \times d$ orthogonal matrix $T = (T_{ij})$ with rational entries such that*

1. $\|O - T\| = \left(\sum_{i,j} |O_{ij} - T_{ij}|^2\right)^{1/2} < \delta$;
2. *each entry of T has a denominator less than a constant depending only on d and δ.*

We now fix $z \in \partial\Omega$, where Ω is a bounded Lipschitz domain. By translation we may assume that z is the origin. There exist $r_0 > 0$ and an orthogonal matrix O such that

$$\Omega_1 \cap B(0, r_0) = \{(y', y_d) \in \mathbb{R}^d : \ y_d > \psi_1(y')\} \cap B(0, r_0),$$

where

$$\Omega_1 = \{y \in \mathbb{R}^d : \ y = Ox \text{ for some } x \in \Omega\},$$

and ψ_1 is a Lipschitz function in $\mathbb{R}^{d-1}$ such that $\psi_1(0) = 0$ and $\|\nabla\psi_1\|_\infty \le M$. Observe that if T is an orthogonal matrix such that $\|T - O\|_\infty < \delta$, where $\delta > 0$, depending only on d and M, is sufficiently small, then

$$\Omega_2 \cap B(0, r_0/2) = \{(y', y_d) \in \mathbb{R}^d : \ y_d > \psi_2(y')\} \cap B(0, r_0/2),$$

where

$$\Omega_2 = \{y \in \mathbb{R}^d : \ y = Tx \text{ for some } x \in \Omega\}$$

and ψ_2 is a Lipschitz function in $\mathbb{R}^{d-1}$ such that $\psi_2(0) = 0$ and $\|\nabla\psi_2\|_\infty \le 2M$. Since

$$\|O - T\| \le C_d \|O^{-1} - T^{-1}\|,$$

in view of Theorem 5.1.2, the orthogonal matrix T may be chosen in such a way that T^{-1} is an orthogonal matrix with rational entries and NT^{-1} is a matrix with integer entries, where N is a large integer depending only on d and M.

Let $w_\varepsilon(y) = u_\varepsilon(x)$, where $y = N^{-1}Tx$. Then

$$-\text{div}_y\big(H(y/\varepsilon)\nabla_y w_\varepsilon\big) = \widetilde{F}(y) \quad \text{in } \Omega_3,$$

where $\widetilde{F}(y) = N^2 F(NT^{-1}x)$,

$$\Omega_3 = \{y \in \mathbb{R}^d : \ y = N^{-1}Tx \text{ for some } x \in \Omega\},$$

and $H(y) = \big(h_{ij}^{\alpha\beta}(y)\big)$ with

$$h_{ij}^{\alpha\beta}(y) = T_{ik}T_{j\ell}a_{k\ell}^{\alpha\beta}(NT^{-1}y).$$

Observe that since NT^{-1} is a matrix with integer entries, $H(y)$ is 1-periodic if $A(y)$ is 1-periodic. Also, the matrix H satisfies the ellipticity conditions (2.1.2)–(2.1.3) with the same μ as for A. Moreover, H satisfies the same smoothness condition (VMO or Hölder) as A. We further note that

$$\Omega_3 \cap B(0, c_0 r_0) = \big\{(y', y_d) \in \mathbb{R}^d : \ y_d > \psi_3(y')\big\} \cap B(0, c_0 r_0),$$

where $\psi_3(y') = N^{-1}\psi_2(Ny')$ and c_0 depends only on d and M. As a result, in the study of uniform boundary estimates for $\mathcal{L}_\varepsilon$, one may localize the problem to a setting where $z \in \partial\Omega$ is the origin and $B(z, r_0) \cap \Omega$ is given by the region above the graph of a function ψ in $\mathbb{R}^{d-1}$. More precisely, it suffices to consider solutions of $\mathcal{L}_\varepsilon(u_\varepsilon) = F$ in D_r with boundary data given on Δ_r, where D_r and Δ_r are defined by

$$\begin{aligned} D_r &= D(r, \psi) \\ &= \big\{(x', x_d) \in \mathbb{R}^d : \ |x'| < r \text{ and } \psi(x') < x_d < \psi(x') + 10(M_0+1)r\big\}, \\ \Delta_r &= \Delta(r, \psi) = \big\{(x', \psi(x')) \in \mathbb{R}^d : \ |x'| < r\big\}, \end{aligned} \tag{5.1.6}$$

and ψ is a Lipschitz (or C^1, $C^{1,\alpha}$) function in $\mathbb{R}^{d-1}$. This localization procedure is used in the proof of several boundary estimates in the monograph. We point out that even if Ω is smooth, it may not be possible to choose a local coordinate system such that $\nabla\psi(0) = 0$ and A is periodic in the x_d direction.

We end this section with a boundary Caccioppoli inequality.

Theorem 5.1.3. *Let Ω be a bounded Lipschitz domain in $\mathbb{R}^d$. Suppose that A satisfies (2.1.2)–(2.1.3). Let $u_\varepsilon \in H^1(B(x_0, r) \cap \Omega; \mathbb{R}^m)$ be a weak solution of $\mathcal{L}_\varepsilon(u_\varepsilon) = F + \operatorname{div}(f)$ in $B(x_0, r) \cap \Omega$ with $u_\varepsilon = g$ on $B(x_0, r) \cap \partial\Omega$ for some $x_0 \in \partial\Omega$ and $0 < r < r_0$. Then*

$$\begin{aligned} \int_{B(x_0, r/2)\cap\Omega} |\nabla u_\varepsilon|^2\, dx &\le \frac{C}{r^2}\int_{B(x_0,r)\cap\Omega} |u_\varepsilon|^2\, dx + Cr^2 \int_{B(x_0,r)\cap\Omega} |F|^2\, dx \\ &\quad + C\int_{B(x_0,r)\cap\Omega} |f|^2\, dx + C\int_{B(x_0,r)\cap\Omega} |\nabla G|^2\, dx \\ &\quad + \frac{C}{r^2}\int_{B(x_0,r)\cap\Omega} |G|^2\, dx, \end{aligned} \tag{5.1.7}$$

where $G \in H^1(B(x_0, r) \cap \Omega; \mathbb{R}^m)$ and $G = g$ on $B(x_0, r) \cap \partial\Omega$.

Proof. By considering the function $u_\varepsilon - G$, one can reduce the general case to the special case $G = 0$. In the latter case the inequality is proved by using the inequality

$$\mu \int_\Omega |\nabla(\varphi u_\varepsilon)|^2\, dx \le \int_\Omega A(x/\varepsilon)\nabla(\varphi u_\varepsilon) \cdot \nabla(\varphi u_\varepsilon)\, dx$$

and the equation

$$
\begin{aligned}
&A(x/\varepsilon)\nabla(\varphi u_\varepsilon)\cdot\nabla(\varphi u_\varepsilon)\\
&\quad = A(x/\varepsilon)\nabla u_\varepsilon\cdot\nabla(\varphi^2 u_\varepsilon) + A(x/\varepsilon)(\nabla\varphi)u_\varepsilon\cdot\nabla(\varphi u_\varepsilon)\\
&\qquad - A(x/\varepsilon)\nabla(\varphi u_\varepsilon)\cdot(\nabla\varphi)u_\varepsilon + A(x/\varepsilon)(\nabla\varphi)u_\varepsilon\cdot(\nabla\varphi)u_\varepsilon,
\end{aligned}
$$

where $\varphi \in C_0^1(B(x_0,r))$ is a cut-off function satisfying $0\le\varphi\le 1$, $\varphi = 1$ on $B(x_0,r/2)$ and $\|\nabla\varphi\|_\infty \le Cr^{-1}$. The argument is similar to that for the interior Caccioppoli inequality (2.1.11). We leave the details to the reader. □

5.2 Boundary Hölder estimates

In this section we prove the following boundary Hölder estimate.

Theorem 5.2.1. *Suppose that A is 1-periodic and satisfies (2.1.2)–(2.1.3). Also assume that A satisfies (4.3.1). Let Ω be a bounded C^1 domain and $0<\sigma<1$. Suppose that $u_\varepsilon \in H^1(B(x_0,r)\cap\Omega;\mathbb{R}^m)$ satisfies*

$$
\begin{cases}
\mathcal{L}_\varepsilon(u_\varepsilon) = 0 & \text{in } B(x_0,r)\cap\Omega,\\
u_\varepsilon = g & \text{on } B(x_0,r)\cap\partial\Omega
\end{cases}
$$

for some $x_0\in\partial\Omega$ and $0<r<r_0$, where $g\in C^{0,1}(B(x_0,r)\cap\partial\Omega)$. Then

$$
\begin{aligned}
&\|u_\varepsilon\|_{C^{0,\sigma}(B(x_0,r/2)\cap\Omega)}\\
&\quad\le Cr^{-\sigma}\left\{\left(\fint_{B(x_0,r)\cap\Omega}|u_\varepsilon|^2\right)^{1/2} + |g(x_0)| + r\|g\|_{C^{0,1}(B(x_0,r)\cap\partial\Omega)}\right\},
\end{aligned}
\tag{5.2.1}
$$

where C depends only on μ, σ, $\omega(t)$ in (4.3.1), and Ω.

Let D_r and Δ_r be defined by (5.1.6), where $\psi:\mathbb{R}^{d-1}\to\mathbb{R}$ is a C^1 function. To quantify the C^1 condition we assume that

$$
\begin{cases}
\psi(0)=0,\ \operatorname{supp}(\psi)\subset\{x'\in\mathbb{R}^{d-1}:\ |x'|\le 1\},\ \|\nabla\psi\|_\infty\le M_0,\\
|\nabla\psi(x')-\nabla\psi(y')|\le\omega_1(|x'-y'|) \quad \text{for any } x',y'\in\mathbb{R}^{d-1},
\end{cases}
\tag{5.2.2}
$$

where $\omega_1(t)$ is a (fixed) nondecreasing continuous function on $[0,\infty)$ with $\omega_1(0)=0$. We will use the following compactness result: if $\{\psi_\ell\}$ is a sequence of C^1 functions satisfying (5.2.2), then there exists a subsequence $\{\psi_{\ell'}\}$ such that $\psi_{\ell'}\to\psi$ in $C^1(\mathbb{R}^{d-1})$ for some ψ satisfying (5.2.2).

By a change of the coordinate system and Campanato's characterization of Hölder spaces, it suffices to establish the following.

Theorem 5.2.2. *Suppose that A satisfies the same conditions as in Theorem* 5.2.1. *Let $0 < \sigma < 1$. Suppose that $u_\varepsilon \in H^1(D_r; \mathbb{R}^m)$ is a solution of $\mathcal{L}_\varepsilon(u_\varepsilon) = 0$ in D_r with $u_\varepsilon = g$ on Δ_r for some $0 < r \le 1$. Also assume that $g(0) = 0$. Then for any $0 < t < r$,*

$$\left(\fint_{D_t} |u_\varepsilon|^2 \right)^{1/2} \le C \left(\frac{t}{r} \right)^{\sigma} \left\{ \left(\fint_{D_r} |u_\varepsilon|^2 \right)^{1/2} + r \|g\|_{C^{0,1}(\Delta_r)} \right\}, \tag{5.2.3}$$

where C depends only on σ, μ, $\omega(t)$ in (4.3.1), *and $(M_0, \omega_1(t))$ in* (5.2.2).

Theorem 5.2.2 is proved by a compactness argument, which is similar to the argument used for the interior Lipschitz estimate in Section 4.1. However, the correctors are not needed for Hölder estimates.

Lemma 5.2.3. *Let $\{\psi_\ell\}$ be a sequence of C^1 functions satisfying* (5.2.2). *Let $v_\ell \in L^2(D(r, \psi_\ell))$. Suppose that $\psi_\ell \to \psi$ in $C^1(|x'| < r)$ and $\|v_\ell\|_{L^2(D(r,\psi_\ell))} \le C$. Then there exists a subsequence, which we still denote by $\{v_\ell\}$, and $v \in L^2(D(r, \psi))$ such that $v_k \to v$ weakly in $L^2(\Omega)$ for any open set $\mathcal{O}$ such that $\overline{\mathcal{O}} \subset D(r, \psi)$.*

Proof. Consider the function $w_\ell(x', x_d) = v_\ell(x', x_d + \psi_\ell(x'))$, defined in

$$D(r, 0) = \Big\{ (x', x_d) : \ |x'| < r \text{ and } 0 < x_d < 10(M_0 + 1)r \Big\}.$$

Since the sequence $\{\|w_\ell\|_{L^2(D(r,0))}\}$ is bounded, there exists a subsequence, which we still denote by $\{w_\ell\}$, such that $w_\ell \to w$ weakly in $L^2(D(r, 0))$. Let

$$v(x', x_d) = w(x', x_d - \psi(x')).$$

It is not hard to verify that $v \in L^2(D(r, \psi))$ and $v_\ell \to v$ weakly in $L^2(\Omega)$ for any open set $\mathcal{O}$ such that $\overline{\mathcal{O}} \subset D(r, \psi)$. ☐

Next we prove a homogenization result for a sequence of domains.

Lemma 5.2.4. *Let $\{A_\ell(y)\}$ be a sequence of 1-periodic matrices satisfying* (2.1.2)–(2.1.3) *and $\{\psi_\ell\}$ a sequence of C^1 functions satisfying* (5.2.2). *Suppose that*

$$\begin{cases} \operatorname{div}(A_\ell(x/\varepsilon_\ell)\nabla u_\ell) = 0 & \text{in } D(r, \psi_\ell), \\ u_\ell = g_\ell & \text{on } \Delta(r, \psi_\ell), \end{cases}$$

where $\varepsilon_\ell \to 0$, $g_\ell(0) = 0$ and

$$\|u_\ell\|_{H^1(D(r,\psi_\ell))} + \|g_\ell\|_{C^{0,1}(\Delta(r,\psi_\ell))} \le C. \tag{5.2.4}$$

Then there exist subsequences of $\{A_\ell\}$, $\{\psi_\ell\}$, $\{u_\ell\}$ and $\{g_\ell\}$, which we still denote by the same notation, and a function ψ satisfying (5.2.2), *$u \in H^1(D(r, \psi); \mathbb{R}^m)$,*

$g \in C^{0,1}(\Delta(r,\psi);\mathbb{R}^m)$ *and a constant matrix* A, *such that*

$$\begin{cases} \widehat{A_\ell} \to A^0, \\ \psi_\ell \to \psi \text{ in } C^1(\mathbb{R}^{d-1}), \\ g_\ell(x',\psi_\ell(x')) \to g(x',\psi(x')) \text{ uniformly for } |x'|<r, \\ u_\ell(x',x_d-\psi_\ell(x')) \to u(x',x_d-\psi(x')) \text{ weakly in } H^1(D(r,0);\mathbb{R}^m), \end{cases} \tag{5.2.5}$$

and

$$\begin{cases} \operatorname{div}(A^0\nabla u) = 0 & \text{in } D(r,\psi), \\ u = g & \text{on } \Delta(r,\psi). \end{cases} \tag{5.2.6}$$

Moreover, the constant matrix A^0 *satisfies the ellipticity condition* (2.2.28).

Proof. Since $\widehat{A_\ell}$ satisfies (2.2.28), uniformly in ℓ, by passing to a subsequence, we may assume that $\widehat{A_\ell} \to A^0$, which also satisfies (2.2.28). The remaining statements in (5.2.5) follow readily from (5.2.2) and (5.2.4) by passing to subsequences. To prove (5.2.6), we fix $\varphi \in C_0^1(D(r,\psi);\mathbb{R}^m)$. Clearly, if ℓ is sufficiently large, $\varphi \in C_0^1(D(r,\psi_\ell);\mathbb{R}^m)$. In view of Lemma 5.2.3 we may assume that $A_\ell(x/\varepsilon_\ell)\nabla u_\ell$ converges weakly in $L^2(\mathcal{O};\mathbb{R}^{m\times d})$, where $\mathcal{O}$ is any open set such that

$$\operatorname{supp}(\varphi) \subset \mathcal{O} \subset \overline{\mathcal{O}} \subset D(r,\psi).$$

Note that $\{u_\ell\}$ converges to u strongly in $L^2(\mathcal{O};\mathbb{R}^m)$ and weakly in $H^1(\mathcal{O};\mathbb{R}^m)$. By Theorem 2.3.2 we obtain

$$\int_{D(r,\psi)} A^0\nabla u \cdot \nabla\varphi \, dx = 0.$$

Hence, $\operatorname{div}(A^0\nabla u) = 0$ in $D(r,\psi)$.

Finally, let

$$v_\ell(x',x_d) = u_\ell(x',x_d+\psi_\ell(x')) \quad \text{and} \quad v(x',x_d) = u(x',x_d+\psi(x')).$$

That $u=g$ on $\Delta(r,\psi)$ in the sense of trace follows from the fact that $v_\ell \to v$ weakly in $H^1(D(r,0);\mathbb{R}^m)$ and $v_\ell = g_\ell(x',\psi_\ell(x'))$ on $\Delta(r,0)$. □

Lemma 5.2.5 (One-step improvement). *Suppose that* A *satisfies* (2.1.2)–(2.1.3) *and is* 1*-periodic. Fix* $0<\sigma<1$. *There exist constants* $\varepsilon_0 \in (0,1/2)$ *and* $\theta \in (0,1/4)$, *depending only on* μ, σ, *and* $(\omega_1(t), M_0)$ *in* (5.2.2), *such that*

$$\fint_{D(\theta)} |u_\varepsilon|^2 \le \theta^{2\sigma}, \tag{5.2.7}$$

whenever $0<\varepsilon<\varepsilon_0$,

$$\begin{cases} \mathcal{L}_\varepsilon(u_\varepsilon) = 0 & \text{in } D_1, \\ u_\varepsilon = g & \text{on } \Delta_1, \end{cases} \tag{5.2.8}$$

and

$$\begin{cases} g(0) = 0, \quad \|g\|_{C^{0,1}(\Delta_1)} \le 1, \\ \displaystyle\fint_{D_1} |u_\varepsilon|^2 \le 1. \end{cases} \tag{5.2.9}$$

Proof. Let $\rho = (1+\sigma)/2$. The proof uses the following observation:

$$\fint_{D_r} |w|^2 \le C_0\, r^{2\rho} \qquad \text{for } 0 < r < \frac{1}{4}, \tag{5.2.10}$$

whenever

$$\begin{cases} \operatorname{div}(A^0 \nabla w) = 0 & \text{in } D_{1/2}, \\ w = g & \text{on } \Delta_{1/2}, \\ \|g\|_{C^{0,1}(\Delta_1)} \le 1, & g(0) = 0, \\ \displaystyle\int_{D_{1/2}} |w|^2 \le |D_1|, \end{cases} \tag{5.2.11}$$

where A^0 is a constant matrix satisfying the ellipticity condition (2.2.28). This follows from the boundary Hölder estimate in C^1 domains for second-order elliptic systems with constant coefficients:

$$\begin{aligned} \left(\fint_{D_r} |w|^2 \right)^{1/2} &\le C r^\rho \|w\|_{C^{0,\rho}(D_r)} \\ &\le C r^\rho \|w\|_{C^{0,\rho}(D_{1/4})} \\ &\le C r^\rho \left\{ \|w\|_{L^2(D_{1/2})} + \|g\|_{C^{0,1}(\Delta_{1/2})} \right\}. \end{aligned}$$

The constant C_0 in (5.2.10) depends only on μ, σ, and $(\omega_1(t), M_0)$ in (5.2.2).

We now choose $\theta \in (0, 1/4)$ so small that $2C_0\theta^{2\rho} \le \theta^{2\sigma}$. We claim that for this θ, there exists $\varepsilon_0 > 0$, depending only on μ, σ, and $(\omega_1(t), M_0)$, such that (5.2.7) holds if $0 < \varepsilon < \varepsilon_0$ and u_ε satisfies (5.2.8)–(5.2.9).

The claim is proved by contradiction. Suppose that there exist sequences $\{\varepsilon_\ell\} \subset \mathbb{R}_+$, $\{A_\ell\}$ satisfying (2.1.2)–(2.1.3) and (2.0.2), $\{\psi_\ell\}$ satisfying (5.2.2), and $\{u_\ell\} \subset H^1(D(1, \psi_\ell); \mathbb{R}^m)$, such that $\varepsilon_\ell \to 0$,

$$\begin{cases} \operatorname{div}\big(A_\ell(x/\varepsilon_\ell) \nabla u_\ell\big) = 0 & \text{in } D(1, \psi_\ell), \\ u_\ell = g_\ell & \text{on } \Delta(1, \psi_\ell), \\ \|g_\ell\|_{C^{0,1}(\Delta(1,\psi_\ell))} \le 1, & g_\ell(0) = 0, \\ \displaystyle\fint_{D(1,\psi_\ell)} |u_\ell|^2 \le 1, \end{cases} \tag{5.2.12}$$

and

$$\fint_{D(\theta, \psi_\ell)} |u_\ell|^2 > \theta^{2\sigma}. \tag{5.2.13}$$

Note that by Caccioppoli's inequality (5.1.7), the $H^1(D(1/2,\psi_\ell);\mathbb{R}^m)$ norm of $\{u_\ell\}$ is uniformly bounded. This allows us to apply Lemma 5.2.4 and to obtain subsequences that satisfy (5.2.5) and (5.2.6). It follows that

$$\begin{aligned}\int_{D(1/2,\psi)} |u|^2 &= \lim_{\ell\to\infty}\int_{D(1/2,\psi_\ell)} |u_\ell|^2\\ &\le \lim_{\ell\to\infty}|D(1,\psi_\ell)|\\ &= |D(1,\psi)|,\end{aligned}$$

and

$$\fint_{D(\theta,\psi)} |u|^2 = \lim_{\ell\to\infty}\fint_{D(\theta,\psi_\ell)} |u_\ell|^2 \ge \theta^{2\sigma}.$$

In view of (5.2.10)–(5.2.11) we obtain $\theta^{2\sigma}\le C_0\,\theta^{2\rho}$, which is in contradiction with the choice of θ. This completes the proof. □

Lemma 5.2.6 (Iteration). *Assume that A satisfies the same conditions as in Lemma 5.2.5. Fix $0<\sigma<1$. Let ε_0 and θ be the positive constants given by Lemma 5.2.5. Suppose that*

$$\mathcal{L}(u_\varepsilon)=0 \quad \text{in } D(1,\psi) \quad \text{and} \quad u_\varepsilon = g \quad \text{on } \Delta(1,\psi),$$

where $g\in C^{0,1}(\Delta(1,\psi);\mathbb{R}^m)$ and $g(0)=0$. Then, if $\varepsilon<\theta^{k-1}\varepsilon_0$ for some $k\ge 1$,

$$\fint_{D(\theta^k,\psi)} |u_\varepsilon|^2 \le \theta^{2k\sigma}\max\left\{\fint_{D(1,\psi)}|u_\varepsilon|^2,\ \|g\|^2_{C^{0,1}(\Delta(1,\psi))}\right\}. \tag{5.2.14}$$

Proof. The lemma is proved by an induction argument on k. Note that the case $k=1$ is given by Lemma 5.2.5. Suppose that the lemma holds for some $k\ge 1$. Let $\varepsilon<\theta^k\varepsilon_0$. We apply Lemma 5.2.5 to the function $w(x)=u_\varepsilon(\theta^k x)$ in $D(1,\psi_k)$, where $\psi_k(x')=\theta^{-k}\psi(\theta^k x')$. Since

$$\mathcal{L}_{\frac{\varepsilon}{\theta^k}}(w)=0 \quad \text{in } D(1,\psi_k)$$

and $\theta^{-k}\varepsilon<\varepsilon_0$, we obtain

$$\begin{aligned}\fint_{D(\theta^{k+1},\psi)} |u_\varepsilon|^2 &= \fint_{D(\theta,\psi_k)} |w|^2\\ &\le \theta^{2\sigma}\max\left\{\fint_{D(1,\psi_k)}|w|^2,\ \|w\|^2_{C^{0,1}(\Delta(1,\psi_k))}\right\}\\ &= \theta^{2\sigma}\max\left\{\fint_{D(\theta^{2k},\psi)}|u_\varepsilon|^2,\ \theta^{2k}\|g\|^2_{C^{0,1}(\Delta(\theta^k,\psi))}\right\}\\ &\le \theta^{2(k+1)\sigma}\max\left\{\fint_{D(1,\psi)}|u_\varepsilon|^2,\ \|g\|^2_{C^{0,1}(\Delta(1,\psi))}\right\},\end{aligned}$$

where we have used the induction assumption in the last step. The fact that $\{\psi_k\}$ satisfies (5.2.2) uniformly in k is essential here. □

We are now ready to give the proof of Theorem 5.2.2.

Proof of Theorem 5.2.2. By rescaling we may assume that $r = 1$. We may also assume that $0 < \varepsilon < \varepsilon_0$, since the case $\varepsilon \geq \varepsilon_0$ follows directly from the boundary Hölder estimates for second-order elliptic systems in divergence form in C^1 domains with VMO coefficients[2]. We further assume that

$$\|g\|_{C^{0,1}(\Delta(1,\psi))} \leq 1 \quad \text{and} \quad \int_{D(1,\psi)} |u_\varepsilon|^2\, dx \leq 1.$$

Under these assumptions we will show that

$$\fint_{D(t,\psi)} |u_\varepsilon|^2 \leq Ct^{2\sigma} \tag{5.2.15}$$

for any $t \in (0, 1/4)$.

To prove (5.2.15) we first consider the case $t \geq (\varepsilon/\varepsilon_0)$. Choose $k \geq 1$ so that $\theta^k \leq t < \theta^{k-1}$. Then $\varepsilon \leq \varepsilon_0 t < \varepsilon_0 \theta^{k-1}$. It follows from Lemma 5.2.6 that

$$\begin{aligned}
\fint_{D(t,\psi)} |u_\varepsilon|^2 &\leq C \fint_{D(\theta^{k-1},\psi)} |u_\varepsilon|^2 \\
&\leq C\theta^{2k\sigma} \\
&\leq Ct^{2\sigma}.
\end{aligned}$$

Next suppose that $0 < t < (\varepsilon/\varepsilon_0)$. Let $w(x) = u_\varepsilon(\varepsilon x)$. Then $\mathcal{L}_1(w) = 0$ in $D(\varepsilon_0^{-1}, \psi_\varepsilon)$ and $w(0) = 0$, where $\psi_\varepsilon(x') = \varepsilon^{-1}\psi(\varepsilon x')$. By the boundary Hölder estimate in C^1 domains for the elliptic operator $\mathcal{L}_1$, we obtain

$$\fint_{D(t\varepsilon^{-1},\psi_\varepsilon)} |w|^2 \leq C\left(\frac{t}{\varepsilon}\right)^{2\sigma} \left\{ \fint_{D(\varepsilon_0^{-1},\psi_\varepsilon)} |w|^2 + \|w\|^2_{C^{0,1}(\Delta(\varepsilon_0^{-1},\psi_\varepsilon))} \right\}.$$

Hence,

$$\begin{aligned}
\fint_{D(t,\psi)} |u_\varepsilon|^2 &\leq C\left(\frac{t}{\varepsilon}\right)^{2\sigma} \left\{ \fint_{D(\varepsilon/\varepsilon_0,\psi)} |u_\varepsilon|^2 + \varepsilon^2 \right\} \\
&\leq Ct^{2\sigma},
\end{aligned}$$

where we have used the estimate (5.2.15) for the case $t = (\varepsilon/\varepsilon_0)$ in the last step. This finishes the proof. □

[2] This result may be proved by using the real-variable method in Section 4.2. The method reduces the problem to a local boundary $W^{1,p}$ estimate for second-order elliptic systems with constant coefficients near a flat boundary.

Theorem 5.2.1 follows from Theorem 5.2.2 by Campanato's characterization of Hölder spaces: if $\mathcal{O}$ is a bounded Lipschitz domain, then

$$\begin{aligned}\|u\|_{C^{0,\sigma}(\mathcal{O})} &:= \sup\left\{\frac{|u(x)-u(y)|}{|x-y|^\sigma}:\ x,y\in\mathcal{O} \text{ and } x\neq y\right\}\\ &\approx \sup\left\{r^{-\sigma}\left(\fint_{\mathcal{O}(x,r)}\Big|u-\fint_{\mathcal{O}(x,r)}u\Big|^2\right)^{1/2},\ x\in\mathcal{O} \text{ and } 0<r<\operatorname{diam}(\mathcal{O})\right\},\end{aligned} \tag{5.2.16}$$

where $\mathcal{O}(x,r)=\mathcal{O}\cap B(x,r)$ (see [50, pp. 70–72]).

Proof of Theorem 5.2.1. Suppose that $\mathcal{L}_\varepsilon(u_\varepsilon)=0$ in $B(x_0,2r)\cap\Omega$ and $u_\varepsilon=g$ on $B(x_0,2r)\cap\partial\Omega$. Also assume that

$$\left(\fint_{B(x_0,2r)\cap\Omega}|u_\varepsilon|^2\right)^{1/2}+|g(x_0)|+r\|g\|_{C^{0,1}(\Delta(x_0,2r)\cap\partial\Omega)}\le 1,$$

where $x_0\in\partial\Omega$ and $0<r<r_0$. By a change of the coordinate system we may deduce from Theorem 5.2.2 that for any $y\in B(x_0,r)\cap\partial\Omega$ and $0<t<cr$,

$$\fint_{B(y,t)\cap\Omega)}|u_\varepsilon-u_\varepsilon(y)|^2\le C\left(\frac{t}{r}\right)^{2\sigma}.$$

It follows that

$$\fint_{B(y,t)\cap\Omega}\left|u_\varepsilon-\fint_{B(y,t)\cap\Omega}u_\varepsilon\right|^2\le C\left(\frac{t}{r}\right)^{2\sigma} \tag{5.2.17}$$

for any $y\in B(x_0,r)\cap\partial\Omega$ and $0<t<cr$. This, together with the interior Hölder estimate in Theorem 4.3.7, implies that estimate (5.2.17) holds for any $y\in B(x_0,r)\cap\Omega$ and $0<t<cr$. It then follows by the Campanato characterization of Hölder spaces that

$$\|u_\varepsilon\|_{C^{0,\sigma}(B(x_0,r)\cap\Omega)}\le C,$$

which gives the estimate (5.2.1). □

Remark 5.2.7. We may use Lemma 5.2.6 to establish a Liouville property for solutions in a half-space with no smoothness condition on A. Indeed, assume that A satisfies (2.1.2)–(2.1.3) and is 1-periodic. Suppose that $u\in H^1_{\text{loc}}(\mathbb{H}_n(a);\mathbb{R}^m)$,

$$\mathcal{L}_1(u)=0\quad\text{in } \mathbb{H}_n(a)\quad\text{and}\quad u=0\quad\text{on } \partial\mathbb{H}_n(a),$$

where $a\in\mathbb{R}$ and

$$\mathbb{H}_n(a)=\{x\in\mathbb{R}^d:\ x\cdot n<a\}$$

is a half-space with outward unit normal $n\in\mathbb{S}^{d-1}$. Under the growth condition

$$\left(\fint_{B(0,R)\cap\mathbb{H}_n(a)}|u|^2\right)^{1/2}\le CR^\rho$$

for $R\ge 1$, where $\rho\in(0,1)$, we may deduce that $u\equiv 0$ in $\mathbb{H}_n(a)$.

To see this, by translation, we may assume that $a=0$. Let $u_\varepsilon(x)=u(\varepsilon^{-1}x)$, where $\varepsilon=\theta^{k+\ell}$ and $\theta\in(0,1)$ is given by Lemma 5.2.5. Then $\mathcal{L}_\varepsilon(u_\varepsilon)=0$ in $\mathbb{H}_n(0)$ and $u_\varepsilon=0$ on $\partial\mathbb{H}_n(0)$. Choose $\sigma\in(\rho,1)$. It follows from Lemma 5.2.6 that

$$\fint_{B(0,\theta^k)\cap\mathbb{H}_n(0)} |u_\varepsilon|^2 \le C\theta^{2k\sigma} \fint_{B(0,1)\cap\mathbb{H}_n(0)} |u_\varepsilon|^2,$$

if k and ℓ are sufficiently large. By a change of variables, this gives

$$\begin{aligned}\fint_{B(0,\theta^{-\ell})\cap\mathbb{H}_n(0)} |u|^2 &\le C\theta^{2k\sigma} \fint_{B(0,\theta^{-\ell-k})\cap\mathbb{H}_n(0)} |u|^2\\ &\le C\theta^{2k(\sigma-\rho)-2\rho\ell},\end{aligned}$$

where for the last inequality we have used the growth condition. Since $\rho<\sigma$, we may let $k\to\infty$ to conclude that $u\equiv 0$ in $B(0,\theta^{-\ell})\cap\mathbb{H}_n(0)$ for any $\ell>2$. It follows that $u\equiv 0$ in $\mathbb{H}_n(0)$.

5.3 Boundary $W^{1,p}$ estimates

In this section we establish the uniform boundary $W^{1,p}$ estimates in C^1 domains under the assumptions that A is elliptic, periodic, and belongs to $\text{VMO}(\mathbb{R}^d)$. We will use $B^{\alpha,p}(\partial\Omega)$ to denote the Besov space on $\partial\Omega$ with exponent $p\in(1,\infty)$ and of order $\alpha\in(0,1)$. If Ω is Lipschitz, the space $B^{1-\frac{1}{p},p}(\partial\Omega)$ may be identified as the set of functions that are traces on $\partial\Omega$ of $W^{1,p}(\Omega)$ functions [76].

Theorem 5.3.1. *Suppose that A is 1-periodic and satisfies* (2.1.2)–(2.1.3). *Also assume that A satisfies* (4.3.1). *Let Ω be a bounded C^1 domain in $\mathbb{R}^d$ and $1<p<\infty$. Let $u_\varepsilon\in W^{1,p}(\Omega;\mathbb{R}^m)$ be the weak solution to the Dirichlet problem*

$$\mathcal{L}_\varepsilon(u_\varepsilon)=F \quad \text{in } \Omega \quad \text{and} \quad u_\varepsilon=g \quad \text{on } \partial\Omega, \tag{5.3.1}$$

where $F\in W^{-1,p}(\Omega;\mathbb{R}^m)$ and $g\in B^{1-\frac{1}{p},p}(\partial\Omega;\mathbb{R}^m)$. Then

$$\|u_\varepsilon\|_{W^{1,p}(\Omega)} \le C_p\left\{\|F\|_{W^{-1,p}(\Omega)}+\|g\|_{B^{1-\frac{1}{p},p}(\partial\Omega)}\right\}, \tag{5.3.2}$$

where C_p depends only on p, μ, $\omega(t)$ in (4.3.1), *and Ω.*

By the real-variable method in Section 4.2, to prove Theorem 5.3.1, it suffices to establish reverse Hölder estimates for local solutions of $\mathcal{L}_\varepsilon(u_\varepsilon)=0$. Since the interior case is already settled in Section 4.3, the remaining task is to establish the following.

Lemma 5.3.2. *Assume that A and Ω satisfy the same conditions as in Theorem 5.3.1. Let $u_\varepsilon\in H^1(B(x_0,r)\cap\Omega;\mathbb{R}^m)$ be a weak solution of $\mathcal{L}_\varepsilon(u_\varepsilon)=0$ in $B(x_0,r)\cap$*

Ω with $u_\varepsilon = 0$ on $B(x_0, r) \cap \partial\Omega$, for some $x_0 \in \partial\Omega$ and $0 < r < r_0$. Then, for any $2 < p < \infty$,

$$\left(\fint_{B(x_0,r/2)\cap\Omega} |\nabla u_\varepsilon|^p \right)^{1/p} \le C_p \left(\fint_{B(x_0,r)\cap\Omega} |\nabla u_\varepsilon|^2 \right)^{1/2}, \tag{5.3.3}$$

where C_p depends only on p, μ, $\omega(t)$ in (4.3.1), and Ω.

Proof. We prove (5.3.3) by combining the interior $W^{1,p}$ estimates in Section 4.3 with the boundary Hölder estimates in Section 5.2. Let $\delta(x) = \text{dist}(x, \partial\Omega)$. It follows from Theorem 4.3.1 and Caccioppoli's inequality that

$$\fint_{B(y,c\delta(y))} |\nabla u_\varepsilon(x)|^p \, dx \le C \fint_{B(y,2c\delta(y))} \left| \frac{u_\varepsilon(x)}{\delta(x)} \right|^p dx, \tag{5.3.4}$$

for any $y \in B(x_0, r/2) \cap \Omega$, where $p > 2$ and $c > 0$ is sufficiently small. Observe that if $|x - y| < c\,\delta(y)$ for some $c \in (0, 1/2)$, then

$$\delta(x) \le |x - y| + \delta(y) \le 2\,\delta(y).$$

Also, since

$$\delta(y) \le |x - y| + \delta(x) \le c\,\delta(y) + \delta(x),$$

we have $\delta(y) \le 2\,\delta(x)$. By integrating both sides of (5.3.4) in y over $B(x_0, r/2)\cap\Omega$, we obtain

$$\int_{B(x_0,r/2)\cap\Omega} |\nabla u_\varepsilon|^p \, dx \le C \int_{B(x_0,3r/4)\cap\Omega} \left| \frac{u_\varepsilon(x)}{\delta(x)} \right|^p dx. \tag{5.3.5}$$

Finally, since $u_\varepsilon = 0$ on $B(x_0, r) \cap \partial\Omega$, the boundary Hölder estimate in Theorem (5.2.1) gives

$$|u_\varepsilon(x)| \le C_\sigma \left(\frac{\delta(x)}{r} \right)^\sigma \left(\fint_{B(x_0,r)\cap\Omega} |u_\varepsilon|^2 \right)^{1/2} \tag{5.3.6}$$

for any $x \in B(x_0, 3r/4) \cap \Omega$, where $\sigma \in (0, 1)$. By choosing σ close to 1 so that $p(1 - \sigma) < 1$ and substituting estimate (5.3.6) into (5.3.5), we see that

$$\left(\fint_{B(x_0,r/2)\cap\Omega} |\nabla u_\varepsilon|^p \right)^{1/p} \le \frac{C}{r} \left(\fint_{B(x_0,r)\cap\Omega} |u_\varepsilon|^2 \right)^{1/2}.$$

Since $u_\varepsilon = 0$ on $B(x_0, r) \cap \partial\Omega$, by the Poincaré inequality, this yields (5.3.3). □

Proof of Theorem 5.3.1. First, since $g \in B^{1-\frac{1}{p},p}(\partial\Omega; \mathbb{R}^m)$, there exists a function $G \in W^{1,p}(\Omega; \mathbb{R}^m)$ such that $G = g$ on $\partial\Omega$ and

$$\|G\|_{W^{1,p}(\Omega)} \le C \, \|g\|_{B^{1-\frac{1}{p},p}(\partial\Omega)}.$$

By considering the function $u_\varepsilon - G$, we reduce the general case to the case $G = 0$.

Next, note that if $u_\varepsilon \in W_0^{1,p}(\Omega;\mathbb{R}^m)$ is a weak solution of $\mathcal{L}_\varepsilon(u_\varepsilon) = F$ in Ω, and $\widetilde{u}_\varepsilon \in W_0^{1,p'}(\Omega;\mathbb{R}^m)$ is a weak solution of $\mathcal{L}_\varepsilon^*(\widetilde{u}_\varepsilon) = \widetilde{F}$ in Ω, then

$$\begin{aligned}\langle F, \widetilde{u}_\varepsilon\rangle_{W^{-1,p}(\Omega)\times W_0^{1,p'}(\Omega)} &= \int_\Omega A(x/\varepsilon)\nabla u_\varepsilon \cdot \nabla \widetilde{u}_\varepsilon \, dx \\ &= \langle \widetilde{F}, u_\varepsilon\rangle_{W^{-1,p'}(\Omega)\times W_0^{1,p}(\Omega)}.\end{aligned}$$

Thus, by a duality argument, it suffices to prove the estimate (5.3.2) with $g = 0$ for $p > 2$. We point out that A^* satisfies the same conditions as A.

Finally, let $p > 2$ and $F \in W^{-1,p}(\Omega;\mathbb{R}^m)$. There exist $\{f_0, f_1, \ldots, f_d\} \subset L^p(\Omega;\mathbb{R}^m)$ such that

$$F = f_0 + \frac{\partial f_i}{\partial x_i} \quad \text{and} \quad \sum_{i=0}^d \|f_i\|_{L^p(\Omega)} \le C\,\|F\|_{W^{-1,p}(\Omega)}.$$

We show that if $u_\varepsilon \in W_0^{1,2}(\Omega;\mathbb{R}^m)$ and $\mathcal{L}_\varepsilon(u_\varepsilon) = F$ in Ω, then

$$\|\nabla u_\varepsilon\|_{L^p(\Omega)} \le C \sum_{i=0}^d \|f_i\|_{L^p(\Omega)}. \tag{5.3.7}$$

This will be done by applying the real-variable argument given by Theorem 4.2.6 (with $\eta = 0$). Let $q = p+1$. Consider the two functions

$$H(x) = |\nabla u_\varepsilon(x)| \quad \text{and} \quad h(x) = \sum_{i=0}^d |f_i(x)| \quad \text{in } \Omega.$$

For each ball B with the property that $|B| \le c_0|\Omega|$ and either $4B \subset \Omega$ or B is centered on $\partial\Omega$, we need to construct two measurable functions H_B and R_B that satisfy $H \le |H_B| + |R_B|$ on $\Omega \cap 2B$ and condition (4.2.25). We will only deal with the case where B is centered on $\partial\Omega$. The other case is already treated in the proof of the interior $W^{1,p}$ estimates in Section 4.3.

Let $B = B(x_0, r)$ for some $x_0 \in \partial\Omega$ and $0 < r < r_0/16$. Write $u_\varepsilon = v_\varepsilon + w_\varepsilon$ in $4B \cap \Omega$, where $v_\varepsilon \in W_0^{1,2}(4B\cap\Omega;\mathbb{R}^m)$ and $\mathcal{L}_\varepsilon(v_\varepsilon) = F$ in $4B \cap \Omega$. Let

$$H_B = |\nabla v_\varepsilon| \quad \text{and} \quad R_B = |\nabla w_\varepsilon|.$$

Then $H \le H_B + R_B$ in $2B \cap \Omega$, and by Theorem 2.1.5,

$$\begin{aligned}\left(\fint_{\Omega\cap 2B} |H_B|^2\right)^{1/2} &\le C\left(\fint_{\Omega\cap 4B} |\nabla v_\varepsilon|^2\right)^{1/2} \\ &\le C\left(\fint_{\Omega\cap 4B} |h|^2\right)^{1/2}.\end{aligned}$$

Note that $\mathcal{L}_\varepsilon(w_\varepsilon)=0$ in $\Omega\cap 4B$ and $w_\varepsilon=0$ on $4B\cap\partial\Omega$. In view of Lemma 5.3.2 we obtain

$$
\begin{aligned}
\left(\fint_{\Omega\cap 2B}|R_B|^q\right)^{1/q} &= \left(\fint_{\Omega\cap 2B}|\nabla w_\varepsilon|^q\right)^{1/q} \le C\left(\fint_{\Omega\cap 4B}|\nabla w_\varepsilon|^2\right)^{1/2}\\
&\le C\left(\fint_{\Omega\cap 4B}|\nabla u_\varepsilon|^2\right)^{1/2} + C\left(\fint_{\Omega\cap 4B}|\nabla v_\varepsilon|^2\right)^{1/2}\\
&\le C\left(\fint_{\Omega\cap 4B}|H_B|^2\right)^{1/2} + C\left(\fint_{\Omega\cap 4B}|h|^2\right)^{1/2}.
\end{aligned}
$$

Thus we have verified all conditions in Theorem 4.2.6 with $\eta=0$. Consequently,

$$
\begin{aligned}
\left(\int_\Omega |\nabla u_\varepsilon|^p\right)^{1/p} &\le C\left\{\left(\int_\Omega|\nabla u_\varepsilon|^2\right)^{1/2} + \left(\int_\Omega |h|^p\right)^{1/p}\right\}\\
&\le C\sum_{i=0}^d \|f_i\|_{L^p(\Omega)}\\
&\le C\,\|F\|_{W^{-1,p}(\Omega)}.
\end{aligned}
$$

This completes the proof. □

Let Ω be a bounded Lipschitz domain in $\mathbb{R}^d$. Let $x_0\in\partial\Omega$ and $0<r<r_0$. Suppose that $\mathcal{L}_\varepsilon(u_\varepsilon)=0$ in $B(x_0,r)\cap\Omega$ and $u_\varepsilon=0$ on $B(x_0,r)\cap\partial\Omega$. It follows from the proof of Theorem 5.1.3 that

$$
\left(\fint_{B(x_0,sr)\cap\Omega}|\nabla u_\varepsilon|^2\right)^{1/2} \le \frac{C}{(t-s)r}\left(\fint_{B(x_0,tr)\cap\Omega}|u_\varepsilon|^2\right)^{1/2}, \tag{5.3.8}
$$

where $(1/2)\le s<t\le 1$ and C depends only on μ and Ω. By the Sobolev–Poincaré inequality, it follows that

$$
\left(\fint_{B(x_0,sr)\cap\Omega}|\nabla u_\varepsilon|^2\right)^{1/2} \le \frac{C}{t-s}\left(\fint_{B(x_0,tr)\cap\Omega}|\nabla u_\varepsilon|^q\right)^{1/2}, \tag{5.3.9}
$$

where $q=\frac{2d}{d+2}$ for $d\ge 3$, and $1<q<2$ for $d=2$. Consequently, as in the interior case, by a real-variable argument [50], there exists $\bar{p}>2$, depending only on μ and Ω, such that

$$
\left(\fint_{B(x_0,r/4)\cap\Omega}|\nabla u_\varepsilon|^{\bar{p}}\right)^{1/\bar{p}} \le C\left(\fint_{B(x_0,r/2)\cap\Omega}|\nabla u_\varepsilon|^2\right)^{1/2}. \tag{5.3.10}
$$

By the proof of Theorem 5.3.1, the boundary reverse Hölder inequality (5.3.10) and its interior counterpart (2.1.12) yield the following.

Theorem 5.3.3. *Suppose that A satisfies* (2.1.2)–(2.1.3). *Let Ω be a bounded Lipschitz domain. Then there exists $\delta \in (0, 1/2)$, depending only on μ and Ω, such that for any $F \in W^{-1,p}(\Omega; \mathbb{R}^m)$ and $g \in B^{1-\frac{1}{p},p}(\partial\Omega; \mathbb{R}^m)$ with $|\frac{1}{p} - \frac{1}{2}| < \delta$, there exists a unique solution in $W^{1,p}(\Omega; \mathbb{R}^m)$ to the Dirichlet problem: $\mathcal{L}_\varepsilon(u_\varepsilon) = F$ in Ω and $u_\varepsilon = g$ on $\partial\Omega$. Moreover, the solution satisfies the estimate* (5.3.2) *with constant C depending only on p, μ and Ω.*

Theorem 5.3.3 is due to N. Meyers [72] in the case where Ω is smooth. We point out that neither a smoothness nor the periodicity assumption on A is needed.

5.4 Green functions and Dirichlet correctors

The $m \times m$ matrix $G_\varepsilon(x, y) = \big(G_\varepsilon^{\alpha\beta}(x, y)\big)$ of Green functions for the operator $\mathcal{L}_\varepsilon$ in Ω is defined, at least formally, by

$$\begin{cases} \mathcal{L}_\varepsilon\big(G_\varepsilon^\beta(\cdot, y)\big) = e^\beta \delta_y(\cdot) & \text{in } \Omega, \\ G_\varepsilon^\beta(\cdot, y) = 0 & \text{on } \partial\Omega, \end{cases} \tag{5.4.1}$$

where $G_\varepsilon^\beta(x, y) = (G_\varepsilon^{1\beta}(x, y), \dots, G_\varepsilon^{m\beta}(x, y))$, $e^\beta = (0, \dots, 1, \dots, 0)$ with 1 in the βth position, and $\delta_y(\cdot)$ denotes the Dirac delta function supported at y. More precisely, if $F \in C_0^\infty(\Omega; \mathbb{R}^m)$, then

$$u_\varepsilon(x) = \int_\Omega G_\varepsilon(x, y) F(y)\, dy$$

is the weak solution in $H_0^1(\Omega; \mathbb{R}^m)$ to the Dirichlet problem: $\mathcal{L}_\varepsilon(u_\varepsilon) = F$ in Ω and $u_\varepsilon = 0$ on $\partial\Omega$.

In the case $m = 1$ or $d = 2$, it is known that if Ω is Lipschitz and A satisfies conditions (2.1.2)–(2.1.3), the Green functions exist and satisfy the estimate

$$|G_\varepsilon(x, y)| \le \begin{cases} C\,|x - y|^{2-d} & \text{if } d \ge 3, \\ C\left\{1 + \ln(r_0 |x - y|^{-1})\right\} & \text{if } d = 2, \end{cases} \tag{5.4.2}$$

for any $x, y \in \Omega$ and $x \ne y$, where $r_0 = \operatorname{diam}(\Omega)$. See, e.g., [55, 101]. If $d \ge 3$ and $m \ge 2$, the matrix of Green functions can be constructed and the estimate $|G_\varepsilon(x, y)| \le C\,|x - y|^{2-d}$ continues to hold as long as local solutions of $\mathcal{L}_\varepsilon(u_\varepsilon) = 0$ and $\mathcal{L}_\varepsilon^*(v_\varepsilon) = 0$ satisfy the De Giorgi–Nash Hölder estimates [59]. As a result, in view of the interior and boundary Hölder estimates in Sections 4.3 and 5.2, we obtain the following.

Theorem 5.4.1. *Suppose that A is 1-periodic and satisfies* (2.1.2)–(2.1.3). *Also assume that A satisfies the* VMO *condition* (4.3.1). *Let Ω be a bounded C^1 domain* *in* $\mathbb{R}^d$. *Then the matrix* $G_\varepsilon(x,y)$ *of Green functions exists and satisfies the estimate* (5.4.2). *Moreover,*

$$\begin{cases} |G_\varepsilon(x,y)| \le \dfrac{C[\delta(x)]^\sigma}{|x-y|^{d-2+\sigma}} & \textit{if } \ \delta(x) < \dfrac{1}{2}|x-y|, \\ |G_\varepsilon(x,y)| \le \dfrac{C[\delta(y)]^{\sigma_1}}{|x-y|^{d-2+\sigma_1}} & \textit{if } \ \delta(y) < \dfrac{1}{2}|x-y|, \\ |G_\varepsilon(x,y)| \le \dfrac{C[\delta(x)]^\sigma[\delta(y)]^{\sigma_1}}{|x-y|^{d-2+\sigma+\sigma_1}} & \textit{if } \ \delta(x) < \dfrac{1}{2}|x-y| \textit{ or } \delta(y) < \dfrac{1}{2}|x-y|, \end{cases} \tag{5.4.3}$$

for any $x, y \in \Omega$ *and* $x \neq y$, *where* $0 < \sigma, \sigma_1 < 1$ *and* $\delta(x) = \text{dist}(x, \partial\Omega)$. *The constants* C *depend at most on* μ, σ, σ_1, $\omega(t)$ *in* (4.3.1), *and* Ω.

Proof. As we mentioned above, the existence of Green functions and estimate (5.4.2) follow from the interior and boundary Hölder estimates in Theorems 4.3.7 and 5.2.1, by the general results in [59, 101]. Suppose that $d \ge 3$. To see the first inequality in (5.4.3), we fix $x_0, y_0 \in \mathbb{R}^d$ with $\delta(x_0) < \frac{1}{2}|x_0 - y_0|$, and consider $u_\varepsilon(x) = G_\varepsilon^\beta(x, y_0)$. Let $r = |x_0 - y_0|$. Since

$$\mathcal{L}_\varepsilon(u_\varepsilon) = 0 \quad \text{in } B(z_0, r/2) \cap \Omega \quad \text{and} \quad u_\varepsilon = 0 \quad \text{on } \partial\Omega,$$

by Theorem 5.2.1, we obtain

$$|u_\varepsilon(x)| \le C_\sigma \left(\frac{\delta(x)}{r}\right)^\sigma \left(\fint_{B(z_0,r/2)\cap\Omega} |u_\varepsilon|^2\right)^{1/2} \tag{5.4.4}$$

for any $x \in B(z_0, r/4) \cap \Omega$, where $z_0 \in \partial\Omega$ and $\delta(x_0) = |x_0 - z_0|$. This gives

$$|G_\varepsilon(x_0, y_0)| \le C\,(\delta(x_0))^\sigma\, |x_0 - y_0|^{2-d-\sigma}.$$

The second inequality in (5.4.3) follows from the first and the fact that

$$G_\varepsilon^{*\alpha\beta}(x,y) = G_\varepsilon^{\beta\alpha}(y,x) \tag{5.4.5}$$

for any $x, y \in \Omega$, where $G_\varepsilon^*(x,y) = \big(G_\varepsilon^{*\alpha\beta}(x,y)\big)$ denotes the matrix of Green functions for the operator $\mathcal{L}_\varepsilon^*$ in Ω. To prove the third inequality, we assume that $\delta(x) < \frac{1}{4}|x-y|$ and $\delta(y) < \frac{1}{4}|x-y|$, for otherwise the estimate follows from the first two. We repeat the argument for the first inequality, but using the second inequality to estimate the RHS of (5.4.4).

Finally, we note that if $d = 2$, the argument given above works equally well, provided the estimate $|G_\varepsilon(x,y)| \le C$ holds in the case $\delta(x) < (1/2)|x-y|$. The latter is indeed true in the general case (see [101]) □

An argument similar to that used in the proof of Theorem 5.4.1 gives

$$|G_\varepsilon(x,y) - G_\varepsilon(z,y)| \le \frac{C_\sigma\, |x-z|^\sigma}{|x-y|^{d-2+\sigma}} \quad \text{if } |x-z| < (1/2)|x-y|, \tag{5.4.6}$$

and

$$|G_\varepsilon(x,y) - G_\varepsilon(x,z)| \le \frac{C_\sigma\, |y-z|^\sigma}{|x-y|^{d-2+\sigma}} \quad \text{if } |y-z| < (1/2)|x-y|, \tag{5.4.7}$$

for any $\sigma \in (0,1)$.

Theorem 5.4.2. *Assume that A and Ω satisfy the same conditions as in Theorem 5.4.1. Then, for any $\sigma \in (0,1)$,*

$$\int_\Omega |\nabla_y G_\varepsilon(x,y)| \big[\delta(y)\big]^{\sigma-1}\, dy \le C_\sigma \big[\delta(x)\big]^\sigma, \tag{5.4.8}$$

where C_σ depends only on σ, μ, $\omega(t)$ in (4.3.1), and Ω.

Proof. Fix $x \in \Omega$ and let $r = \delta(x)/2$. By Hölder's inequality, Caccioppoli's inequality, and the estimate (5.4.3),

$$\begin{aligned}
\int_{B(x,r)} |\nabla G_\varepsilon(x,y)|\, dy &= \sum_{j=0}^\infty \int_{2^{-j-1}\le|y-x|<2^{-j}r} |\nabla_y G_\varepsilon(x,y)|\, dy\\
&\le C\sum_{j=0}^\infty \left(\fint_{2^{-j-2}r\le|y-x|\le 2^{-j+1}r} |G_\varepsilon(x,y)|^2\, dy\right)^{1/2} (2^{-j}r)^{d-1}\\
&\le Cr.
\end{aligned}$$

It follows that

$$\int_{B(x,r)} |\nabla_y G_\varepsilon(x,y)| \big[\delta(y)\big]^{\sigma-1}\, dy \le Cr^\sigma. \tag{5.4.9}$$

Next, to estimate the integral on $\Omega\setminus B(x,r)$, we observe that if Q is a cube in $\mathbb{R}^d$ with the property that $3Q \subset \Omega\setminus\{x\}$ and its side length $\ell(Q) \sim \operatorname{dist}(Q,\partial\Omega)$, then

$$\begin{aligned}
&\int_Q |\nabla_y G_\varepsilon(x,y)| \big[\delta(y)\big]^{\sigma-1}\, dy\\
&\qquad \le C\big[\ell(Q)\big]^{\sigma-1} |Q| \left(\fint_Q |\nabla_y G_\varepsilon(x,y)|^2\, dy\right)^{1/2}\\
&\qquad \le C\big[\ell(Q)\big]^{\sigma-2} |Q| \left(\fint_{2Q} |G_\varepsilon(x,y)|^2\, dy\right)^{1/2},
\end{aligned} \tag{5.4.10}$$

where we have used Caccioppoli's inequality for the last step.

This, together with the third inequality in (5.4.3), gives

$$
\begin{aligned}
&\int_Q |\nabla_y G_\varepsilon(x,y)| \big[\delta(y)\big]^{\sigma-1}\, dy \\
&\le C r^{\sigma_1} [\ell(Q)]^{\sigma+\sigma_2-2} |Q| \left(\fint_{2Q} \frac{dy}{|x-y|^{2(d-2+\sigma_1+\sigma_2)}} \right)^{1/2} \\
&\le C r^{\sigma_1} [\ell(Q)]^{\sigma+\sigma_2-2} \int_{2Q} \frac{dy}{|x-y|^{(d-2+\sigma_1+\sigma_2)}} \\
&\le C r^{\sigma_1} \int_Q \frac{[\delta(y)]^{\sigma+\sigma_2-2}}{|x-y|^{d-2+\sigma_1+\sigma_2}}\, dy,
\end{aligned}
\tag{5.4.11}
$$

where $0 < \sigma_1, \sigma_2 < 1$, and we have used the observation that $\delta(y) \sim \ell(Q)$ for $y \in (3/2)Q$ and $|x-y| \sim |x-z|$ for any $y, z \in (3/2)Q$.

Finally, we perform a Whitney decomposition on Ω (see [95, pp. 167–170]). This gives $\Omega = \bigcup_j Q_j$, where $\{Q_j\}$ is a sequence of (closed) non-overlapping cubes with the property that

$$
\operatorname{diam}(Q_j) \le \operatorname{dist}(Q_j, \partial\Omega) \le 4 \operatorname{diam}(Q_j).
$$

Note that $3Q_j \subset \Omega$. Let

$$
\mathcal{O} = \bigcup_{3Q_j \subset \Omega \setminus \{x\}} Q_j.
$$

If $y \in \Omega \setminus \mathcal{O}$, then $y \in Q_j$ for some Q_j such that $x \in 3Q_j$. It follows that $|y-x| \le C\,\ell(Q_j) \le C\delta(x)$. Hence,

$$
\int_{\Omega\setminus\mathcal{O}} |\nabla_y G_\varepsilon(x,y)| \big[\delta(y)\big]^{\sigma-1}\, dy \le C r^{\sigma-1}, \tag{5.4.12}
$$

by the proof of (5.4.9). By summation, the estimate(5.4.11) leads to

$$
\begin{aligned}
&\int_{\mathcal{O}} |\nabla_y G_\varepsilon(x,y)| \big[\delta(y)\big]^{\sigma-1}\, dy \\
&\qquad \le C r^{\sigma_1} \int_\Omega \frac{[\delta(y)]^{\sigma+\sigma_2-2}}{(|x-y|+r)^{d-2+\sigma_1+\sigma_2}}\, dy.
\end{aligned}
\tag{5.4.13}
$$

Since Ω is Lipschitz, the integral in the RHS of (5.4.13) is bounded by

$$
\begin{aligned}
C \int_0^\infty \int_{\mathbb{R}^{d-1}} & \frac{t^{\sigma+\sigma_2-2}}{(|y'|+|t-r|+r)^{d-2+\sigma_1+\sigma_2}}\, dy'\, dt \\
&\le C \int_0^\infty \frac{t^{\sigma+\sigma_2-2}}{(|t-r|+r)^{\sigma_1+\sigma_2-1}}\, dt \\
&\le C r^{\sigma-\sigma_1} \int_0^\infty \frac{t^{\sigma+\sigma_2-2}}{(|t-1|+1)^{\sigma_1+\sigma_2-1}}\, dt \\
&\le C r^{\sigma-\sigma_1},
\end{aligned}
$$

where we have chosen $\sigma_1, \sigma_2 \in (0,1)$ so that $\sigma_1 + \sigma_2 > 1$ and $\sigma < \sigma_1 < 1$. This, together with (5.4.13) and (5.4.12), completes the proof. □

Definition 5.4.3. The Dirichlet corrector $\Phi_\varepsilon = \big(\Phi_{\varepsilon,j}^\beta\big)$ for the operator $\mathcal{L}_\varepsilon$ in Ω is defined by

$$\begin{cases} \mathcal{L}_\varepsilon(\Phi_{\varepsilon,j}^\beta) = 0 & \text{in } \Omega, \\ \Phi_{\varepsilon,j}^\beta = P_j^\beta & \text{on } \partial\Omega, \end{cases} \tag{5.4.14}$$

where $P_j^\beta(x) = x_j e^\beta$ for $1 \le j \le d$ and $1 \le \beta \le m$.

In the study of the boundary Lipschitz estimates for solutions with the Dirichlet condition, the function $\Phi_{\varepsilon,j}^\beta(x) - P_j^\beta(x)$ plays a role similar to $\varepsilon\chi_j^\beta(x/\varepsilon)$ for the interior Lipschitz estimates. Note that $\Phi_{\varepsilon,j}^\beta - P_j^\beta \in H_0^1(\Omega;\mathbb{R}^m)$ satisfies

$$\mathcal{L}_\varepsilon\big\{\Phi_{\varepsilon,j}^\beta - P_j^\beta\big\} = \mathcal{L}_\varepsilon\big\{\varepsilon\chi_j^\beta(x/\varepsilon)\big\} \quad \text{in } \Omega.$$

Our goal in the rest of this section is to establish the following Lipschitz estimate of Φ_ε.

Theorem 5.4.4. *Suppose that A is 1-periodic and satisfies* (2.1.2)–(2.1.3). *Also assume that A satisfies the Hölder continuity condition* (4.0.2). *Let Ω be a bounded $C^{1,\alpha}$ domain. Then*

$$\|\nabla\Phi_\varepsilon\|_{L^\infty(\Omega)} \le C, \tag{5.4.15}$$

where C depends only on μ, λ, τ and Ω.

The proof of Theorem 5.4.4 uses the estimate of Green functions in Theorem 5.4.2 and a blow-up argument.

Lemma 5.4.5. *Suppose that A and Ω satisfy the same conditions as in Theorem* 5.4.1. *For $g \in C^{0,1}(\Omega;\mathbb{R}^m)$, let $u_\varepsilon \in H^1(\Omega;\mathbb{R}^m)$ be the solution of the Dirichlet problem $\mathcal{L}_\varepsilon(u_\varepsilon) = 0$ in Ω and $u_\varepsilon = g$ on $\partial\Omega$. Then for any $x_0 \in \partial\Omega$ and $c\,\varepsilon \le r <$* $\operatorname{diam}(\Omega)$,

$$\left(\fint_{B(x_0,r)\cap\Omega} |\nabla u_\varepsilon|^2\right)^{1/2} \le C\Big\{\|\nabla g\|_{L^\infty(\Omega)} + \varepsilon^{-1}\|g\|_{L^\infty(\Omega)}\Big\}, \tag{5.4.16}$$

where C depends only on μ, $\omega(t)$ in (4.3.1), *and Ω.*

Proof. Let $v_\varepsilon = g\varphi_\varepsilon$, where φ_ε is a cut-off function in $C_0^\infty(\mathbb{R}^d)$ such that $0 \le \varphi_\varepsilon \le 1$, $\varphi_\varepsilon(x) = 1$ if $\delta(x) \le (\varepsilon/4)$, $\varphi_\varepsilon(x) = 0$ if $\delta(x) \ge (\varepsilon/2)$, and $|\nabla\varphi_\varepsilon| \le C\,\varepsilon^{-1}$. Note that

$$\|\nabla v_\varepsilon\|_{L^\infty(\Omega)} \le C\Big\{\|\nabla g\|_{L^\infty(\Omega)} + \varepsilon^{-1}\|g\|_{L^\infty(\Omega)}\Big\}. \tag{5.4.17}$$

Thus it suffices to estimate $w_\varepsilon = u_\varepsilon - v_\varepsilon$.

To this end, we observe that $\mathcal{L}_\varepsilon(w_\varepsilon) = -\mathcal{L}_\varepsilon(v_\varepsilon)$ in Ω and $w_\varepsilon = 0$ on $\partial\Omega$. It follows that

$$w_\varepsilon^\alpha(x) = -\int_\Omega \frac{\partial}{\partial y_i}\Big\{G_\varepsilon^{\alpha\beta}(x,y)\Big\} \cdot a_{ij}^{\beta\gamma}(y/\varepsilon) \cdot \frac{\partial v_\varepsilon^\gamma}{\partial y_j}\, dy.$$

Hence,

$$|w_\varepsilon(x)| \le C\,\|\nabla v_\varepsilon\|_{L^\infty(\Omega)} \int_{\delta(y)\le\varepsilon} |\nabla_y G_\varepsilon(x,y)|\, dy.$$

By Theorem 5.4.2,

$$\int_{\delta(y)\le\varepsilon} |\nabla_y G_\varepsilon(x,y)|\, dy \le C_\sigma\,[\delta(x)]^\sigma \varepsilon^{1-\sigma} \tag{5.4.18}$$

for any $\sigma \in (0,1)$. This implies that

$$\begin{aligned}\|w_\varepsilon\|_{L^\infty(B(Q,2r)\cap\Omega)} &\le C_\sigma r^\sigma \varepsilon^{1-\sigma}\|\nabla v_\varepsilon\|_{L^\infty(\Omega)}\\ &\le Cr\|\nabla v_\varepsilon\|_{L^\infty(\Omega)},\end{aligned}$$

where we have used the assumption $r \ge c\varepsilon$. By Caccioppoli's inequality we obtain

$$\begin{aligned}\fint_{B(Q,r)\cap\Omega} |\nabla w_\varepsilon|^2 &\le \frac{C}{r^2}\fint_{B(Q,2r)\cap\Omega} |w_\varepsilon|^2 + C\,\|\nabla v_\varepsilon\|^2_{L^\infty(\Omega)}\\ &\le C\|\nabla v_\varepsilon\|^2_{L^\infty(\Omega)},\end{aligned}$$

which, in view of (5.4.17), gives the desired estimate. □

Proof of Theorem 5.4.4. Let $r_0 = \operatorname{diam}(\Omega)$. The case $\varepsilon \ge cr_0$ follows from the boundary Lipschitz estimates in $C^{1,\alpha}$ domains for elliptic systems in divergence form with Hölder continuous coefficients [50]. Suppose that $0 < \varepsilon < cr_0$. Fix $1 \le j \le d$ and $1 \le \beta \le m$, and let

$$u_\varepsilon = \Phi_{\varepsilon,j}^\beta(x) - P_j^\beta(x) - \varepsilon\chi_j^\beta(x/\varepsilon).$$

Then $\mathcal{L}_\varepsilon(u_\varepsilon) = 0$ in Ω and $u_\varepsilon = -\varepsilon\chi_j^\beta(x/\varepsilon)$ on $\partial\Omega$. Under the assumption that A is Hölder continuous, $\nabla\chi$ is bounded. It then follows from Lemma 5.4.5 that

$$\fint_{B(x_0,r)\cap\Omega} |\nabla u_\varepsilon|^2 \le C \tag{5.4.19}$$

for any $x_0 \in \partial\Omega$ and $c\varepsilon < r < r_0$. By the interior Lipschitz estimates in Section 4.1, this implies that $|\nabla u_\varepsilon(x)| \le C$ if $\delta(x) \ge \varepsilon$. Indeed, suppose $x \in \Omega$ and $\delta(x) \ge \varepsilon$. Let $r = \delta(x) = |x - x_0|$, where $x_0 \in \partial\Omega$. Then

$$\begin{aligned}|\nabla u_\varepsilon(x)| &\le C\left(\fint_{B(x,r/2)} |\nabla u_\varepsilon|^2\right)^{1/2}\\ &\le C\left(\fint_{B(x_0,2r)\cap\Omega} |\nabla u_\varepsilon|^2\right)^{1/2}\\ &\le C.\end{aligned}$$

Consequently, we obtain

$$|\nabla \Phi_\varepsilon(x)| \le C + C\, \|\nabla \chi\|_\infty \le C,$$

if $\delta(x) \ge \varepsilon$.

The case $\delta(x) < \varepsilon$ can be handled by a blow-up argument, observing that

$$\fint_{B(x_0, 2\varepsilon)\cap\Omega} |\nabla \Phi_\varepsilon|^2 \le C, \tag{5.4.20}$$

which follows from (5.4.19). Without loss of generality, suppose $x_0 = 0 \in \partial\Omega$. Consider $w(x) = \varepsilon^{-1}\Phi^\beta_{\varepsilon,j}(\varepsilon x)$ in $\Omega^\varepsilon = \{x \in \mathbb{R}^d : \ \varepsilon x \in \Omega\}$. Then $\mathcal{L}_1(w) = 0$ in Ω^ε and $w = P^\beta_j(x)$ on $\partial\Omega^\varepsilon$. It follows from the boundary Lipschitz estimates for $\mathcal{L}_1$ that

$$\|\nabla w\|_{L^\infty(B(0,1)\cap\Omega^\varepsilon)} \le C \left\{ 1 + \left(\fint_{B(0,2)\cap\Omega^\varepsilon} |\nabla w|^2 \right)^{1/2} \right\}.$$

By a change of variables, this yields

$$\begin{aligned}\|\nabla \Phi_\varepsilon\|_{L^\infty(B(0,\varepsilon)\cap\Omega)} &\le C \left\{ 1 + \left(\fint_{B(0,2\varepsilon)\cap\Omega} |\nabla \Phi_\varepsilon|^2 \right)^{1/2} \right\}\\ &\le C,\end{aligned}$$

which completes the proof of Theorem 5.4.4. □

Remark 5.4.6. Let $\psi : \mathbb{R}^{d-1} \to \mathbb{R}$ be a function such that

$$\begin{cases} \psi(0) = 0, \quad \text{supp}(\psi) \subset \{x' : \ |x'| \le 3\}, \\ \|\nabla \psi\|_\infty \le M_0 \quad \text{ and } \quad \|\nabla \psi\|_{C^{0,\eta}(\mathbb{R}^{d-1})} \le M_0, \end{cases} \tag{5.4.21}$$

where $\eta \in (0,1)$ and $M_0 > 0$ are fixed. We construct a bounded $C^{1,\eta}$ domain Ω_ψ in $\mathbb{R}^d$ with the following properties:

$$\begin{cases} D(4,\psi) \subset \Omega_\psi \subset \big\{(x', x_d) : \ |x'| < 8 \text{ and } |x_d| < 100(M_0+1)\big\}, \\ \big\{(x', \psi(x')) : \ |x'| < 4\big\} \subset \partial\Omega_\psi, \\ \Omega_\psi \setminus \{(x', \psi(x')) : \ |x'| \le 4\} \text{ depends only on } M_0. \end{cases} \tag{5.4.22}$$

Let $\Phi_\varepsilon = \Phi_\varepsilon(x, \Omega_\psi, A)$ be the matrix of Dirichlet correctors for $\mathcal{L}_\varepsilon$ in Ω_ψ. It follows from the proof of Theorem 5.4.4 that

$$\|\nabla \Phi_\varepsilon(\cdot, \Omega_\psi, A)\|_{L^\infty(\Omega_\psi)} \le C,$$

where C depends only on μ, λ, τ, and (η, M_0) in (5.4.21). Also note that by (3.2.9),

$$\|\Phi^\beta_{\varepsilon,j}(\cdot, \Omega_\psi, A) - P^\beta_j(\cdot)\|_{L^2(\Omega_\psi)} \le C\sqrt{\varepsilon}.$$

5.5 Boundary Lipschitz estimates

In this section we establish uniform boundary Lipschitz estimates in $C^{1,\alpha}$ domains for solutions with the Dirichlet condition under the assumptions that A is elliptic, periodic, and Hölder continuous.

Let $\nabla_{\tan} g$ denote the tangential gradient of g on the boundary.

Theorem 5.5.1. *Suppose that A is 1-periodic and satisfies (2.1.2)–(2.1.3). Also assume A satisfies (4.0.2). Let Ω be a bounded $C^{1,\eta}$ domain for some $\eta \in (0,1)$. Suppose $u_\varepsilon \in H^1(B(x_0,r)\cap\Omega;\mathbb{R}^m)$ is a solution of*

$$\begin{cases} \mathcal{L}_\varepsilon(u_\varepsilon) = 0 & \text{in } B(x_0,r)\cap\Omega, \\ u_\varepsilon = g & \text{on } B(x_0,r)\cap\partial\Omega \end{cases}$$

for some $g \in C^{1,\eta}(B(x_0,r)\cap\partial\Omega;\mathbb{R}^m)$, where $x_0 \in \partial\Omega$ and $0<r<r_0$. Then

$$\begin{aligned} &\|\nabla u_\varepsilon\|_{L^\infty(B(x_0,r/2)\cap\Omega)} \\ &\le C\Big\{ r^{-1}\left(\fint_{B(x_0,r)\cap\Omega} |u_\varepsilon|^2\right)^{1/2} + r^\eta \|\nabla_{\tan} g\|_{C^{0,\eta}(B(x_0,r)\cap\partial\Omega)} \\ &\qquad + \|\nabla_{\tan} g\|_{L^\infty(B(x_0,r)\cap\partial\Omega)} + r^{-1}\|g\|_{L^\infty(B(x_0,r)\cap\partial\Omega)}\Big\}, \end{aligned} \tag{5.5.1}$$

where C depends only on μ, λ, τ, and Ω.

As in the case of the interior Lipschitz estimates, Theorem 5.5.1 is proved by a compactness argument. However, we need to replace $\varepsilon\chi(x/\varepsilon)$ by the function $\Phi_\varepsilon(x) - P(x)$, where Φ_ε is the matrix of Dirichlet correctors constructed in Section 5.4.

Let $D_r = D(r,\psi)$ and $\Delta_r = \Delta(r,\psi)$ be defined by (5.1.6) with ψ satisfying (5.4.21).

Lemma 5.5.2 (One-step improvement). *Let $\sigma = \eta/4$. Let the boundary corrector $\Phi^\beta_{\varepsilon,j} = \Phi^\beta_{\varepsilon,j}(x,\Omega_\psi,A)$ be defined as in Remark 5.4.6. There exist constants $\varepsilon_0 \in (0,1/2)$ and $\theta \in (0,1/4)$, depending only on μ, λ, τ, and (η, M_0), such that*

$$\left(\fint_{D_\theta} \left| u_\varepsilon - \Phi^\beta_{\varepsilon,j} n_j(0) n_i(0) \fint_{D_\theta} \frac{\partial u^\beta_\varepsilon}{\partial x_i}\right|^2\right)^{1/2} \le \theta^{1+\sigma}, \tag{5.5.2}$$

whenever $0<\varepsilon<\varepsilon_0$,

$$\begin{cases} \mathcal{L}_\varepsilon(u_\varepsilon) = 0 & \text{in } D_1, \\ u = g & \text{on } \Delta_1, \end{cases}$$

and

$$\begin{cases} \left(\fint_{D_1} |u_\varepsilon|^2\right)^{1/2} \le 1, \quad \|\nabla_{\tan} g\|_{C^{0,2\sigma}(\Delta_1)} \le 1, \\ g(0) = 0, \quad |\nabla_{\tan} g(0)| = 0. \end{cases}$$

Proof. The lemma is proved by contradiction, using Lemma 5.2.4 and the following observation: if $\mathrm{div}(A^0\nabla w) = 0$ in $D_{1/2}$ and $w = g$ on $\Delta_{1/2}$, where A^0 is a constant matrix satisfying (2.2.28) and $|g(0)| = |\nabla_{\tan} g(0)| = 0$, then

$$\begin{aligned}&\|w - x_j n_j(0) n_i(0) \fint_{D_r} \frac{\partial w}{\partial x_i}\|_{L^\infty(D_r)} \\ &\qquad\le C_0\, r^{1+2\sigma}\Big\{\|w\|_{L^2(D_{1/2})} + \|\nabla_{\tan} g\|_{C^{0,2\sigma}(\Delta_{1/2})}\Big\}\end{aligned} \tag{5.5.3}$$

for any $r \in (0, 1/4)$, where C_0 depends only on μ and (η, M_0) in (5.4.21). To show (5.5.3), we use the boundary $C^{1,2\sigma}$ estimates in $C^{1,\eta}$ domains for second-order elliptic systems with constant coefficients satisfying the Legendre–Hadamard ellipticity condition (2.2.28) [50],

$$\|\nabla w\|_{C^{0,2\sigma}(D_{1/4})} \le C\Big\{\|w\|_{L^2(D_{1/2})} + \|\nabla_{\tan} g\|_{C^{0,2\sigma}(\Delta_{1/2})}\Big\},$$

to obtain

$$\begin{aligned}&\|w - x_j \fint_{D_r} \frac{\partial w}{\partial x_j}\|_{L^\infty(D_r)} \\ &\le C r^{1+2\sigma}\|\nabla w\|_{C^{0,2\sigma}(D_r)} \\ &\le C r^{1+2\sigma}\|\nabla w\|_{C^{0,2\sigma}(D_{1/4})} \\ &\le C r^{1+2\sigma}\Big\{\|w\|_{L^2(D_{1/2})} + \|\nabla_{\tan} g\|_{C^{0,2\sigma}(\Delta_{1/2})}\Big\}.\end{aligned} \tag{5.5.4}$$

Since $\nabla_{\tan} w(0) = 0$, we have

$$n_i(0)\frac{\partial w}{\partial x_j}(0) - n_j(0)\frac{\partial w}{\partial x_i}(0) = 0.$$

Hence, for $x \in D_r$,

$$\begin{aligned}&\Big|x_j \fint_{D_r} \frac{\partial w}{\partial x_j} - x_j n_j(0) n_i(0) \fint_{D_r} \frac{\partial w}{\partial x_i}\Big| \\ &= \Big|x_j n_i(0)\Big\{n_i(0) \fint_{D_r} \frac{\partial w}{\partial x_j} - n_j(0) \fint_{D_r} \frac{\partial w}{\partial x_i}\Big\}\Big| \\ &= \Big|x_j n_i(0)\Big\{n_i(0) \fint_{D_r} \Big(\frac{\partial w}{\partial x_j} - \frac{\partial w}{\partial x_j}(0)\Big) - n_j(0) \fint_{D_r} \Big(\frac{\partial w}{\partial x_i} - \frac{\partial w}{\partial x_i}(0)\Big)\Big\}\Big| \\ &\le C|x| \fint_{D_r} |\nabla w - \nabla w(0)| \\ &\le C r^{1+2\sigma}\|\nabla w\|_{C^{0,2\sigma}(D_r)}.\end{aligned}$$

This, together with (5.5.4), gives (5.5.3).

Next we choose $\theta \in (0, 1/4)$, depending only on μ, η and M_0, such that

$$2C_0\theta^\sigma\{1 + |D(1,\psi)|^{1/2}\} \le 1.$$

We will show that for this θ, there exists $\varepsilon_0 > 0$, depending only on μ, λ, τ, and (η, M_0), such that the estimate (5.5.2) holds.

Suppose this is not the case. Then there exist sequences $\{\varepsilon_\ell\} \subset \mathbb{R}_+$, $\{A^\ell\}$ satisfying (2.0.2), (2.1.2)–(2.1.3) and (4.0.2), $\{u_\ell\}$, $\{g_\ell\}$ and $\{\psi_\ell\}$, such that $\varepsilon_\ell \to 0$, ψ_ℓ satisfies (5.4.21),

$$\begin{cases} \operatorname{div}(A^\ell(x/\varepsilon_\ell)\nabla u_\ell) = 0 & \text{in } D(1,\psi_\ell), \\ u_\ell = g_\ell & \text{on } \Delta(1,\psi_\ell), \end{cases}$$

$$\left(\fint_{D(\theta,\psi_\ell)} \left| u_\ell - \Phi^{\beta,\ell}_{\varepsilon_\ell,j} n^\ell_j(0) n^\ell_i(0) \fint_{D(\theta,\psi_\ell)} \frac{\partial u^\beta_\ell}{\partial x_i} \right|^2 \right)^{1/2} > \theta^{1+\sigma}, \tag{5.5.5}$$

and

$$\begin{aligned} &|g_\ell(0)| = |\nabla_{\tan} g_\ell(0)| = 0, \\ &\|u_\ell\|_{L^2(D(1,\psi_\ell))} \le 1, \quad \|\nabla_{\tan} g_\ell\|_{C^{0,2\sigma}(\Delta(1,\psi_\ell))} \le 1, \end{aligned} \tag{5.5.6}$$

where $n^\ell = (n^\ell_1, \dots, n^\ell_d)$ denotes the outward unit normal to $\partial D(1,\psi_\ell)$ and

$$\Phi^{\beta,\ell}_{\varepsilon_\ell}(x) = \Phi^\beta_{\varepsilon_\ell}(x, \Omega_{\psi_\ell}, A^\ell).$$

In view of Lemma 5.2.4, by passing to subsequences we may assume that as $\ell \to \infty$,

$$\begin{cases} \widehat{A^\ell} \to A, \\ \psi_\ell \to \psi \quad \text{in } C^1(|x'| < 4), \\ g_\ell(x', \psi_\ell(x')) \to g(x', \psi(x')) \quad \text{in } C^1(|x'| < 1), \\ u_\ell(x', x_d - \psi_\ell(x')) \to u(x', x_d - \psi(x')) \text{ weakly in } L^2(D(1,0); \mathbb{R}^m), \\ u_\ell(x', x_d - \psi_\ell(x')) \to u(x', x_d - \psi(x')) \text{ weakly in } H^1(D(1/2,0); \mathbb{R}^m). \end{cases} \tag{5.5.7}$$

Moreover, u is a weak solution of $\operatorname{div}(A\nabla w) = 0$ in $D(1/2,\psi)$ with $u = g$ on $\Delta(1/2,\psi)$, and A is a constant matrix satisfying (2.2.28).

By Remark 5.4.6,

$$\|\Phi^{\beta,\ell}_{\varepsilon_\ell,j} - P^\beta_j\|_{L^2(\Omega_{\psi_\ell})} \le C\sqrt{\varepsilon_\ell}. \tag{5.5.8}$$

This, together with (5.5.7), allows us to take the limit ($\ell \to \infty$) in (5.5.5) to arrive at

$$\left(\fint_{D_\theta} \left| u - x_j n_j(0) n_i(0) \fint_{D_\theta} \frac{\partial u}{\partial x_i} \right|^2 dx \right)^{1/2} \ge \theta^{1+\sigma}, \tag{5.5.9}$$

where $D_\theta = D(\theta,\psi)$. Similarly, by (5.5.6) and (5.5.7), we obtain

$$\|u\|_{L^2(D_1)} \le 1 \quad \text{and} \quad \|\nabla_{\tan} g\|_{C^{0,2\sigma}(\Delta_1)} \le 1.$$

It then follows from (5.5.9) and (5.5.3) that

$$\theta^{1+\sigma} \le C_0\, \theta^{1+2\sigma}\{1 + |D(1,\psi)|^{1/2}\},$$

which is in contradiction with the choice of θ. This completes the proof. □

Let $\psi_k(x') = \theta^{-k}\psi(\theta^k x')$, where $k \ge 1$, and $\psi : \mathbb{R}^{d-1} \to \mathbb{R}$ be a $C^{1,\eta}$ function satisfying (5.4.21). Let

$$\Pi^{\beta,k}_{\varepsilon,j}(x) = \theta^k \Phi^{\beta}_{\frac{\varepsilon}{\theta^k},j}(\theta^{-k}x, \Omega_{\psi_k}, A). \tag{5.5.10}$$

Then

$$\mathcal{L}_\varepsilon\left(\Pi^{\beta,k}_{\varepsilon,j}\right) = 0 \quad \text{in } D(1,\psi) \quad \text{and} \quad \Pi^{\beta,k}_{\varepsilon,j} = P^\beta_j \quad \text{on } \Delta(1,\psi). \tag{5.5.11}$$

This will be used in the next lemma.

Lemma 5.5.3 (Iteration). *Let σ, ε_0 and θ be constants given by Lemma* 5.5.2. *Suppose that*

$$\mathcal{L}_\varepsilon(u_\varepsilon) = 0 \quad \text{in } D(1,\psi) \quad \text{and} \quad u_\varepsilon = g \quad \text{on } \Delta(1,\psi),$$

where $g \in C^{1,2\sigma}(\Delta(1,\psi);\mathbb{R}^m)$ and $|g(0)| = |\nabla_{\tan} g(0)| = 0$. Assume that $\varepsilon < \theta^{\ell-1}\varepsilon_0$ for some $\ell \ge 1$. Then there exist constants $E^k_\varepsilon = \left(E^{\beta,k}_\varepsilon\right) \in \mathbb{R}^m$ for $k = 0,\dots,\ell-1$, such that

$$|E^k_\varepsilon| \le C\theta^{-1} J \tag{5.5.12}$$

and

$$\left(\fint_{D(\theta^\ell,\psi)} \left| u_\varepsilon - \sum_{k=0}^{\ell-1} \theta^{\sigma k}\Pi^{\beta,k}_{\varepsilon,j}(x)\, n_j(0)\, E^{\beta,k}_\varepsilon \right|^2 dx \right)^{1/2} \le \theta^{\ell(1+\sigma)} J, \tag{5.5.13}$$

where $\Pi^{\beta,k}_{\varepsilon,j}$ is defined by (5.5.10),

$$J = \max\left\{ \|\nabla_{\tan} g\|_{C^{0,2\sigma}(\Delta(1,\psi))}, \left(\fint_{D(1,\psi)} |u_\varepsilon|^2\right)^{1/2} \right\}, \tag{5.5.14}$$

and $\psi_k(x') = \theta^{-k}\psi(\theta^k x')$.

Proof. The lemma is proved by induction on ℓ. The case $\ell = 1$ follows by applying Lemma 5.5.2 to u_ε/J, with

$$E^{\beta,0}_\varepsilon = n_i(0) \fint_{D(\theta,\psi)} \frac{\partial u^\beta_\varepsilon}{\partial x_i}.$$

Suppose that Lemma 5.5.3 holds for some $\ell \ge 1$. Consider the function

$$w(x) = \theta^{-\ell} \left\{ u_\varepsilon(\theta^\ell x) - \sum_{k=0}^{\ell-1} \theta^{\sigma k}\, \Pi^{\beta,k}_{\varepsilon,j}(\theta^\ell x)\, n_j(0)\, E^{\beta,k}_\varepsilon \right\}$$

on $D(1,\psi_\ell)$. Note that $\mathcal{L}_{\theta^{-\ell}\varepsilon}(w) = 0$ in $D(1,\psi_\ell)$ and $w(0) = 0$. Also, since $\nabla_{\tan} u_\varepsilon(0) = 0$ and

$$\nabla_{\tan} \Pi^{\beta,k}_{\varepsilon,j}(0)\, n_j(0) = \nabla_{\tan} P^\beta_j(0)\, n_j(0) = 0,$$

we see that $\nabla_{\tan} w(0) = 0$. It then follows from Lemma 5.5.2 that

$$\begin{aligned} &\left(\fint_{D(\theta,\psi_\ell)} \left| w - \Phi^\beta_{\frac{\varepsilon}{\theta^\ell},j}(x,\Omega_{\psi_\ell},A)\, n_j(0)\, n_i(0) \fint_{D(\theta,\psi_\ell)} \frac{\partial w^\beta}{\partial x_i} \right|^2 dx \right)^{1/2} \\ &\le \theta^{1+\sigma} \max\left\{ \left(\fint_{D(1,\psi_\ell)} |w|^2 \right)^{1/2},\ \|\nabla_{\tan} w\|_{C^{0,2\sigma}(\Delta(1,\psi_\ell))} \right\}. \end{aligned} \tag{5.5.15}$$

By a change of variables this yields

$$\begin{aligned} &\left(\fint_{D(\theta^{\ell+1},\psi)} \left| u_\varepsilon - \sum_{k=0}^{\ell} \theta^{\sigma k} \Pi^{\beta,k}_{\varepsilon,j}(x)\, n_j(0)\, E^{\beta,k}_\varepsilon \right|^2 dx \right)^{1/2} \\ &\le \theta^{\ell+1+\sigma} \max\left\{ \left(\fint_{D(1,\psi_\ell)} |w|^2 \right)^{1/2},\ \|\nabla_{\tan} w\|_{C^{0,2\sigma}(\Delta(1,\psi_\ell))} \right\}, \end{aligned} \tag{5.5.16}$$

where

$$E^{\beta,\ell}_\varepsilon = \theta^{-\sigma\ell} n_i(0) \fint_{D(\theta,\psi_\ell)} \frac{\partial w^\beta}{\partial x_i}.$$

Note that by the induction assumption,

$$\left(\fint_{D(1,\psi_\ell)} |w|^2 \right)^{1/2} \le \theta^{\ell\sigma} J. \tag{5.5.17}$$

Also, note that

$$\begin{aligned} &\|\nabla_{\tan} w\|_{C^{0,2\sigma}(\Delta(1,\psi_\ell))} \\ &= \theta^{2\ell\sigma} \|\nabla_{\tan} f - \sum_{k=0}^{\ell-1} \theta^{\sigma k} \nabla_{\tan} \Pi^{\beta,k}_{\varepsilon,j}\, n_j(0)\, E^{\beta,k}_\varepsilon \|_{C^{0,2\sigma}(\Delta(\theta^\ell,\psi))} \\ &\le J\theta^{2\ell\sigma} \left\{ 1 + C\theta^{-1} \sum_{k=0}^{\ell-1} \theta^{\sigma k} \|\nabla_{\tan} \Pi^k_{\varepsilon,j} n_j(0)\|_{C^{0,2\sigma}(\Delta(\theta^\ell,\psi))} \right\} \end{aligned}$$

$$
\begin{aligned}
&\leq \theta^{\ell\sigma} J \left\{ \theta^{\ell\sigma} + C\theta^{-1} \sum_{k=0}^{\ell-1} \theta^{\sigma k} \|\nabla_{\tan} P_j n_j(0)\|_{C^{0,2\sigma}(\Delta(\theta^\ell,\psi))} \right\} \\
&\leq \theta^{\ell\sigma} J \left\{ \theta^{\sigma} + \frac{C\|n\|_{C^{0,2\sigma}(\Delta(\theta^\ell,\psi))}}{\theta(1-\theta^\sigma)} \right\}.
\end{aligned}
$$

Since $\|\nabla\psi\|_{C^{0,4\sigma}(\mathbb{R}^{d-1})} \leq M_0$,

$$
\|n\|_{C^{0,2\sigma}(\Delta(t,\psi))} \leq Ct^{2\sigma} \to 0 \qquad \text{as } t \to 0.
$$

Thus, by making an initial dilation of the independent variables, if necessary, we may assume that ψ is such that

$$
\theta^\sigma + \frac{C\|n\|_{C^{0,2\sigma}(\Delta(1,\psi))}}{\theta(1-\theta^\sigma)} < 1.
$$

This, together with (5.5.16) and (5.5.17), gives the estimate (5.5.13).

Finally, we note that by Caccioppoli's inequality,

$$
\begin{aligned}
|E_\varepsilon^\ell| &\leq C\,\theta^{-\sigma\ell-1} \left\{ \left(\fint_{D(1,\psi_\ell)} |w|^2 \right)^{1/2} + \|\nabla_{\tan} w\|_{L^\infty(\Delta(1,\psi_\ell))} \right\} \\
&\leq C\,\theta^{-\sigma\ell-1} \Big\{ \theta^\sigma J + \|\nabla_{\tan} w\|_{C^{0,2\sigma}(\Delta(1,\psi_\ell))} \Big\} \\
&\leq C\theta^{-1} J,
\end{aligned}
$$

where we have used the fact that $\nabla_{\tan} w(0) = 0$. The induction argument is now complete. □

Lemma 5.5.4. *Suppose that $\mathcal{L}_\varepsilon(u_\varepsilon) = 0$ in D_1 and $u_\varepsilon = g$ on Δ_1. Then*

$$
\left(\fint_{D_r} |\nabla u_\varepsilon|^2 \right)^{1/2} \leq C \left\{ \left(\fint_{D_1} |u_\varepsilon|^2 \right)^{1/2} + \|g\|_{C^{1,\eta}(\Delta_1)} \right\} \tag{5.5.18}
$$

for any $0 < r < (1/2)$, where C depends only on μ, λ, τ, and (η, M_0).

Proof. Let $\Phi_\varepsilon = \Phi_\varepsilon(x, \Omega_\psi, A) = (\Phi_{\varepsilon,j}^\beta(x))$ and

$$
b_j^\beta = \frac{\partial}{\partial x_j} \{ g^\beta(x', \psi(x')) \}(0).
$$

By considering the function

$$
u_\varepsilon(x) - \left\{ u_\varepsilon(0) + \sum_{j=1}^d \Phi_{\varepsilon,j}^\beta(x) b_j^\beta \right\},
$$

we may assume that $|u_\varepsilon(0)| = |\nabla_{\tan} u_\varepsilon(0)| = 0$. Under these assumptions we will show that

$$\left(\fint_{D_r} |u_\varepsilon|^2\right)^{1/2} \le Cr\left\{\left(\fint_{D_1} |u_\varepsilon|^2\right)^{1/2} + \|\nabla_{\tan} g\|_{C^{0,\eta}(\Delta_1)}\right\} \tag{5.5.19}$$

for $0 < r < (1/2)$. Estimate (5.5.18) follows from (5.5.19) by Caccioppoli's inequality.

Let σ, θ and ε_0 be the constants given by Lemma 5.5.3. Let $0 < \varepsilon < \theta\varepsilon_0$ (the case $\varepsilon \ge \theta\varepsilon_0$ follows from the Lipschitz estimates for elliptic systems with Hölder continuous coefficients [50]). Suppose that

$$\theta^{i+1} \le \frac{\varepsilon}{\varepsilon_0} < \theta^i \quad \text{for some } i \ge 1.$$

We may assume that $0 < r < \theta$ (the case $\theta \le r < 1/2$ is trivial). We first consider the case $\frac{\varepsilon}{\varepsilon_0} \le r < \theta$. Since $\theta^{\ell+1} \le r < \theta^\ell$ for some $\ell = 1, \ldots, i$, it follows from Lemma 5.5.3 that

$$\begin{aligned}
\left(\fint_{D_r} |u_\varepsilon|^2\right)^{1/2} &\le C\left(\fint_{D_{\theta^\ell}} |u_\varepsilon|^2\right)^{1/2} \\
&\le C\left(\fint_{D_{\theta^\ell}} \left| u_\varepsilon - \sum_{k=1}^{\ell-1} \theta^{\sigma k} \Pi_{\varepsilon,j}^{\beta,k} n_j(0) E_\varepsilon^{\beta,k}\right|^2\right)^{1/2} \\
&\quad + C\sum_{k=0}^{\ell-1} \theta^{\sigma k} |E_\varepsilon^k| \|\Pi_\varepsilon^k\|_{L^\infty(D_{\theta^\ell})} \\
&\le \theta^{\ell(1+\sigma)} J + CJ \sum_{k=0}^{\ell-1} \theta^{\sigma k} \|\Pi_\varepsilon^k\|_{L^\infty(D_{\theta^\ell})},
\end{aligned} \tag{5.5.20}$$

where

$$J = \max\left\{\|u_\varepsilon\|_{L^2(D_1)},\ \|g\|_{C^{1,\eta}(\Delta_1)}\right\}.$$

Note that $\Pi_\varepsilon^k(0) = 0$, and by Remark 5.4.6,

$$\|\nabla \Pi_\varepsilon^k\|_{L^\infty(D_{\theta^k})} \le C.$$

This implies that

$$\|\Pi_\varepsilon^k\|_{L^\infty(D_{\theta^\ell})} \le C\theta^\ell \quad \text{for } k < \ell.$$

In view of (5.5.20) we obtain the estimate (5.5.19) for any $r \in (\varepsilon/\varepsilon_0, 1)$.

Finally, to treat the case $0 < r < (\varepsilon/\varepsilon_0)$, we use a familiar blow-up argument. Let $w(x) = \varepsilon^{-1} u_\varepsilon(\varepsilon x)$. Then

$$\mathcal{L}_1(w) = 0 \quad \text{in } D(2\varepsilon_0^{-1}, \psi_\varepsilon) \quad \text{and} \quad w = \varepsilon^{-1} g(\varepsilon x) \quad \text{on } \Delta(2\varepsilon_0^{-1}, \psi_\varepsilon),$$

where $\psi_\varepsilon(x') = \varepsilon^{-1}\psi(\varepsilon x')$. Since $|w(0)| = |\nabla_{\tan} w(0)| = 0$, by the boundary Lipschitz estimates for $\mathcal{L}_1$,

$$\|w\|_{L^\infty(D(s,\psi_\varepsilon))} \le Cs\left\{\left(\fint_{D(2\varepsilon_0^{-1},\psi_\varepsilon))} |w|^2\right)^{1/2} + \|\nabla_{\tan} w\|_{C^{0,\eta}(\Delta(2\varepsilon_0^{-1},\psi_\varepsilon))}\right\}$$

for $0 < s < (2/\varepsilon_0)$. By a change of variables this yields

$$\begin{aligned}\|u_\varepsilon\|_{L^\infty(D(r,\psi))} &\le Cr\left\{\varepsilon^{-1}\left(\fint_{D\left(\frac{2\varepsilon}{\varepsilon_0},\psi\right)} |u_\varepsilon|^2\right)^{1/2} + \|\nabla_{\tan} g\|_{C^{0,\eta}(\Delta(1,\psi))}\right\}\\ &\le CrJ,\end{aligned}$$

where we have used the estimate (5.5.19) for the case $r = (2\varepsilon/\varepsilon_0)$ in the last inequality. The proof is now complete. □

We are now in a position to give the proof of Theorem 5.5.1.

Proof of Theorem 5.5.1. By rescaling we may assume that $x_0 = 0$ and $r = 1$. By a change of the coordinate system we can deduce from Lemma 5.5.4 that if $y \in \partial\Omega$, $|y| < (1/2)$ and $0 < t < (1/4)$, then

$$\left(\fint_{B(y,t)\cap\Omega} |\nabla u_\varepsilon|^2\right)^{1/2} \le C\left\{\left(\fint_{B(0,1)\cap\Omega} |u_\varepsilon|^2\right)^{1/2} + \|g\|_{C^{1,\eta}(B(0,1)\cap\partial\Omega)}\right\},$$

where C depends only on μ, λ, τ, η, and Ω. This, together with the interior Lipschitz estimate, gives (5.5.1). Indeed, let $x \in B(0,1/2)\cap\Omega$ and $t = \text{dist}(x,\partial\Omega)$. Choose $z \in B(0,1/2)\cap\partial\Omega$ such that $|x - z| \le Ct$. Then

$$\begin{aligned}|\nabla u_\varepsilon(x)| &\le C\left(\fint_{B(x,t/2)} |\nabla u_\varepsilon|^2\right)^{1/2}\\ &\le C\left(\fint_{B(z,(C+1)t)\cap\Omega} |\nabla u_\varepsilon|^2\right)^{1/2}\\ &\le C\left\{\left(\fint_{B(0,1)\cap\Omega} |u_\varepsilon|^2\right)^{1/2} + \|g\|_{C^{1,\eta}(B(0,1)\cap\partial\Omega)}\right\},\end{aligned}$$

which completes the proof. □

5.6 The Dirichlet problem in C^1 and $C^{1,\alpha}$ domains

In this section we establish uniform Hölder estimates and Lipschitz estimates as well as the nontangential-maximal-function estimates for the Dirichlet problem,

$$\begin{cases} \mathcal{L}_\varepsilon(u_\varepsilon) = F & \text{in } \Omega, \\ u_\varepsilon = g & \text{on } \partial\Omega. \end{cases} \tag{5.6.1}$$

The results for the $W^{1,p}$ estimates were already given in Section 5.3.

We begin with improved estimates for Green functions.

Theorem 5.6.1. *Suppose that A is 1-periodic and satisfies (2.1.2)–(2.1.3). Also assume that A satisfies (4.0.2). Let Ω be a bounded $C^{1,\alpha}$ domain for some $\alpha \in (0,1)$. Then the matrix of Green functions satisfies*

$$|\nabla_x G_\varepsilon(x,y)| + |\nabla_y G_\varepsilon(x,y)| \le C\,|x-y|^{1-d}, \tag{5.6.2}$$

$$|\nabla_x G_\varepsilon(x,y)| \le \frac{C\delta(y)}{|x-y|^d}, \qquad |\nabla_y G_\varepsilon(x,y)| \le \frac{C\delta(x)}{|x-y|^d}, \tag{5.6.3}$$

and

$$|\nabla_x \nabla_y G_\varepsilon(x,y)| \le C\,|x-y|^{-d} \tag{5.6.4}$$

for any $x, y \in \Omega$ and $x \ne y$, where C depends only on μ, λ, τ, and Ω.

Proof. Recall that if $d \ge 3$, $|G_\varepsilon(x,y)| \le C|x-y|^{2-d}$. Using $\mathcal{L}_\varepsilon\big(G_\varepsilon(\cdot,y)\big) = 0$ in $\Omega \setminus \{y\}$ and $G_\varepsilon^{*\alpha\beta}(x,y) = G_\varepsilon^{\beta\alpha}(y,x)$, estimates (5.6.2), (5.6.3), and (5.6.4) follow readily from the interior and boundary Lipschitz estimates in Sections 4.1 and 5.5. If $d = 2$ one should replace the size estimate on $|G_\varepsilon(x,y)|$ by

$$|G_\varepsilon(x,y) - G_\varepsilon(z,y)| \le C \quad \text{if } |x-z| < (1/2)|x-y|,$$

and apply the Lipschitz estimates to $u_\varepsilon(x) = G_\varepsilon(x,y) - G_\varepsilon(z,y)$. □

Theorem 5.6.2 (Lipschitz estimate)**.** *Suppose that A and Ω satisfy the same conditions as in Theorem 5.6.1. Let $g \in C^{1,\sigma}(\partial\Omega;\mathbb{R}^m)$ and $F \in L^p(\Omega;\mathbb{R}^m)$, where $\sigma \in (0,\alpha)$ and $p > d$. Then the unique solution in $C^{0,1}(\Omega;\mathbb{R}^m)$ to the Dirichlet problem (5.6.1) satisfies the Lipschitz estimate*

$$\|\nabla u_\varepsilon\|_{L^\infty(\Omega)} \le C\Big\{ \|g\|_{C^{1,\sigma}(\partial\Omega)} + \|F\|_{L^p(\Omega)} \Big\}, \tag{5.6.5}$$

where C depends only on μ, λ, τ, σ, p, and Ω.

Proof. Let

$$v_\varepsilon(x) = \int_\Omega G_\varepsilon(x,y) F(y)\, dy.$$

Then $\mathcal{L}_\varepsilon(v_\varepsilon) = F$ in Ω and $v_\varepsilon = 0$ on $\partial\Omega$. Note that by (5.6.2),

$$|\nabla v_\varepsilon(x)| \le C \int_\Omega \frac{|F(y)|}{|x-y|^{d-1}}\, dy. \tag{5.6.6}$$

By Hölder's inequality we obtain $\|\nabla v_\varepsilon\|_{L^\infty(\Omega)} \le C\|F\|_{L^p(\Omega)}$ for $p > d$. Thus, by subtracting v_ε from u_ε, we may now assume that $F = 0$ in Theorem 5.6.2. In this case, by covering Ω with balls of radius cr_0, we deduce from Theorems 5.5.1 and 4.1.2 that

$$\begin{aligned}\|\nabla u_\varepsilon\|_{L^\infty(\Omega)} &\le C\Big\{\|\nabla u_\varepsilon\|_{L^2(\Omega)} + \|g\|_{C^{1,\sigma}(\partial\Omega)}\Big\}\\ &\le C\,\|g\|_{C^{1,\sigma}(\partial\Omega)},\end{aligned}$$

where we have used the energy estimate

$$\|\nabla u_\varepsilon\|_{L^2(\Omega)} \le C\,\|g\|_{H^{1/2}(\partial\Omega)} \le C\,\|g\|_{C^1(\partial\Omega)}. \qquad \square$$

Remark 5.6.3. Let v_ε be the same as in the proof of Theorem 5.6.2. Using the estimate (5.6.6) and writing

$$\Omega = \bigcup_{j=j_0}^\infty B(x, 2^{-j}r) \cap \Omega,$$

where $2^{-j_0}r_0 \approx \operatorname{diam}(\Omega)$, it is not hard to deduce that

$$\|\nabla v_\varepsilon\|_{L^\infty(\Omega)} \le C_\rho \sup_{\substack{x\in\Omega\\ 0<r<r_0}} r^{1-\rho} \fint_{B(x,r)\cap\Omega} |F|,$$

where $\rho \in (0,1)$. This allows us to replace (5.6.5) by

$$\|\nabla u_\varepsilon\|_{L^\infty(\Omega)} \le C \left\{ \|g\|_{C^{1,\sigma}(\partial\Omega)} + \sup_{\substack{x\in\Omega\\ 0<r<r_0}} r^{1-\rho} \fint_{B(x,r)\cap\Omega} |F| \right\} \tag{5.6.7}$$

for any $\sigma, \rho \in (0,1)$.

Remark 5.6.4. Consider the Dirichlet problem,

$$\mathcal{L}_\varepsilon(u_\varepsilon) = \operatorname{div}(f) \quad \text{in } \Omega \quad \text{and} \quad u_\varepsilon = g \quad \text{on } \partial\Omega, \tag{5.6.8}$$

where $f = (f_i^\beta)$. Let

$$w_\varepsilon^\alpha(x) = -\int_\Omega \frac{\partial}{\partial y_i}\Big\{G_\varepsilon^{\alpha\beta}(x,y)\Big\} f_i^\beta(y)\, dy. \tag{5.6.9}$$

Then $\mathcal{L}_\varepsilon(w_\varepsilon) = \operatorname{div}(f)$ in Ω and $w_\varepsilon = 0$ on $\partial\Omega$. Since

$$w_\varepsilon^\alpha(x) = -\int_\Omega \frac{\partial}{\partial y_i}\Big\{G_\varepsilon^{\alpha\beta}(x,y)\Big\}\cdot \big(f_i^\beta(y) - f_i^\beta(x)\big)\, dy,$$

it follows that

$$
\begin{aligned}
|\nabla w_\varepsilon(x)| &\le C \int_\Omega |\nabla_x \nabla_y G_\varepsilon(x,y)|\, |f(y)-f(x)|\, dy \\
&\le C\, \|f\|_{C^{0,\rho}(\Omega)} \int_\Omega \frac{dy}{|x-y|^{d-\rho}} \\
&\le C\, \|f\|_{C^{0,\rho}(\Omega)},
\end{aligned}
$$

where we have used the estimate (5.6.4). As a result, we obtain the following estimate for the solutions of (5.6.8),

$$
\|\nabla u_\varepsilon\|_{L^\infty(\Omega)} \le C\Big\{ \|g\|_{C^{1,\sigma}(\partial\Omega)} + \|f\|_{C^\rho(\Omega)} \Big\}, \tag{5.6.10}
$$

under the assumptions that A is periodic, elliptic and Hölder continuous, and that Ω is $C^{1,\alpha}$.

Recall that for a continuous function u in Ω, the nontangential maximal function is defined by

$$
(u)^*(x) = \sup\Big\{ |u(y)| : \ y \in \Omega \text{ and } |y-x| < C_0 \operatorname{dist}(y, \partial\Omega) \Big\} \tag{5.6.11}
$$

for $x \in \partial\Omega$, where $C_0 = C_0(\Omega) > 1$ is sufficiently large.

Theorem 5.6.5 (Nontangential-maximal-function estimates). *Suppose that A and Ω satisfy the same conditions as in Theorem* 5.6.1. *Let $1 < p \le \infty$. For $g \in L^p(\partial\Omega; \mathbb{R}^m)$, let u_ε be the unique solution to the Dirichlet problem,*

$$
\mathcal{L}_\varepsilon(u_\varepsilon) = 0 \quad \text{in } \Omega \qquad \text{and } u_\varepsilon = g \quad \text{on } \partial\Omega,
$$

with the property $(u_\varepsilon)^* \in L^p(\partial\Omega)$. Then

$$
\|(u_\varepsilon)^*\|_{L^p(\partial\Omega)} \le C_p\, \|g\|_{L^p(\partial\Omega)}, \tag{5.6.12}
$$

where C_p depends only on p, μ, λ, τ, and Ω.

Proof. Write

$$
u_\varepsilon(x) = \int_{\partial\Omega} P_\varepsilon(x,y) g(y)\, dy,
$$

where the Poisson kernel $P_\varepsilon(x,y) = (P_\varepsilon^{\alpha\beta}(x,y))$ for $\mathcal{L}_\varepsilon$ on Ω is given by

$$
P_\varepsilon^{\alpha\beta}(x,y) = -n_i(y) a_{ji}^{\gamma\beta}(y/\varepsilon) \frac{\partial}{\partial y_j}\Big\{ G_\varepsilon^{\alpha\gamma}(x,y) \Big\} \tag{5.6.13}
$$

for $x \in \Omega$ and $y \in \partial\Omega$, and $n = (n_1, \dots, n_d)$ denotes the outward unit normal to $\partial\Omega$. In view of the estimate (5.6.3), we have

$$
|P_\varepsilon(x,y)| \le \frac{C\delta(x)}{|x-y|^d}, \tag{5.6.14}
$$

and hence

$$|u_\varepsilon(x)| \le C\,\delta(x) \int_{\partial\Omega} \frac{|g(y)|}{|x-y|^d}\, dy \qquad \text{for } x \in \Omega. \tag{5.6.15}$$

It follows from (5.6.15) that if $|x-z| < C_0\,\delta(x)$ for some $z \in \partial\Omega$,

$$(u_\varepsilon)^*(z) \le C\mathcal{M}_{\partial\Omega}(g)(z), \tag{5.6.16}$$

where

$$\mathcal{M}_{\partial\Omega}(g)(z) = \sup\left\{ ⨍_{B(z,r)\cap\partial\Omega} |g| : 0 < r < \operatorname{diam}(\partial\Omega) \right\} \tag{5.6.17}$$

is the Hardy–Littlewood maximal function of g on $\partial\Omega$. Since

$$\|\mathcal{M}_{\partial\Omega}(g)\|_{L^p(\partial\Omega)} \le C_p\,\|g\|_{L^p(\partial\Omega)} \qquad \text{for } 1 < p \le \infty,$$

the desired estimate $\|\mathcal{M}(u_\varepsilon)\|_{L^p(\partial\Omega)} \le C_p\,\|g\|_{L^p(\partial\Omega)}$ follows readily from (5.6.16). ◻

Remark 5.6.6 (Agmon–Miranda maximum principle). In the case $p = \infty$, Theorem 5.6.5 gives

$$\|u_\varepsilon\|_{L^\infty(\Omega)} \le C\,\|u_\varepsilon\|_{L^\infty(\partial\Omega)}, \tag{5.6.18}$$

where $\mathcal{L}_\varepsilon(u_\varepsilon) = 0$ in Ω and C is independent of ε. In particular, let

$$u_\varepsilon = \Phi^\beta_{\varepsilon,j} - P^\beta_j - \varepsilon\chi^\beta_j(x/\varepsilon),$$

where $(\Phi^\beta_{\varepsilon,j})$ are the Dirichlet correctors for $\mathcal{L}_\varepsilon$ in Ω. Then

$$\mathcal{L}_\varepsilon(u_\varepsilon) = 0 \quad \text{in } \Omega \quad \text{and} \quad u_\varepsilon = -\varepsilon\chi^\beta_j(x/\varepsilon) \quad \text{on } \partial\Omega.$$

It follows from (5.6.18) that $\|u_\varepsilon\|_{L^\infty(\Omega)} \le C\varepsilon$. This yields that

$$\|\Phi^\beta_{\varepsilon,j} - P^\beta_j\|_{L^\infty(\Omega)} \le C\varepsilon. \tag{5.6.19}$$

We end this section with some sharp Hölder estimates in C^1 domains under the assumptions that A is elliptic, periodic, and belongs to $\text{VMO}(\mathbb{R}^d)$.

Theorem 5.6.7. *Suppose that A satisfies* (2.1.2)–(2.1.3), *is* 1*-periodic and belongs to* $VMO(\mathbb{R}^d)$. *Let Ω be a bounded C^1 domain. Let u_ε be the solution to the Dirichlet problem* (5.6.1). *Then*

$$\|u_\varepsilon\|_{C^\rho(\overline{\Omega})} \le C\Big\{ \|g\|_{C^\rho(\partial\Omega)} + \|F\|_{W^{-1,p}(\Omega)} \Big\}, \tag{5.6.20}$$

where $\rho \in (0,1)$ *and* $\rho = 1 - \frac{d}{p}$.

Proof. By supposition of solutions it suffices to consider two separate cases: (1) $g = 0$; (2) $F = 0$. In the first case we use the $W^{1,p}$ estimates in Theorem 5.3.1 to obtain

$$\|\nabla u_\varepsilon\|_{L^p(\Omega)} \le C\, \|F\|_{W^{-1,p}(\Omega)}.$$

By the Sobolev embedding, this implies that

$$\|u_\varepsilon\|_{C^\rho(\overline{\Omega})} \le C\, \|F\|_{W^{-1,p}(\Omega)},$$

where $\rho = 1 - \frac{d}{p}$.

To treat the second case we assume, without loss of generality, that

$$\|g\|_{C^\rho(\partial\Omega)} = 1.$$

We need to show that

$$|u_\varepsilon(x) - u_\varepsilon(y)| \le C\, |x-y|^\rho \qquad \text{for any } x, y \in \Omega. \tag{5.6.21}$$

To this end we choose a harmonic function v in Ω such that $v = g$ on $\partial\Omega$. Since Ω is C^1, it is known that such v exists and is unique. Moreover,

$$\|v\|_{C^\rho(\overline{\Omega})} \le C\, \|g\|_{C^\rho(\partial\Omega)} = C.$$

Furthermore, by the boundary Hölder estimates for harmonic functions,

$$|\nabla v(x)| \le C\big[\delta(x)\big]^{\rho-1}. \tag{5.6.22}$$

Let $w_\varepsilon = u_\varepsilon - v(x)$. Then $\mathcal{L}_\varepsilon(w_\varepsilon) = -\mathcal{L}_\varepsilon(v)$ in Ω and $w_\varepsilon = 0$ on $\partial\Omega$. Representing w_ε as

$$w_\varepsilon^\alpha(x) = -\int_\Omega \frac{\partial}{\partial y_i}\Big\{ G_\varepsilon^{\alpha\beta}(x,y)\Big\} a_{ij}^{\beta\gamma}(y/\varepsilon) \frac{\partial v^\gamma}{\partial y_j}\, dy,$$

we obtain

$$\begin{aligned} |w_\varepsilon(x)| &\le C \int_\Omega |\nabla_y G_\varepsilon(x,y)|\big[\delta(y)\big]^{\rho-1} dy \\ &\le C\big[\delta(x)\big]^\rho, \end{aligned} \tag{5.6.23}$$

where we have used the estimate (5.6.22) for the first inequality and Theorem 5.4.2 for the second. By Caccioppoli's inequality, we also have

$$\begin{aligned} &\left(\fint_{B(x,\delta(x)/2)} |\nabla w_\varepsilon|^2 \right)^{1/2} \\ &\le \frac{C}{\delta(x)} \left(\fint_{B(x,3\delta(x)/4)} |w_\varepsilon|^2 \right)^{1/2} + C \left(\fint_{B(x,3\delta(x)/4)} |\nabla v|^2 \right)^{1/2} \\ &\le \big[\delta(x)\big]^{\rho-1}, \end{aligned} \tag{5.6.24}$$

where we have used (5.6.23) and (5.6.22) for the last inequality.

Finally, to show (5.6.21), we consider three sub-cases: (i) $|y-x| \le \delta(x)/4$; (ii) $|y-x| \le \delta(y)/4$; (iii) $|y-x| > \delta(x)/4$ and $|y-x| > \delta(y)/4$. In the first sub-case, we use the interior Hölder estimate to obtain

$$
\begin{aligned}
|u_\varepsilon(x) - u_\varepsilon(y)| &\le Cr\left(\frac{|x-y|}{r}\right)^\rho \left(\fint_{B(x,r)} |\nabla u_\varepsilon|^2\right)^{1/2}\\
&\le C\,|x-y|^\rho,
\end{aligned}
$$

where $r = \delta(x)/2$ and we have used (5.6.24) and (5.6.22). The second sub-case can be handled in the same manner. In the third sub-case we choose $x_0, y_0 \in \partial\Omega$ so that $|x - x_0| = \delta(x)$ and $|y - y_0| = \delta(y)$. Note that

$$
|x_0 - y_0| \le |x - x_0| + |x - y| + |y - y_0| \le 9\,|x - y|.
$$

Hence,

$$
\begin{aligned}
|u_\varepsilon(x) - u_\varepsilon(y)| &\le |u_\varepsilon(x) - g(x_0)| + |g(x_0) - g(y_0)| + |u_\varepsilon(y) - g(y_0)|\\
&\le C\Big\{|w_\varepsilon(x)| + |v(x) - g(x_0)| + |x_0 - y_0|^\rho + |w_\varepsilon(y)| + |v(y) - g(y_0)|\Big\}\\
&\le C\Big\{\big[\delta(x)\big]^\rho + |x-y|^\rho + \big[\delta(y)\big]^\rho\Big\}\\
&\le C\,|x-y|^\rho,
\end{aligned}
$$

where we have used estimates (5.6.23) and $\|v\|_{C^\rho(\overline{\Omega})} \le C$. This finishes the proof. □

5.7 Notes

Under the assumptions that A is elliptic, periodic, and Hölder continuous, the uniform boundary Hölder, $W^{1,p}$, and Lipschitz estimates with Dirichlet boundary conditions as well as nontangential-maximal-function estimates are due to M. Avellaneda and F. Lin [11] (also see [12, 13, 16]). Our exposition on Hölder, Lipschitz, and nontangential-maximal-function estimates follows closely [11]. The uniform $W^{1,p}$ estimates in C^1 domains for operators with periodic VMO coefficients were established in [88]. See also [46, 45]. Theorem 5.4.2 on Green functions is taken from [89].

Chapter 6

Regularity for the Neumann Problem

In this chapter we study uniform regularity estimates for the Neumann boundary value problem

$$\begin{cases} \mathcal{L}_\varepsilon(u_\varepsilon) = F & \text{in } \Omega, \\ \dfrac{\partial u_\varepsilon}{\partial \nu_\varepsilon} = g & \text{on } \partial\Omega, \end{cases} \tag{6.0.1}$$

where $\mathcal{L}_\varepsilon = -\text{div}\big(A(x/\varepsilon)\nabla\big)$ and $\frac{\partial u_\varepsilon}{\partial \nu_\varepsilon}$ denotes the conormal derivative of u_ε, defined by

$$\left(\frac{\partial u_\varepsilon}{\partial \nu_\varepsilon}\right)^\alpha = n_i a_{ij}^{\alpha\beta}(x/\varepsilon)\frac{\partial u_\varepsilon^\beta}{\partial x_j}. \tag{6.0.2}$$

Our approach is based on a general scheme for establishing large-scale regularity estimates in homogenization, developed by S.N. Armstrong and C. Smart [7] in the study of stochastic homogenization. Roughly speaking, the scheme states that if a function u_ε is well approximated by $C^{1,\alpha}$ functions at every scale greater than ε, then u_ε is Lipschitz continuous at every scale greater than ε. The approach relies on a certain (very weak) result on convergence rates and does not involve correctors in a direct manner. In comparison with the compactness method used in Chapters 4 and 5, when applied to boundary Lipschitz estimates, it does not require a priori Lipschitz estimates for boundary correctors.

We start out in Section 6.1 by establishing a result on the approximation of solutions of $\mathcal{L}_\varepsilon(u_\varepsilon) = F$ with (partial) Neumann data by solutions of $\mathcal{L}_0(u) = F$ at the large scale. In Section 6.2 we test the scheme in the simple case of boundary Hölder estimates in C^1 domains. As in the case of the Dirichlet condition, the boundary $W^{1,p}$ estimates in C^1 domains are obtained in Section 6.3 by combining the boundary Hölder estimates with the interior $W^{1,p}$ estimates in Chapter 4. In

Z. Shen, *Periodic Homogenization of Elliptic Systems*, Operator Theory: Advances and Applications 269, https://doi.org/10.1007/978-3-319-91214-1_6

Section 6.4 we prove the boundary Lipschitz estimates in $C^{1,\alpha}$ domains, assuming that A is elliptic, periodic, and Hölder continuous. Section 6.5 is devoted to the study of the matrix of Neumann functions $N_\varepsilon(x,y)$. We obtain uniform size estimates for $N_\varepsilon(x,y)$ and its derivatives $\nabla_x N_\varepsilon(x,y)$, $\nabla_y N_\varepsilon(x,y)$, $\nabla_x\nabla_y N_\varepsilon(x,y)$. Throughout Sections 6.1–6.5 we shall assume that A satisfies the Legendre ellipticity condition (2.1.20). In Section 6.6 we discuss boundary estimates for elliptic systems of linear elasticity with Neumann conditions.

We will be working with local solutions with partial Neumann conditions. Let $g \in L^2(B(x_0,r)\cap\partial\Omega;\mathbb{R}^m)$ and $F \in L^2(B(x_0,r)\cap\Omega;\mathbb{R}^m)$ for some $x_0 \in \partial\Omega$ and $0 < r < r_0$. We say that $u_\varepsilon \in H^1(B(x_0,r)\cap\Omega;\mathbb{R}^m)$ is a weak solution to

$$\mathcal{L}_\varepsilon(u_\varepsilon) = F \quad \text{in } B(x_0,r)\cap\Omega \quad \text{and} \quad \frac{\partial u_\varepsilon}{\partial \nu_\varepsilon} = g \quad \text{on } B(x_0,r)\cap\partial\Omega,$$

where $F \in L^2(B(x_0,r)\cap\Omega;\mathbb{R}^m)$ and $g \in L^2(B(x_0,r)\cap\partial\Omega;\mathbb{R}^m)$, if

$$\begin{aligned}\int_{B(x_0,r)\cap\Omega} & A(x/\varepsilon)\nabla u_\varepsilon\cdot\nabla\varphi\,dx \\ &= \int_{B(x_0,r)\cap\Omega} F\cdot\varphi\,dx + \int_{B(x_0,r)\cap\partial\Omega} g\cdot\varphi\,d\sigma\end{aligned} \tag{6.0.3}$$

for any $\varphi \in C_0^\infty(B(x_0,r);\mathbb{R}^m)$.

6.1 Approximation of solutions at the large scale

Throughout this section we assume that A is 1-periodic and satisfies (2.1.20). No smoothness condition on A is needed. Let $D_r = D(r,\psi)$ and $\Delta_r = D(r,\psi)$ be defined by (5.1.6), where $\psi : \mathbb{R}^{d-1} \to \mathbb{R}$ is a Lipschitz function such that $\psi(0) = 0$ and $\|\nabla\psi\|_\infty \le M_0$.

The goal of this section is to establish the following.

Theorem 6.1.1. *Suppose that A is 1-periodic and satisfies (2.1.20). Let $u_\varepsilon \in H^1(D_{2r},\mathbb{R}^m)$ be a weak solution to*

$$\mathcal{L}_\varepsilon(u_\varepsilon) = F \quad \text{in } D_{2r} \quad \text{and} \quad \frac{\partial u_\varepsilon}{\partial \nu_\varepsilon} = g \quad \text{on } \Delta_{2r},$$

where $F \in L^2(D_{2r};\mathbb{R}^m)$ and $g \in L^2(\Delta_{2r};\mathbb{R}^m)$. Assume that $r \ge \varepsilon$. Then there exists $w \in H^1(D_r;\mathbb{R}^m)$ such that

$$\mathcal{L}_0(w) = F \quad \text{in } D_r \quad \text{and} \quad \frac{\partial w}{\partial \nu_0} = g \quad \text{on } \Delta_r,$$

and

$$\left(\fint_{D_r} |u_\varepsilon - w|^2\right)^{1/2} \tag{6.1.1}$$
$$\le C\left(\frac{\varepsilon}{r}\right)^\alpha \left\{\left(\fint_{D_{2r}} |u_\varepsilon|^2\right)^{1/2} + r^2\left(\fint_{D_{2r}} |F|^2\right)^{1/2} + r\left(\fint_{\Delta_{2r}} |g|^2\right)^{1/2}\right\},$$

where $C > 0$ and $\alpha \in (0, 1/2)$ depend only on μ and M_0.

We start with a Caccioppoli inequality for solutions with Neumann conditions.

Lemma 6.1.2. *Suppose that $\mathcal{L}_\varepsilon(u_\varepsilon) = F$ in D_{2r} and $\frac{\partial u_\varepsilon}{\partial \nu_\varepsilon} = g$ on Δ_{2r}. Then*

$$\fint_{D_{3r/2}} |\nabla u_\varepsilon|^2\, dx \le \frac{C}{r^2}\fint_{D_{2r}} |u_\varepsilon|^2\, dx + C\, r\fint_{D_{2r}} |F|^2\, dx + C\fint_{\Delta_{2r}} |g|^2\, d\sigma, \tag{6.1.2}$$

where C depends only on μ and M_0.

Proof. By rescaling we may assume that $r = 1$. We first consider the special case $g = 0$, in which the proof is similar to that of (2.1.11). Choose $\psi \in C_0^\infty(\mathbb{R}^d)$ such that $0 \le \psi \le 1$, $\psi = 1$ in $D_{3/2}$ and $\psi = 0$ in $D_2 \setminus D_{7/4}$. Note that

$$\int_{D_2} A(x/\varepsilon)\nabla u_\varepsilon \cdot \nabla(\psi^2 u_\varepsilon)\, dx = \int_{D_2} F \cdot \psi^2 u_\varepsilon\, dx.$$

This, together with the equation (2.1.9), leads to

$$\int_{D_{3/2}} |\nabla u_\varepsilon|^2\, dx \le C\int_{D_2} |u_\varepsilon|^2\, dx + C\int_{D_2} |F|^2\, dx,$$

by using the inequality (2.1.10). The general case may be reduced to the special case by considering $u_\varepsilon - v_\varepsilon$, where $\mathcal{L}_\varepsilon(v_\varepsilon) = F$ in D_2, $\frac{\partial v_\varepsilon}{\partial \nu_\varepsilon} = \widetilde{g}$ on ∂D_2, and $\int_{D_2} v_\varepsilon\, dx = 0$. Here $\widetilde{g} = g$ on Δ_2 and $\widetilde{g}$ is a constant on $\partial D_2 \setminus \Delta_2$, chosen so that $\int_{\partial D_2} \widetilde{g}\, d\sigma + \int_{D_2} F\, dx = 0$. Note that by Theorem 2.1.7,

$$\|v_\varepsilon\|_{H^1(D_2)} \le C\Big\{\|\widetilde{g}\|_{L^2(\partial D_2)} + \|F\|_{L^2(D_2)}\Big\} \le C\Big\{\|g\|_{L^2(\Delta_2)} + \|F\|_{L^2(D_2)}\Big\}.$$

It follows that

$$\begin{aligned}
\int_{D_{3/2}} |\nabla u_\varepsilon|^2 &\le 2\int_{D_{3/2}} |\nabla v_\varepsilon|^2 + 2\int_{D_{3/2}} |\nabla(u_\varepsilon - v_\varepsilon)|^2\\
&\le C\,\|g\|^2_{L^2(\Delta_2)} + C\int_{D_2} |u_\varepsilon - v_\varepsilon|^2 + C\,\|F\|^2_{L^2(D_2)}\\
&\le C\int_{D_2} |u_\varepsilon|^2 + C\,\|F\|^2_{L^2(D_2)} + C\,\|g\|^2_{L^2(\Delta_2)},
\end{aligned}$$

where we have used (6.1.2) for the special case. □

Remark 6.1.3. Suppose that $\mathcal{L}_\varepsilon(u_\varepsilon) = 0$ in D_{2r} and $\frac{\partial u_\varepsilon}{\partial \nu_\varepsilon} = 0$ on Δ_{2r}. It follows from the proof of Lemma 6.1.2 that

$$\int_{D_{sr}} |\nabla u_\varepsilon|^2 \, dx \le \frac{C}{(t-s)^2 r^2} \int_{D_{tr}} |u_\varepsilon - E|^2 \, dx,$$

where $E \in \mathbb{R}^m$, $1 \le s < t \le 2$, and C depends only on μ and M_0. By the Poincaré–Sobolev inequality we obtain

$$\left(\fint_{D_{sr}} |\nabla u_\varepsilon|^2 \right)^{1/2} \le \frac{C}{(t-s)^2} \left(\fint_{D_{tr}} |\nabla u_\varepsilon|^q \right)^{1/q},$$

where $\frac{1}{q} = \frac{1}{2} + \frac{1}{d}$. As in the interior case, by a real-variable argument (see [50, Theorem 6.38]), this leads to the reverse Hölder inequality,

$$\left(\fint_{D_r} |\nabla u_\varepsilon|^{\bar{p}} \right)^{1/\bar{p}} \le C \left(\fint_{D_{2r}} |\nabla u_\varepsilon|^2 \right)^{1/2}, \tag{6.1.3}$$

where $C > 0$ and $\bar{p} > 2$ depend only on μ and M_0. Note that the periodicity of A is not needed for (6.1.2) and (6.1.3). It is also not needed for the next lemma, which gives a Meyers estimate [72].

Lemma 6.1.4. *Let $\Omega = D_r$ for some $1 \le r \le 2$. Let $u_\varepsilon \in H^1(\Omega; \mathbb{R}^m)$ $(\varepsilon \ge 0)$ be a weak solution to the Neumann problem $\mathcal{L}_\varepsilon(u_\varepsilon) = F$ in Ω and $\frac{\partial u_\varepsilon}{\partial \nu_\varepsilon} = g$ on $\partial\Omega$, where $F \in L^2(\Omega; \mathbb{R}^m)$, $g \in L^2(\partial\Omega; \mathbb{R}^m)$, and $\int_\Omega F \, dx + \int_{\partial\Omega} g \, d\sigma = 0$. Then there exists $p > 2$, depending only on μ and M_0, such that*

$$\|\nabla u_\varepsilon\|_{L^p(\Omega)} \le C \Big\{ \|F\|_{L^2(\Omega)} + \|g\|_{L^2(\partial\Omega)} \Big\}, \tag{6.1.4}$$

where $C > 0$ depend only on μ and M_0.

Proof. Consider the Neumann problem $\mathcal{L}_\varepsilon(v_\varepsilon) = \operatorname{div}(f)$ in Ω and $\frac{\partial v_\varepsilon}{\partial \nu_\varepsilon} = 0$ on $\partial\Omega$, where $f \in C_0^1(\Omega; \mathbb{R}^{m \times d})$. Clearly, $\|\nabla v_\varepsilon\|_{L^2(\Omega)} \le C \|f\|_{L^2(\Omega)}$. In view of the reverse Hölder inequality (6.1.3), using the real-variable argument in Section 4.2, one can deduce that

$$\|\nabla v_\varepsilon\|_{L^p(\Omega)} \le C \, \|f\|_{L^p(\Omega)}, \tag{6.1.5}$$

for any $2 < p < \bar{p}$. Since this is also true for the adjoint operator $\mathcal{L}_\varepsilon^*$, by a duality argument (see Remark 6.3.6), it follows that

$$\|\nabla u_\varepsilon\|_{L^p(\Omega)} \le C \Big\{ \|F\|_{L^q(\Omega)} + \|g\|_{B^{-1/p,p}(\partial\Omega)} \Big\}, \tag{6.1.6}$$

for $2 < p < \bar{p}$, where $\frac{1}{q} = \frac{1}{p} + \frac{1}{d}$ and $B^{-1/p,p}(\partial\Omega)$ denotes the dual of the Besov space $B^{1/p,p'}(\partial\Omega)$. By choosing $p \in (2, \bar{p})$ close to 2 so that $L^2(\Omega) \subset L^q(\Omega)$ and $L^2(\partial\Omega) \subset B^{-1/p,p}(\partial\Omega)$, we see that the estimate (6.1.4) follows from (6.1.6). $\square$

It follows by Theorem 3.3.4 that

$$\|u_\varepsilon - u_0\|_{L^2(\Omega)} \le C\sqrt{\varepsilon}\Big\{\|F\|_{L^2(\Omega)} + \|g\|_{L^2(\partial\Omega)}\Big\}, \tag{6.1.7}$$

where Ω is a bounded Lipschitz domain. However, this was proved under the symmetry condition $A^* = A$. The next lemma provides a very weak rate of convergence without the symmetry condition.

Lemma 6.1.5. *Let $u_\varepsilon \in H^1(\Omega;\mathbb{R}^m)$ $(\varepsilon \ge 0)$ be the weak solution to the Neumann problem (6.0.1) with $\int_\Omega u_\varepsilon\, dx = 0$, where $\Omega = D_r$ for some $1 \le r \le 2$. Then*

$$\|u_\varepsilon - u_0\|_{L^2(\Omega)} \le C\varepsilon^\sigma\Big\{\|F\|_{L^2(\Omega)} + \|g\|_{L^2(\partial\Omega)}\Big\} \tag{6.1.8}$$

for any $0 < \varepsilon < 2$, where $\sigma > 0$ and $C > 0$ depend only on μ and M_0.

Proof. It follows from (3.3.1) that

$$\|u_\varepsilon - u_0\|_{L^2(\Omega)} \le C\Big\{\varepsilon\,\|\nabla^2 u_0\|_{L^2(\Omega\setminus\Omega_\varepsilon)} + \varepsilon\,\|\nabla u_0\|_{L^2(\Omega)} + \|\nabla u_0\|_{L^2(\Omega_{5\varepsilon})}\Big\}, \tag{6.1.9}$$

where $\Omega_t = \{x \in \Omega : \text{dist}(x,\partial\Omega) < t\}$. To bound the RHS of (6.1.9), we first use interior estimates for $\mathcal{L}_0$ to obtain

$$\fint_{B(x,\delta(x)/8)} |\nabla^2 u_0|^2 \le \frac{C}{[\delta(x)]^2}\fint_{B(x,\delta(x)/4)} |\nabla u_0|^2 + C\fint_{B(x,\delta(x)/4)} |F|^2$$

for any $x \in \Omega$, where $\delta(x) = \text{dist}(x,\partial\Omega)$. We then integrate both sides of the inequality above over the set $\Omega\setminus\Omega_\varepsilon$. Observe that if $|x-y| < \delta(x)/4$, then $|\delta(x)-\delta(y)| \le |x-y| < \delta(x)/4$, which gives

$$(4/5)\delta(y) < \delta(x) < (4/3)\delta(y).$$

It follows that

$$\begin{aligned}
&\int_{\Omega\setminus\Omega_\varepsilon} |\nabla^2 u_0(y)|^2\, dy \\
&\le C\int_{\Omega\setminus\Omega_{\varepsilon/2}} [\delta(y)]^{-2}|\nabla u_0(y)|^2\, dy + C\int_\Omega |F(y)|^2\, dy \\
&\le C\left(\int_\Omega |\nabla u_0|^{2s}\, dy\right)^{1/s}\left(\int_{\Omega\setminus\Omega_{\varepsilon/2}} [\delta(y)]^{-2s'}\, dy\right)^{1/s'} + C\int_\Omega |F(y)|^2\, dy \\
&\le C\varepsilon^{-1-\frac{1}{s}}\|\nabla u_0\|^2_{L^{2s}(\Omega)} + +C\int_\Omega |F(y)|^2\, dy,
\end{aligned}$$

where $s > 1$ and we have used Hölder's inequality for the second step. Let $p = 2s > 2$. We see that

$$\begin{aligned}
\varepsilon\,\|\nabla^2 u_0\|_{L^2(\Omega\setminus\Omega_\varepsilon)} &\le C\varepsilon^{\frac{1}{2}-\frac{1}{p}}\|\nabla u_0\|_{L^p(\Omega)} + C\varepsilon\,\|F\|_{L^2(\Omega)} \\
&\le C\varepsilon^{\frac{1}{2}-\frac{1}{p}}\Big\{\|F\|_{L^2(\Omega)} + \|g\|_{L^2(\partial\Omega)}\Big\},
\end{aligned} \tag{6.1.10}$$

where we have used (6.1.4) for the last inequality. Also, by Hölder's inequality,

$$\begin{aligned}\|\nabla u_0\|_{L^2(\Omega_{5\varepsilon})} &\leq C\varepsilon^{\frac{1}{2}-\frac{1}{p}}\|\nabla u_0\|_{L^p(\Omega)}\\ &\leq C\varepsilon^{\frac{1}{2}-\frac{1}{p}}\Big\{\|F\|_{L^2(\Omega)}+\|g\|_{L^2(\partial\Omega)}\Big\}.\end{aligned} \tag{6.1.11}$$

The inequality (6.1.8) with $\sigma = \frac{1}{2} - \frac{1}{p} > 0$ follows from (6.1.9), (6.1.10) and (6.1.11). □

We are now in a position to give the proof of Theorem 6.1.1.

Proof of Theorem 6.1.1. By rescaling we may assume that $r = 1$. It follows from the Caccioppoli inequality (6.1.2) and the co-area formula that there exists some $t \in (1, 3/2)$ such that

$$\int_{\partial D_t\setminus\Delta_2} |\nabla u_\varepsilon|^2\, d\sigma \leq C\int_{D_2} |u_\varepsilon|^2\, dx + C\int_{D_2} |F|^2\, dx + C\int_{\Delta_2} |g|^2\, d\sigma. \tag{6.1.12}$$

For otherwise, we could integrate the reverse inequality,

$$\int_{\partial D_t\setminus\Delta_2} |\nabla u_\varepsilon|^2\, d\sigma > C\int_{D_2} |u_\varepsilon|^2\, dx + C\int_{D_2} |F|^2\, dx + C\int_{\Delta_2} |g|^2\, d\sigma,$$

in t over the interval $(1, 3/2)$ to obtain

$$\int_{D_{3/2}\setminus D_1} |\nabla u_\varepsilon|^2\, dx > C\int_{D_2} |u_\varepsilon|^2\, dx + C\int_{D_2} |F|^2\, dx + C\int_{\Delta_2} |g|^2\, d\sigma,$$

which is in contradiction with (6.1.2). We now let w be the unique solution of the Neumann problem

$$\mathcal{L}_0(w) = F \quad \text{in } D_t \quad \text{and} \quad \frac{\partial w}{\partial \nu_0} = \frac{\partial u_\varepsilon}{\partial \nu_\varepsilon} \quad \text{on } \partial D_t,$$

with $\int_{D_t} w\, dx = \int_{D_t} u_\varepsilon\, dx$. Note that $\frac{\partial w}{\partial \nu_0} = g$ on Δ_1, and that, by Lemma 6.1.5,

$$\begin{aligned}\|u_\varepsilon - w\|_{L^2(D_1)} &\leq \|u_\varepsilon - w\|_{L^2(D_t)}\\ &\leq C\varepsilon^\sigma\Big\{\|F\|_{L^2(D_2)} + \|g\|_{L^2(\Delta_2)} + \|\nabla u_\varepsilon\|_{L^2(\partial D_t\setminus\Delta_2)}\Big\},\end{aligned}$$

which, together with (6.1.12), yields (6.1.1). □

6.2 Boundary Hölder estimates

In this section we establish uniform boundary Hölder estimates in C^1 domains for $\mathcal{L}_\varepsilon$ with Neumann conditions. Throughout this section we assume that $D_r = D(r, \psi)$, $\Delta_r = \Delta(r, \psi)$, and $\psi : \mathbb{R}^{d-1} \to \mathbb{R}$ is a C^1 function satisfying (5.2.2).

Theorem 6.2.1 (Large-scale Hölder estimate). *Suppose that A satisfies (2.1.20) and is 1-periodic. Let $u_\varepsilon \in H^1(D_2; \mathbb{R}^m)$ be a weak solution of $\mathcal{L}_\varepsilon(u_\varepsilon) = 0$ in D_2 and $\frac{\partial u_\varepsilon}{\partial \nu_\varepsilon} = g$ on Δ_2. Then, for any $\rho \in (0,1)$ and $\varepsilon \le r \le 1$,*

$$\left(\fint_{D_r} |\nabla u_\varepsilon|^2 \right)^{1/2} \le C r^{\rho - 1} \left\{ \left(\fint_{D_2} |u_\varepsilon|^2 \right)^{1/2} + \|g\|_\infty \right\}, \tag{6.2.1}$$

where C depends only on ρ, μ, and $(\omega_1(t), M_0)$ in (5.2.2).

Proof. Fix $\gamma \in (\rho, 1)$. For each $r \in [\varepsilon, 1]$, let $w = w_r$ be the function given by Theorem 6.1.1. By the boundary C^γ estimate in C^1 domains for the operator $\mathcal{L}_0$,

$$\begin{aligned}
\inf_{q \in \mathbb{R}^m} \left(\fint_{D_{\theta r}} |w - q|^2 \right)^{1/2} &\le (\theta r)^\gamma \|w\|_{C^{0,\gamma}(D_{\theta r})} \\
&\le (\theta r)^\gamma \|w\|_{C^{0,\gamma}(D_{r/2})} \\
&\le C_0 \theta^\gamma \inf_{q \in \mathbb{R}^m} \left(\fint_{D_r} |w - q|^2 \right)^{1/2} + C r \|g\|_\infty
\end{aligned}$$

for any $\theta \in (0, 1/2)$, where C_0 depends only on μ, γ and $(\omega_1(t), M_0)$. This, together with Theorem 6.1.1, gives

$$\begin{aligned}
&\inf_{q \in \mathbb{R}^m} \left(\fint_{D_{\theta r}} |u_\varepsilon - q|^2 \right)^{1/2} \\
&\le \inf_{q \in \mathbb{R}^m} \left(\fint_{D_{\theta r}} |w - q|^2 \right)^{1/2} + \left(\fint_{D_{\theta r}} |w_\varepsilon - w|^2 \right)^{1/2} \\
&\le C_0 \theta^\gamma \inf_{q \in \mathbb{R}^m} \left(\fint_{D_r} |w - q|^2 \right)^{1/2} + C_\theta \left(\fint_{D_r} |u_\varepsilon - w|^2 \right)^{1/2} + C r \|g\|_\infty \\
&\le C_0 \theta^\gamma \inf_{q \in \mathbb{R}^m} \left(\fint_{D_r} |u_\varepsilon - q|^2 \right)^{1/2} + C_\theta \left(\fint_{D_r} |u_\varepsilon - w|^2 \right)^{1/2} + C r \|g\|_\infty \\
&\le C_0 \theta^\gamma \inf_{q \in \mathbb{R}^m} \left(\fint_{D_r} |u_\varepsilon - q|^2 \right)^{1/2} + C_\theta \left(\frac{\varepsilon}{r} \right)^\alpha \left(\fint_{D_{2r}} |u_\varepsilon|^2 \right)^{1/2} + C r \|g\|_\infty,
\end{aligned}$$

where the constant C_θ also depends on θ. Replacing u_ε with $u_\varepsilon - q$, we obtain

$$\phi(\theta r) \le C_0 \theta^{\gamma - \rho} \phi(r) + C_\theta \left(\frac{\varepsilon}{r} \right)^\alpha \phi(2r) + C_\theta r^{1-\rho} \|g\|_\infty \tag{6.2.2}$$

for any $r \in [\varepsilon, 1]$, where

$$\phi(r) = r^{-\rho} \inf_{q \in \mathbb{R}^m} \left(\fint_{D_r} |u_\varepsilon - q|^2 \right)^{1/2}.$$

We now choose $\theta \in (0, 1/2)$ so small that

$$C_0 \theta^{\gamma - \rho} < (1/4),$$

which is possible, since $\gamma - \rho > 0$. With θ fixed, we choose $N > 1$ so large that

$$C_\theta N^{-\alpha} < (1/4).$$

It follows that if $1 \ge r \ge N\varepsilon$, then

$$\phi(\theta r) \le \frac{1}{4}\{\phi(r) + \phi(2r)\} + C r^{1-\rho} \|g\|_\infty. \tag{6.2.3}$$

Finally, we divide both sides of (6.2.3) by r and integrate the resulting inequality in r over the interval $(N\varepsilon, 1)$. This gives

$$\begin{aligned} \int_{\theta N\varepsilon}^{\theta} \phi(r) \frac{dr}{r} &\le \frac{1}{4} \int_{N\varepsilon}^{1} \phi(r) \frac{dr}{r} + \frac{1}{4} \int_{2N\varepsilon}^{2} \phi(r) \frac{dr}{r} + C \|g\|_\infty \\ &\le \frac{1}{2} \int_{\theta N\varepsilon}^{2} \phi(r) \frac{dr}{r} + C \|g\|_\infty. \end{aligned}$$

It follows that

$$\int_{\theta N\varepsilon}^{\theta} \phi(r) \frac{dr}{r} \le 2 \int_{\theta}^{2} \phi(r) \frac{dr}{r} + C \|g\|_\infty. \tag{6.2.4}$$

Using the observation that $\phi(r) \le C\phi(t)$ for $t \in [r, 2r]$, we may deduce from (6.2.4) that

$$\phi(r) \le C\phi(2) + C \|g\|_\infty$$

for any $r \in [\varepsilon, 2]$. The estimate (6.2.1) now follows by Caccioppoli's inequality. □

No smoothness condition on A is needed for (6.2.1). Under the additional VMO condition (4.3.1), we obtain the following.

Theorem 6.2.2 (Boundary Hölder estimate). *Suppose that A satisfies the ellipticity condition (2.1.20) and is 1-periodic. Also assume A satisfies (4.3.1). Let Ω be a bounded C^1 domain and $0 < \rho < 1$. Let $u_\varepsilon \in H^1(B(x_0, r) \cap \Omega; \mathbb{R}^m)$ be a solution of $\mathcal{L}_\varepsilon(u_\varepsilon) = 0$ in $B(x_0, r) \cap \Omega$ with $\frac{\partial u_\varepsilon}{\partial \nu_\varepsilon} = g$ on $B(x_0, r) \cap \partial\Omega$ for some $x_0 \in \partial\Omega$ and $0 < r < r_0$. Then, for any $x, y \in B(x_0, r/2) \cap \Omega$,*

$$|u_\varepsilon(x) - u_\varepsilon(y)| \le C \left(\frac{|x - y|}{r} \right)^\rho \left\{ \left(\fint_{B(x_0, r) \cap \Omega} |u_\varepsilon|^2 \right)^{1/2} + r\|g\| \right\}, \tag{6.2.5}$$

where C depends at most on ρ, μ, $\omega(t)$ in (4.3.1), and Ω.

Proof. By a change of the coordinate system it suffices to show that

$$|u_\varepsilon(x) - u_\varepsilon(y)| \le C \left(\frac{|x-y|}{r}\right)^\rho \left\{ \left(\fint_{D_{2r}} |u_\varepsilon|^2 \right)^{1/2} + r\|g\|_\infty \right\} \tag{6.2.6}$$

for any $x, y \in D_r$, where $0 < r \le 1$, $\mathcal{L}_\varepsilon(u_\varepsilon) = 0$ in D_{2r} and $\frac{\partial u_\varepsilon}{\partial \nu_\varepsilon} = g$ on Δ_{2r}.

By rescaling we may assume $r = 1$. We may also assume $\varepsilon < 1$, since the case $\varepsilon \ge 1$ follows readily from boundary Hölder estimates in C^1 domains for elliptic systems with VMO coefficients (see, e.g., [10, 23]). Under these assumptions we will show that

$$\left(\fint_{D_t} \left|u_\varepsilon - \fint_{D_t} u_\varepsilon\right|^2 \right)^{1/2} \le C t^\rho \Big\{ \|u_\varepsilon\|_{L^2(D_2)} + \|g\|_\infty \Big\} \tag{6.2.7}$$

for any $t \in (0, 1/4)$. As in the case of the Dirichlet condition, the desired estimate follows from the interior Hölder estimate and (6.2.7), using Campanato's characterization of Hölder spaces.

To prove (6.2.7), we first consider the case $t \ge \varepsilon$. It follows from the Poincaré inequality and Theorem 6.2.1 that

$$\begin{aligned} \left(\fint_{D_t} \left|u_\varepsilon - \fint_{D_t} u_\varepsilon\right|^2 \right)^{1/2} &\le Ct \left(\fint_{D_t} |\nabla u_\varepsilon|^2 \right)^{1/2} \\ &\le C t^\rho \Big\{ \|u_\varepsilon\|_{L^2(D_2)} + \|g\|_\infty \Big\}. \end{aligned}$$

Next, suppose that $t < \varepsilon$. Let $w(x) = u_\varepsilon(\varepsilon x)$. Then $\mathcal{L}_1(w) = 0$ in $D(1, \psi_\varepsilon)$, where $\psi_\varepsilon(x') = \varepsilon^{-1}\psi(\varepsilon x')$. By the boundary Hölder estimates in C^1 domains for $\mathcal{L}_1$, we obtain

$$\begin{aligned} \left(\fint_{D(t,\psi)} \left|u_\varepsilon - \fint_{D(t,\psi)} u_\varepsilon\right|^2 \right)^{1/2} &= \left(\fint_{D(\frac{t}{\varepsilon},\psi_\varepsilon)} \left|w - \fint_{D(\frac{t}{\varepsilon},\psi_\varepsilon)} w\right|^2 \right)^{1/2} \\ &\le C \left(\frac{t}{\varepsilon}\right)^\rho \left\{ \left(\fint_{D(1,\psi_\varepsilon)} \left|w - \fint_{D(1,\psi_\varepsilon)} w\right|^2 \right)^{1/2} + \|g\|_\infty \right\} \\ &= C \left(\frac{t}{\varepsilon}\right)^\rho \left\{ \left(\fint_{D(\varepsilon,\psi)} \left|u_\varepsilon - \fint_{D(\varepsilon,\psi)} u_\varepsilon\right|^2 \right)^{1/2} + \|g\|_\infty \right\} \\ &\le C \left(\frac{t}{\varepsilon}\right)^\rho \varepsilon^\rho \Big\{ \|u_\varepsilon\|_{L^2(D_2)} + \|g\|_\infty \Big\} \\ &= C t^\rho \Big\{ \|u_\varepsilon\|_{L^2(D_2)} + \|g\|_\infty \Big\}, \end{aligned}$$

where the last inequality follows from the previous case with $t = \varepsilon$. This finishes the proof of (6.2.7) and thus of Theorem 6.2.2. □

6.3 Boundary $W^{1,p}$ estimates

In this section we establish uniform $W^{1,p}$ estimates for the Neumann problem

$$\begin{cases} \mathcal{L}_\varepsilon(u_\varepsilon) = \operatorname{div}(f) + F & \text{in } \Omega, \\ \dfrac{\partial u_\varepsilon}{\partial \nu_\varepsilon} = g - n \cdot f & \text{on } \partial\Omega. \end{cases} \tag{6.3.1}$$

Throughout the section we assume that Ω is C^1 and that A satisfies (2.1.20), (2.0.2) and (4.3.1).

Let $B^{-\frac{1}{p},p}(\partial\Omega; \mathbb{R}^m)$ denote the dual of the Besov space $B^{\frac{1}{p},p'}(\partial\Omega; \mathbb{R}^m)$ for $1 < p < \infty$.

Definition 6.3.1. We call $u_\varepsilon \in W^{1,p}(\Omega; \mathbb{R}^m)$ a weak solution of (6.3.1), if

$$\begin{aligned} &\int_\Omega A(x/\varepsilon)\nabla u_\varepsilon \cdot \nabla \phi \, dx \\ &= \int_\Omega \left\{ -f_i^\alpha \frac{\partial \phi^\alpha}{\partial x_i} + F^\alpha \phi^\alpha \right\} dx + \langle g, \phi \rangle_{B^{-1/p,p}(\partial\Omega) \times B^{1/p,p'}(\partial\Omega)} \end{aligned} \tag{6.3.2}$$

for any $\phi = (\phi^\alpha) \in C_0^\infty(\mathbb{R}^d; \mathbb{R}^m)$, where $f = (f_i^\alpha)$ and $F = (F^\alpha)$.

Theorem 6.3.2. *Let $1 < p < \infty$. Let $g = (g^\alpha) \in B^{-1/p,p}(\partial\Omega; \mathbb{R}^m)$, $f = (f_i^\alpha) \in L^p(\Omega; \mathbb{R}^{m\times d})$, and $F = (F^\alpha) \in L^q(\Omega; \mathbb{R}^m)$, where $q = \frac{pd}{p+d}$ for $p > \frac{d}{d-1}$, $q > 1$ for $p = \frac{d}{d-1}$, and $q = 1$ for $1 < p < \frac{d}{d-1}$. Then, if F and g satisfy the compatibility condition*

$$\int_\Omega F \cdot b \, dx + \langle g, b \rangle_{B^{-1/p,p}(\partial\Omega) \times B^{1/p,p'}(\partial\Omega)} = 0 \tag{6.3.3}$$

for any $b \in \mathbb{R}^m$, the Neumann problem (6.3.1) has a unique (up to constants) weak solution. Moreover, the solution satisfies the estimate,

$$\|\nabla u_\varepsilon\|_{L^p(\Omega)} \le C_p \left\{ \|f\|_{L^p(\Omega)} + \|F\|_{L^q(\Omega)} + \|g\|_{B^{-1/p,p}(\partial\Omega)} \right\}, \tag{6.3.4}$$

where $C_p > 0$ depends only on p, μ, $\omega(t)$ in (4.3.1), and Ω.

The proof of Theorem 6.3.2 is divided into several steps.

Lemma 6.3.3. *Let $f \in L^p(\Omega; \mathbb{R}^{m\times d})$ for some $1 < p < \infty$. Then there exists a unique (up to constants) u_ε in $W^{1,p}(\Omega; \mathbb{R}^m)$ such that*

$$\mathcal{L}_\varepsilon(u_\varepsilon) = \operatorname{div}(f) \quad \text{in } \Omega \quad \text{and} \quad \frac{\partial u_\varepsilon}{\partial \nu_\varepsilon} = -n \cdot f \quad \text{on } \partial\Omega.$$

Moreover, the solution satisfies the estimate,

$$\|\nabla u_\varepsilon\|_{L^p(\Omega)} \le C_p \|f\|_{L^p(\Omega)}, \tag{6.3.5}$$

where C_p depends only on p, μ, $\omega(t)$ in (4.3.1), and Ω.

Proof. The proof is parallel to that of the same estimate for the Dirichlet problem (see Section 5.3). Let $p > 2$. We first observe that if $\mathcal{L}_\varepsilon(u_\varepsilon) = 0$ in $\Omega \cap 2B$ and $\frac{\partial u_\varepsilon}{\partial \nu_\varepsilon} = 0$ in $\partial\Omega \cap 2B$, where B is a ball in $\mathbb{R}^d$ with the property that $|B| \le c_0|\Omega|$ and either $2B \subset \Omega$ or B is centered on $\partial\Omega$, then

$$\left(\fint_{\Omega\cap B} |\nabla u_\varepsilon|^p \right)^{1/p} \le C \left(\fint_{\Omega\cap 2B} |\nabla u_\varepsilon|^2 \right)^{1/2}. \tag{6.3.6}$$

The interior case where $2B \subset \Omega$ follows directly from the interior $W^{1,p}$ estimate in Section 4.3. To handle the boundary case where $x_0 \in \partial\Omega$, one uses a line of argument similar to that used in the Dirichlet problem, by combining the interior $W^{1,p}$ estimates with the boundary Hölder estimate in Theorem 6.2.2. More precisely, let $B = B(x_0, r)$, where $x_0 \in \partial\Omega$ and $0 < r < r_0$. Note that if $x \in B(x_0, r)$, then

$$\begin{aligned} \fint_{B(x,\delta(x)/8)} |\nabla u_\varepsilon|^p &\le \frac{C}{[\delta(x)]^p} \left(\fint_{B(x,\delta(x)/4)} \left| u_\varepsilon - \fint_{B(x,\delta(x)/4)} u_\varepsilon \right|^2 \right)^{p/2} \\ &\le \frac{C}{[\delta(x)]^p} \left(\frac{\delta(x)}{r} \right)^{\rho p} \left(\fint_{\Omega\cap 2B} |u_\varepsilon|^2 \right)^{p/2}, \end{aligned}$$

where $\rho \in (0,1)$ is close to 1 so that $p(1-\rho) < 1$. By integrating both sides of the inequality above over the set $B \cap \Omega$, we obtain

$$\int_{\Omega\cap B} |\nabla u_\varepsilon|^p \le C r^{d-p} \left(\fint_{\Omega\cap 2B} |u_\varepsilon|^2 \right)^{p/2}.$$

Let E be the L^1 average of u_ε over the set $\Omega \cap 2B$. By replacing u_ε with $u_\varepsilon - E$ in the inequality above and applying the Poincaré inequality, we obtain (6.3.6).

With the reverse Hölder inequality (6.3.6) at our disposal, we may deduce the estimate (6.3.5) by Theorem 4.2.7. Indeed, for $f \in L^2(\Omega; \mathbb{R}^{m\times d})$, let $T_\varepsilon(f) = \nabla u_\varepsilon$, where $u_\varepsilon \in W^{1,2}(\Omega; \mathbb{R}^m)$ is the unique weak solution to $\mathcal{L}_\varepsilon(u_\varepsilon) = \operatorname{div}(f)$ in Ω, $\frac{\partial u_\varepsilon}{\partial \nu_\varepsilon} = -n \cdot f$ on $\partial\Omega$, and $\int_\Omega u_\varepsilon \, dx = 0$. Clearly,

$$\|T_\varepsilon(f)\|_{L^2(\Omega)} \le C \, \|f\|_{L^2(\Omega)},$$

where C depends only on μ and Ω. Suppose now that $f = 0$ in $\Omega \cap 2B$, where $|B| \le c_0|\Omega|$ and either $2B \subset \Omega$ or B is centered on $\partial\Omega$. Then

$$\mathcal{L}_\varepsilon(u_\varepsilon) = 0 \quad \text{in } \Omega \cap 2B \quad \text{and} \quad \frac{\partial u_\varepsilon}{\partial \nu_\varepsilon} = 0 \quad \text{on } \partial\Omega \cap 2B.$$

In view of (6.3.6) we obtain

$$\left(\fint_{\Omega\cap B} |T_\varepsilon(f)|^p \right)^{1/p} \le C \left(\fint_{\Omega\cap 2B} |T_\varepsilon(f)|^2 \right)^{1/2}. \tag{6.3.7}$$

As a result, the operator T_ε satisfies the assumptions in Theorem 4.2.7 with constants depending at most on μ, p, $\omega(t)$ in (4.3.1), and Ω. This gives (6.3.5) for $2 < p < \infty$. Note that the uniqueness for $p > 2$ follows from the uniqueness for $p = 2$.

The case $1 < p < 2$ may be handled by duality. Let $g = (g_i^\alpha) \in C_0^1(\Omega; \mathbb{R}^{m\times d})$ and v_ε be a weak solution in $W^{1,2}(\Omega; \mathbb{R}^m)$ of $\mathcal{L}_\varepsilon^*(v_\varepsilon) = \operatorname{div}(g)$ in Ω with $\frac{\partial v_\varepsilon}{\partial \nu_\varepsilon^*} = 0$ on $\partial\Omega$, where $\mathcal{L}_\varepsilon^*$ denotes the adjoint of $\mathcal{L}_\varepsilon$. Since A^* satisfies the same conditions as A and $p' > 2$, we have

$$\|\nabla v_\varepsilon\|_{L^{p'}(\Omega)} \le C\, \|g\|_{L^{p'}(\Omega)}.$$

Also, note that if $f = (f_i^\alpha) \in C_0^1(\Omega; \mathbb{R}^{m\times d})$ and u_ε is a weak solution in $W^{1,2}(\Omega; \mathbb{R}^m)$ of $\mathcal{L}_\varepsilon(u_\varepsilon) = \operatorname{div}(f)$ in Ω with $\frac{\partial u_\varepsilon}{\partial \nu_\varepsilon} = 0$ on $\partial\Omega$, then

$$\int_\Omega f_i^\alpha \cdot \frac{\partial v_\varepsilon^\alpha}{\partial x_i}\, dx = -\int_\Omega A(x/\varepsilon)\nabla u_\varepsilon \cdot \nabla v_\varepsilon\, dx = \int_\Omega g_i^\alpha \cdot \frac{\partial u_\varepsilon^\alpha}{\partial x_i}\, dx. \tag{6.3.8}$$

It follows from (6.3.8) by duality that

$$\|\nabla u_\varepsilon\|_{L^p(\Omega)} \le C\, \|f\|_{L^p(\Omega)}.$$

By a density argument, this gives the existence of solutions in $W^{1,p}(\Omega; \mathbb{R}^m)$ for general f in $L^p(\Omega; \mathbb{R}^{m\times d})$. Observe that the duality argument above in fact shows that any solution in $W^{1,p}(\Omega; \mathbb{R}^m)$ with data $f \in L^p(\Omega; \mathbb{R}^{d\times m})$ satisfies (6.3.5). As a consequence we obtain the uniqueness for $1 < p < 2$. □

Lemma 6.3.4. *Let $1 < p < \infty$ and $g \in B^{-1/p,p}(\partial\Omega; \mathbb{R}^m)$ with*

$$\langle g, b\rangle_{B^{-1/p,p}(\partial\Omega)\times B^{1/p,p'}(\partial\Omega)} = 0 \tag{6.3.9}$$

for any $b \in \mathbb{R}^m$. Then there exists a unique (up to constants) $u_\varepsilon \in W^{1,p}(\Omega; \mathbb{R}^m)$ such that $\mathcal{L}_\varepsilon(u_\varepsilon) = 0$ in Ω and $\frac{\partial u_\varepsilon}{\partial \nu_\varepsilon} = g$ on $\partial\Omega$. Moreover, the solution satisfies

$$\|\nabla u_\varepsilon\|_{L^p(\Omega)} \le C\, \|g\|_{B^{-1/p,p}(\partial\Omega)}, \tag{6.3.10}$$

where C depends only on μ, p, $\omega(t)$ in (4.3.1), and Ω.

Proof. The uniqueness is contained in Lemma 6.3.3. To establish the existence as well as the estimate (6.3.10), we first assume that

$$g \in B^{-1/p,p}(\partial\Omega; \mathbb{R}^m) \cap B^{-1/2,2}(\partial\Omega; \mathbb{R}^m)$$

and show that the estimate (6.3.10) holds for solutions in $W^{1,2}(\Omega; \mathbb{R}^m)$, given by Theorem 2.1.7. By a density argument, this gives the existence of solutions in $W^{1,p}(\Omega; \mathbb{R}^m)$ for general $g \in B^{-1/p,p}(\partial\Omega; \mathbb{R}^m)$ satisfying (6.3.9).

Let $f = (f_i^\alpha) \in C_0^1(\Omega; \mathbb{R}^{m\times d})$ and v_ε be a weak solution in $W^{1,2}(\Omega; \mathbb{R}^m)$ to

$$\mathcal{L}_\varepsilon^*(v_\varepsilon) = \text{div}(f) \quad \text{in } \Omega \quad \text{and} \quad \frac{\partial v_\varepsilon}{\partial \nu_\varepsilon^*} = 0 \quad \text{on } \partial\Omega.$$

Since A^* satisfies the same conditions as A, by Lemma 6.3.3, we have

$$\|\nabla v_\varepsilon\|_{L^{p'}(\Omega)} \le C\, \|f\|_{L^{p'}(\Omega)}.$$

Note that

$$\begin{aligned}\int_\Omega f_i^\alpha \cdot \frac{\partial u_\varepsilon^\alpha}{\partial x_i}\, dx &= -\int_\Omega A(x/\varepsilon)\nabla u_\varepsilon \cdot \nabla v_\varepsilon\, dx \\ &= -\langle g, v_\varepsilon\rangle_{B^{-1/p,p}(\partial\Omega)\times B^{1/p,p'}(\partial\Omega)}.\end{aligned} \tag{6.3.11}$$

Let E denote the L^1 average of v_ε over Ω. Then

$$\begin{aligned}|\langle g, v_\varepsilon\rangle_{B^{-1/p,p}(\partial\Omega)\times B^{1/p,p'}(\partial\Omega)}| &= |\langle g, v_\varepsilon - E\rangle_{B^{-1/p,p}(\partial\Omega)\times B^{1/p,p'}(\partial\Omega)}| \\ &\le \|g\|_{B^{-1/p,p}(\partial\Omega)} \|v_\varepsilon - E\|_{B^{1/p,p'}(\partial\Omega)} \\ &\le C\, \|g\|_{B^{-1/p,p}(\partial\Omega)} \|v_\varepsilon - E\|_{W^{1,p'}(\Omega)} \\ &\le C\, \|g\|_{B^{-1/p,p}(\partial\Omega)} \|\nabla v_\varepsilon\|_{L^{p'}(\Omega)} \\ &\le C\, \|g\|_{B^{-1/p,p}(\partial\Omega)} \|f\|_{L^{p'}(\Omega)},\end{aligned} \tag{6.3.12}$$

where we have used a trace theorem for the second inequality and the Poincaré inequality for the third. The estimate (6.3.10) follows from (6.3.11)–(6.3.12) by duality. □

Let $1 < q < d$ and $\frac{1}{p} = \frac{1}{q} - \frac{1}{d}$. In the proof of the next lemma we will need the following Sobolev inequality:

$$\left(\int_\Omega |u|^p\, dx\right)^{1/p} \le C\left(\int_\Omega |\nabla u|^q\, dx\right)^{1/q}, \tag{6.3.13}$$

where $u \in W^{1,q}(\Omega)$ and $\int_{\partial\Omega} u = 0$. Note that by the Sobolev embedding, (6.3.13) also holds for $q > d$ and $p = \infty$. If $q = d$, it holds for any $1 < p < \infty$. To see (6.3.13), by the Poincaré–Sobolev inequality, it suffices to show that

$$\left|\int_\Omega u\, dx\right| \le C\left|\int_{\partial\Omega} u\, d\sigma\right| + C\left(\int_\Omega |\nabla u|^q\, dx\right)^{1/q} \tag{6.3.14}$$

for $q > 1$. This may be done by using a proof by contradiction.

Lemma 6.3.5. *Let $1 < p < \infty$ and $1 \le q < \infty$, where*

$$q = \begin{cases} \dfrac{pd}{p+d} & \text{if } p > \dfrac{d}{d-1}, \\ q > 1 & \text{if } p = \dfrac{d}{d-1}, \\ q = 1 & \text{if } 1 < p < \dfrac{d}{d-1}. \end{cases} \tag{6.3.15}$$

Then, for any function $F \in L^q(\Omega; \mathbb{R}^m)$, there exists a unique (up to constants) solution u_ε in $W^{1,p}(\Omega; \mathbb{R}^m)$ to $\mathcal{L}_\varepsilon(u_\varepsilon) = F$ in Ω and $\frac{\partial u_\varepsilon}{\partial \nu_\varepsilon} = -b$ on $\partial\Omega$, where $b = \frac{1}{\partial\Omega}\int_\Omega F$. Moreover, the solution satisfies

$$\|\nabla u_\varepsilon\|_{L^p(\Omega)} \le C\, \|F\|_{L^q(\Omega)}. \tag{6.3.16}$$

Proof. The uniqueness is contained in Lemma 6.3.3. To establish the existence as well as the estimate (6.3.16), we first assume $F \in C_0^1(\Omega; \mathbb{R}^m)$. By Theorem 2.4.4, there exists a unique (up to constants) solution in $W^{1,2}(\Omega; \mathbb{R}^m)$ to

$$\mathcal{L}_\varepsilon(u_\varepsilon) = F \quad \text{in } \Omega \qquad \text{and} \qquad \frac{\partial u_\varepsilon}{\partial \nu_\varepsilon} = -\frac{1}{|\partial\Omega|}\int_\Omega F\, dx \qquad \text{on } \partial\Omega.$$

We will show the solution satisfies (6.3.16). By a density argument this would give the existence of solutions in $W^{1,p}(\Omega; \mathbb{R}^m)$ for general $F \in L^q(\Omega; \mathbb{R}^m)$.

Let $f = (f_i^\alpha) \in C_0^1(\Omega; \mathbb{R}^m)$ and $v_\varepsilon \in W^{1,2}(\Omega; \mathbb{R}^m)$ be a weak solution to the Neumann problem $\mathcal{L}_\varepsilon^*(v_\varepsilon) = \text{div}(f)$ in Ω and $\frac{\partial v_\varepsilon}{\partial \nu_\varepsilon} = 0$ on $\partial\Omega$. By Lemma 6.3.3, we have

$$\|\nabla v_\varepsilon\|_{L^{p'}(\Omega)} \le C\, \|f\|_{L^{p'}(\Omega)}. \tag{6.3.17}$$

Note that

$$\begin{aligned}
-\int_\Omega \frac{\partial u_\varepsilon^\alpha}{\partial x_i} \cdot f_i^\alpha\, dx &= \int_\Omega A(x/\varepsilon)\nabla u_\varepsilon \cdot \nabla v_\varepsilon\, dx \\
&= \int_\Omega F \cdot v_\varepsilon\, dx - \int_{\partial\Omega} b \cdot v_\varepsilon\, d\sigma \\
&= \int_\Omega F(v_\varepsilon - E)\, dx,
\end{aligned} \tag{6.3.18}$$

where $b = \frac{1}{|\partial\Omega|}\int_\Omega F$ and E is the L^1 average of v_ε over $\partial\Omega$. Moreover,

$$\|v_\varepsilon - E\|_{L^{q'}(\Omega)} \le C\, \|\nabla v_\varepsilon\|_{L^{p'}(\Omega)} \le C\, \|f\|_{L^{p'}(\Omega)}$$

by (6.3.13) and (6.3.17). In view of (6.3.18), we obtain

$$\begin{aligned}
\left| \int_\Omega \frac{\partial u_\varepsilon^\alpha}{\partial x_i} \cdot f_i^\alpha\, dx \right| &\le \|F\|_{L^q(\Omega)} \|v_\varepsilon - E\|_{L^{q'}(\Omega)} \\
&\le C\, \|F\|_{L^q(\Omega)} \|\nabla v_\varepsilon\|_{L^{p'}(\Omega)} \\
&\le C\, \|F\|_{L^q(\Omega)} \|f\|_{L^{p'}(\Omega)}.
\end{aligned}$$

By duality this gives the estimate (6.3.16). □

We are now in a position to give the proof of Theorem 6.3.2.

Proof of Theorem 6.3.2. The uniqueness is contained in Lemma 6.3.3. To establish the existence as well as the estimate (6.3.4), we let $v_\varepsilon \in W^{1,p}(\Omega; \mathbb{R}^m)$ be a weak solution to

$$\mathcal{L}_\varepsilon(v_\varepsilon) = \text{div}(f) \quad \text{in } \Omega \qquad \text{and} \qquad \frac{\partial v_\varepsilon}{\partial \nu_\varepsilon} = -n \cdot f \qquad \text{on } \partial\Omega.$$

Also, let w_ε be a weak solution to

$$\mathcal{L}_\varepsilon(w_\varepsilon) = F \quad \text{in } \Omega \quad \text{and} \quad \frac{\partial w_\varepsilon}{\partial \nu_\varepsilon} = b \quad \text{on } \partial\Omega,$$

where $b = -\frac{1}{|\partial\Omega|}\int_\Omega F$, and $z_\varepsilon \in W^{1,p}(\Omega; \mathbb{R}^m)$ be a weak solution to

$$\mathcal{L}_\varepsilon(z_\varepsilon) = 0 \quad \text{in } \Omega \quad \text{and} \quad \frac{\partial z_\varepsilon}{\partial \nu_\varepsilon} = g - b \quad \text{on } \partial\Omega.$$

Let $u_\varepsilon = v_\varepsilon + w_\varepsilon + z_\varepsilon$. Note that

$$\mathcal{L}_\varepsilon(u_\varepsilon) = \operatorname{div}(f) + F \quad \text{in } \Omega \quad \text{and} \quad \frac{\partial u_\varepsilon}{\partial \nu_\varepsilon} = g - n \cdot f \quad \text{on } \partial\Omega.$$

In view of Lemmas 6.3.3, 6.3.4, and 6.3.5, we obtain

$$\begin{aligned}
\|\nabla u_\varepsilon\|_{L^p(\Omega)} &\le \|\nabla v_\varepsilon\|_{L^p(\Omega)} + \|\nabla w_\varepsilon\|_{L^p(\Omega)} + \|\nabla z_\varepsilon\|_{L^p(\Omega)} \\
&\le C\Big\{ \|f\|_{L^p(\Omega)} + \|F\|_{L^q(\Omega)} + \|g\|_{B^{-1/p,p}(\partial\Omega)} \Big\},
\end{aligned}$$

where $q = \frac{pd}{p+d}$ for $p > \frac{d}{d-1}$, $q > 1$ for $p = \frac{d}{d-1}$, and $q = 1$ for $1 < p < \frac{d}{d-1}$. □

Remark 6.3.6. Let Ω be a fixed Lipschitz domain in $\mathbb{R}^d$. We also fix $2 < p < \infty$. Suppose that for any $f \in C_0^1(\Omega; \mathbb{R}^{m\times d})$, weak solutions in $W^{1,2}(\Omega; \mathbb{R}^m)$ to the Neumann problem,

$$\mathcal{L}_\varepsilon(v_\varepsilon) = \operatorname{div}(f) \quad \text{in } \Omega \quad \text{and} \quad \frac{\partial v_\varepsilon}{\partial \nu_\varepsilon} = 0 \quad \text{on } \partial\Omega,$$

satisfy the $W^{1,p}$ estimate

$$\|\nabla v_\varepsilon\|_{L^p(\Omega)} \le C_0 \, \|f\|_{L^p(\Omega)}$$

for some $C_0 > 0$. Then, for any $g \in B^{-\frac{1}{p},p}(\partial\Omega; \mathbb{R}^m)$ satisfying the compatibility condition (6.3.9), weak solutions to the Neumann problem,

$$\mathcal{L}_\varepsilon(u_\varepsilon) = 0 \quad \text{in } \Omega \quad \text{and} \quad \frac{\partial u_\varepsilon}{\partial \nu_\varepsilon} = g \quad \text{on } \partial\Omega,$$

satisfy the estimate

$$\|\nabla u_\varepsilon\|_{L^p(\Omega)} \le C C_0 \, \|g\|_{B^{-1/p,p}(\partial\Omega)},$$

where C depends only on Ω. This follows from the duality argument used in the proof of Lemma 6.3.4. Similarly, by the duality argument used in the proof of Lemma 6.3.5, for any $F \in L^q(\Omega; \mathbb{R}^m)$, where $\frac{1}{q} = \frac{1}{p} + \frac{1}{d}$, weak solutions in $W^{1,2}(\Omega; \mathbb{R}^m)$ to the Neumann problem,

$$\mathcal{L}^*_\varepsilon(u_\varepsilon) = F \quad \text{in } \Omega \quad \text{and} \quad \frac{\partial u_\varepsilon}{\partial \nu^*_\varepsilon} = b \quad \text{on } \partial\Omega,$$

satisfy the $W^{1,p}$ estimate

$$\|\nabla u_\varepsilon\|_{L^p(\Omega)} \le CC_0 \, \|F\|_{L^q(\Omega)},$$

where $b = -\frac{1}{|\partial\Omega|}\int_\Omega F$ and C depends only on Ω.

Remark 6.3.7. Let $B = B(x_0, r)$ for some $x_0 \in \partial\Omega$ and $0 < r < r_0$. Let $u_\varepsilon \in H^1(2B \cap \Omega; \mathbb{R}^m)$ be a weak solution of $\mathcal{L}_\varepsilon(u_\varepsilon) = 0$ in $2B \cap \Omega$ with $\frac{\partial u_\varepsilon}{\partial \nu_\varepsilon} = g$ on $2B \cap \partial\Omega$. Then for $2 < p < \infty$,

$$\begin{aligned} &\left(\fint_{B\cap\Omega} |\nabla u_\varepsilon|^p \, dx \right)^{1/p} \\ &\le \frac{C}{r} \left(\fint_{2B\cap\Omega} |u_\varepsilon|^2 \, dx \right)^{1/2} + C \left(\fint_{2B\cap\partial\Omega} |g|^t \, d\sigma \right)^{1/t}, \end{aligned} \tag{6.3.19}$$

where $t = p(d-1)/d$. To see this, we apply the estimate (6.3.4) to the function ψu_ε, where $\psi \in C_0^\infty(2B)$ is a cut-off function such that $\psi = 1$ in B and $|\nabla\psi| \le Cr^{-1}$. A bootstrap argument as well as the Sobolev embedding $B^{1/p,p'}(\partial\Omega) \subset L^s(\partial\Omega)$, where $\frac{1}{s} = \frac{1}{p'} - \frac{1}{p(d-1)}$, are also needed.

6.4 Boundary Lipschitz estimates

The goal of this section is to establish uniform boundary Lipschitz estimates in $C^{1,\eta}$ domains for solutions with Neumann conditions. Throughout the section we assume that $D_r = D(r, \psi)$, $\Delta_r = \Delta(r, \psi)$, and $\psi : \mathbb{R}^{d-1} \to \mathbb{R}$ is a $C^{1,\eta}$ function satisfying the condition (5.4.21).

Theorem 6.4.1 (Large-scale Lipschitz estimate). *Suppose that A is 1-periodic and satisfies (2.1.20). Let $u_\varepsilon \in H^1(D_2; \mathbb{R}^m)$ be a weak solution of $\mathcal{L}_\varepsilon(u_\varepsilon) = F$ in D_2 with $\frac{\partial u_\varepsilon}{\partial \nu_\varepsilon} = g$ on Δ_2, where $F \in L^p(D_2; \mathbb{R}^m)$, $g \in C^\rho(\Delta_2; \mathbb{R}^m)$ for some $p > d$ and $\rho \in (0, \eta)$. Then, for $\varepsilon \le r \le 1$,*

$$\begin{aligned} &\left(\fint_{D_r} |\nabla u_\varepsilon|^2 \right)^{1/2} \\ &\le C \left\{ \left(\fint_{D_2} |\nabla u_\varepsilon|^2 \right)^{1/2} + \|F\|_{L^p(D_2)} + \|g\|_{L^\infty(\Delta_2)} + \|g\|_{C^{0,\rho}(\Delta_2)} \right\}, \end{aligned} \tag{6.4.1}$$

where C depends only on μ, p, ρ, and (M_0, η) in (5.4.21).

No smoothness condition on A is needed for the large-scale estimate (6.4.1). Under the additional Hölder continuity condition (4.0.2), we may deduce the full-scale Lipschitz estimate from Theorem 6.4.1.

Theorem 6.4.2. *Suppose that A satisfies (2.1.20), (2.0.2) and (4.0.2). Let Ω be a bounded $C^{1,\eta}$ domain. Let $u_\varepsilon \in H^1(B(x_0,r)\cap\Omega;\mathbb{R}^m)$ be a weak solution of $\mathcal{L}_\varepsilon(u_\varepsilon)=F$ in $B(x_0,r)\cap\Omega$ with $\frac{\partial u_\varepsilon}{\partial \nu_\varepsilon}=g$ on $B(x_0,r)\cap\partial\Omega$, for some $x_0\in\partial\Omega$ and $0<r<r_0$. Then*

$$
\begin{aligned}
&\|\nabla u_\varepsilon\|_{L^\infty(B(x_0,r/2)\cap\Omega)}\\
&\le C\left\{\left(\fint_{B(x_0,r)\cap\Omega}|\nabla u_\varepsilon|^2\right)^{1/2}+r\left(\fint_{B(x_0,r)\cap\Omega}|F|^p\right)^{1/p}\right.\\
&\qquad\left.+\|g\|_{L^\infty(B(x_0,r)\cap\partial\Omega)}+r^\rho\|g\|_{C^{0,\rho}(B(x_0,r)\cap\partial\Omega)}\right\},
\end{aligned}
\tag{6.4.2}
$$

where $\rho\in(0,\eta)$, $p>d$, and C depends only on ρ, p, μ, (λ,τ) in (4.0.2), and Ω.

Proof. We give the proof of Theorem 6.4.2, assuming Theorem 6.4.1. By a change of the coordinate system it suffices to prove that if $p>d$ and $\rho\in(0,\eta)$,

$$
\begin{aligned}
\|\nabla u_\varepsilon\|_{L^\infty(D_r)}\le C&\left\{\left(\fint_{D_{2r}}|\nabla u_\varepsilon|^2\right)^{1/2}+r\left(\fint_{D_{2r}}|F|^p\right)^{1/p}\right.\\
&\qquad\left.+\|g\|_{L^\infty(\Delta_{2r})}+r^\rho\|g\|_{C^{0,\rho}(\Delta_{2r})}\right\},
\end{aligned}
\tag{6.4.3}
$$

for $0<r\le 1$, where $\mathcal{L}_\varepsilon(u_\varepsilon)=F$ in D_{2r}, $\frac{\partial u_\varepsilon}{\partial\nu_\varepsilon}=g$ on Δ_{2r}, and C depends only on μ, p, ρ and (M_0,η) in (5.4.21). By rescaling we may assume that $r=1$. Note that if $\varepsilon\ge 1$, the matrix $A(x/\varepsilon)$ is uniformly Hölder continuous in ε. Consequently, the case $\varepsilon\ge 1$ follows from the standard boundary Lipschitz estimates in $C^{1,\eta}$ domains for elliptic systems with Hölder continuous coefficients.

We thus assume that $r=1$ and $0<\varepsilon<1$. Let $w(x)=\varepsilon^{-1}u_\varepsilon(\varepsilon x)$. Then

$$
\mathcal{L}_1(w)=\widetilde{F}\quad\text{in } \widetilde{D}_2\quad\text{and}\quad \frac{\partial w}{\partial\nu_1}=\widetilde{g}\quad\text{on } \widetilde{\Delta}_2,
$$

where $\widetilde{F}(x)=\varepsilon F(\varepsilon x)$, $\widetilde{g}(x)=g(\varepsilon x)$, and

$$
\widetilde{D}_r=D(r,\widetilde{\psi}),\quad \widetilde{\Delta}_r=\Delta(r,\widetilde{\psi}),\quad \widetilde{\psi}(x')=\varepsilon^{-1}\psi(\varepsilon x').
$$

Since $0<\varepsilon<1$, the function $\widetilde{\psi}$ satisfies the condition (5.4.21) with the same (M_0,η). It follows from the boundary Lipschitz estimates for the operator $\mathcal{L}_1$ that

$$
\|\nabla w\|_{L^\infty(\widetilde{D}_1)}\le C\left\{\|\nabla w\|_{L^2(\widetilde{D}_2)}+\|\widetilde{F}\|_{L^p(\widetilde{D}_2)}+\|\widetilde{g}\|_{L^\infty(\widetilde{\Delta}_2)}+\|\widetilde{g}\|_{C^{0,\rho}(\widetilde{\Delta}_2)}\right\}.
$$

By a change of variables this leads to

$$
\begin{aligned}
&\|\nabla u_\varepsilon\|_{L^\infty(D_\varepsilon)} \\
&\le C\Big\{ \Big(\fint_{D_{2\varepsilon}} |\nabla u_\varepsilon|^2 \Big)^{1/2} + \varepsilon \Big(\fint_{D_{2\varepsilon}} |F|^p \Big)^{1/p} \\
&\qquad\qquad + \|g\|_{L^\infty(\Delta_{2\varepsilon})} + \varepsilon^\rho \|g\|_{C^{0,\rho}(\Delta_{2\varepsilon})} \Big\} \\
&\le C\Big\{ \Big(\fint_{D_{2\varepsilon}} |\nabla u_\varepsilon|^2 \Big)^{1/2} + \|F\|_{L^p(D_2)} + \|g\|_{L^\infty(\Delta_2)} + \|g\|_{C^{0,\rho}(\Delta_2)} \Big\} \\
&\le C\Big\{ \Big(\fint_{D_2} |\nabla u_\varepsilon|^2 \Big)^{1/2} + \|F\|_{L^p(D_2)} + \|g\|_{L^\infty(\Delta_2)} + \|g\|_{C^{0,\rho}(\Delta_2)} \Big\},
\end{aligned}
\tag{6.4.4}
$$

where we have used the fact that $p > d$ and $\varepsilon < 1$ for the second inequality and (6.4.1) for the last.

Using (6.4.4) and translation, we may bound $|\nabla u_\varepsilon(x)|$ by the RHS of (6.4.4) for any $x \in D_1$ with $\text{dist}(x, \Delta_1) \le c\varepsilon$. Similarly, by combining interior Lipschitz estimates for $\mathcal{L}_1$ with (6.4.1), we may dominate $|\nabla u_\varepsilon(x)|$ by the RHS of (6.4.4) for any $x \in D_1$ with $\text{dist}(x, \Delta_1) \ge c\varepsilon$. This finishes the proof. □

Corollary 6.4.3. *Suppose that A satisfies* (2.0.2), (2.1.20) *and* (4.0.2). *Let Ω be a bounded $C^{1,\eta}$ domain in $\mathbb{R}^d$ for some $\eta \in (0,1)$. Let $u_\varepsilon \in H^1(\Omega; \mathbb{R}^m)$ be a weak solution to the Neumann problem*

$$
\mathcal{L}_\varepsilon(u_\varepsilon) = F \quad \text{in } \Omega \quad \text{and} \quad \frac{\partial u_\varepsilon}{\partial \nu_\varepsilon} = g \quad \text{on } \partial\Omega, \tag{6.4.5}
$$

where $F \in L^p(\Omega; \mathbb{R}^m)$, $g \in C^\rho(\partial\Omega; \mathbb{R}^m)$ for some $p > d$ and $\rho \in (0, \eta)$, and

$$
\int_\Omega F\, dx + \int_{\partial\Omega} g\, d\sigma = 0.
$$

Then

$$
\|\nabla u_\varepsilon\|_{L^\infty(\Omega)} \le C\Big\{ \|F\|_{L^p(\Omega)} + \|g\|_{C^\rho(\partial\Omega)} \Big\}, \tag{6.4.6}
$$

where C depends only on p, ρ, μ, (λ, τ) and Ω.

Proof. By covering $\partial\Omega$ with a finite number of balls $\{B(x_\ell, r_0/4)\}$, where $x_\ell \in \partial\Omega$, we can deduce from Theorem 6.4.2 and the interior Lipschitz estimate in Theorem 4.1.2 that

$$
\|\nabla u_\varepsilon\|_{L^\infty(\Omega)} \le C\Big\{ \|\nabla u_\varepsilon\|_{L^2(\Omega)} + \|F\|_{L^p(\Omega)} + \|g\|_{C^\rho(\partial\Omega)} \Big\},
$$

which, together with the energy estimate for $\|\nabla u_\varepsilon\|_{L^2(\Omega)}$, gives (6.4.6). □

The rest of this section is devoted to the proof of Theorem 6.4.1.

Lemma 6.4.4. *Suppose that* $\mathcal{L}_0(w) = F$ *in* D_r *and* $\frac{\partial w}{\partial \nu_0} = g$ *on* Δ_r *for some* $0 < r \le 1$. *Let*

$$I(t) = \frac{1}{t} \inf_{\substack{E\in\mathbb{R}^{m\times d}\\ q\in\mathbb{R}^m}} \left\{ \left(\fint_{D_t} |w - Ex - q|^2 \right)^{1/2} + t^2 \left(\fint_{D_t} |F|^p \right)^{1/p} + t \left\| \frac{\partial}{\partial \nu_0}(w - Ex) \right\|_{L^\infty(\Delta_t)} + t^{1+\rho} \left\| \frac{\partial}{\partial \nu_0}(w - Ex) \right\|_{C^{0,\rho}(\Delta_t)} \right\}$$

for $0 < t \le r$, *where* $p > d$ *and* $0 < \rho < \min\{\eta, 1 - \frac{d}{p}\}$. *Then there exists* $\theta \in (0, 1/4)$, *depending only on* ρ, p, μ *and* (η, M_0), *such that*

$$I(\theta r) \le (1/2) I(r). \tag{6.4.7}$$

Proof. The proof uses the boundary $C^{1,\rho}$ estimate in $C^{1,\eta}$ domains with Neumann conditions for second-order elliptic systems with constant coefficients. By rescaling we may assume $r = 1$. Choosing $q = w(0)$ and $E = \nabla w(0)$, we see that for any $\theta \in (0, 1/4)$,

$$\begin{aligned} I(\theta) &\le C\theta^\rho \|\nabla w\|_{C^{0,\sigma}(D_\theta)} + C\theta^{1-\frac{d}{p}} \left(\fint_{D_1} |F|^p \right)^{1/p} \\ &\le C\theta^\rho \left\{ \|\nabla w\|_{C^{0,\sigma}(D_{1/4})} + \left(\fint_{D_1} |F|^p \right)^{1/p} \right\}, \end{aligned}$$

where we have used the assumption $\rho < 1 - \frac{d}{p}$. It follows from the boundary $C^{1,\rho}$ estimate for $\mathcal{L}_0$ that

$$\|\nabla w\|_{C^{0,\rho}(D_{1/4})} \le C \left\{ \left(\fint_{D_1} |w|^2 \right)^{1/2} + \left(\fint_{D_1} |F|^p \right)^{1/p} + \left\| \frac{\partial w}{\partial \nu_0} \right\|_{L^\infty(\Delta_1)} + \left\| \frac{\partial w}{\partial \nu_0} \right\|_{C^{0,\rho}(\Delta_1)} \right\}.$$

Hence, for any $\theta \in (0, 1/4)$,

$$I(\theta) \le C\theta^\rho \left\{ \left(\fint_{D_1} |w|^2 \right)^{1/2} + \left(\fint_{D_1} |F|^p \right)^{1/p} + \left\| \frac{\partial w}{\partial \nu_0} \right\|_{L^\infty(\Delta_1)} + \left\| \frac{\partial w}{\partial \nu_0} \right\|_{C^{0,\rho}(\Delta_1)} \right\},$$

where C depends only on μ, ρ, p and (η, M_0) in (5.4.21). Finally, since

$$\mathcal{L}_0(w - Ex - q) = F \quad \text{in } D_2$$

for any $E \in \mathbb{R}^{m\times d}$ and $q \in \mathbb{R}^m$, the inequality above gives

$$I(\theta) \le C\,\theta^\rho I(1).$$

The estimate (6.4.7) follows by choosing $\theta \in (0, 1/4)$ so small that $C\theta^\rho \le (1/2)$. ∎

Lemma 6.4.5. *Suppose that $\mathcal{L}_\varepsilon(u_\varepsilon) = F$ in D_2 and $\frac{\partial u_\varepsilon}{\partial \nu_\varepsilon} = g$ on Δ_2, where $0 < \varepsilon < 1$. Let*

$$H(t) = \frac{1}{t} \inf_{\substack{E\in\mathbb{R}^{m\times d}\\ q\in\mathbb{R}^m}} \Bigg\{ \left(\fint_{D_t} |u_\varepsilon - Ex - q|^2 \right)^{1/2} + t^2 \left(\fint_{D_t} |F|^p \right)^{1/p} + t \left\| g - \frac{\partial}{\partial \nu_0}(Ex) \right\|_{L^\infty(\Delta_t)} + t^{1+\sigma} \left\| g - \frac{\partial}{\partial \nu_0}(Ex) \right\|_{C^{0,\sigma}(\Delta_t)} \Bigg\},$$

where $0 < t \le 1$ and $0 < \rho < \min\{\eta, 1 - \frac{d}{p}\}$. Then, for $\varepsilon < t \le 1$,

$$H(\theta t) \le \frac{1}{2} H(t) + C \left(\frac{\varepsilon}{t}\right)^\alpha \Bigg\{ \frac{1}{t} \inf_{q\in\mathbb{R}^m} \left(\fint_{D_{2t}} |u_\varepsilon - q|^2 \right)^{1/2} + t \left(\fint_{D_{2t}} |F|^p \right)^{1/p} + \|g\|_{L^\infty(\Delta_{2t})} \Bigg\}, \tag{6.4.8}$$

where $\theta \in (0, 1/4)$ is given by Lemma 6.4.4, *$\alpha \in (0, 1/2)$ is given by Theorem* 6.1.1, *and C depends only on ρ, p, μ, and (η, M_0).*

Proof. For each $t \in (\varepsilon, 1]$, let $w = w_t$ be the solution of $\mathcal{L}_0(w) = F$ in D_t with $\frac{\partial w}{\partial \nu_0} = g$ on Δ_t, given by Theorem 6.1.1. Since

$$\left(\fint_{D_{\theta t}} |u_\varepsilon - Ex - q|^2 \right)^{1/2} \le \left(\fint_{D_{\theta t}} |u_\varepsilon - w|^2 \right)^{1/2} + \left(\fint_{D_{\theta t}} |w - Ex - q|^2 \right)^{1/2}$$

for any $E \in \mathbb{R}^{m\times d}$ and $q \in \mathbb{R}^m$, we deduce that

$$H(\theta t) \le I(\theta t) + \frac{1}{\theta t} \left(\fint_{D_{\theta t}} |u_\varepsilon - w|^2 \right)^{1/2}. \tag{6.4.9}$$

Similarly, since

$$\left(\fint_{D_t} |w - Ex - q|^2 \right)^{1/2} \le \left(\fint_{D_t} |u_\varepsilon - w|^2 \right)^{1/2} + \left(\fint_{D_t} |u_\varepsilon - Ex - q|^2 \right)^{1/2},$$

we obtain

$$I(t) \le H(t) + \frac{1}{t} \left(\fint_{D_t} |u_\varepsilon - w|^2 \right)^{1/2}.$$

This, together with (6.4.9) and the estimate $I(\theta t) \le (1/2) I(t)$ in Lemma 6.4.4, gives

$$H(\theta t) \le \frac{1}{2} H(t) + \frac{C}{t} \left(\fint_{D_t} |u_\varepsilon - w|^2 \right)^{1/2}, \tag{6.4.10}$$

which, by Theorem 6.1.1, yields (6.4.8). □

The proof of the next lemma will be given at the end of this section.

Lemma 6.4.6. *Let $H(r)$ and $h(r)$ be two nonnegative, continuous functions on the interval $(0,1]$. Let $0 < \varepsilon < (1/4)$. Suppose that there exists a constant C_0 such that*

$$\max_{r\le t\le 2r} H(t) \le C_0\, H(2r) \quad \text{and} \quad \max_{r\le t,s\le 2r} |h(t)-h(s)| \le C_0\, H(2r) \tag{6.4.11}$$

for any $r \in [\varepsilon, 1/2]$. Suppose further that

$$H(\theta r) \le \frac{1}{2} H(r) + C_0\, \beta(\varepsilon/r)\Big\{H(2r) + h(2r)\Big\} \tag{6.4.12}$$

for any $r \in [\varepsilon, 1/2]$, where $\theta \in (0, 1/4)$ and $\beta(t)$ is a nonnegative, nondecreasing function on $[0,1]$ such that $\beta(0) = 0$ and

$$\int_0^1 \frac{\beta(t)}{t}\, dt < \infty. \tag{6.4.13}$$

Then

$$\max_{\varepsilon\le r\le 1} \{H(r) + h(r)\} \le C\{H(1) + h(1)\}, \tag{6.4.14}$$

where C depends only on C_0, θ, and the function $\beta(t)$.

We now give the proof of Theorem 6.4.1, using Lemmas 6.4.5 and 6.4.6.

Proof of Theorem 6.4.1. Let u_ε be a solution of $\mathcal{L}_\varepsilon(u_\varepsilon) = F$ in D_2 with $\frac{\partial u_\varepsilon}{\partial \nu_\varepsilon} = g$ on Δ_2. We define the function $H(t)$ by (6.4.8). It is not hard to see that

$$H(t) \le CH(2r) \quad \text{if } t \in [r, 2r] \tag{6.4.15}$$

Next, we define $h(t) = |E_t|$, where E_t is an $m \times d$ matrix such that

$$\begin{aligned} H(t) = \frac{1}{t} \inf_{q\in\mathbb{R}^m} \Bigg\{ \left(\fint_{D_t} |u_\varepsilon - E_t x - q|^2 \right)^{1/2} + t^2 \left(\fint_{D_t} |F|^p \right)^{1/p} \\ + t \left\| g - \frac{\partial}{\partial \nu_0}(E_t x) \right\|_{L^\infty(\Delta_t)} + t^{1+\rho} \left\| g - \frac{\partial}{\partial \nu_0}(E_t x) \right\|_{C^{0,\rho}(\Delta_t)} \Bigg\}. \end{aligned}$$

Let $t, s \in [r, 2r]$. Using that

$$\begin{aligned} |E_t - E_s| &\le \frac{C}{r} \inf_{q\in\mathbb{R}^m} \left(\fint_{D_r} |(E_t - E_s)x - q|^2 \right)^{1/2} \\ &= \frac{C}{r} \inf_{q_1,q_2\in\mathbb{R}^m} \left(\fint_{D_r} |(E_t - E_s)x - q_1 + q_2|^2 \right)^{1/2} \end{aligned}$$

$$
\begin{aligned}
&\le \frac{C}{t} \inf_{q\in\mathbb{R}^m} \left(\fint_{D_t} |u_\varepsilon - E_t x - q|^2 \right)^{1/2} \\
&\qquad\qquad + \frac{C}{s} \inf_{q\in\mathbb{R}^m} \left(\fint_{D_s} |u_\varepsilon - E_s x - q|^2 \right)^{1/2} \\
&\le C\{H(t) + H(s)\} \\
&\le CH(2r),
\end{aligned}
$$

we obtain

$$
\max_{r\le t,s\le 2r} |h(t) - h(s)| \le CH(2r). \tag{6.4.16}
$$

Furthermore, by (6.4.8),

$$
H(\theta r) \le \frac{1}{2} H(r) + C\left(\frac{\varepsilon}{r}\right)^\alpha \Phi(2r) \tag{6.4.17}
$$

for $r \in [\varepsilon, 1]$, where $\alpha \in (0, 1/2)$ and

$$
\Phi(t) = \left\{ \frac{1}{t} \inf_{q\in\mathbb{R}^m} \left(\fint_{D_t} |u_\varepsilon - q|^2 \right)^{1/2} + t \left(\fint_{D_t} |F|^p \right)^{1/p} + \|g\|_{L^\infty(\Delta_t)} \right\}.
$$

It is easy to see that

$$
\Phi(t) \le C\{H(t) + h(t)\},
$$

which, together with (6.4.17), leads to

$$
H(\theta r) \le \frac{1}{2} H(r) + C\left(\frac{\varepsilon}{r}\right)^\alpha \Big\{ H(2r) + h(2r) \Big\}. \tag{6.4.18}
$$

Thus the functions $H(r)$ and $h(r)$ satisfy the conditions (6.4.11), (6.4.12) and (6.4.13). As a result, we obtain that for $r \in [\varepsilon, 1]$,

$$
\begin{aligned}
\inf_{q\in\mathbb{R}^m} \frac{1}{r} \left(\fint_{D_r} |u_\varepsilon - q|^2 \right)^{1/2} &\le C\{H(r) + h(r)\} \\
&\le C\{H(1) + h(1)\}.
\end{aligned}
$$

By taking $E = 0$ and $q = 0$, we see that

$$
H(1) \le C \left\{ \left(\fint_{D_1} |u_\varepsilon|^2 \right)^{1/2} + \|F\|_{L^p(D_1)} + \|g\|_{L^\infty(\Delta_1)} + \|g\|_{C^{0,\rho}(\Delta_1)} \right\}.
$$

Also, note that

$$
\begin{aligned}
h(1) &\le C \inf_{q\in\mathbb{R}^m} \left(\fint_{D_1} |E_1 x + q|^2 \right)^{1/2} \\
&\le C \left\{ H(1) + \left(\fint_{D_1} |u_\varepsilon|^2 \right)^{1/2} \right\}.
\end{aligned}
$$

Hence, we have proved that for $\varepsilon \le r \le 1$,

$$\begin{aligned} &\inf_{q\in\mathbb{R}^m} \frac{1}{r}\left(\fint_{D_r} |u_\varepsilon - q|^2\right)^{1/2} \\ &\le C\left\{\left(\fint_{D_2} |u_\varepsilon|^2\right)^{1/2} + \|F\|_{L^p(D_2)} + \|g\|_{L^\infty(\Delta_2)} + \|g\|_{C^{0,\rho}(\Delta_2)}\right\}. \end{aligned} \tag{6.4.19}$$

Replacing u_ε by $u_\varepsilon - \fint_{D_2} u_\varepsilon$ in the estimate above and using the Poincaré inequality, we obtain

$$\begin{aligned} &\inf_{q\in\mathbb{R}^m} \frac{1}{r}\left(\fint_{D_r} |u_\varepsilon - q|^2\right)^{1/2} \\ &\le C\left\{\left(\fint_{D_2} |\nabla u_\varepsilon|^2\right)^{1/2} + \|F\|_{L^p(D_2)} + \|g\|_{L^\infty(\Delta_2)} + \|g\|_{C^{0,\rho}(\Delta_2)}\right\}. \end{aligned} \tag{6.4.20}$$

This, together with Caccioppoli's inequality (6.1.2), gives (6.4.1). □

We end this section with the proof of Lemma 6.4.6.

Proof of Lemma 6.4.6. It follows from the second inequality in (6.4.11) that

$$h(r) \le h(2r) + C_0 H(2r)$$

for any $r \in [\varepsilon, 1/2]$. Hence,

$$\begin{aligned} \int_a^{1/2} \frac{h(r)}{r}\,dr &\le \int_a^{1/2} \frac{h(2r)}{r}\,dr + C_0\int_a^{1/2} \frac{H(2r)}{r}\,dr \\ &= \int_{2a}^{1} \frac{h(r)}{r}\,dr + C_0\int_a^{1/2} \frac{H(2r)}{r}\,dr, \end{aligned}$$

where $a \in [\varepsilon, 1/4]$. This implies that

$$\begin{aligned} \int_a^{2a} \frac{h(r)}{r}\,dr &\le \int_{1/2}^{1} \frac{h(r)}{r}\,dr + C_0\int_a^{1/2} \frac{H(2r)}{r}\,dr \\ &\le C_0\Big\{H(1) + h(1)\Big\} + C_0\int_a^{1/2} \frac{H(2r)}{r}\,dr, \end{aligned}$$

which, together with (6.4.11), leads to

$$H(a) + h(a) \le C\left\{H(2a) + h(1) + H(1) + \int_{2a}^{1} \frac{H(r)}{r}\,dr\right\}$$

for any $a \in [\varepsilon, 1/4]$. By the first inequality in (6.4.11) we may further deduce that

$$H(a) + h(a) \le C\left\{H(1) + h(1) + \int_a^1 \frac{H(r)}{r}\,dr\right\} \tag{6.4.21}$$

for any $a \in [\varepsilon, 1]$.

To bound the integral in the RHS of (6.4.21), we use (6.4.12) and (6.4.21) to obtain

$$H(\theta r) \leq \frac{1}{2} H(r) + C\beta(\varepsilon/r)\Big\{H(1) + h(1)\Big\} + C\beta(\varepsilon/r)\int_r^1 \frac{H(t)}{t}\, dt$$

for $r \in [\varepsilon, 1/2]$. It follows that

$$\begin{aligned}\int_{\alpha\theta\varepsilon}^{\theta} \frac{H(r)}{r} dr \leq \frac{1}{2}\int_{\alpha\varepsilon}^{1} \frac{H(r)}{r} dr + C\Big\{H(1) + h(1)\Big\} \\ + C\int_{\alpha\varepsilon}^{1} \beta(\varepsilon/r)\left\{\int_r^1 \frac{H(t)}{t} dt\right\}\frac{dr}{r},\end{aligned} \tag{6.4.22}$$

where $\alpha > 1$ and we have used the condition (6.4.13) on $\beta(t)$ for

$$\int_{\alpha\varepsilon}^{1} \beta(\varepsilon/r)\frac{dr}{r} = \int_{\varepsilon}^{\frac{1}{\alpha}} \frac{\beta(t)}{t} dt \leq \int_0^1 \frac{\beta(t)}{t} dt < \infty.$$

Note that, by Fubini's Theorem,

$$\begin{aligned}\int_{\alpha\varepsilon}^{1} \beta(\varepsilon/r)\left\{\int_r^1 \frac{H(t)}{t} dt\right\}\frac{dr}{r} &= \int_{\alpha\varepsilon}^{1}\left\{\int_{\alpha\varepsilon}^{t} \beta(\varepsilon/r)\frac{dr}{r}\right\}\frac{H(t)}{t} dt \\ &= \int_{\alpha\varepsilon}^{1} H(t)\left\{\int_{\varepsilon/t}^{1/\alpha} \frac{\beta(s)}{s} ds\right\}\frac{dt}{t} \\ &\leq \int_0^{1/\alpha} \frac{\beta(s)}{s} ds \int_{\alpha\varepsilon}^{1} \frac{H(t)}{t} dt \\ &\leq \frac{1}{4C}\int_{\alpha\varepsilon}^{1} \frac{H(t)}{t} dt,\end{aligned}$$

if $\alpha > 1$, which only depends on C_0 and the function β, is sufficiently large. In view of (6.4.22) this gives

$$\int_{\alpha\theta\varepsilon}^{\theta} \frac{H(r)}{r} dr \leq \frac{1}{2}\int_{\alpha\varepsilon}^{1} \frac{H(r)}{r} dr + C\Big\{H(1) + h(1)\Big\} + \frac{1}{4}\int_{\alpha\varepsilon}^{1} \frac{H(t)}{t} dt.$$

It follows that

$$\int_{\alpha\theta\varepsilon}^{\theta} \frac{H(r)}{r} dr \leq C\Big\{H(1) + h(1)\Big\},$$

which, by (6.4.11) and (6.4.21), yields

$$\begin{aligned}H(r) + h(r) &\leq C\left\{H(1) + h(1) + \int_r^1 \frac{H(t)}{t} dt\right\} \\ &\leq C\{H(1) + h(1)\}\end{aligned}$$

for any $r \in [\varepsilon, 1]$. □

6.5 Matrix of Neumann functions

Assume that A is 1-periodic and satisfies the ellipticity condition (2.1.20) and the VMO condition (4.3.1). Suppose that either $\mathcal{L}_\varepsilon(u_\varepsilon) = 0$ or $\mathcal{L}_\varepsilon^*(u_\varepsilon) = 0$ in $2B = B(x_0, 2r)$. It follows from the interior Hölder estimate (4.3.11) that

$$\|u_\varepsilon\|_{C^{0,\rho}(B)} \le Cr^{-\rho} \left(\fint_{2B} |u_\varepsilon|^2 \right)^{1/2}$$

for any $\rho \in (0,1)$, where C depends only on μ, ρ, and the function $\omega(t)$ in (4.3.1). This allows one to construct an $m \times m$ matrix of Neumann functions in a bounded Lipschitz domain Ω in $\mathbb{R}^d$,

$$N_\varepsilon(x,y) = \big(N_\varepsilon^{\alpha\beta}(x,y)\big),$$

with the following properties:

- For $d \ge 3$, one has
$$|N_\varepsilon(x,y)| \le C\, |x-y|^{2-d},$$
for $x, y \in \Omega$ with $|x-y| < \frac{1}{2}\text{dist}(y, \partial\Omega)$. If $d = 2$, then
$$|N_\varepsilon(x,y)| \le C\big\{1 + \ln[r_0 |x-y|^{-1}]\big\},$$
for any $x, y \in \Omega$, where $r_0 = \text{diam}(\Omega)$.
- If $F \in L^p(\Omega; \mathbb{R}^m)$ for some $p > \frac{d}{2}$ and $g \in L^2(\partial\Omega; \mathbb{R}^m)$ satisfy the compatibility condition $\int_\Omega F\, dx + \int_{\partial\Omega} g = 0$, then
$$u_\varepsilon(x) = \int_\Omega N_\varepsilon(x,y) F(y)\, dy + \int_{\partial\Omega} N_\varepsilon(x,y) g(y)\, d\sigma(y) \tag{6.5.1}$$
is the unique weak solution in $H^1(\Omega; \mathbb{R}^m)$ of the Neumann problem $\mathcal{L}_\varepsilon(u_\varepsilon) = F$ in Ω and $\frac{\partial u_\varepsilon}{\partial \nu_\varepsilon} = g$ on $\partial\Omega$, with $\int_{\partial\Omega} u_\varepsilon\, d\sigma = 0$.
- Let $N_\varepsilon^*(x,y)$ denote the matrix of Neumann functions for the adjoint operator $\mathcal{L}_\varepsilon^*$ in Ω. Then
$$N_\varepsilon^*(x,y) = \big(N_\varepsilon(y,x)\big)^T \quad \text{for any } x, y \in \Omega. \tag{6.5.2}$$

Theorem 6.5.1. *Suppose that A is 1-periodic and satisfies conditions* (2.1.20) *and* (4.3.1). *Let Ω be a bounded C^1 domain in $\mathbb{R}^d$. Then, if $d \ge 3$,*

$$|N_\varepsilon(x,y)| \le C\, |x-y|^{2-d}, \tag{6.5.3}$$

for any $x, y \in \Omega$ and $x \ne y$. Moreover, if $d \ge 2$,

$$|N_\varepsilon(x,y) - N(z,y)| + |N_\varepsilon(y,x) - N_\varepsilon(y,z)| \le \frac{C|x-z|^\rho}{|x-y|^{d-2+\rho}}, \tag{6.5.4}$$

for any $x, y, z \in \Omega$ such that $|x-z| < \frac{1}{2}|x-y|$. The constant C depends at most on μ, ρ, $\omega(t)$ and Ω.

Proof. Since Ω is C^1, solutions of $\mathcal{L}_\varepsilon(u_\varepsilon) = 0$ in $B(x_0, r) \cap \Omega$ with $\frac{\partial u_\varepsilon}{\partial \nu_\varepsilon} = 0$ on $B(x_0, r) \cap \partial\Omega$, where $x_0 \in \partial\Omega$ and $0 < r < r_0$, satisfy the estimate

$$\|u_\varepsilon\|_{C^{0,\rho}(B(x_0,r/2)\cap\Omega)} \le C r^{-\rho} \left(\fint_{B(x_0,r)\cap\Omega} |u_\varepsilon|^2 \right)^{1/2} \tag{6.5.5}$$

for any $\rho \in (0,1)$, where C depends only on ρ, μ, $\omega(t)$ and Ω. This is a consequence of Theorem 6.2.2, which also gives (6.5.5) for solutions of $\mathcal{L}_\varepsilon^*(u_\varepsilon) = 0$ in $B(x_0, r) \cap \Omega$ with $\frac{\partial u_\varepsilon}{\partial \nu_\varepsilon^*} = 0$ on $B(x_0, r) \cap \partial\Omega$. The estimates (6.5.3)–(6.5.4) now follow from general results in [101] for $d = 2$ and in [27] for $d \ge 3$. □

Using interior and boundary Lipschitz estimates, stronger estimates can be proved in $C^{1,\alpha}$ domains under the assumption that A is Hölder continuous.

Theorem 6.5.2. *Assume that A is 1-periodic and satisfies conditions (2.1.20) and (4.0.2). Let Ω be a bounded $C^{1,\alpha}$ domain in $\mathbb{R}^d$ for some $\alpha > 0$. Then for any $x, y \in \Omega$ and $x \ne y$,*

$$|\nabla_x N_\varepsilon(x, y)| + |\nabla_y N_\varepsilon(x, y)| \le C\,|x - y|^{1-d}, \tag{6.5.6}$$

and

$$|\nabla_y \nabla_x N_\varepsilon(x, y)| \le C\,|x - y|^{-d}, \tag{6.5.7}$$

where C depends only on μ, (λ, τ) and Ω.

Proof. Suppose $d \ge 3$. Fix $x_0, y_0 \in \Omega$ and let $r = |x_0 - y_0|/8$. Let $u_\varepsilon(x) = N_\varepsilon(x, y_0)$. Then $\mathcal{L}_\varepsilon(u_\varepsilon) = 0$ in $B(x_0, 4r) \cap \Omega$ and

$$\frac{\partial u_\varepsilon}{\partial \nu_\varepsilon} = -\frac{1}{|\partial\Omega|} I_{m\times m} \quad \text{on } B(x_0, 4r) \cap \partial\Omega,$$

where $I_{m\times m}$ denotes the $m \times m$ identity matrix. By the boundary Lipschitz estimate (6.4.2) it follows that

$$\begin{aligned}
|\nabla u_\varepsilon(x_0)| &\le \frac{C}{r} \left(\fint_{B(x_0,4r)\cap\Omega} |u_\varepsilon|^2 \right)^{1/2} + C \\
&\le \frac{C}{r^{d-1}},
\end{aligned}$$

where we have used the size estimate (6.5.3) for the last inequality. This shows that $|\nabla_x N_\varepsilon(x_0, y_0)| \le C r^{1-d}$. Thus we have proved that $|\nabla_x N_\varepsilon(x, y)| \le C|x - y|^{1-d}$. Since the same argument also yields $|\nabla_x N_\varepsilon^*(x, y)| \le C|x - y|^{1-d}$, in view of (6.5.2), we obtain

$$|\nabla_y N_\varepsilon(x, y)| = |\nabla_y N_\varepsilon^*(y, x)| \le C\,|x - y|^{1-d}.$$

To see (6.5.7), we note that the boundary Lipschitz estimate gives

$$
\begin{aligned}
|\nabla_x N_\varepsilon(x_0, y_1) - \nabla_x N_\varepsilon(x_0, y_2)\}| &\le C r^{-1} \max_{z \in B(x_0, r)\cap\Omega} |N_\varepsilon(z, y_1) - N_\varepsilon(z, y_2)| \\
&\le \frac{C|y_1 - y_2|}{r^d},
\end{aligned}
$$

where $y_1, y_2 \in B(y_0, r)$. It follows that $|\nabla_y \nabla_x N_\varepsilon(x_0, y_0)| \le C r^{-d}$.

Finally, in the case $d = 2$, we apply the Lipschitz estimate to

$$
u_\varepsilon(x) = N_\varepsilon(x, y_0) - N_\varepsilon(x_0, y_0)
$$

and use the fact that

$$
|N_\varepsilon(x, y_0) - N_\varepsilon(x_0, y_0)| \le C
$$

if $|x - x_0| < (1/2)|x_0 - y_0|$. □

6.6 Elliptic systems of linear elasticity

A careful inspection of the proof in the previous sections in this chapter shows that all results, with a few minor modifications, hold for the elliptic system of elasticity. In particular, we obtain uniform boundary Hölder and $W^{1,p}$ estimates in C^1 domains as well as uniform Lipschitz estimate in $C^{1,\alpha}$ domains.

We start with a Caccioppoli inequality for elliptic systems of elasticity with Neumann conditions.

Lemma 6.6.1. *Suppose that $A \in E(\kappa_1, \kappa_2)$. Let $u_\varepsilon \in H^1(D_{2r}; \mathbb{R}^d)$ be a solution of $\mathcal{L}_\varepsilon(u_\varepsilon) = F$ in D_{2r} with $\frac{\partial u_\varepsilon}{\partial \nu_\varepsilon} = g$ on Δ_{2r}. Then the estimate* (6.1.2) *holds with constant C depending only on κ_1, κ_2 and M_0.*

Proof. We follow the same line of argument as in the proof of Lemma 6.1.2. The special case where $g = 0$ may be handled in the same manner with help of the second Korn inequality. To deal with the general case, in view of the proof of Lemma 6.1.2, it suffices to construct a function $\widetilde{g} \in L^2(\partial D_2; \mathbb{R}^d)$ such that $\widetilde{g} = g$ on Δ_2,

$$
\|\widetilde{g}\|_{L^2(\partial D_2)} \le C\big\{\|g\|_{L^2(\Delta_2)} + \|F\|_{L^2(D_2)}\big\},
$$

and $\widetilde{g}$ satisfies the compatibility condition,

$$
\int_{\partial D_2} \widetilde{g} \cdot \phi \, d\sigma + \int_{D_2} F \cdot \phi \, dx = 0 \tag{6.6.1}
$$

for any $\phi \in \mathcal{R}$. To this end, we let

$$
\widetilde{g} = \alpha_1 \phi_1 + \alpha_2 \phi_2 + \cdots + \alpha_N \phi_N \quad \text{on } \partial D_2 \setminus \Delta_2,
$$

where $N = d(d+1)/2$, $\{\phi_1, \phi_2, \ldots, \phi_N\}$ is an orthonormal basis of $\mathcal{R}$ in $L^2(D_2; \mathbb{R}^d)$, and $(\alpha_1, \alpha_2, \ldots, \alpha_N) \in \mathbb{R}^N$ is to be determined by solving the $N \times N$ system of linear equations,

$$\int_{\partial D_2 \setminus \Delta_2} (\alpha_1 \phi_1 + \alpha_2 \phi_2 + \cdots + \alpha_N \phi_N) \cdot \phi_j \, d\sigma = -\int_{D_2} F \cdot \phi_j \, dx - \int_{\Delta_2} g \cdot \phi_j \, d\sigma$$

for $j = 1, 2, \ldots, N$. This linear system is uniquely solvable, provided that

$$\det \left(\int_{\partial D_2 \setminus \Delta_2} \phi_i \cdot \phi_j \, d\sigma \right) \neq 0. \tag{6.6.2}$$

To see that (6.6.2) holds, let us assume the contrary. Then there exists

$$(\beta_1, \beta_2, \ldots, \beta_N) \in \mathbb{R}^N \setminus \{0\}$$

such that

$$\int_{\partial D_2 \setminus \Delta_2} |\beta_i \phi_i|^2 \, d\sigma = \beta_i \beta_j \int_{\partial D_2 \setminus \Delta_2} \phi_i \cdot \phi_j \, d\sigma = 0,$$

which implies that $\beta_i \phi_i = 0$ on $\partial D_2 \setminus \Delta_2$ (the repeated indices are summed from 1 to N). Since $\beta_i \phi_i$ is a linear function and $\partial D_2 \setminus \Delta_2$ cannot be a hyperplane, we conclude that $\beta_i \phi_i \equiv 0$ in $\mathbb{R}^d$. Consequently, $\beta_1 = \beta_2 = \cdots = \beta_N = 0$, which gives us a contradiction. □

The next theorem is an analog of Theorem 6.1.1.

Theorem 6.6.2. *Suppose that $A \in E(\kappa_1, \kappa_2)$ and is 1-periodic. Let $u_\varepsilon \in H^1(D_{2r}, \mathbb{R}^d)$ be a weak solution of $\mathcal{L}_\varepsilon(u_\varepsilon) = F$ in D_{2r} with $\frac{\partial u_\varepsilon}{\partial \nu_\varepsilon} = g$ on Δ_{2r}, where $F \in L^2(D_{2r}; \mathbb{R}^d)$ and $g \in L^2(\Delta_{2r}; \mathbb{R}^d)$. Assume that $r \geq \varepsilon$. Then there exists $w \in H^1(D_r; \mathbb{R}^d)$ such that $\mathcal{L}_0(w) = F$ in D_r, $\frac{\partial w}{\partial \nu_0} = g$ on Δ_r, and*

$$\begin{aligned} &\left(\fint_{D_r} |u_\varepsilon - w|^2 \right)^{1/2} \\ &\leq C \left(\frac{\varepsilon}{r} \right)^{1/2} \left\{ \left(\fint_{D_{2r}} |u_\varepsilon|^2 \right)^{1/2} + r^2 \left(\fint_{D_{2r}} |F|^2 \right)^{1/2} + r \left(\fint_{\Delta_{2r}} |g|^2 \right)^{1/2} \right\}, \end{aligned} \tag{6.6.3}$$

where C depends only on μ and M_0.

Proof. The proof is similar to that of Theorem 6.1.1, using Lemma 6.6.1 and estimate (6.1.8). Note that by (3.6.6), the estimate (6.1.8) holds for $\sigma = 1/2$, where $\mathcal{L}_\varepsilon(u_\varepsilon) = \mathcal{L}_0(u_0)$ in Ω, $\frac{\partial u_\varepsilon}{\partial \nu_\varepsilon} = \frac{\partial u_0}{\partial \nu_0}$ on $\partial\Omega$, and

$$\int_\Omega (u_\varepsilon - u_0) \cdot \phi \, dx = 0,$$

for any $\phi \in \mathcal{R}$. □

Theorem 6.6.3. *Suppose that $A \in E(\kappa_1, \kappa_2)$ is 1-periodic and satisfies (4.3.1). Let Ω be a bounded C^1 domain in $\mathbb{R}^d$ and $0 < \rho < 1$. Let $u_\varepsilon \in H^1(B(x_0, r) \cap \Omega; \mathbb{R}^d)$ be a solution of $\mathcal{L}_\varepsilon(u_\varepsilon) = 0$ in $B(x_0, r) \cap \Omega$ with $\frac{\partial u_\varepsilon}{\partial \nu_\varepsilon} = g$ on $B(x_0, r) \cap \partial\Omega$ for some $x_0 \in \partial\Omega$ and $0 < r < r_0$. Then the Hölder estimate (6.2.5) holds for any $x, y \in B(x_0, r/2) \cap \Omega$, where the constant C depends only on μ, ρ, κ_1, κ_2, and Ω.*

Proof. With Theorem 6.6.2 at our disposal, the proof is the same as that for Theorem 6.2.2. □

Theorem 6.6.4. *Suppose that $A \in E(\kappa_1, \kappa_2)$ is 1-periodic and satisfies (4.3.1). Let Ω be a bounded C^1 domain in $\mathbb{R}^d$ and $1 < p < \infty$. Let $F \in L^q(\Omega; \mathbb{R}^d)$, $f \in L^p(\Omega; \mathbb{R}^{d\times d})$ and $g \in B^{-1/p,p}(\partial\Omega; \mathbb{R}^d)$ satisfy the compatibility condition (2.4.9), where q given by (6.3.15). Then the Neumann problem (6.3.1) has a solution u_ε in $W^{1,p}(\Omega; \mathbb{R}^d)$ such that the $W^{1,p}$ estimate (6.3.4) holds for some constant C_p depending only on p, κ_1, κ_2, $\omega(t)$ in (4.3.1), and Ω. The solution is unique in $W^{1,p}(\Omega; \mathbb{R}^d)$, up to an element of $\mathcal{R}$.*

Proof. The proof follows the same line of argument used for Theorem 6.3.2. Due to the compatibility condition (2.4.9), some modifications are needed.

Step 1. Consider the Neumann problem:

$$\mathcal{L}_\varepsilon(u_\varepsilon) = \operatorname{div}(f) \quad \text{in } \Omega \quad \text{and} \quad \frac{\partial u_\varepsilon}{\partial \nu_\varepsilon} = -n \cdot f \quad \text{on } \partial\Omega, \tag{6.6.4}$$

which has a unique solution in $H^1(\Omega; \mathbb{R}^d)$ such that $u_\varepsilon \perp \mathcal{R}$ in $L^2(\Omega; \mathbb{R}^d)$, provided that $f = (f_i^\alpha) \in L^2(\Omega; \mathbb{R}^{d\times d})$ satisfies the compatibility condition

$$\int_\Omega f_i^\alpha \frac{\partial \phi^\alpha}{\partial x_i}\, dx = 0$$

for any $\phi = (\phi^\alpha) \in \mathcal{R}$. This condition is equivalent to

$$\int_\Omega (f_i^\alpha - f_\alpha^i)\, dx = 0. \tag{6.6.5}$$

As a result, Theorem 4.2.7 cannot be applied directly. Rather, we use Theorem 4.2.6 to show that

$$\|\nabla u_\varepsilon\|_{L^p(\Omega)} \le C_p \|f\|_{L^p(\Omega)} \tag{6.6.6}$$

for $1 < p < \infty$.

To this end we fix a ball B with $|B| \le c_0|\Omega|$. Assume that either $4B \subset \Omega$ or B is centered on $\partial\Omega$. We write $u_\varepsilon = v_\varepsilon + w_\varepsilon$ in Ω, where v_ε is the unique solution in $H^1(\Omega; \mathbb{R}^d)$ to the Neumann problem,

$$\begin{cases} \mathcal{L}_\varepsilon(v_\varepsilon) = \operatorname{div}\big((f - E)\chi_{4B\cap\Omega}\big) & \text{in } \Omega, \\ \dfrac{\partial v_\varepsilon}{\partial \nu_\varepsilon} = -n \cdot (f - E)\chi_{4B\cap\Omega} & \text{on } \partial\Omega, \end{cases} \tag{6.6.7}$$

such that $v_\varepsilon \perp \mathcal{R}$ in $L^2(\Omega; \mathbb{R}^d)$, where $E = (E_i^\alpha)$ is a constant with

$$E_i^\alpha = \frac{1}{2} \fint_{4B\cap\Omega} (f_i^\alpha - f_\alpha^i).$$

It is easy to verify that the function $(f - E)\chi_{4B\cap\Omega}$ satisfies the compatibility condition (6.6.5). Thus,

$$\|\nabla v_\varepsilon\|_{L^2(\Omega)} \le C\, \|f - E\|_{L^2(4B\cap\Omega)} \le C\, \|f\|_{L^2(4B\cap\Omega)},$$

which leads to

$$\left(\fint_{4B\cap\Omega} |\nabla v_\varepsilon|^2\right)^{1/2} \le C \left(\fint_{4B\cap\Omega} |f|^2\right)^{1/2}. \tag{6.6.8}$$

To estimate ∇w_ε in $2B \cap \Omega$, we first consider the case where $4B \subset \Omega$. Since $\mathcal{L}_\varepsilon(w_\varepsilon) = 0$ in $4B$, we may use the interior $W^{1,p}$ estimate in Theorem 4.3.1 to obtain

$$\begin{aligned}
\left(\fint_{2B} |\nabla w_\varepsilon|^p\right)^{1/p} &\le C \left(\fint_{4B} |\nabla w_\varepsilon|^2\right)^{1/2}\\
&\le C \left(\fint_{4B} |\nabla u_\varepsilon|^2\right)^{1/2} + \left(\fint_{4B} |\nabla v_\varepsilon|^2\right)^{1/2}\\
&\le C \left(\fint_{4B} |\nabla u_\varepsilon|^2\right)^{1/2} + C \left(\fint_{4B} |f|^2\right)^{1/2},
\end{aligned}$$

for $2 < p < \infty$, where we have used (6.6.8) for the last step. If B is centered on $\partial\Omega$, we observe that w_ε satisfies

$$\mathcal{L}_\varepsilon(w_\varepsilon) = 0 \quad \text{in } 4B \cap \Omega \quad \text{and} \quad \frac{\partial w_\varepsilon}{\partial \nu_\varepsilon} = n \cdot E \quad \text{on } 4B \cap \partial\Omega.$$

It follows by Theorem 6.6.3 that

$$|w_\varepsilon(x) - w_\varepsilon(y)| \le Cr \left(\frac{|x-y|}{r}\right)^\rho \left\{ \left(\fint_{4B\cap\Omega} |\nabla w_\varepsilon|^2\right)^{1/2} + |E| \right\}$$

for any $x, y \in 2B \cap \Omega$, where $0 < \rho < 1$. As in the proof of Theorem 6.3.2, this, together with the interior $W^{1,p}$ estimate, yields that

$$\begin{aligned}
\left(\fint_{2B\cap\Omega} |\nabla w_\varepsilon|^p\right)^{1/p} &\le C \left\{ \left(\fint_{4B\cap\Omega} |\nabla w_\varepsilon|^2\right)^{1/2} + |E| \right\}\\
&\le C \left\{ \left(\fint_{4B\cap\Omega} |\nabla u_\varepsilon|^2\right)^{1/2} + \left(\fint_{4B\cap\Omega} |\nabla v_\varepsilon|^2\right)^{1/2} + |E| \right\}\\
&\le C \left\{ \left(\fint_{4B\cap\Omega} |\nabla u_\varepsilon|^2\right)^{1/2} + \left(\fint_{4B\cap\Omega} |f|^2\right)^{1/2} \right\}.
\end{aligned}$$

By Theorem 4.2.6, it follows that

$$\left(\fint_\Omega |\nabla u_\varepsilon|^p\right)^{1/p} \le C\left(\fint_\Omega |\nabla u_\varepsilon|^2\right)^{1/2} + C\left(\fint_\Omega |f|^p\right)^{1/p}$$

for any $2 < p < \infty$. Using

$$\|\nabla u_\varepsilon\|_{L^2(\Omega)} \le C\,\|f\|_{L^2(\Omega)} \le C\,\|f\|_{L^p(\Omega)},$$

we obtain (6.6.6) for $2 < p < \infty$.

Step 2. To prove the estimate (6.6.6) for $1 < p < 2$, we use a duality argument. Let $f = (f_i^\alpha) \in L^2(\Omega; \mathbb{R}^{d\times d})$, $g = (g_i^\alpha) \in L^{p'}(\Omega; \mathbb{R}^{d\times d})$ satisfy the compatibility condition (6.6.5). Let $u_\varepsilon, v_\varepsilon$ be weak solutions of (6.6.4) with data f, g, respectively. Then

$$\int_\Omega g_i^\alpha \frac{\partial u_\varepsilon^\alpha}{\partial x_i}\, dx = -\int_\Omega A(x/\varepsilon)\nabla u_\varepsilon \cdot \nabla v_\varepsilon\, dx = \int_\Omega f_i^\alpha \frac{\partial v_\varepsilon^\alpha}{\partial x_i}\, dx.$$

Assume that $v_\varepsilon \perp \mathcal{R}$ in $L^2(\Omega; \mathbb{R}^{d\times d})$. By Step 1, $\|\nabla v_\varepsilon\|_{L^{p'}(\Omega)} \le C\|g\|_{L^{p'}(\Omega)}$. It follows that

$$\left|\int_\Omega g_i^\alpha \frac{\partial u_\varepsilon^\alpha}{\partial x_i}\, dx\right| \le C\,\|g\|_{L^{p'}(\Omega)}\|f\|_{L^p(\Omega)}. \tag{6.6.9}$$

Now, for any $h = (h_i^\alpha) \in L^{p'}(\Omega; \mathbb{R}^{d\times d})$, let $g = (g_i^\alpha)$, where

$$g_i^\alpha = 2h_i^\alpha - \fint_\Omega h_i^\alpha\, dx + \fint_\Omega h_\alpha^i\, dx.$$

Then g satisfies (6.6.5). As a result, we may deduce from (6.6.9) that

$$\left|\int_\Omega h_i^\alpha \frac{\partial u_\varepsilon^\alpha}{\partial x_i}\, dx\right| \le C\left\{\|f\|_{L^p(\Omega)} + \left|\int_\Omega \left(\nabla u_\varepsilon - (\nabla u_\varepsilon)^T\right) dx\right|\right\}\|h\|_{L^{p'}(\Omega)}. \tag{6.6.10}$$

By duality this implies that

$$\|\nabla u_\varepsilon\|_{L^p(\Omega)} \le C\,\|f\|_{L^p(\Omega)} + C\left|\int_\Omega \left(\nabla u_\varepsilon - (\nabla u_\varepsilon)^T\right) dx\right|. \tag{6.6.11}$$

We may eliminate the last term in the inequality above by subtracting an element of $\mathcal{R}$ from u_ε. The duality argument above also gives the uniqueness in $W^{1,p}(\Omega; \mathbb{R}^d)$, up to an element in $\mathcal{R}$. Consequently, by a density argument, for any $f \in L^p(\Omega; \mathbb{R}^d)$ satisfying (6.6.6), there exists a weak solution u_ε of (6.6.4) in $W^{1,p}(\Omega; \mathbb{R}^d)$ such that $\|\nabla u_\varepsilon\|_{L^p(\Omega)} \le C\|f\|_{L^p(\Omega)}$. The solution is unique in $W^{1,p}(\Omega; \mathbb{R}^d)$, up to an element of $\mathcal{R}$.

Step 3. Let $1 < p < \infty$. Consider the Neumann problem

$$\mathcal{L}_\varepsilon(u_\varepsilon) = 0 \quad \text{in } \Omega \quad \text{and} \quad \frac{\partial u_\varepsilon}{\partial \nu_\varepsilon} = g \quad \text{on } \partial\Omega, \tag{6.6.12}$$

where $g \in B^{-1/p,p}(\partial\Omega;\mathbb{R}^d)$ satisfies the compatibility condition

$$\langle g,\phi\rangle_{B^{-1/p,p}(\partial\Omega)\times B^{1/p,p'}(\partial\Omega)} = 0 \tag{6.6.13}$$

for any $\phi \in \mathcal{R}$. Then there exists a weak solution of (6.6.12) in $W^{1,p}(\Omega;\mathbb{R}^d)$, unique up to an element of $\mathcal{R}$, such that

$$\|\nabla u_\varepsilon\|_{L^p(\Omega)} \le C\,\|g\|_{B^{-1/p,p}(\partial\Omega)}. \tag{6.6.14}$$

This follows from Steps 1 and 2 by a duality argument, similar to that in the proof of Lemma 6.3.4.

Step 4. Let $1 < p < \infty$. Consider the Neumann problem

$$\mathcal{L}_\varepsilon(u_\varepsilon) = F \quad \text{in } \Omega \quad \text{and} \quad \frac{\partial u_\varepsilon}{\partial \nu_\varepsilon} = h \quad \text{on } \partial\Omega, \tag{6.6.15}$$

where $F \in L^q(\Omega;\mathbb{R}^d)$, q is given by (6.3.15), and $h \in \mathcal{R}$ satisfies the compatibility condition

$$\int_\Omega F\cdot\phi\,dx + \int_{\partial\Omega} h\cdot\phi\,d\sigma = 0 \tag{6.6.16}$$

for any $\phi \in \mathcal{R}$. We first show that for any $F \in L^1(\Omega;\mathbb{R}^d)$, there exists a unique $h \in \mathcal{R}$ such that h satisfies (6.6.16) and

$$\|h\|_{C^1(\Omega)} \le C\,\|F\|_{L^1(\Omega)}, \tag{6.6.17}$$

where C depends only on Ω. Indeed, let $\{\phi_1,\phi_2,\dots,\phi_N\}$ be an orthonormal basis of $\mathcal{R}$ in $L^2(\Omega;\mathbb{R}^d)$, where $N = \frac{d(d+1)}{2}$, and let

$$h = \alpha_1\phi_1 + \alpha_2\phi_2 + \cdots + \alpha_N\phi_N,$$

where $(\alpha_1,\alpha_2,\dots,\alpha_N) \in \mathbb{R}^N$. To find $(\alpha_1,\alpha_2,\dots,\alpha_N)$, we solve the $N\times N$ system of linear equations

$$\alpha_i \int_{\partial\Omega} \phi_i\cdot\phi_j\,d\sigma = -\int_\Omega F\cdot\phi_j\,dx, \quad j = 1,2,\dots,N,$$

which is uniquely solvable, provided that

$$\det\left(\int_{\partial\Omega} \phi_i\cdot\phi_j\,d\sigma\right) \neq 0. \tag{6.6.18}$$

The proof of (6.6.18) is similar to that of (6.6.2). Suppose (6.6.18) does not hold. Then there exists $(\beta_1,\beta_2,\dots,\beta_N) \in \mathbb{R}^N$ such that

$$\beta_i \int_{\partial\Omega} \phi_i\cdot\phi_j\,d\sigma = 0.$$

Let $v = \beta_1\phi_1 + \cdots + \beta_N\phi_N \in \mathcal{R}$. Then

$$\int_{\partial\Omega} |v|^2 \, d\sigma = 0,$$

which implies that $v = 0$ in $\partial\Omega$. Since v is a linear function and $\partial\Omega$ is not a hyperplane, we obtain $v \equiv 0$ in $\mathbb{R}^d$. By the linearly independence of $\phi_1, \phi_2, \dots, \phi_N$ in $L^2(\Omega; \mathbb{R}^d)$, this leads to $\beta_1 = \beta_2 = \cdots = \beta_N = 0$ and gives us a contradiction.

With the construction of h, the argument in the proof of Lemma 6.3.4 shows that there exists a weak solution of (6.6.15) in $W^{1,p}(\Omega; \mathbb{R}^d)$, unique up to an element of $\mathcal{R}$, such that

$$\|\nabla u_\varepsilon\|_{L^p(\Omega)} \le C \, \|F\|_{L^q(\Omega)}. \tag{6.6.19}$$

Step 5. Theorem 6.6.4 follows from steps 1-5 by writing

$$u_\varepsilon = u_\varepsilon^{(1)} + u_\varepsilon^{(2)} + u_\varepsilon^{(3)},$$

where $u_\varepsilon^{(1)}$ is a solution of (6.6.4) with datum f, $u_\varepsilon^{(2)}$ is a solution of (6.6.14) with datum $g - h$, and $u_\varepsilon^{(3)}$ is a solution of (6.6.15) with datum F and h. In view of (6.6.16), the boundary datum $g - h$ satisfies the compatibility condition (6.6.13), and by (6.6.19),

$$\|g - h\|_{B^{-1/p,p}(\partial\Omega)} \le C\Big\{ \|g\|_{B^{-1/p,p}(\partial\Omega)} + \|F\|_{L^1(\Omega)} \Big\}.$$

It follows that

$$\begin{aligned}
\|\nabla u_\varepsilon\|_{L^p(\Omega)} &\le \|\nabla u_\varepsilon^{(1)}\|_{L^p(\Omega)} + \|\nabla u_\varepsilon^{(2)}\|_{L^p(\Omega)} + \|\nabla u_\varepsilon^{(3)}\|_{L^p(\Omega)} \\
&\le C\Big\{ \|f\|_{L^p(\Omega)} + \|F\|_{L^q(\Omega)} + \|g\|_{B^{-1/p,p}(\partial\Omega)} \Big\}.
\end{aligned}$$

This completes the proof of Theorem 6.6.4. □

The next theorem gives the boundary Lipschitz estimate for the Neumann problem in a $C^{1,\alpha}$ domain.

Theorem 6.6.5. *Suppose that $A \in E(\kappa_1, \kappa_2)$ is 1-periodic and satisfies (4.0.2). Let Ω be a bounded $C^{1,\alpha}$ domain for some $\alpha \in (0,1)$. Let $u_\varepsilon \in H^1(B(x_0, r) \cap \Omega; \mathbb{R}^d)$ be a weak solution of $\mathcal{L}_\varepsilon(u_\varepsilon) = F$ in $B(x_0, r) \cap \Omega; \mathbb{R}^d)$ with $\frac{\partial u_\varepsilon}{\partial \nu_\varepsilon} = g$ on $B(x_0, r) \cap \partial\Omega$, for some $x_0 \in \partial\Omega$ and $0 < r < r_0$. Then the estimate (6.4.2) holds for $\rho \in (0,1)$ and $p > d$, where C depends only on ρ, p, κ_1, κ_2, (λ, τ) in (4.0.2), and Ω.*

Proof. With Theorem 6.6.2, the proof is the same as that of Theorem 6.4.2. □

Corollary 6.6.6. *Assume that A and Ω satisfy the same conditions as in Theorem 6.6.5. Let $p > d$ and $\rho \in (0,1)$. Let $u_\varepsilon \in H^1(\Omega; \mathbb{R}^d)$ be a weak solution to the Neumann problem,*

$$\mathcal{L}_\varepsilon(u_\varepsilon) = F \quad \text{in } \Omega \quad \text{and} \quad \frac{\partial u_\varepsilon}{\partial \nu_\varepsilon} = g \quad \text{on } \partial\Omega,$$

where $F \in L^p(\Omega; \mathbb{R}^d)$ *and* $g \in C^\rho(\partial\Omega; \mathbb{R}^d)$ *satisfy the compatibility condition* (2.4.9). *Assume that* $u_\varepsilon \perp \mathcal{R}$. *Then*

$$\|\nabla u_\varepsilon\|_{L^\infty(\Omega)} \le C\Big\{ \|F\|_{L^p(\Omega)} + \|g\|_{C^\rho(\partial\Omega)} \Big\}, \tag{6.6.20}$$

where C *depends only on* p, ρ, κ_1, κ_2, (λ, τ) *and* Ω.

Proof. The proof is the same as that of Corollary 6.4.3. □

6.7 Notes

The boundary Hölder and $W^{1,p}$ estimates for solutions with Neumann conditions were proved by C. Kenig, F. Lin, and Z. Shen in [65]. Under the additional symmetry condition $A^* = A$, the boundary Lipschitz estimate for Neumann problems was also established in [65]. This was achieved by using a compactness method, similar to that used in Chapter 5 for the boundary Lipschitz estimate for solutions with Dirichlet conditions. As we pointed out earlier, the compactness method reduces the problem to the Lipschitz estimate for the correctors. In the case of Neumann correctors, the Lipschitz estimate was obtained in [65] by utilizing the nontangential-maximal-function estimates in [67, 68] by C. Kenig and Z. Shen for solutions to Neumann problems with L^2 boundary data (see Chapter 8). As results in [67, 68] were proved in Lipschitz domains under the symmetry condition, so were the main estimates in [65].

The symmetry condition was removed by S.N. Armstrong and Z. Shen in [6], where the interior and boundary Lipschitz estimates were established for second-order elliptic systems with uniformly almost-periodic coefficients. The approach used in [6] was developed by S.N. Armstrong and C. Smart in [7] for the study of large-scale regularity in stochastic homogenization. Also see related work in [51, 5, 2, 3, 52] and references therein.

The presentation in this chapter follows closely [90] by Z. Shen, where the boundary regularity estimates were studied for elliptic systems of elasticity with periodic coefficients. In particular, Lemma 6.4.6, which improves the analogous results in [7, 6], is taken from [90].

Chapter 7

Convergence Rates, Part II

In Chapter 3 we establish the $O(\sqrt{\varepsilon})$ error estimates for some two-scale expansions in H^1 and the $O(\varepsilon)$ convergence rate for solutions u_ε in L^2. The results are obtained without any smoothness assumption on the coefficient matrix A. In this chapter we return to the problem of convergence rates and prove various results under some additional smoothness assumptions, using uniform regularity estimates obtained in Chapters 4–6. We shall be mainly interested in the sharp $O(\varepsilon)$ or near sharp rates of convergence.

We start out in Section 7.1 with an error estimate in H^1 for a two-scale expansion involving boundary correctors. The result is used in Section 7.2 to study the convergence rates for the Dirichlet eigenvalues of $\mathcal{L}_\varepsilon$. In Section 7.3 we derive asymptotic expansions, as $\varepsilon \to 0$, of Green functions $G_\varepsilon(x,y)$ as well as their derivatives $\nabla_x G_\varepsilon(x,y)$, $\nabla_y G_\varepsilon(x,y)$, and $\nabla_x \nabla_y G_\varepsilon(x,y)$, using the Dirichlet correctors. As a corollary, we also obtain an asymptotic expansion of the Poisson kernel $P_\varepsilon(x,y)$ for $\mathcal{L}_\varepsilon$ in a $C^{2,\alpha}$ domain Ω. Analogous expansions are obtained for Neumann functions $N_\varepsilon(x,y)$ and their derivatives in Section 7.4. Results in Sections 7.3 and 7.4 are used in Section 7.5 to establish convergence rates of $u_\varepsilon - u_0$ in L^p and $u_\varepsilon - u_0 - v_\varepsilon$ in $W^{1,p}$ for $p \neq 2$, where v_ε is a first-order corrector.

7.1 Convergence rates in H^1 and L^2

For solutions of $\mathcal{L}_\varepsilon(u_\varepsilon) = F$ in Ω subject to the Dirichlet condition $u_\varepsilon = f$ or Neumann condition $\frac{\partial u_\varepsilon}{\partial \nu_\varepsilon} = g$ on $\partial\Omega$, it is proved in Chapter 3 that

$$\|u_\varepsilon - u_0 - \varepsilon\chi(x/\varepsilon)\nabla u_0\|_{H^1(\Omega)} \leq C\sqrt{\varepsilon}\,\|u_0\|_{W^{2,d}(\Omega)}, \tag{7.1.1}$$

where Ω is a bounded Lipschitz domain in $\mathbb{R}^d$. If $d \geq 3$ and the corrector χ is bounded, the estimate (7.1.1) is improved to

$$\|u_\varepsilon - u_0 - \varepsilon\chi(x/\varepsilon)\nabla u_0\|_{H^1(\Omega)} \leq C\sqrt{\varepsilon}\,\|u_0\|_{H^2(\Omega)}. \tag{7.1.2}$$

Z. Shen, *Periodic Homogenization of Elliptic Systems*, Operator Theory: Advances and Applications 269, https://doi.org/10.1007/978-3-319-91214-1_7

See Theorems 3.2.7 and 3.3.5. Furthermore, the sharp $O(\varepsilon)$ rate in $L^2(\Omega)$

$$\|u_\varepsilon - u_0\|_{L^2(\Omega)} \le C\varepsilon \, \|u_0\|_{H^2(\Omega)}, \tag{7.1.3}$$

is established in Sections 3.4 and 3.5, if Ω is $C^{1,1}$. See Theorems 3.4.5 and 3.5.4.

Recall that the Dirichlet corrector $\Phi_\varepsilon = \big(\Phi_{\varepsilon,j}^\beta\big)$, with $1 \le j \le d$ and $1 \le \beta \le m$, for $\mathcal{L}_\varepsilon$ in Ω is defined by

$$\mathcal{L}_\varepsilon\big(\Phi_{\varepsilon,j}^\beta\big) = 0 \quad \text{in } \Omega \quad \text{ and } \quad \Phi_{\varepsilon,j}^\beta = P_j^\beta \quad \text{on } \partial\Omega, \tag{7.1.4}$$

where $P_j^\beta(x) = x_j e^\beta$. The following theorem gives the $O(\varepsilon)$ rate of convergence in $H_0^1(\Omega)$ for the Dirichlet problem.

Theorem 7.1.1. *Suppose that A is 1-periodic and satisfies (2.1.2)–(2.1.3). Let Ω be a bounded Lipschitz domain in $\mathbb{R}^d$. Assume that $\chi = (\chi_j^\beta)$ is Hölder continuous and that $\Phi_\varepsilon = \big(\Phi_{\varepsilon,j}^\beta\big)$ is bounded (if $m \ge 2$). Let $u_\varepsilon \in H^1(\Omega; \mathbb{R}^m)$ $(\varepsilon \ge 0)$ be the weak solution of the Dirichlet problem $\mathcal{L}_\varepsilon(u_\varepsilon) = F$ in Ω and $u_\varepsilon = f$ on $\partial\Omega$. Then, if $u_0 \in H^2(\Omega; \mathbb{R}^m)$,*

$$\Big\| u_\varepsilon - u_0 - \Big\{ \Phi_{\varepsilon,j}^\beta - P_j^\beta \Big\} \frac{\partial u_0^\beta}{\partial x_j} \Big\|_{H_0^1(\Omega)} \le C \Big\{ \varepsilon + \|\Phi_\varepsilon - P\|_\infty \Big\} \|\nabla^2 u_0\|_{L^2(\Omega)}, \tag{7.1.5}$$

where $P = (P_j^\beta)$ and C depends only on A and Ω.

The proof of Theorem 7.1.1 uses energy estimates and the formula in the following lemma.

Lemma 7.1.2. *Suppose that $u_\varepsilon \in H^1(\Omega; \mathbb{R}^m)$ and $u_0 \in H^2(\Omega; \mathbb{R}^m)$. Let*

$$w_\varepsilon(x) = u_\varepsilon(x) - u_0(x) - \Big\{ V_{\varepsilon,j}^\beta(x) - P_j^\beta(x) \Big\} \frac{\partial u_0^\beta}{\partial x_j}, \tag{7.1.6}$$

where $V_{\varepsilon,j}^\beta = (V_{\varepsilon,j}^{1\beta}, \dots, V_{\varepsilon,j}^{m\beta}) \in H^1(\Omega; \mathbb{R}^m)$ and $\mathcal{L}_\varepsilon\big(V_{\varepsilon,j}^\beta\big) = 0$ in Ω for each $1 \le j \le d$ and $1 \le \beta \le m$. Assume that $\mathcal{L}_\varepsilon(u_\varepsilon) = \mathcal{L}_0(u_0)$ in Ω. Then

$$\begin{aligned}
\big(\mathcal{L}_\varepsilon(w_\varepsilon)\big)^\alpha = & -\varepsilon \frac{\partial}{\partial x_i} \left\{ \phi_{jik}^{\alpha\gamma}(x/\varepsilon) \frac{\partial^2 u_0^\gamma}{\partial x_j \partial x_k} \right\} \\
& + \frac{\partial}{\partial x_i} \left\{ a_{ij}^{\alpha\beta}(x/\varepsilon) \Big[V_{\varepsilon,k}^{\beta\gamma}(x) - x_k \delta^{\beta\gamma} \Big] \frac{\partial^2 u_0^\gamma}{\partial x_j \partial x_k} \right\} \\
& + a_{ij}^{\alpha\beta}(x/\varepsilon) \frac{\partial}{\partial x_j} \Big[V_{\varepsilon,k}^{\beta\gamma}(x) - x_k \delta^{\beta\gamma} - \varepsilon \chi_k^{\beta\gamma}(x/\varepsilon) \Big] \frac{\partial^2 u_0^\gamma}{\partial x_i \partial x_k},
\end{aligned} \tag{7.1.7}$$

where $\phi = \big(\phi_{kij}^{\alpha\beta}(y)\big)$ is the flux corrector, given by Lemma 3.1.1.

Proof. Note that

$$
\begin{aligned}
a_{ij}^{\alpha\beta}\left(\frac{x}{\varepsilon}\right)\frac{\partial w_\varepsilon^\beta}{\partial x_j} = a_{ij}^{\alpha\beta}\left(\frac{x}{\varepsilon}\right)\frac{\partial u_\varepsilon^\beta}{\partial x_j} &- a_{ij}^{\alpha\beta}\left(\frac{x}{\varepsilon}\right)\frac{\partial u_0^\beta}{\partial x_j}\\
&- a_{ij}^{\alpha\beta}\left(\frac{x}{\varepsilon}\right)\frac{\partial}{\partial x_j}\left\{V_{\varepsilon,k}^{\beta\gamma} - x_k\delta^{\beta\gamma}\right\}\cdot\frac{\partial u_0^\gamma}{\partial x_k}\\
&- a_{ij}^{\alpha\beta}\left(\frac{x}{\varepsilon}\right)\left\{V_{\varepsilon,k}^{\beta\gamma} - x_k\delta^{\beta\gamma}\right\}\frac{\partial^2 u_0^\gamma}{\partial x_k\partial x_j},
\end{aligned}
$$

and

$$
\begin{aligned}
\left(\mathcal{L}_\varepsilon(w_\varepsilon)\right)^\alpha = \left(\mathcal{L}_\varepsilon(u_\varepsilon)\right)^\alpha - \left(\mathcal{L}_0(u_0)\right)^\alpha &- \frac{\partial}{\partial x_i}\left\{\left[\widehat{a}_{ij}^{\alpha\beta} - a_{ij}^{\alpha\beta}(x/\varepsilon)\right]\frac{\partial u_0^\beta}{\partial x_j}\right\}\\
&+ \left\{\mathcal{L}_\varepsilon(V_{\varepsilon,k}^\gamma - P_k^\gamma)\right\}^\alpha\cdot\frac{\partial u_0^\gamma}{\partial x_k}\\
&+ a_{ij}^{\alpha\beta}(x/\varepsilon)\frac{\partial}{\partial x_j}\left\{V_{\varepsilon,k}^{\beta\gamma} - x_k\delta^{\beta\gamma}\right\}\cdot\frac{\partial^2 u_0^\gamma}{\partial x_i\partial x_k}\\
&+ \frac{\partial}{\partial x_i}\left\{a_{ij}^{\alpha\beta}(x/\varepsilon)\left[V_{\varepsilon,k}^{\beta\gamma} - x_k\delta^{\beta\gamma}\right]\frac{\partial^2 u_0^\gamma}{\partial x_k\partial x_j}\right\}.
\end{aligned}
$$

Using the fact that

$$
\mathcal{L}_\varepsilon\big(V_{\varepsilon,k}^\gamma - P_k^\gamma\big) = -\mathcal{L}_\varepsilon\big(P_k^\gamma\big) = \mathcal{L}_\varepsilon\{\varepsilon\chi_k^\gamma(x/\varepsilon)\},
$$

and $\mathcal{L}_\varepsilon(u_\varepsilon) = \mathcal{L}_0(u_0)$, we obtain

$$
\begin{aligned}
\left(\mathcal{L}_\varepsilon(w_\varepsilon)\right)^\alpha = &\frac{\partial}{\partial x_i}\left\{b_{ij}^{\alpha\beta}(x/\varepsilon)\frac{\partial u_0^\beta}{\partial x_j}\right\}\\
&+ a_{ij}^{\alpha\beta}(x/\varepsilon)\frac{\partial}{\partial x_j}\left\{V_{\varepsilon,k}^{\beta\gamma}(x) - x_k\delta^{\beta\gamma} - \varepsilon\chi_k^{\beta\gamma}(x/\varepsilon)\right\}\cdot\frac{\partial^2 u_0^\gamma}{\partial x_i\partial x_k}\\
&+ \frac{\partial}{\partial x_i}\left\{a_{ij}^{\alpha\beta}(x/\varepsilon)\left[V_{\varepsilon,k}^{\beta\gamma} - x_k\delta^{\beta\gamma}\right]\cdot\frac{\partial^2 u_0^\gamma}{\partial x_k\partial x_j}\right\},
\end{aligned}
\tag{7.1.8}
$$

where the 1-periodic function $b_{ij}^{\alpha\beta}(y)$ is defined by (3.1.1). The formula (7.1.7) now follows from the identity

$$
\frac{\partial}{\partial x_i}\left\{b_{ij}^{\alpha\beta}(x/\varepsilon)\frac{\partial u_0^\beta}{\partial x_j}\right\} = -\varepsilon\frac{\partial}{\partial x_i}\left\{\phi_{kij}^{\alpha\beta}(x/\varepsilon)\frac{\partial^2 u_0^\beta}{\partial x_k\partial x_j}\right\}, \tag{7.1.9}
$$

which is a consequence of (3.1.5). □

Proof of Theorem 7.1.1. Let

$$w_\varepsilon = u_\varepsilon - u_0 - \left\{ \Phi^\beta_{\varepsilon,j} - P^\beta_j \right\} \frac{\partial u^\beta_0}{\partial x_j}.$$

Since Φ_ε is bounded, $\mathcal{L}_\varepsilon(\Phi_\varepsilon) = 0$ and $u_0 \in H^2(\Omega; \mathbb{R}^m)$, we can use the argument in the proof of Lemma 3.2.8 to show that $|\nabla \Phi_\varepsilon||\nabla u_0| \in L^2(\Omega)$. This implies that $w_\varepsilon \in H^1_0(\Omega; \mathbb{R}^m)$. It follows from Lemma 7.1.2 that

$$\begin{aligned}
\big(\mathcal{L}_\varepsilon(w_\varepsilon)\big)^\alpha = & -\varepsilon \frac{\partial}{\partial x_i} \left\{ \phi^{\alpha\gamma}_{jik}(x/\varepsilon) \frac{\partial^2 u^\gamma_0}{\partial x_j \partial x_k} \right\} \\
& + \frac{\partial}{\partial x_i} \left\{ a^{\alpha\beta}_{ij}(x/\varepsilon) \left[\Phi^{\beta\gamma}_{\varepsilon,k}(x) - x_k \delta^{\beta\gamma} \right] \frac{\partial u^\gamma_0}{\partial x_j \partial x_k} \right\} \\
& + a^{\alpha\beta}_{ij}(x/\varepsilon) \frac{\partial}{\partial x_j} \left[\Phi^{\beta\gamma}_{\varepsilon,k}(x) - x_k \delta^{\beta\gamma} - \varepsilon \chi^{\beta\gamma}_k(x/\varepsilon) \right] \frac{\partial^2 u^\gamma_0}{\partial x_i \partial x_k}.
\end{aligned}$$

Hence,

$$\begin{aligned}
\int_\Omega |\nabla w_\varepsilon|^2 \, dx \le C\varepsilon & \int_\Omega |\phi(x/\varepsilon)| \, |\nabla^2 u_0| \, |\nabla w_\varepsilon| \, dx \\
& + C \int_\Omega |\Phi_\varepsilon - P| \, |\nabla^2 u_0| \, |\nabla w_\varepsilon| \, dx \\
& + C \int_\Omega |\nabla \big(\Phi_\varepsilon - P - \varepsilon \chi(x/\varepsilon) \big)| \, |w_\varepsilon| \, |\nabla^2 u_0| \, dx,
\end{aligned} \tag{7.1.10}$$

where $\phi = \big(\phi^{\alpha\beta}_{jik}\big)$. Since χ is Hölder continuous, the flux corrector ϕ is bounded. Recall that

$$\mathcal{L}_\varepsilon\big(\Phi_\varepsilon - P - \varepsilon\chi(x/\varepsilon)\big) = 0 \quad \text{in } \Omega.$$

Thus, by (2.1.7),

$$\begin{aligned}
& \int_\Omega |\nabla \big(\Phi_\varepsilon - P - \varepsilon \chi(x/\varepsilon) \big)|^2 |w_\varepsilon|^2 \, dx \\
& \qquad \le C \int_\Omega |\Phi_\varepsilon - P - \varepsilon \chi(x/\varepsilon)|^2 |\nabla w_\varepsilon|^2 \, dx.
\end{aligned}$$

This, together with (7.1.10) and the inequality (2.1.10), gives (7.1.5). □

Corollary 7.1.3. *Let $m = 1$ and Ω be a bounded Lipschitz domain in $\mathbb{R}^d$. Let u_ε $(\varepsilon \ge 0)$ be the same as in Theorem* 7.1.1. *Then*

$$\left\| u_\varepsilon - u_0 - \left\{ \Phi_{\varepsilon,j} - x_j \right\} \frac{\partial u_0}{\partial x_j} \right\|_{H^1_0(\Omega)} \le C\varepsilon \, \|\nabla^2 u_0\|_{L^2(\Omega)}. \tag{7.1.11}$$

Consequently,

$$\|u_\varepsilon - u_0\|_{L^2(\Omega)} \le C\varepsilon \, \|u_0\|_{H^2(\Omega)}, \tag{7.1.12}$$

where C depends only on μ and Ω.

Proof. In the scalar case $m = 1$, the corrector χ is Hölder continuous. Also note that if

$$v_\varepsilon = \Phi_{\varepsilon,j}(x) - x_j - \varepsilon\chi_j(x/\varepsilon),$$

then $\mathcal{L}_\varepsilon(v_\varepsilon) = 0$ in Ω and $v_\varepsilon = -\varepsilon\chi_j(x/\varepsilon)$ on $\partial\Omega$. Thus, by the maximum principle,

$$\|v_\varepsilon\|_{L^\infty(\Omega)} \le \|v_\varepsilon\|_{L^\infty(\partial\Omega)} \le C\varepsilon,$$

which gives

$$\|\Phi_{\varepsilon,j} - x_j\|_{L^\infty(\Omega)} \le C\varepsilon. \tag{7.1.13}$$

Thus the estimate (7.1.11) follows from (7.1.5). Finally, since

$$\Big\|\big(\Phi_{\varepsilon,j} - x_j\big)\frac{\partial u_0}{\partial x_j}\Big\|_{L^2(\Omega)} \le C\varepsilon\,\|\nabla u_0\|_{L^2(\Omega)},$$

the estimate (7.1.12) follows readily from (7.1.11). □

Corollary 7.1.4. *Suppose that $m \ge 2$ and Ω is $C^{1,\alpha}$ for some $\alpha \in (0,1)$. Assume that A is Hölder continuous. Let u_ε $(\varepsilon \ge 0)$ be the same as in Theorem 7.1.1. Then,*

$$\Big\|u_\varepsilon - u_0 - \Big\{\Phi^\beta_{\varepsilon,j} - P^\beta_j\Big\}\frac{\partial u^\beta_0}{\partial x_j}\Big\|_{H^1_0(\Omega)} \le C\varepsilon\,\|\nabla^2 u_0\|_{L^2(\Omega)}, \tag{7.1.14}$$

where C depends only on A and Ω.

Proof. Under the assumptions that A is Hölder continuous and Ω is $C^{1,\alpha}$, it follows by the Agmon–Miranda maximum principle in Remark 5.6.6 that

$$\|\Phi^\beta_{\varepsilon,j} - P^\beta_j\|_{L^\infty(\Omega)} \le C\varepsilon, \tag{7.1.15}$$

which, together with (7.1.5), gives the estimate (7.1.14). □

Remark 7.1.5. Since

$$\begin{aligned}
&\frac{\partial}{\partial x_i}\left\{u_\varepsilon - u_0 - \Big\{\Phi^\beta_{\varepsilon,j} - P^\beta_j\Big\}\frac{\partial u^\beta_0}{\partial x_j}\right\}\\
&\quad= \frac{\partial u_\varepsilon}{\partial x_i} - \frac{\partial}{\partial x_i}\Big\{\Phi^\beta_{\varepsilon,j}\Big\}\cdot\frac{\partial u^\beta_0}{\partial x_j} - \Big\{\Phi^\beta_{\varepsilon,j} - P^\beta_j\Big\}\frac{\partial^2 u^\beta_0}{\partial x_i \partial x_j},
\end{aligned}$$

it follows from (7.1.11) and (7.1.14) that

$$\Big\|\frac{\partial u_\varepsilon}{\partial x_i} - \frac{\partial}{\partial x_i}\Big\{\Phi^\beta_{\varepsilon,j}\Big\}\cdot\frac{\partial u^\beta_0}{\partial x_j}\Big\|_{L^2(\Omega)} \le C\varepsilon\,\|\nabla^2 u_0\|_{L^2(\Omega)}, \tag{7.1.16}$$

where C depends only on A and Ω.

7.2 Convergence rates of eigenvalues

In this section we study the convergence rate of Dirichlet eigenvalues for the operator $\mathcal{L}_\varepsilon$. Throughout the section we assume that A is 1-periodic and satisfies the ellipticity condition (2.1.2)–(2.1.3) and the symmetry condition $A^* = A$.

For $f \in L^2(\Omega; \mathbb{R}^m)$, under the ellipticity condition (2.1.2)–(2.1.3), the elliptic system $\mathcal{L}_\varepsilon(u_\varepsilon) = f$ in Ω has a unique (weak) solution in $H_0^1(\Omega; \mathbb{R}^m)$. Define $T_\varepsilon^{\mathrm{D}}(f) = u_\varepsilon$. Note that

$$\langle T_\varepsilon^{\mathrm{D}}(f), f\rangle = \langle u_\varepsilon, f\rangle = \int_\Omega A(x/\varepsilon)\nabla u_\varepsilon \cdot \nabla u_\varepsilon \, dx \tag{7.2.1}$$

(if $\varepsilon = 0$, $A(x/\varepsilon)$ is replaced by $\widehat{A}$), where $\langle\, , \rangle$ denotes the inner product in $L^2(\Omega; \mathbb{R}^m)$. Since

$$\|u_\varepsilon\|_{H_0^1(\Omega)} \le C\, \|f\|_{L^2(\Omega)},$$

where C depends only on μ and Ω, the linear operator $T_\varepsilon^{\mathrm{D}}$ is bounded, positive, and compact on $L^2(\Omega; \mathbb{R}^m)$. With the symmetry condition $A^* = A$, the operator T_ε is also self-adjoint. Let

$$\sigma_{\varepsilon,1} \ge \sigma_{\varepsilon,2} \ge \cdots \ge \sigma_{\varepsilon,k} \ge \cdots > 0 \tag{7.2.2}$$

denote the sequence of eigenvalues in a decreasing order of $T_\varepsilon^{\mathrm{D}}$. Recall that λ_ε is called a Dirichlet eigenvalue of $\mathcal{L}_\varepsilon$ in Ω if there exists a nonzero $u_\varepsilon \in H_0^1(\Omega; \mathbb{R}^m)$ such that $\mathcal{L}_\varepsilon(u_\varepsilon) = \lambda_\varepsilon u_\varepsilon$ in Ω. Thus, if A is elliptic and symmetric, for each $\varepsilon > 0$,

$$\{\lambda_{\varepsilon,k} = (\sigma_{\varepsilon,k})^{-1}\}$$

forms the sequence of Dirichlet eigenvalues in an increasing order of $\mathcal{L}_\varepsilon$ in Ω.

By the mini-max principle,

$$\sigma_{\varepsilon,k} = \min_{\substack{f_1,\cdots,f_{k-1}\\ \in L^2(\Omega;\mathbb{R}^m)}} \max_{\substack{\|f\|_{L^2(\Omega)}=1\\ f\perp f_i\\ i=1,\dots,k-1}} \langle T_\varepsilon^{\mathrm{D}}(f), f\rangle. \tag{7.2.3}$$

Let $\{\phi_{\varepsilon,k}\}$ be an orthonormal basis of $L^2(\Omega; \mathbb{R}^m)$, where each $\phi_{\varepsilon,k}$ is an eigenfunction associated with $\sigma_{\varepsilon,k}$. Let $V_{\varepsilon,0} = \{0\}$ and $V_{\varepsilon,k}$ be the subspace of $L^2(\Omega; \mathbb{R}^m)$ spanned by $\{\phi_{\varepsilon,1}, \dots, \phi_{\varepsilon,k}\}$ for $k \ge 1$. Then

$$\sigma_{\varepsilon,k} = \max_{\substack{f\perp V_{\varepsilon,k-1}\\ \|f\|_{L^2(\Omega)}=1}} \langle T_\varepsilon^{\mathrm{D}}(f), f\rangle. \tag{7.2.4}$$

Lemma 7.2.1. *For any* $\varepsilon > 0$,

$$|\sigma_{\varepsilon,k} - \sigma_{0,k}| \le$$

$$\max\left\{ \max_{\substack{f\perp V_{0,k-1}\\ \|f\|_{L^2(\Omega)}=1}} \langle (T_\varepsilon^{\mathrm{D}} - T_0^{\mathrm{D}})f, f\rangle|, \max_{\substack{f\perp V_{\varepsilon,k-1}\\ \|f\|_{L^2(\Omega)}=1}} \langle (T_\varepsilon^{\mathrm{D}} - T_0^{\mathrm{D}})f, f\rangle| \right\}.$$

Proof. It follows from (7.2.3) that

$$\begin{aligned}
\sigma_{\varepsilon,k} &\le \max_{\substack{f \perp V_{0,k-1} \\ \|f\|_{L^2(\Omega)}=1}} \langle T_\varepsilon^{\mathrm{D}}(f), f\rangle \\
&\le \max_{\substack{f \perp V_{0,k-1} \\ \|f\|_{L^2(\Omega)}=1}} \langle (T_\varepsilon^{\mathrm{D}} - T_0^{\mathrm{D}})(f), f\rangle + \max_{\substack{f \perp V_{0,k-1} \\ \|f\|_{L^2(\Omega)}=1}} \langle T_0^{\mathrm{D}}(f), f\rangle \\
&= \max_{\substack{f \perp V_{0,k-1} \\ \|f\|_{L^2(\Omega)}=1}} \langle (T_\varepsilon^{\mathrm{D}} - T_0^{\mathrm{D}})(f), f\rangle + \sigma_{0,k},
\end{aligned}$$

where we have used (7.2.4). Hence,

$$\sigma_{\varepsilon,k} - \sigma_{0,k} \le \max_{\substack{f \perp V_{0,k-1} \\ \|f\|_{L^2(\Omega)}=1}} \langle (T_\varepsilon^{\mathrm{D}} - T_0^{\mathrm{D}})(f), f\rangle. \tag{7.2.5}$$

Similarly, one can show that

$$\sigma_{0,k} - \sigma_{\varepsilon,k} \le \max_{\substack{f \perp V_{\varepsilon,k-1} \\ \|f\|_{L^2(\Omega)}=1}} \langle (T_0^{\mathrm{D}} - T_\varepsilon^{\mathrm{D}})(f), f\rangle. \tag{7.2.6}$$

The desired estimate follows from (7.2.5) and (7.2.6). □

By Theorem 3.4.3,

$$\|T_\varepsilon^{\mathrm{D}}(f) - T_0^{\mathrm{D}}(f)\|_{L^2(\Omega)} \le C\varepsilon \, \|T_\varepsilon^{\mathrm{D}}(f)\|_{H^2(\Omega)},$$

where Ω is a bounded Lipschitz domain. If Ω is $C^{1,1}$ (or convex in the case $m = 1$), the H^2 estimate

$$\|T_0^{\mathrm{D}}(f)\|_{H^2(\Omega)} \le C \, \|f\|_{L^2(\Omega)}$$

holds for $\mathcal{L}_0$. It follows that

$$\|T_\varepsilon^{\mathrm{D}} - T_0^{\mathrm{D}}\|_{L^2 \to L^2} \le C\varepsilon. \tag{7.2.7}$$

In view of Lemma 7.2.1, this gives

$$|\sigma_{\varepsilon.k} - \sigma_{0,k}| \le C\varepsilon,$$

where C depends only on Ω and μ. Thus,

$$|\lambda_{\varepsilon,k} - \lambda_{0,k}| \le C\varepsilon \, (\lambda_{0,k})^2. \tag{7.2.8}$$

We will see that the error estimate in $H_0^1(\Omega)$ in Corollaries 7.1.3 and 7.1.4 allows us to improve the estimate (7.2.8) by a factor of $(\lambda_{0,k})^{1/2}$. Note that the smoothness condition on A is not needed in the scalar case $m = 1$ in the following theorem.

Theorem 7.2.2. *Suppose that A is 1-periodic, symmetric, and satisfies the ellipticity condition (2.1.2)–(2.1.3). If $m \ge 2$, we also assume that A is Hölder continuous. Let Ω be a bounded $C^{1,1}$ domain or convex domain in the case $m = 1$. Then*

$$|\lambda_{\varepsilon,k} - \lambda_{0,k}| \le C\varepsilon\, (\lambda_{0,k})^{3/2}, \tag{7.2.9}$$

where C is independent of ε and k.

Proof. We will use Lemma 7.2.1, Corollaries 7.1.3 and 7.1.4 to show that

$$|\sigma_{\varepsilon,k} - \sigma_{0,k}| \le C\varepsilon\, (\sigma_{0,k})^{1/2}, \tag{7.2.10}$$

where C is independent of ε and k. Since $\lambda_{\varepsilon,k} = (\sigma_{\varepsilon,k})^{-1}$ for $\varepsilon \ge 0$ and $\lambda_{\varepsilon,k} \approx \lambda_{0,k}$, this gives the desired estimate.

Let $u_\varepsilon = T_\varepsilon^{\mathrm{D}}(f)$ and $u_0 = T_0^{\mathrm{D}}(f)$, where $\|f\|_{L^2(\Omega)} = 1$ and $f \perp V_{0,k-1}$. In view of (7.2.4) for $\varepsilon = 0$, we have $\langle u_0, f\rangle \le \sigma_{0,k}$. Hence,

$$c\, \|\nabla u_0\|^2_{L^2(\Omega)} \le \langle u_0, f\rangle \le \sigma_{0,k},$$

where c depends only on μ. It follows that

$$\|f\|_{H^{-1}(\Omega)} \le C\, \|\nabla u_0\|_{L^2(\Omega)} \le C\, (\sigma_{0,k})^{1/2}. \tag{7.2.11}$$

Now, write

$$\langle u_\varepsilon - u_0, f\rangle = \Big\langle u_\varepsilon - u_0 - \{\Phi^\beta_{\varepsilon,\ell} - P^\beta_\ell\}\frac{\partial u_0^\beta}{\partial x_\ell}, f\Big\rangle + \Big\langle \{\Phi^\beta_{\varepsilon,\ell} - P^\beta_\ell\}\frac{\partial u_0^\beta}{\partial x_\ell}, f\Big\rangle.$$

This implies that for any $f \perp V_{0,k-1}$ with $\|f\|_{L^2(\Omega)} = 1$,

$$\begin{aligned}
|\langle u_\varepsilon - u_0, f\rangle| &\le \Big\| u_\varepsilon - u_0 - \Big\{\Phi^\beta_{\varepsilon,\ell} - P^\beta_\ell\Big\}\frac{\partial u_0^\beta}{\partial x_\ell}\Big\|_{H_0^1(\Omega)} \|f\|_{H^{-1}(\Omega)}\\
&\quad + \Big\|\Big\{\Phi^\beta_{\varepsilon,\ell} - P^\beta_\ell\Big\}\frac{\partial u_0^\beta}{\partial x_\ell}\Big\|_{L^2(\Omega)} \|f\|_{L^2(\Omega)}\\
&\le C\varepsilon\, \|f\|_{L^2(\Omega)} \|f\|_{H^{-1}(\Omega)} + C\varepsilon\, \|\nabla u_0\|_{L^2(\Omega)} \|f\|_{L^2(\Omega)}\\
&\le C\varepsilon\, \|\nabla u_0\|_{L^2(\Omega)}\\
&\le C\varepsilon\, (\sigma_{0,k})^{1/2},
\end{aligned} \tag{7.2.12}$$

where we have used Corollaries 7.1.3 and 7.1.4 as well as the estimate

$$\|\Phi^\beta_{\varepsilon,\ell} - P^\beta_\ell\|_\infty \le C\varepsilon$$

for the second inequality, and (7.2.11) for the third and fourth.

Next we consider the case $f \perp V_{\varepsilon,k-1}$ and $\|f\|_{L^2(\Omega)} = 1$. In view of (7.2.4), we have $\langle u_\varepsilon, f\rangle \le \sigma_{\varepsilon,k}$. Hence,

$$c\, \|\nabla u_\varepsilon\|^2_{L^2(\Omega)} \le \langle u_\varepsilon, f\rangle \le \sigma_{\varepsilon,k}.$$

It follows that

$$\|f\|_{H^{-1}(\Omega)} \le C \|\nabla u_\varepsilon\|_{L^2(\Omega)} \le C \left(\sigma_{\varepsilon,k}\right)^{1/2}, \tag{7.2.13}$$

and

$$\|\nabla u_0\|_{L^2(\Omega)} \le C \|f\|_{H^{-1}(\Omega)} \le C \left(\sigma_{\varepsilon,k}\right)^{1/2}, \tag{7.2.14}$$

where C depends only on μ. As before, this implies that for any $f \perp V_{\varepsilon,k-1}$ with $\|f\|_{L^2(\Omega)} = 1$,

$$\begin{aligned}
|\langle u_\varepsilon - u_0, f\rangle| &\le \left\| u_\varepsilon - u_0 - \left\{\Phi^\beta_{\varepsilon,\ell} - P^\beta_\ell\right\}\frac{\partial u_0^\beta}{\partial x_\ell}\right\|_{H^1_0(\Omega)} \|f\|_{H^{-1}(\Omega)} \\
&\quad + \left\| \left\{\Phi^\beta_{\varepsilon,\ell} - P^\beta_\ell\right\}\frac{\partial u_0^\beta}{\partial x_\ell}\right\|_{L^2(\Omega)} \|f\|_{L^2(\Omega)} \\
&\le C\varepsilon \|f\|_{H^{-1}(\Omega)} + C\varepsilon \|\nabla u_0\|_{L^2(\Omega)} \\
&\le C\varepsilon \left(\sigma_{\varepsilon,k}\right)^{1/2} \\
&\le C\varepsilon \left(\sigma_{0,k}\right)^{1/2},
\end{aligned} \tag{7.2.15}$$

where we have used the fact $\sigma_{\varepsilon,k} \approx \sigma_{0,k}$. In view of Lemma 7.2.1, the estimate (7.2.10) follows from (7.2.12) and (7.2.15). □

7.3 Asymptotic expansions of Green functions

Assume A satisfies the ellipticity condition (2.1.2)–(2.1.3). Let

$$G_\varepsilon(x, y) = \left(G_\varepsilon^{\alpha\beta}(x, y)\right)$$

denote the $m \times m$ matrix of Green functions for $\mathcal{L}_\varepsilon$ in Ω. Recall that in the scalar case $m = 1$,

$$|G_\varepsilon(x, y)| \le \begin{cases} C\,|x-y|^{2-d} & \text{if } d \ge 3, \\ C\left\{1 + \ln\left(r_0|x-y|^{-1}\right)\right\} & \text{if } d = 2, \end{cases} \tag{7.3.1}$$

for any $x, y \in \Omega$ and $x \ne y$, where Ω is a bounded Lipschitz domain in $\mathbb{R}^d$, $r_0 = \operatorname{diam}(\Omega)$, and C depends only on μ and Ω. The estimate in (7.3.1) for $d = 2$ also holds for $m \ge 2$. No smoothness or periodicity condition is needed in both cases. If A is 1-periodic and belongs to $\text{VMO}(\mathbb{R}^d)$, it follows from the interior and boundary Hölder estimates that the estimate (7.3.1) for $d \ge 3$ holds if $m \ge 2$ and Ω is C^1. Furthermore, if A satisfies the Hölder continuity condition (4.0.2) and Ω is $C^{1,\alpha}$, it is proved in Chapter 5 that

$$\begin{aligned}
|\nabla_x G_\varepsilon(x, y)| + |\nabla_y G_\varepsilon(x, y)| &\le C\,|x-y|^{1-d}, \\
|\nabla_x \nabla_y G_\varepsilon(x, y)| &\le C\,|x-y|^{-d}
\end{aligned} \tag{7.3.2}$$

for any $x, y \in \Omega$ and $x \ne y$, where C depends only on μ, λ, τ, and Ω.

In this section we study the asymptotic behavior, as $\varepsilon \to 0$, of $G_\varepsilon(x,y)$, $\nabla_x G_\varepsilon(x,y)$, $\nabla_y G_\varepsilon(x,y)$, and $\nabla_x\nabla_y G_\varepsilon(x,y)$. We shall use $G_0(x,y) = \big(G_0^{\alpha\beta}(x,y)\big)$ to denote the $m\times m$ matrix of Green functions for the homogenized operator $\mathcal{L}_0$ in Ω.

We begin with a size estimate for $|G_\varepsilon(x,y) - G_0(x,y)|$.

Theorem 7.3.1. *Suppose that A satisfies (2.1.2)–(2.1.3) and is 1-periodic. If $m \ge 2$, we also assume that A is Hölder continuous. Let Ω be a bounded $C^{1,1}$ domain. Then*

$$|G_\varepsilon(x,y) - G_0(x,y)| \le \frac{C\,\varepsilon}{|x-y|^{d-1}} \quad \text{for any } x,y\in\Omega \text{ and } x\ne y, \tag{7.3.3}$$

where C depends only on μ and Ω as well as (λ,τ) (if $m\ge 2$).

The proof of Theorem 7.3.1, which follows the same line of argument for the estimate of $|\Gamma_\varepsilon(x,y) - \Gamma_0(x,y)|$ in Section 4.3, relies on some boundary L^∞ estimates. Define

$$T_r = T(x,r) = B(x,r)\cap\Omega \quad\text{and}\quad I_r = I(x,r) = B(x,r)\cap\partial\Omega \tag{7.3.4}$$

for some $x\in\overline{\Omega}$ and $0<r<r_0 = c_0\,\mathrm{diam}(\Omega)$.

Lemma 7.3.2. *Suppose that A satisfies the same conditions as in Theorem 7.3.1. Assume that Ω is Lipschitz if $m=1$, and $C^{1,\alpha}$ if $m\ge 2$. Then*

$$\|u_\varepsilon\|_{L^\infty(T_r)} \le C\,\|f\|_{L^\infty(I_{3r})} + C\fint_{T_{3r}} |u_\varepsilon|, \tag{7.3.5}$$

where $\mathcal{L}_\varepsilon(u_\varepsilon) = 0$ in T_{3r} and $u_\varepsilon = f$ on I_{3r}.

Proof. By rescaling we may assume that $r=1$. If $f=0$, the estimate is a consequence of (5.2.1). To treat the general case, let v_ε be the solution to $\mathcal{L}_\varepsilon(v_\varepsilon)=0$ in $\widetilde{\Omega}$ with the Dirichlet condition $v_\varepsilon = f$ on $\partial\widetilde{\Omega}\cap\partial\Omega$ and $v_\varepsilon = 0$ on $\partial\widetilde{\Omega}\setminus\partial\Omega$, where $\widetilde{\Omega}$ is a $C^{1,\alpha}$ domain such that $T_2\subset\widetilde{\Omega}\subset T_3$. By the Agmon–Miranda maximum principle in Remark 5.6.6, $\|v_\varepsilon\|_{L^\infty(\widetilde{\Omega})} \le C\|f\|_{L^\infty(I_3)}$. This, together with

$$\begin{aligned}\|u_\varepsilon - v_\varepsilon\|_{L^\infty(T_1)} &\le C\fint_{T_2} |u_\varepsilon - v_\varepsilon|\\ &\le C\fint_{T_3} |u_\varepsilon| + C\,\|f\|_{L^\infty(I_3)},\end{aligned}$$

gives (7.3.5) for the case $m\ge 2$. Finally, we observe that if $m=1$, the L^∞ estimate and the maximum principle used above hold for Lipschitz domains without any smoothness (and periodicity) condition on A. □

Lemma 7.3.3. *Assume that A and Ω satisfy the same conditions as in Lemma 7.3.2. Let $u_\varepsilon\in H^1(T_{4r};\mathbb{R}^m)$ and $u_0\in W^{2,p}(T_{4r};\mathbb{R}^m)$ for some $d<p<\infty$. Suppose that*

$$\mathcal{L}_\varepsilon(u_\varepsilon) = \mathcal{L}_0(u_0) \quad \text{in } T_{4r} \quad\text{and}\quad u_\varepsilon = u_0 \quad \text{on } I_{4r}.$$

Then,

$$\|u_\varepsilon - u_0\|_{L^\infty(T_r)} \le C \fint_{T_{4r}} |u_\varepsilon - u_0| + C\varepsilon \, \|\nabla u_0\|_{L^\infty(T_{4r})} \\ + C_p \, \varepsilon \, r^{1-\frac{d}{p}} \|\nabla^2 u_0\|_{L^p(T_{4r})}. \tag{7.3.6}$$

Proof. By rescaling we may assume that $r = 1$. Choose a domain $\widetilde{\Omega}$ which is Lipschitz for $m = 1$ and $C^{1,\alpha}$ for $m \ge 2$, such that $T_3 \subset \widetilde{\Omega} \subset T_4$. Consider

$$w_\varepsilon = u_\varepsilon - u_0 - \varepsilon \chi_j^\beta(x/\varepsilon) \frac{\partial u_0^\beta}{\partial x_j} = w_\varepsilon^{(1)} + w_\varepsilon^{(2)} \quad \text{in } \widetilde{\Omega},$$

where

$$\mathcal{L}_\varepsilon\big(w_\varepsilon^{(1)}\big) = \mathcal{L}_\varepsilon(w_\varepsilon) \quad \text{in } \widetilde{\Omega} \quad \text{and} \quad w_\varepsilon^{(1)} \in H_0^1(\widetilde{\Omega}; \mathbb{R}^m), \tag{7.3.7}$$

and

$$\mathcal{L}_\varepsilon\big(w_\varepsilon^{(2)}\big) = 0 \quad \text{in } \widetilde{\Omega} \quad \text{and} \quad w_\varepsilon^{(2)} = w_\varepsilon \quad \text{on} \quad \partial\widetilde{\Omega}. \tag{7.3.8}$$

Since $w_\varepsilon^{(2)} = w_\varepsilon = -\varepsilon\chi(x/\varepsilon)\nabla u_0$ on I_3 and $\|\chi\|_\infty \le C$, it follows from Lemma 7.3.2 that

$$\begin{aligned} \|w_\varepsilon^{(2)}\|_{L^\infty(T_1)} &\le C\varepsilon \, \|\nabla u_0\|_{L^\infty(I_3)} + C \fint_{T_3} |w_\varepsilon^{(2)}| \\ &\le C\varepsilon \, \|\nabla u_0\|_{L^\infty(I_3)} + C \fint_{T_3} |w_\varepsilon| + C \fint_{T_3} |w_\varepsilon^{(1)}| \\ &\le C \fint_{T_3} |u_\varepsilon - u_0| + C\varepsilon \, \|\nabla u_0\|_{L^\infty(T_3)} + C \, \|w_\varepsilon^{(1)}\|_{L^\infty(T_3)}. \end{aligned}$$

This gives

$$\|u_\varepsilon - u_0\|_{L^\infty(T_1)} \le C \fint_{T_3} |u_\varepsilon - u_0| + C\varepsilon \, \|\nabla u_0\|_{L^\infty(T_3)} + C \, \|w_\varepsilon^{(1)}\|_{L^\infty(T_3)}. \tag{7.3.9}$$

To estimate $w_\varepsilon^{(1)}$ on T_3, we use the Green function representation

$$w_\varepsilon^{(1)}(x) = \int_{\widetilde{\Omega}} \widetilde{G}_\varepsilon(x, y) \mathcal{L}_\varepsilon(w_\varepsilon)(y) \, dy,$$

where $\widetilde{G}_\varepsilon(x, y)$ denotes the matrix of Green functions for $\mathcal{L}_\varepsilon$ in $\widetilde{\Omega}$. In view of (4.4.14), we obtain

$$w_\varepsilon^{(1)}(x) = \varepsilon \int_{\widetilde{\Omega}} \frac{\partial}{\partial y_i} \Big\{ \widetilde{G}_\varepsilon(x, y) \Big\} \cdot \Big[\phi_{jik}(y/\varepsilon) - a_{ij}(y/\varepsilon)\chi_k(y/\varepsilon) \Big] \cdot \frac{\partial^2 u_0}{\partial y_j \partial y_k} \, dy,$$

where we have suppressed the superscripts for notational simplicity.

Since $\|\phi_{jik}\|_\infty \le C$ and $p > d$, it follows that

$$\begin{aligned}|w_\varepsilon^{(1)}(x)| &\le C\varepsilon \int_{\widetilde{\Omega}} |\nabla_y \widetilde{G}_\varepsilon(x,y)|\, |\nabla^2 u_0(y)|\, dy \\ &\le C\varepsilon\, \|\nabla^2 u_0\|_{L^p(T_4)} \left(\int_{\widetilde{\Omega}} |\nabla_y \widetilde{G}_\varepsilon(x,y)|^{p'}\, dy \right)^{1/p'} \\ &\le C_p\, \varepsilon\, \|\nabla^2 u_0\|_{L^p(T_4)},\end{aligned}$$

where we have used Hölder's inequality and the observation that

$$\|\nabla_y \widetilde{G}_\varepsilon(x,\cdot)\|_{L^{p'}(\widetilde{\Omega})} \le C. \tag{7.3.10}$$

This, together with (7.3.9), gives the estimate (7.3.6). We point out that the estimate (7.3.10) follows from the size estimate (7.3.1) and Caccioppoli's inequality by decomposing Ω as a union of the subsets $\Omega \cap \{y : |y-x| \sim 2^{-\ell}\}$. □

Proof of Theorem 7.3.1. We first note that under the assumptions on A and Ω in the theorem, the size estimate (7.3.1) and $|\nabla_x G_0(x,y)| \le C|x-y|^{1-d}$ hold for any $x, y \in \Omega$ and $x \ne y$. We now fix $x_0, y_0 \in \Omega$ and $r = |x_0 - y_0|/8 > 0$. For $F \in C_0^\infty(T(y_0,r); \mathbb{R}^m)$, let

$$u_\varepsilon(x) = \int_\Omega G_\varepsilon(x,y) F(y)\, dy \quad \text{and} \quad u_0(x) = \int_\Omega G_0(x,y) F(y)\, dy.$$

Then $\mathcal{L}_\varepsilon(u_\varepsilon) = \mathcal{L}_0(u_0) = F$ in Ω and $u_\varepsilon = u_0 = 0$ on $\partial\Omega$. Note that since Ω is $C^{1,1}$,

$$\begin{aligned}&\|\nabla^2 u_0\|_{L^p(\Omega)} \le C_p\, \|F\|_{L^p(\Omega)} && \text{for } 1 < p < \infty, \\ &\|\nabla u_0\|_{L^\infty(\Omega)} \le C_p\, r^{1-\frac{d}{p}} \|F\|_{L^p(T(y_0,r))} && \text{for } p > d.\end{aligned} \tag{7.3.11}$$

The first inequality in (7.3.11) is the $W^{2,p}$ estimate in $C^{1,1}$ domains for second-order elliptic systems with constant coefficients[3], while the second follows from the estimate $|\nabla_x G_0(x,y)| \le C\, |x-y|^{1-d}$ by Hölder's inequality.

Next, let

$$w_\varepsilon = u_\varepsilon - u_0 - \varepsilon \chi_j^\beta(x/\varepsilon) \frac{\partial u_0^\beta}{\partial x_j} = \theta_\varepsilon(x) + z_\varepsilon(x),$$

where $\theta_\varepsilon \in H_0^1(\Omega; \mathbb{R}^m)$ and $\mathcal{L}_\varepsilon(\theta_\varepsilon) = \mathcal{L}_\varepsilon(w_\varepsilon)$ in Ω. Observe that, by the formula (4.4.14) for $\mathcal{L}_\varepsilon(w_\varepsilon)$,

$$\|\nabla \theta_\varepsilon\|_{L^2(\Omega)} \le C\varepsilon\, \|\nabla^2 u_0\|_{L^2(\Omega)} \le C\varepsilon\, \|F\|_{L^2(T(y_0,r))},$$

[3]The proof for $p = 2$ can be found in [50, 76]. The proof for $1 < p < \infty$ is similar and uses the $W^{1,p}$ estimates in the place of energy estimates.

where we have used the fact that χ and ϕ are bounded. By Hölder and Sobolev inequalities, this implies that if $d \geq 3$,

$$\begin{aligned}\|\theta_\varepsilon\|_{L^2(T(x_0,r))} &\leq Cr\,\|\theta_\varepsilon\|_{L^q(\Omega)}\\ &\leq Cr\,\|\nabla\theta_\varepsilon\|_{L^2(\Omega)}\\ &\leq C\varepsilon\, r^{1+\frac{d}{2}-\frac{d}{p}}\,\|F\|_{L^p(T(y_0,r))},\end{aligned}\tag{7.3.12}$$

where $\frac{1}{q}=\frac{1}{2}-\frac{1}{d}$ and $p>d$. We point out that if $d=2$, one has

$$\|\theta_\varepsilon\|_{L^2(T(x_0,r))} \leq C\varepsilon\, r\,\|F\|_{L^2(T(y_0,r))}$$

in place of (7.3.12). To see this, we use the fact that the $W^{1,p}$ estimate holds for $\mathcal{L}_\varepsilon$ for p close to 2, even without the smoothness assumption on A (see Remark 4.3.6). Thus there exists some $\bar{p}<2$ such that

$$\|\nabla\theta_\varepsilon\|_{L^{\bar{p}}(\Omega)} \leq C\varepsilon\,\|\nabla^2 u_0\|_{L^{\bar{p}}(\Omega)} \leq C\varepsilon\,\|F\|_{L^{\bar{p}}(T(y_0,r))},$$

which, by the Hölder inequality and the Sobolev inequality, leads to

$$\begin{aligned}\|\theta_\varepsilon\|_{L^2(T(x_0,r))} &\leq Cr^{1-\frac{2}{q}}\|\theta_\varepsilon\|_{L^q(T(x_0,r))} \leq Cr^{1-\frac{2}{q}}\|\nabla\theta_\varepsilon\|_{L^{\bar{p}}(\Omega)}\\ &\leq Cr^{2-\frac{2}{\bar{p}}}\|\nabla\theta_\varepsilon\|_{L^{\bar{p}}(\Omega)} \quad \leq C\varepsilon\, r^{2-\frac{2}{\bar{p}}}\|F\|_{L^{\bar{p}}(T(y_0,r))}\\ &\leq C\varepsilon\, r\|F\|_{L^2(T(y_0,r))},\end{aligned}\tag{7.3.13}$$

where $\frac{1}{q}=\frac{1}{\bar{p}}-\frac{1}{2}$.

Observe that since $\mathcal{L}_\varepsilon(z_\varepsilon)=0$ in Ω and $z_\varepsilon=w_\varepsilon$ on $\partial\Omega$, by the maximum principle (5.6.18),

$$\|z_\varepsilon\|_{L^\infty(\Omega)} \leq C\,\|z_\varepsilon\|_{L^\infty(\partial\Omega)} \leq C\varepsilon\,\|\nabla u_0\|_{L^\infty(\partial\Omega)}.\tag{7.3.14}$$

In view of (7.3.11)–(7.3.14), we obtain

$$\begin{aligned}\|u_\varepsilon-u_0\|_{L^2(T(x_0,r))} &\leq \|\theta_\varepsilon\|_{L^2(T(x_0,r))}+\|z_\varepsilon\|_{L^2(T(x_0,r))}+C\varepsilon\, r^{\frac{d}{2}}\,\|\nabla u_0\|_{L^\infty(\Omega)}\\ &\leq \|\theta_\varepsilon\|_{L^2(T(x_0,r))}+C\varepsilon\, r^{\frac{d}{2}}\,\|\nabla u_0\|_{L^\infty(\Omega)}\\ &\leq C\varepsilon\, r^{1+\frac{d}{2}-\frac{d}{p}}\,\|f\|_{L^p(T(y_0,r))},\end{aligned}$$

where $p>d$. This, together with Lemma 7.3.3 and (7.3.11), gives

$$|u_\varepsilon(x_0)-u_0(x_0)| \leq C\varepsilon\, r^{1-\frac{d}{p}}\|f\|_{L^p(T(y_0,r))}.$$

It then follows by duality that

$$\left(\int_{T(y_0,r)} |G_\varepsilon(x_0,y)-G_0(x_0,y)|^{p'}\,dy\right)^{1/p'} \leq C_p\,\varepsilon\, r^{1-\frac{d}{p}} \quad \text{for any } p>d.$$

Finally, since $\mathcal{L}_\varepsilon^*\big(G_\varepsilon(x_0,\cdot)\big) = \mathcal{L}_0^*\big(G_0(x_0,\cdot)\big) = 0$ in $T(y_0,r)$, we may invoke Lemma 7.3.3 again to conclude that

$$
\begin{aligned}
|G_\varepsilon(x_0,y_0) - G_0(x_0,y_0)| &\le \fint_{T(y_0,r)} |G_\varepsilon(x_0,y) - G_0(x_0,y)|\, dy \\
&\qquad + C\varepsilon\, \|\nabla_y G_0(x_0,\cdot)\|_{L^\infty(T(y_0,r))} \\
&\qquad + C_p\, \varepsilon\, r^{1-\frac{d}{p}}\, \|\nabla_y^2 G_0(x_0,\cdot)\|_{L^p(T(y_0,r))} \\
&\le C\varepsilon\, r^{1-d},
\end{aligned}
$$

where we have used that

$$
\begin{aligned}
\left(\fint_{T(y_0,r)} |\nabla_y^2 G_0(x_0,y)|^p\, dy \right)^{1/p} &\le C_p\, r^{-2} \|G_0(x_0,\cdot)\|_{L^\infty(T(y_0,2r))} \\
&\le C_p\, r^{-d},
\end{aligned}
$$

which in turn follows by using the boundary $W^{2,p}$estimates on $C^{1,1}$ domains for $\mathcal{L}_0^*$. □

The next theorem gives an asymptotic expansion of $\nabla_x G_\varepsilon(x,y)$. Recall that $\big(\Phi_{\varepsilon,j}^{\alpha\beta}(x)\big)$ denotes the matrix of Dirichlet correctors for $\mathcal{L}_\varepsilon$ in Ω.

Theorem 7.3.4. *Suppose that A is 1-periodic and satisfies (2.1.2)–(2.1.3). Also assume that A is Hölder continuous. Let Ω be a bounded $C^{2,\eta}$ domain for some $\eta \in (0,1)$. Then*

$$
\begin{aligned}
\Big| \frac{\partial}{\partial x_i} \Big\{ G_\varepsilon^{\alpha\gamma}(x,y) \Big\} - \frac{\partial}{\partial x_i} \Big\{ \Phi_{\varepsilon,j}^{\alpha\beta}(x) \Big\} \cdot \frac{\partial}{\partial x_j} \Big\{ G_0^{\beta\gamma}(x,y) \Big\} \Big| & \\
\le \frac{C\varepsilon \ln[\varepsilon^{-1}|x-y| + 2]}{|x-y|^d} &
\end{aligned}
\tag{7.3.15}
$$

for any $x, y \in \Omega$ and $x \ne y$, where C depends only on μ, (λ,τ), and Ω.

The proof of Theorem 7.3.4 relies on a boundary Lipschitz estimate. The argument is similar to that for the estimate of $\nabla_x \Gamma_\varepsilon(x,y) - \nabla\chi(x/\varepsilon)\cdot\nabla_x\Gamma_\varepsilon(x,y)$ in Section 4.3.

Lemma 7.3.5. *Suppose that A and Ω satisfy the same conditions as in Theorem 7.3.4. Let $u_\varepsilon \in H^1(T_{4r};\mathbb{R}^m)$ and $u_0 \in C^{2,\rho}(T_{4r};\mathbb{R}^m)$ for some $0<\rho<\eta$. Assume that $\mathcal{L}_\varepsilon(u_\varepsilon) = \mathcal{L}_0(u_0)$ in T_{4r} and $u_\varepsilon = u_0$ on I_{4r}. Then, if $0<\varepsilon<r$,*

$$
\begin{aligned}
&\Big\| \frac{\partial u_\varepsilon^\alpha}{\partial x_i} - \frac{\partial}{\partial x_i}\Big\{\Phi_{\varepsilon,j}^{\alpha\beta}\Big\} \cdot \frac{\partial u_0^\beta}{\partial x_j} \Big\|_{L^\infty(T_r)} \\
&\le \frac{C}{r} \fint_{\Omega_{4r}} |u_\varepsilon - u_0| + C\varepsilon\, r^{-1} \|\nabla u_0\|_{L^\infty(T_{4r})} \\
&\quad + C\varepsilon \ln[\varepsilon^{-1} r + 2] \|\nabla^2 u_0\|_{L^\infty(T_{4r})} + C\varepsilon\, r^\rho\, \|\nabla^2 u_0\|_{C^{0,\rho}(T_{4r})}.
\end{aligned}
\tag{7.3.16}
$$

Proof. We start out by choosing a $C^{2,\eta}$ domain $\widetilde{\Omega}$ such that $T_{3r} \subset \widetilde{\Omega} \subset T_{4r}$. Let

$$w_\varepsilon = u_\varepsilon - u_0 - \left\{ \Phi^\beta_{\varepsilon,j} - P^\beta_j \right\} \frac{\partial u_0^\beta}{\partial x_j}.$$

Note that $w_\varepsilon = 0$ on I_{4r}. Write $w_\varepsilon = \theta_\varepsilon + z_\varepsilon$ in $\widetilde{\Omega}$, where $\theta_\varepsilon \in H^1_0(\widetilde{\Omega}; \mathbb{R}^m)$ and $\mathcal{L}_\varepsilon(\theta_\varepsilon) = \mathcal{L}_\varepsilon(w_\varepsilon)$ in $\widetilde{\Omega}$. Since $\mathcal{L}_\varepsilon(z_\varepsilon) = 0$ in $\widetilde{\Omega}$ and $z_\varepsilon = w_\varepsilon = 0$ on I_{3r}, it follows from the boundary Lipschitz estimate (5.5.1) that

$$\begin{aligned}
\|\nabla z_\varepsilon\|_{L^\infty(T_r)} &\le \frac{C}{r} \fint_{T_{2r}} |z_\varepsilon| \\
&\le \frac{C}{r} \fint_{T_{2r}} |w_\varepsilon| + C r^{-1} \|\theta_\varepsilon\|_{L^\infty(T_{2r})} \\
&\le \frac{C}{r} \fint_{T_{2r}} |u_\varepsilon - u_0| + C\varepsilon\, r^{-1} \|\nabla u_0\|_{L^\infty(T_{2r})} + C r^{-1} \|\theta_\varepsilon\|_{L^\infty(T_{2r})},
\end{aligned}$$

where we have used the estimate $\|\Phi^\beta_{\varepsilon,j} - P^\beta_j\|_{L^\infty(\Omega)} \le C\varepsilon$. This implies that

$$\|\nabla w_\varepsilon\|_{L^\infty(T_r)} \le \frac{C}{r} \fint_{T_{2r}} |u_\varepsilon - u_0| + C\varepsilon\, r^{-1} \|\nabla u_0\|_{L^\infty(T_{2r})} + C\, \|\nabla \theta_\varepsilon\|_{L^\infty(T_{2r})},$$

where we have used $\|\theta_\varepsilon\|_{L^\infty(T_{2r})} \le C r \|\nabla \theta_\varepsilon\|_{L^\infty(T_{2r})}$. Thus,

$$\begin{aligned}
&\left\| \frac{\partial u_\varepsilon^\alpha}{\partial x_i} - \frac{\partial}{\partial x_i} \left\{ \Phi^{\alpha\beta}_{\varepsilon,j} \right\} \cdot \frac{\partial u_0^\beta}{\partial x_j} \right\|_{L^\infty(T_r)} \\
&\quad \le \frac{C}{r} \fint_{T_{2r}} |u_\varepsilon - u_0| \, dx + C\varepsilon r^{-1} \|\nabla u_0\|_{L^\infty(T_{2r})} \\
&\qquad + C\varepsilon\, \|\nabla^2 u_0\|_{L^\infty(T_{2r})} + C\, \|\nabla \theta_\varepsilon\|_{L^\infty(T_{2r})}.
\end{aligned} \tag{7.3.17}$$

It remains to estimate $\nabla \theta_\varepsilon$ on T_{2r}. To this end we use the Green function representation

$$\theta_\varepsilon(x) = \int_{\widetilde{\Omega}} \widetilde{G}_\varepsilon(x,y) \mathcal{L}_\varepsilon(w_\varepsilon)(y)\, dy,$$

where $\widetilde{G}_\varepsilon(x,y)$ is the matrix of Green functions for $\mathcal{L}_\varepsilon$ in the $C^{2,\eta}$ domain $\widetilde{\Omega}$. Let

$$f_i(x) = -\varepsilon \phi_{kij}\,(x/\varepsilon)\, \frac{\partial^2 u_0}{\partial x_j \partial x_k} + a_{ij}\,(x/\varepsilon) \left[\Phi_{\varepsilon,k} - P_k \right] \cdot \frac{\partial^2 u_0}{\partial x_j \partial x_k},$$

where we have suppressed the superscripts for notational simplicity. In view of (7.1.7), we obtain

$$\begin{aligned}
\theta_\varepsilon(x) = &- \int_{\widetilde{\Omega}} \frac{\partial}{\partial y_i} \left\{ \widetilde{G}_\varepsilon(x,y) \right\} \cdot \left\{ f_i(y) - f_i(x) \right\} dy \\
&+ \int_{\widetilde{\Omega}} \widetilde{G}_\varepsilon(x,y) a_{ij}\,(y/\varepsilon)\, \frac{\partial}{\partial y_j} \Big[\Phi_{\varepsilon,k} - P_k - \varepsilon \chi_k\,(y/\varepsilon) \Big] \cdot \frac{\partial^2 u_0}{\partial y_i \partial y_k}\, dy.
\end{aligned}$$

It follows that

$$\begin{aligned}|\nabla\theta_\varepsilon(x)| &\le \int_{\widetilde{\Omega}} |\nabla_x\nabla_y\widetilde{G}_\varepsilon(x,y)|\,|f(y)-f(x)|\,dy\\ &\quad + C\,\|\nabla^2 u_0\|_{L^\infty(\Omega_{4r})}\int_{\widetilde{\Omega}} |\nabla_x\widetilde{G}_\varepsilon(x,y)|\,\big|\nabla_y\big[\Phi_\varepsilon - P - \varepsilon\chi(y/\varepsilon)\big]\big|\,dy.\end{aligned} \tag{7.3.18}$$

To handle the first term in the RHS of (7.3.18), we use

$$|\nabla_x\nabla_y\widetilde{G}_\varepsilon(x,y)| \le C\,|x-y|^{-d}$$

and the observation that

$$\begin{aligned}&\|f\|_{L^\infty(T_{4r})} \le C\varepsilon\,\|\nabla^2 u_0\|_{L^\infty(T_{4r})},\\ &|f(x)-f(y)| \le C\,|x-y|^\rho\Big\{\varepsilon^{1-\rho}\|\nabla^2 u_0\|_{L^\infty(T_{4r})} + \varepsilon\,\|\nabla^2 u_0\|_{C^{0,\rho}(T_{4r})}\Big\}.\end{aligned}$$

This yields that

$$\begin{aligned}&\int_{\widetilde{\Omega}} |\nabla_x\nabla_y\widetilde{G}_\varepsilon(x,y)||f(y)-f(x)|\,dy\\ &\le C\varepsilon\,\|\nabla^2 u_0\|_{L^\infty(T_{4r})}\int_{\widetilde{\Omega}\setminus B(x,\varepsilon)}\frac{dy}{|x-y|^d}\\ &\qquad + C\Big\{\varepsilon^{1-\rho}\|\nabla^2 u_0\|_{L^\infty(T_{4r})} + \varepsilon\,\|\nabla^2 u_0\|_{C^{0,\rho}(T_{4r})}\Big\}\int_{\widetilde{\Omega}\cap B(x,\varepsilon)}\frac{dy}{|x-y|^{d-\rho}}\\ &\le C\varepsilon\ln[\varepsilon^{-1}r+2]\|\nabla^2 u_0\|_{L^\infty(T_{4r})} + C\varepsilon^{1+\rho}\|\nabla^2 u_0\|_{C^{0,\rho}(T_{4r})}.\end{aligned}$$

Finally, using the estimates

$$|\nabla_x\widetilde{G}_\varepsilon(x,y)| \le C\,\mathrm{dist}(y,\partial\widetilde{\Omega})\,|x-y|^{-d}$$

and $|\nabla_x\widetilde{G}_\varepsilon(x,y)| \le C\,|x-y|^{1-d}$ as well as the observation that

$$\Big|\nabla\Big\{\Phi_{\varepsilon,j} - P_j - \varepsilon\chi_j(x/\varepsilon)\Big\}\Big| \le C\,\min\Big\{1, \varepsilon\,\big[\mathrm{dist}(x,\partial\widetilde{\Omega})\big]^{-1}\Big\}, \tag{7.3.19}$$

we can bound the second term in the RHS of (7.3.18) by

$$\begin{aligned}&C\,\|\nabla^2 u_0\|_{L^\infty(T_{4r})}\left\{\varepsilon\int_{\widetilde{\Omega}\setminus B(x,\varepsilon)}\frac{dy}{|x-y|^d} + \int_{\widetilde{\Omega}\cap B(x,\varepsilon)}\frac{dy}{|x-y|^{d-1}}\right\}\\ &\quad\le C\varepsilon\ln[\varepsilon^{-1}r+2]\|\nabla^2 u_0\|_{L^\infty(T_{4r})}.\end{aligned}$$

We remark that the inequality (7.3.19) follows from the estimate

$$\|\Phi_{\varepsilon,j} - P_j - \varepsilon\chi_j(x/\varepsilon)\|_{L^\infty(\Omega)} \le C\varepsilon$$

and the interior Lipschitz estimate for $\mathcal{L}_\varepsilon$. As a result, we have proved that

$$\|\nabla\theta_\varepsilon\|_{L^\infty(T_{3r})} \le C\varepsilon\ln[\varepsilon^{-1}r+2]\|\nabla^2 u_0\|_{L^\infty(T_{4r})} + C\varepsilon^{1+\rho}\|\nabla^2 u_0\|_{C^{0,\rho}(T_{4r})}.$$

This, together with (7.3.17), completes the proof of (7.3.16). ☐

Proof of Theorem 7.3.4. Fix $x_0, y_0 \in \Omega$ and $r = |x_0 - y_0|/8$. We may assume that $0 < \varepsilon < r$, since the case $\varepsilon \geq r$ is trivial and follows directly from the size estimates of $|\nabla_x G_\varepsilon(x,y)|$, $|\nabla_x G_0(x,y)|$ and $\|\nabla \Phi_\varepsilon\|_{L^\infty(\Omega)} \leq C$.

Let $u_\varepsilon(x) = G_\varepsilon(x, y_0)$ and $u_0(x) = G_0(x, y_0)$. Observe that

$$\mathcal{L}_\varepsilon(u_\varepsilon) = \mathcal{L}_0(u_0) = 0 \quad \text{ in } T_{4r} = T(x_0, 4r),$$

and $u_\varepsilon = u_0 = 0$ on $I_{4r} = I(x_0, 4r)$. By Theorem 7.3.1,

$$\|u_\varepsilon - u_0\|_{L^\infty(T_{4r})} \leq C\varepsilon\, r^{1-d}.$$

Also, since Ω is $C^{2,\eta}$, we have $\|\nabla u_0\|_{L^\infty(T_{4r})} \leq C r^{1-d}$,

$$\|\nabla^2 u_0\|_{L^\infty(T_{4r})} \leq C r^{-d} \quad \text{ and } \quad \|\nabla^2 u_0\|_{C^{0,\rho}(T_{4r})} \leq C r^{-d-\rho}.$$

Hence, by Lemma 7.3.5, we obtain

$$\left\| \frac{\partial u_\varepsilon^\alpha}{\partial x_i} - \frac{\partial}{\partial x_i}\left\{\Phi_{\varepsilon,j}^{\alpha\beta}\right\} \cdot \frac{\partial u_0^\beta}{\partial x_j} \right\|_{L^\infty(T_r)} \leq C\varepsilon\, r^{-d} \ln[\varepsilon^{-1} r + 2].$$

This finishes the proof. □

Let $G_\varepsilon^*(x,y) = \left(G_\varepsilon^{*\alpha\beta}(x,y)\right)_{m\times m}$ denote the matrix of Green functions for $\mathcal{L}_\varepsilon^*$, the adjoint of $\mathcal{L}_\varepsilon$. Since A^* satisfies the same conditions as A, by Theorem 7.3.4,

$$\begin{aligned} \left| \frac{\partial}{\partial x_i}\left\{G_\varepsilon^{*\alpha\gamma}(x,y)\right\} - \frac{\partial}{\partial x_i}\left\{\Phi_{\varepsilon,j}^{*\alpha\beta}(x)\right\} \cdot \frac{\partial}{\partial x_j}\left\{G_0^{*\beta\gamma}(x,y)\right\} \right| \\ \leq \frac{C\varepsilon \ln[\varepsilon^{-1}|x-y| + 2]}{|x-y|^d}, \end{aligned} \tag{7.3.20}$$

where $\Phi_\varepsilon^* = \left(\Phi_{\varepsilon,j}^{*\alpha\beta}(x)\right)_{m\times m}$ denotes the matrix of Dirichlet correctors for $\mathcal{L}_\varepsilon^*$ in Ω. Using $G_\varepsilon^{*\alpha\beta}(x,y) = G_\varepsilon^{\beta\alpha}(y,x)$, we obtain

$$\begin{aligned} \left| \frac{\partial}{\partial y_i}\left\{G_\varepsilon^{\gamma\alpha}(x,y)\right\} - \frac{\partial}{\partial y_i}\left\{\Phi_{\varepsilon,j}^{*\alpha\beta}(y)\right\} \cdot \frac{\partial}{\partial y_j}\left\{G_0^{\gamma\beta}(x,y)\right\} \right| \\ \leq \frac{C\varepsilon \ln[\varepsilon^{-1}|x-y| + 2]}{|x-y|^d}. \end{aligned} \tag{7.3.21}$$

This leads to an asymptotic expansion of the Poisson kernel for $\mathcal{L}_\varepsilon$ on Ω.

Let $(h^{\alpha\beta}(y))$ denote the inverse matrix of $\left(n_i(y) n_j(y) \widehat{a}_{ij}^{\alpha\beta}\right)_{m\times m}$.

Theorem 7.3.6. *Suppose that A satisfies the same conditions as in Theorem* 7.3.4. *Let $P_\varepsilon(x,y) = \left(P_\varepsilon^{\alpha\beta}(x,y)\right)_{m\times m}$ denote the Poisson kernel for $\mathcal{L}_\varepsilon$ in a bounded $C^{2,\eta}$ domain Ω. Then*

$$P_\varepsilon^{\alpha\beta}(x,y) = P_0^{\alpha\gamma}(x,y)\,\omega_\varepsilon^{\gamma\beta}(y) + R_\varepsilon^{\alpha\beta}(x,y), \tag{7.3.22}$$

where

$$\omega_\varepsilon^{\gamma\beta}(y) = h^{\gamma\sigma}(y) \cdot \frac{\partial}{\partial n(y)} \Big\{ \Phi_{\varepsilon,k}^{*\rho\sigma}(y) \Big\} \cdot n_k(y) \cdot n_i(y) n_j(y) a_{ij}^{\rho\beta}(y/\varepsilon), \tag{7.3.23}$$

and

$$|R_\varepsilon^{\alpha\beta}(x,y)| \le \frac{C\varepsilon \ln[\varepsilon^{-1}|x-y|+2]}{|x-y|^d} \tag{7.3.24}$$

for any $x \in \Omega$ and $y \in \partial\Omega$. The constant C depends only on μ, λ, τ, and Ω.

Proof. Note that for $x \in \Omega$ and $y \in \partial\Omega$,

$$\begin{aligned} P_\varepsilon^{\alpha\beta}(x,y) &= -n_i(y) a_{ji}^{\gamma\beta}(y/\varepsilon) \frac{\partial}{\partial y_j} \Big\{ G_\varepsilon^{\alpha\gamma}(x,y) \Big\} \\ &= -\frac{\partial}{\partial n(y)} \Big\{ G_\varepsilon^{\alpha\gamma}(x,y) \Big\} \cdot n_i(y) n_j(y) a_{ij}^{\gamma\beta}(y/\varepsilon), \end{aligned} \tag{7.3.25}$$

where the second equality follows from the fact $G_\varepsilon(x,\cdot) = 0$ on $\partial\Omega$. By (7.3.21), we obtain

$$\begin{aligned} \Big| P_\varepsilon^{\alpha\beta}(x,y) + \frac{\partial}{\partial n(y)} \Big\{ G_0^{\alpha\sigma}(x,y) \Big\} \cdot \frac{\partial}{\partial n(y)} \Big\{ \Phi_{\varepsilon,k}^{*\gamma\sigma}(y) \Big\} \cdot n_i(y) n_j(y) a_{ij}^{\gamma\beta}(y/\varepsilon) n_k(y) \Big| \\ \le \frac{C\varepsilon \ln[\varepsilon^{-1}|x-y|+2]}{|x-y|^d}. \end{aligned} \tag{7.3.26}$$

In view of (7.3.25) (with $\varepsilon = 0$), we have

$$P_0^{\alpha\beta}(x,y) h^{\beta\sigma}(y) = -\frac{\partial}{\partial n(y)} \Big\{ G_0^{\alpha\sigma}(x,y) \Big\}.$$

This, together with (7.3.26), gives

$$|P_\varepsilon^{\alpha\beta}(x,y) - P_0^{\alpha\gamma}(x,y) \omega_\varepsilon^{\gamma\beta}(y)| \le \frac{C\varepsilon \ln[\varepsilon^{-1}|x-y|+2]}{|x-y|^d},$$

for any $x \in \Omega$ and $y \in \partial\Omega$, where $\omega_\varepsilon(y)$ is defined by (7.3.23). □

We end this section with an asymptotic expansion for $\nabla_x \nabla_y G_\varepsilon(x,y)$.

Theorem 7.3.7. *Let A satisfy the same conditions as in Theorem 7.3.4. Let Ω be a bounded $C^{2,\eta}$ domain for some $\eta \in (0,1)$. Then*

$$\begin{aligned} \Big| \frac{\partial^2}{\partial x_i \partial y_j} \Big\{ G_\varepsilon^{\alpha\beta}(x,y) \Big\} - \frac{\partial}{\partial x_i} \Big\{ \Phi_{\varepsilon,k}^{\alpha\gamma}(x) \Big\} \cdot \frac{\partial^2}{\partial x_k \partial y_\ell} \Big\{ G_0^{\gamma\sigma}(x,y) \Big\} \cdot \frac{\partial}{\partial y_j} \Big\{ \Phi_{\varepsilon,\ell}^{*\beta\sigma}(y) \Big\} \Big| \\ \le \frac{C\varepsilon \ln\big[\varepsilon^{-1}|x-y|+2\big]}{|x-y|^{d+1}} \end{aligned} \tag{7.3.27}$$

for any $x, y \in \Omega$ and $x \ne y$, where C depends only on μ, λ, τ, and Ω.

Proof. Fix $x_0, y_0 \in \Omega$. Let $r = |x_0 - y_0|/8$. Since

$$|\nabla_x \nabla_y G_\varepsilon(x,y)| \le C\,|x-y|^{-d},$$

it suffices to consider the case $0 < \varepsilon < r$. Fix $1 \le \beta \le m$ and $1 \le j \le d$, let

$$\begin{cases} u_\varepsilon^\alpha(x) = \dfrac{\partial G_\varepsilon^{\alpha\beta}}{\partial y_j}(x, y_0), \\ u_0^\alpha(x) = \dfrac{\partial}{\partial y_j}\Big\{\Phi_{\varepsilon,\ell}^{*\beta\sigma}\Big\}(y_0)\cdot \dfrac{\partial G_0^{\alpha\sigma}}{\partial y_\ell}(x,y_0) \end{cases}$$

in $T_{4r} = \Omega \cap B(x_0, 4r)$. Observe that $\mathcal{L}_\varepsilon(u_\varepsilon) = \mathcal{L}_0(u_0) = 0$ in T_{4r} and $u_\varepsilon = u_0 = 0$ on I_{4r}. It follows from (7.3.21) that

$$\|u_\varepsilon - u_0\|_{L^\infty(T_{4r})} \le C\varepsilon\, r^{-d} \ln\big[\varepsilon^{-1} r + 2\big]. \tag{7.3.28}$$

Since Ω is $C^{2,\eta}$, we have $\|\nabla u_0\|_{L^\infty(T_{4r})} \le C r^{-d}$,

$$\|\nabla^2 u_0\|_{L^\infty(T_{4r})} \le C r^{-d-1} \quad \text{and} \quad \|\nabla^2 u_0\|_{C^{0,\eta}(T_{4r})} \le C r^{-d-1-\eta}. \tag{7.3.29}$$

By Lemma 7.3.5, estimates (7.3.28) and (7.3.29) imply that

$$\left\| \frac{\partial u_\varepsilon^\alpha}{\partial x_i} - \frac{\partial}{\partial x_i}\Big\{\Phi_{\varepsilon,k}^{\alpha\gamma}\Big\} \cdot \frac{\partial u_0^\gamma}{\partial x_k} \right\|_{L^\infty(T_r)} \le \frac{C\varepsilon \ln\big[\varepsilon^{-1} r + 2\big]}{r^{d+1}}.$$

This gives the desired estimate (7.3.27). □

7.4 Asymptotic expansions of Neumann functions

Throughout this section we will assume that A is 1-periodic and satisfies the ellipticity condition (2.1.20) and the Hölder continuity condition (4.0.2). Let

$$N_\varepsilon(x,y) = \big(N_\varepsilon^{\alpha\beta}(x,y)\big)$$

denote the $m \times m$ matrix of Neumann functions for $\mathcal{L}_\varepsilon$ in Ω. Under the assumption that Ω is $C^{1,\alpha}$, it is proved in Chapter 6 that if $d \ge 3$, then

$$|N_\varepsilon(x,y)| \le C\,|x-y|^{2-d}, \tag{7.4.1}$$

and

$$\begin{aligned} |\nabla_x N_\varepsilon(x,y)| + |\nabla_y N_\varepsilon(x,y)| &\le C\,|x-y|^{1-d}, \\ |\nabla_x \nabla_y N_\varepsilon(x,y)| &\le C\,|x-y|^{-d}, \end{aligned} \tag{7.4.2}$$

for any $x, y \in \Omega$ and $x \ne y$, where C depends only on μ, (λ, τ), and Ω. In the case $d = 2$, the estimates in (7.4.2) continue to hold, while (7.4.1) is replaced by

$$|N_\varepsilon(x,y)| \le C\{1 + \ln(r_0|x-y|^{-1})\}, \tag{7.4.3}$$

where $r_0 = \operatorname{diam}(\Omega)$. In this section we investigate the asymptotic behavior, as $\varepsilon \to 0$, of $N_\varepsilon(x,y)$, $\nabla_x N_\varepsilon(x,y)$, $\nabla_y N_\varepsilon(x,y)$, and $\nabla_x\nabla_y N_\varepsilon(x,y)$. We shall use $N_0(x,y) = \big(N_0^{\alpha\beta}(x,y)\big)$ to denote the $m\times m$ matrix of Neumann functions for $\mathcal{L}_0$ in Ω.

Theorem 7.4.1. *Let Ω be a bounded $C^{1,1}$ domain in $\mathbb{R}^d$. Then*

$$|N_\varepsilon(x,y) - N_0(x,y)| \le \frac{C\varepsilon \ln[\varepsilon^{-1}|x-y|+2]}{|x-y|^{d-1}} \tag{7.4.4}$$

for any $x, y \in \Omega$ and $x \ne y$, where C depends only on μ, (λ,τ), and Ω.

As in the case of Green functions, Theorem 7.4.1 is a consequence of an L^∞ estimate for local solutions. Recall that $T_r = B(x_0,r)\cap\Omega$ and $I(r) = B(x_0,r)\cap\partial\Omega$, where $x_0 \in \overline{\Omega}$ and $0 < r < c_0 \operatorname{diam}(\Omega)$.

Lemma 7.4.2. *Let Ω be a bounded $C^{1,\eta}$ domain for some $\eta \in (0,1)$. Let $u_\varepsilon \in H^1(T_{3r};\mathbb{R}^m)$ and $u_0 \in W^{2,p}(T_{3r};\mathbb{R}^m)$ for some $p > d$. Suppose that*

$$\mathcal{L}_\varepsilon(u_\varepsilon) = \mathcal{L}_0(u_0) \quad \text{in } T_{3r} \quad \text{and} \quad \frac{\partial u_\varepsilon}{\partial \nu_\varepsilon} = \frac{\partial u_0}{\partial \nu_0} \quad \text{on } I_{3r}.$$

Then, if $0 < \varepsilon < (r/2)$,

$$\begin{aligned} \|u_\varepsilon - u_0\|_{L^\infty(T_r)} \le C \fint_{T_{3r}} |u_\varepsilon - u_0| &+ C\varepsilon \ln[\varepsilon^{-1} r + 2] \|\nabla u_0\|_{L^\infty(T_{3r})} \\ &+ C_p\, \varepsilon\, r^{1-\frac{d}{p}} \|\nabla^2 u_0\|_{L^p(T_{3r})}. \end{aligned} \tag{7.4.5}$$

Proof. By rescaling we may assume that $r = 1$. Choose a $C^{1,\eta}$ domain $\widetilde{\Omega}$ such that $T_2 \subset \widetilde{\Omega} \subset T_3$. Let

$$w_\varepsilon = u_\varepsilon(x) - u_0(x) - \varepsilon \chi_j^\beta(x/\varepsilon) \frac{\partial u_0^\beta}{\partial x_j}.$$

Recall that

$$(\mathcal{L}(w_\varepsilon))^\alpha = -\varepsilon \frac{\partial}{\partial x_i} \left\{ \left[\phi_{jik}^{\alpha\beta}(x/\varepsilon) - a_{ij}^{\alpha\gamma}(x/\varepsilon)\chi_k^{\gamma\beta}(x/\varepsilon) \right] \frac{\partial^2 u_0^\beta}{\partial x_j \partial x_k} \right\}. \tag{7.4.6}$$

Using (3.1.3), one can verify that

$$\begin{aligned} \left(\frac{\partial w_\varepsilon}{\partial \nu_\varepsilon}\right)^\alpha &= \left(\frac{\partial u_\varepsilon}{\partial \nu_\varepsilon}\right)^\alpha - \left(\frac{\partial u_0}{\partial \nu_0}\right)^\alpha \\ &\quad - \frac{\varepsilon}{2}\left(n_i \frac{\partial}{\partial x_j} - n_j \frac{\partial}{\partial x_i}\right) \left\{ \phi_{jik}^{\alpha\gamma}(x/\varepsilon) \frac{\partial u_0^\gamma}{\partial x_k} \right\} \\ &\quad + \varepsilon n_i \left[\phi_{jik}^{\alpha\beta}(x/\varepsilon) - a_{ij}^{\alpha\gamma}(x/\varepsilon)\chi_k^{\gamma\beta}(x/\varepsilon) \right] \frac{\partial^2 u_0^\beta}{\partial x_j \partial x_k}. \end{aligned} \tag{7.4.7}$$

We point out that $n_i \frac{\partial}{\partial x_j} - n_j \frac{\partial}{\partial x_i}$ is a tangential derivative on the boundary and possesses the following property (integration by parts on $\partial\Omega$),

$$\int_{\partial\Omega} \left(n_i \frac{\partial}{\partial x_j} - n_j \frac{\partial}{\partial x_i} \right) u \cdot v \, d\sigma = - \int_{\partial\Omega} u \cdot \left(n_i \frac{\partial}{\partial x_j} - n_j \frac{\partial}{\partial x_i} \right) v \, d\sigma \qquad (7.4.8)$$

for any $u, v \in C^1(\overline{\Omega})$. The equality (7.4.8) may be proved by using the divergence theorem and a simple approximation argument.

Let $w_\varepsilon = \theta_\varepsilon + z_\varepsilon$, where

$$(\theta_\varepsilon(x))^\alpha = \varepsilon \int_{\widetilde{\Omega}} \frac{\partial}{\partial y_i} \left\{ \widetilde{N}_\varepsilon^{\alpha\beta}(x,y) \right\} \cdot \left[\phi_{jik}^{\beta\gamma}(y/\varepsilon) - a_{ij}^{\beta\sigma}(y/\varepsilon) \chi_k^{\sigma\gamma}(y/\varepsilon) \right] \frac{\partial^2 u_0^\gamma}{\partial y_j \partial y_k} \, dy \qquad (7.4.9)$$

and $\widetilde{N}_\varepsilon(x,y)$ denotes the matrix of Neumann functions for $\mathcal{L}_\varepsilon$ in $\widetilde{\Omega}$. Since

$$|\nabla_y \widetilde{N}_\varepsilon(x,y)| \le C \, |x-y|^{1-d},$$

it follows by Hölder's inequality that

$$\|\theta_\varepsilon\|_{L^\infty(T_2)} \le C_p \, \varepsilon \, \|\nabla^2 u_0\|_{L^p(T_3)} \qquad \text{for any } p > d. \qquad (7.4.10)$$

To estimate z_ε, we observe that $\mathcal{L}_\varepsilon(z_\varepsilon) = 0$ in $\widetilde{\Omega}$ and

$$\left(\frac{\partial z_\varepsilon}{\partial \nu_\varepsilon} \right)^\alpha = \left(\frac{\partial u_\varepsilon}{\partial \nu_\varepsilon} \right)^\alpha - \left(\frac{\partial u_0}{\partial \nu_0} \right)^\alpha - \frac{\varepsilon}{2} \left(n_i \frac{\partial}{\partial x_j} - n_j \frac{\partial}{\partial x_i} \right) \left\{ \phi_{jik}^{\alpha\gamma}(x/\varepsilon) \frac{\partial u_0^\gamma}{\partial x_k} \right\} \qquad (7.4.11)$$

on $\partial\widetilde{\Omega}$. Let $z_\varepsilon = z_\varepsilon^{(1)} + z_\varepsilon^{(2)}$, where

$$(z_\varepsilon^{(1)})^\alpha(x) = \frac{\varepsilon}{2} \int_{\partial\widetilde{\Omega}} \left(n_i \frac{\partial}{\partial y_j} - n_j \frac{\partial}{\partial y_i} \right) \left\{ \widetilde{N}_\varepsilon^{\alpha\beta}(x,y) \right\} \cdot \phi_{jik}^{\beta\gamma}(y/\varepsilon) \frac{\partial u_0^\gamma}{\partial y_k} \, d\sigma(y). \qquad (7.4.12)$$

For each $x \in \widetilde{\Omega}$, choose $\hat{x} \in \partial\widetilde{\Omega}$ such that $|\hat{x} - x| = \text{dist}(x, \partial\widetilde{\Omega})$. Note that for $y \in \partial\widetilde{\Omega}$,

$$|y - \hat{x}| \le |y - x| + |x - \hat{x}| \le 2|y - x|.$$

Hence, $|\nabla_y \widetilde{N}_\varepsilon(x,y)| \le C|y - \hat{x}|^{1-d}$ and

$$\begin{aligned} |(z_\varepsilon^{(1)})^\alpha(x)| &= \frac{\varepsilon}{2} \left| \int_{\partial\widetilde{\Omega}} \left(n_i \frac{\partial}{\partial y_j} - n_j \frac{\partial}{\partial y_i} \right) \left\{ \widetilde{N}_\varepsilon^{\alpha\beta}(x,y) \right\} \cdot \{ f_{ji}^\beta(y) - f_{ji}^\beta(\hat{x}) \} \, d\sigma(y) \right| \\ &\le C\varepsilon \int_{\partial\widetilde{\Omega}} \frac{|f(y) - f(\hat{x})|}{|y - \hat{x}|^{d-1}} \, d\sigma(y), \end{aligned}$$

where $f(y) = (f_{ji}^\beta(y)) = (\phi_{jik}^{\beta\gamma}(y/\varepsilon) \frac{\partial u_0^\gamma}{\partial y_k}(y))$. Since

$$\|f\|_{L^\infty(T_3)} \le C \, \|\nabla u_0\|_{L^\infty(T_3)}$$

and

$$|f(y) - f(\hat{x})| \le C\varepsilon^{-1}|y-\hat{x}| \|\nabla u_0\|_{L^\infty(T_3)} + C|y-\hat{x}|^\rho \|\nabla u_0\|_{C^{0,\rho}(T_3)},$$

where $0 < \rho < \eta$ and we have used the fact that $\|\phi_{jik}^{\beta\gamma}\|_{C^1(Y)} \le C$, we get

$$\begin{aligned} |z_\varepsilon^{(1)}(x)| &\le C\varepsilon \, \|\nabla u_0\|_{L^\infty(T_3)} \int_{\partial\widetilde{\Omega}\setminus B(\hat{x},\varepsilon)} |\hat{x}-y|^{1-d}\, d\sigma(y) \\ &\qquad + C\, \|\nabla u_0\|_{L^\infty(T_3)} \int_{B(\hat{x},\varepsilon)\cap\partial\widetilde{\Omega}} |y-\hat{x}|^{2-d}\, d\sigma(y) \\ &\qquad + C\varepsilon \, \|\nabla u_0\|_{C^{0,\rho}(T_3)} \int_{B(\hat{x},\varepsilon)\cap\partial\widetilde{\Omega}} |y-\hat{x}|^{1-d+\rho}\, d\sigma(y) \\ &\le C\varepsilon \ln[\varepsilon^{-1}+2] \|\nabla u_0\|_{L^\infty(T_3)} + C\varepsilon^{1+\rho} \|\nabla u_0\|_{C^{0,\rho}(T_3)}. \end{aligned}$$

By the Sobolev embedding, this implies that

$$\|z_\varepsilon^{(1)}\|_{L^\infty(T_2)} \le C\varepsilon \ln[\varepsilon^{-1}+2] \|\nabla u_0\|_{L^\infty(T_3)} + C_p\, \varepsilon \, \|\nabla^2 u_0\|_{L^p(T_3)} \tag{7.4.13}$$

for any $p > d$.

Finally, to estimate $z_\varepsilon^{(2)}$, we note that $\mathcal{L}_\varepsilon(z_\varepsilon^{(2)}) = 0$ in T_2 and

$$\frac{\partial}{\partial \nu_\varepsilon}\left\{z_\varepsilon^{(2)}\right\} = \frac{\partial u_\varepsilon}{\partial \nu_\varepsilon} - \frac{\partial u_0}{\partial \nu_0} = 0 \quad \text{on } I_2.$$

It follows from the boundary Hölder estimate (6.2.6) as well as interior estimates that

$$\begin{aligned} \|z_\varepsilon^{(2)}\|_{L^\infty(T_1)} &\le C \fint_{T_2} |z_\varepsilon^{(2)}| \\ &\le C \fint_{T_2} |u_\varepsilon - u_0| + C\varepsilon \, \|\nabla u_0\|_{L^\infty(T_2)} + C\, \|\theta_\varepsilon\|_{L^\infty(T_2)} + C\, \|z_\varepsilon^{(1)}\|_{L^\infty(T_2)}. \end{aligned}$$

Hence,

$$\begin{aligned} &\|u_\varepsilon - u_0\|_{L^\infty(T_1)} \\ &\quad \le C \fint_{T_2} |u_\varepsilon - u_0| + C\varepsilon \, \|\nabla u_0\|_{L^\infty(T_2)} + C\, \|\theta_\varepsilon\|_{L^\infty(T_2)} + C\, \|z_\varepsilon^{(1)}\|_{L^\infty(T_2)}. \end{aligned}$$

This, together with (7.4.10) and (7.4.13), gives the estimate (7.4.5). □

Lemma 7.4.3. *Let Ω be a bounded Lipschitz domain in $\mathbb{R}^d$. Let*

$$u(x) = \int_\Omega \frac{g(y)}{|x-y|^{d-1}}\, dy \quad \text{and} \quad v(x) = \int_{\partial\Omega} \frac{f(y)}{|x-y|^{d-1}}\, d\sigma(y).$$

Then

$$\begin{aligned} \|u\|_{L^2(\partial\Omega)} &\le C\, \|g\|_{L^q(\Omega)}, \\ \|v\|_{L^p(\Omega)} &\le C\, \|f\|_{L^2(\partial\Omega)}, \end{aligned} \tag{7.4.14}$$

where $p = \frac{2d}{d-1}$, $q = p' = \frac{2d}{d+1}$, and C depends only on Ω.

Proof. It follows from (3.1.15) that

$$\|u\|_{L^2(\partial\Omega)} \le C\,\|u\|_{W^{1,q}(\Omega)} \le C\,\|g\|_{L^q(\Omega)}.$$

The estimate for v follows from that for u by a duality argument. □

Proof of Theorem 7.4.1. By rescaling we may assume that $\text{diam}(\Omega) = 1$. Fix $x_0, y_0 \in \Omega$ and let $r = |x_0 - y_0|/8$. Since $|N_\varepsilon(x_0, y_0)| \le Cr^{2-d}$, we may also assume that $0 < \varepsilon < r$. For $g \in C_0^\infty(T(y_0, r); \mathbb{R}^m)$ and $\varepsilon \ge 0$, let

$$u_\varepsilon(x) = \int_\Omega N_\varepsilon(x, y) g(y)\, dy.$$

Then $\mathcal{L}_\varepsilon(u_\varepsilon) = g$ in Ω, $\frac{\partial u_\varepsilon}{\partial \nu_\varepsilon} = -\frac{1}{|\partial\Omega|}\int_\Omega g\, dx$ on $\partial\Omega$ and $\int_{\partial\Omega} u_\varepsilon\, d\sigma = 0$. It follows that $\mathcal{L}_\varepsilon(u_\varepsilon) = \mathcal{L}_0(u_0)$ in Ω and $\frac{\partial u_\varepsilon}{\partial \nu_\varepsilon} = \frac{\partial u_0}{\partial \nu_0}$ on $\partial\Omega$. Let

$$w_\varepsilon = u_\varepsilon(x) - u_0(x) - \varepsilon \chi_j^\beta(x/\varepsilon)\frac{\partial u_0^\beta}{\partial x_j}.$$

As in the proof of Lemma 7.4.2, $\mathcal{L}(w_\varepsilon)$ and $\frac{\partial w_\varepsilon}{\partial \nu_\varepsilon}$ are given by (7.4.6) and (7.4.7), respectively. Now, write $w_\varepsilon = \theta_\varepsilon + z_\varepsilon + \rho$, where $\mathcal{L}_\varepsilon(z_\varepsilon) = \mathcal{L}_\varepsilon(w_\varepsilon)$ in Ω, $\int_\Omega z_\varepsilon\, dx = 0$,

$$\frac{\partial z_\varepsilon}{\partial \nu_\varepsilon} = \varepsilon n_i \left[\phi_{jik}^{\alpha\beta}(x/\varepsilon) - a_{ij}^{\alpha\gamma}(x/\varepsilon)\chi_k^{\gamma\beta}(x/\varepsilon)\right] \frac{\partial^2 u_0^\beta}{\partial x_j \partial x_k} \qquad \text{on } \partial\Omega,$$

and $\rho = \fint_{\partial\Omega} \{w_\varepsilon - z_\varepsilon\}$ is a constant. Note that

$$\|\nabla z_\varepsilon\|_{L^2(\Omega)} \le C\varepsilon\,\|\nabla^2 u_0\|_{L^2(\Omega)} \le C\varepsilon\,\|g\|_{L^2(\Omega)}. \tag{7.4.15}$$

Since $\int_\Omega z_\varepsilon\, dx = 0$, by the Poincaré inequality, we obtain $\|z_\varepsilon\|_{L^p(\Omega)} \le C\varepsilon\|g\|_{L^2(\Omega)}$, where $d \ge 3$ and $p = \frac{2d}{d-2}$. It follows by Hölder's inequality that

$$\|z_\varepsilon\|_{L^2(T(x_0,r))} \le Cr^{\frac{d}{2}-\frac{d}{p}}\|z_\varepsilon\|_{L^p(T(x_0,r))} \le C\varepsilon r\|g\|_{L^2(T(y_0,r))}. \tag{7.4.16}$$

For the case $d = 2$, in place of (7.4.15)–(7.4.16) we use a Meyers type estimate to obtain

$$\|\nabla z_\varepsilon\|_{L^{\bar p}(\Omega)} \le C\varepsilon\,\|\nabla^2 u_0\|_{L^{\bar p}(\Omega)} \le C\varepsilon\,\|g\|_{L^{\bar p}(\Omega)} \tag{7.4.17}$$

for some $\bar p < 2$ (see the proof of Lemma 6.1.4). This gives

$$\begin{aligned}\|z_\varepsilon\|_{L^2(T(x_0,r))} &\le Cr^{1-\frac{2}{q}}\|z_\varepsilon\|_{L^q(\Omega)} \le Cr^{1-\frac{2}{q}}\|\nabla z_\varepsilon\|_{L^{\bar p}(\Omega)}\\ &\le C\varepsilon r^{1-\frac{2}{q}}\|g\|_{L^{\bar p}(\Omega)} \le C\varepsilon r\|g\|_{L^2(T(y_0,r))},\end{aligned}$$

where $\frac{1}{q} = \frac{1}{\bar p} - \frac{1}{2}$.

Next, to estimate θ_ε, we observe that $\mathcal{L}_\varepsilon(\theta_\varepsilon)=0$ in Ω, $\int_{\partial\Omega}\theta_\varepsilon\, d\sigma=0$ and

$$\frac{\partial\theta_\varepsilon}{\partial\nu_\varepsilon}=-\frac{\varepsilon}{2}\left(n_i\frac{\partial}{\partial x_j}-n_j\frac{\partial}{\partial x_i}\right)\left\{\phi_{jik}^{\alpha\gamma}(x/\varepsilon)\frac{\partial u_0^\gamma}{\partial x_k}\right\}.$$

It follows that

$$\theta_\varepsilon^\alpha(x)=\frac{\varepsilon}{2}\int_{\partial\Omega}\left(n_i\frac{\partial}{\partial y_j}-n_j\frac{\partial}{\partial y_i}\right)N_\varepsilon^{\alpha\beta}(x,y)\left\{\phi_{jik}^{\beta\gamma}(y/\varepsilon)\frac{\partial u_0^\gamma}{\partial y_k}\right\}d\sigma(y).$$

Using the estimate $|\nabla_y N_\varepsilon(x,y)|\le C|x-y|^{1-d}$, we obtain

$$|\theta_\varepsilon(x)|\le C\varepsilon\int_{\partial\Omega}\frac{|\nabla u_0(y)|}{|x-y|^{d-1}}\,d\sigma(y).$$

In view of Lemma 7.4.3 we see that

$$\|\theta_\varepsilon\|_{L^p(\Omega)}\le C\varepsilon\,\|\nabla u_0\|_{L^2(\partial\Omega)},$$

where $p=\frac{2d}{d-1}$. By Hölder's inequality, this gives

$$\|\theta_\varepsilon\|_{L^2(T(x_0,r))}\le C\varepsilon r^{\frac{1}{2}}\|\nabla u_0\|_{L^2(\partial\Omega)}. \tag{7.4.18}$$

Since

$$|\nabla u_0(x)|\le C\int_\Omega\frac{|g(y)|\,dy}{|x-y|^{d-1}},$$

we can invoke Lemma 7.4.3 again to claim that

$$\|\nabla u_0\|_{L^2(\partial\Omega)}\le C\,\|g\|_{L^q(T(y_0,r))}\le Cr^{1/2}\|g\|_{L^2(T(y_0,r))},$$

where $q=\frac{2d}{d+1}$. In view of (7.4.18) we obtain

$$\|\theta_\varepsilon\|_{L^2(T(x_0,r))}\le C\varepsilon r\|g\|_{L^2(T(y_0,r))}.$$

This, together with (7.4.16) and the observation that

$$|\rho|\le C\int_{\partial\Omega}\{\varepsilon|\nabla u_0|+|z_\varepsilon|\}d\sigma\le C\varepsilon\,\|g\|_{L^2(T(y_0,r))},$$

gives

$$\|w_\varepsilon\|_{L^2(T(x_0,r))}\le C\varepsilon r\|g\|_{L^2(T(y_0,r))}.$$

It follows that

$$\left\{\fint_{T(x_0,r)}|u_\varepsilon-u_0|^2\right\}^{1/2}\le C\varepsilon r^{\frac{2-d}{2}}\|g\|_{L^2(T(y_0,r))}.$$

Since

$$\|\nabla u_0\|_{L^\infty(T(x_0,r))} \le C r^{\frac{2-d}{2}} \|g\|_{L^2(\Omega)} \quad \text{and} \quad \|\nabla^2 u_0\|_{L^p(\Omega)} \le C\, \|g\|_{L^p(\Omega)},$$

by Lemma 7.4.2, we obtain

$$|u_\varepsilon(x_0) - u_0(x_0)| \le C_p\, \varepsilon r^{1-\frac{d}{p}} \ln[\varepsilon^{-1} r + 2] \|g\|_{L^p(T(y_0,r))},$$

where $p > d$. By duality this gives

$$\left\{ \fint_{T(y_0,r)} |N_\varepsilon(x_0,y) - N_0(x_0,y)|^{p'}\, dy \right\}^{1/p'} \le C_p\, \varepsilon r^{1-d} \ln[\varepsilon^{-1} r + 2].$$

Finally, since

$$\frac{\partial}{\partial \nu_\varepsilon(y)} \{N_\varepsilon(x,y)\} = \frac{\partial}{\partial \nu_0(y)} \{N_0(x,y)\} = -\frac{I_{m\times m}}{|\partial\Omega|} \qquad \text{on } \partial\Omega,$$

$|\nabla_y N_0(x,y)| \le C|x-y|^{1-d}$ and $\|\nabla_y^2 N_0(x_0,y)\|_{L^p(T(y_0,r))} \le C r^{\frac{d}{p}-d}$, we may invoke Lemma 7.4.2 again to obtain

$$\begin{aligned} |N_\varepsilon(x_0,y_0) - N_0(x_0,y_0)| &\le \fint_{T(y_0,r)} |N_\varepsilon(x_0,y) - N_0(x_0,y)|\, dy + C\varepsilon r^{1-d} \ln[\varepsilon^{-1} r + 2] \\ &\le C\varepsilon r^{1-d} \ln[\varepsilon^{-1} r + 2]. \end{aligned}$$

This completes the proof. □

Definition 7.4.4. For $1 \le j \le d$ and $1 \le \beta \le m$, a Neumann corrector $\Psi_{\varepsilon,j}^\beta = \big(\Psi_{\varepsilon,j}^{\alpha\beta}\big)$ is defined to be a weak solution of the Neumann problem,

$$\begin{cases} \mathcal{L}_\varepsilon\big(\Psi_{\varepsilon,j}^\beta\big) = 0 & \text{in } \Omega, \\ \dfrac{\partial}{\partial \nu_\varepsilon}\big(\Psi_{\varepsilon,j}^\beta\big) = \dfrac{\partial}{\partial \nu_0}\big(P_j^\beta\big) & \text{on } \partial\Omega, \end{cases} \tag{7.4.19}$$

where $P_j^\beta = x_j e^\beta$.

Since solutions of (7.4.19) are unique up to a constant in $\mathbb{R}^m$, to fix the corrector we assume that $\Psi_{\varepsilon,j}^\beta(x_0) = P_j^\beta(x_0)$ for some $x_0 \in \Omega$. By the boundary Lipschitz estimate in Corollary 6.4.3, we see that

$$\|\nabla \Psi_{\varepsilon,j}^\beta\|_{L^\infty(\Omega)} \le C, \tag{7.4.20}$$

where C depends only on μ, λ, τ, and Ω.

Lemma 7.4.5. *Suppose that A is 1-periodic and satisfies (2.1.20) and (4.0.2). Let Ω be a bounded $C^{1,\eta}$ domain in $\mathbb{R}^d$. Then for any $x \in \Omega$,*

$$|\nabla\{\Psi_{\varepsilon,j}^\beta(x) - P_j^\beta(x) - \varepsilon\chi_j^\beta(x/\varepsilon)\}| \le \frac{C\varepsilon}{\delta(x)}, \tag{7.4.21}$$

where $\delta(x) = \text{dist}(x, \partial\Omega)$ and C depends only on μ, λ, τ, and Ω. Moreover,

$$\|\Psi_{\varepsilon,j}^\beta - P_j^\beta\|_{L^\infty(\Omega)} \le C\varepsilon \ln\left[\varepsilon^{-1}r_0 + 2\right], \tag{7.4.22}$$

where $r_0 = \text{diam}(\Omega)$.

Proof. Fix $1 \le j \le d$ and $1 \le \beta \le m$. Let

$$w_\varepsilon = \Psi_{\varepsilon,j}^\beta - P_j^\beta - \varepsilon\chi_j^\beta(x/\varepsilon).$$

Then $\mathcal{L}_\varepsilon(w_\varepsilon) = 0$ in Ω. Using the formula (7.4.7) with $u_\varepsilon = \Psi_{\varepsilon,j}^\beta$ and $u_0 = P_j^\beta$, we obtain

$$\left(\frac{\partial w_\varepsilon}{\partial \nu_\varepsilon}\right)^\alpha = -\frac{\varepsilon}{2}\left(n_i\frac{\partial}{\partial x_j} - n_j\frac{\partial}{\partial x_i}\right)\left\{\phi_{jik}^{\alpha\gamma}(x/\varepsilon)\frac{\partial u_0^\gamma}{\partial x_k}\right\} \quad \text{on } \partial\Omega.$$

It follows that for any $x, y \in \Omega$,

$$\begin{aligned} w_\varepsilon^\alpha(x) - w_\varepsilon^\alpha(y) = \frac{\varepsilon}{2}\int_{\partial\Omega}\left(n_i\frac{\partial}{\partial z_j} - n_j\frac{\partial}{\partial z_i}\right)&\left(N_\varepsilon^{\alpha\beta}(x,z) - N_\varepsilon^{\alpha\beta}(y,z)\right) \\ &\left\{\phi_{jik}^{\beta\gamma}(z/\varepsilon)\frac{\partial u_0^\gamma}{\partial z_k}\right\} d\sigma(z). \end{aligned}$$

Thus, if $|x - y| < (1/2)\delta(x)$,

$$\begin{aligned} |w_\varepsilon(x) - w_\varepsilon(y)| &\le C\varepsilon\int_{\partial\Omega}|\nabla_z N_\varepsilon(x,z) - \nabla_z N_\varepsilon(y,z)|\, d\sigma(z) \\ &\le C\varepsilon\,|x - y|\int_{\partial\Omega}\frac{d\sigma(z)}{|x-y|^d} \\ &\le C\varepsilon\,|x - y|[\delta(x)]^{-1}, \end{aligned}$$

where we have used the estimate $|\nabla_x\nabla_y N_\varepsilon(x,y)| \le C|x-y|^{-d}$. This gives the estimate (7.4.21).

To see (7.4.22), note that $\|\nabla w_\varepsilon\|_{L^\infty(\Omega)} \le C$, and thus

$$|\nabla w_\varepsilon(x)| \le C\min\left\{\frac{\varepsilon}{\delta(x)}, 1\right\}.$$

This, together with the fact that $|w_\varepsilon(x_0)| = \varepsilon|\chi(x_0/\varepsilon)|$, yields (7.4.22) by a simple integration. $\square$

The next theorem gives an asymptotic expansion of $\nabla_x N_\varepsilon(x,y)$.

Theorem 7.4.6. *Suppose that A is 1-periodic and satisfies conditions (2.1.20) and (4.0.2). Let Ω be a bounded $C^{2,\eta}$ domain for some $\eta \in (0,1)$. Then, for any $\rho \in (0,1)$,*

$$\begin{aligned} \Big| \frac{\partial}{\partial x_i} \left\{ N_\varepsilon^{\alpha\beta}(x,y) \right\} - \frac{\partial}{\partial x_i} \left\{ \Psi_{\varepsilon,j}^{\alpha\gamma}(x) \right\} \cdot \frac{\partial}{\partial x_j} \Big\{ N_0^{\gamma\beta}(x,y) \Big\} \Big| \\ \le \frac{C_\rho\, \varepsilon^{1-\rho} \ln[\varepsilon^{-1} r_0 + 2]}{|x-y|^{d-\rho}} \end{aligned} \tag{7.4.23}$$

for any $x, y \in \Omega$ and $x \neq y$, where $r_0 = \mathrm{diam}(\Omega)$ and C_ρ depends only on μ, ρ, λ, τ, and Ω.

We need two lemmas before we carry out the proof of Theorem 7.4.6.

Lemma 7.4.7. *Let Ω be a bounded $C^{1,\alpha}$ domain for some $\alpha \in (0,1]$ and*

$$u_\varepsilon(x) = \int_\Omega \frac{\partial}{\partial y_j} \{N_\varepsilon(x,y)\} f(y)\, dy$$

for some $1 \le j \le d$. Then

$$\|\nabla u_\varepsilon\|_{L^\infty(\Omega)} \le C \left\{ \ln[\varepsilon^{-1} r_0 + 2] + r_0^\eta \right\} \|f\|_{L^\infty(\Omega)} + C\varepsilon^\eta H_{\varepsilon,\eta}(f), \tag{7.4.24}$$

where $r_0 = \mathrm{diam}(\Omega)$ and

$$H_{\varepsilon,\eta}(f) = \sup \left\{ \frac{|f(x) - f(y)|}{|x-y|^\eta} : \ x, y \in \Omega \text{ and } 0 < |x-y| < \varepsilon \right\}.$$

Proof. For $x \in \Omega$, choose $\hat{x} \in \partial\Omega$ such that $|x - \hat{x}| = \mathrm{dist}(x, \partial\Omega)$. Note that

$$\begin{aligned} \frac{\partial u_\varepsilon}{\partial x_i} &= \int_\Omega \frac{\partial^2}{\partial x_i \partial y_j} \{N_\varepsilon(x,y)\} \cdot \{f(y) - f(x)\}\, dy \\ &\quad + \int_{\partial\Omega} \{n_j(y) - n_j(\hat{x})\} \cdot \frac{\partial}{\partial x_i} \{N_\varepsilon(x,y)\} \cdot f(x)\, d\sigma(y), \end{aligned}$$

where we have used the fact $\int_{\partial\Omega} N_\varepsilon(x,y)\, d\sigma(y) = 0$. This, together with the estimates $|\nabla_x N_\varepsilon(x,y)| \le C|x-y|^{1-d}$ and $|\nabla_x \nabla_y N_\varepsilon(x,y)| \le C|x-y|^{-d}$, gives

$$\begin{aligned} |\nabla u_\varepsilon(x)| &\le C \int_\Omega \frac{|f(y) - f(x)|}{|x-y|^d}\, dy + C\, \|f\|_{L^\infty(\Omega)} \int_{\partial\Omega} \frac{d\sigma(y)}{|y - \hat{x}|^{d-1-\eta}} \\ &\le C \|f\|_{L^\infty(\Omega)} \int_{\Omega \setminus B(x,\varepsilon)} \frac{dy}{|x-y|^d} \\ &\qquad + C H_{\varepsilon,\eta}(f) \int_{|y-x|<\varepsilon} \frac{dy}{|y-x|^{d-\eta}} + C\, \|f\|_{L^\infty(\Omega)} r_0^\eta \\ &\le C \|f\|_{L^\infty(\Omega)} \ln[\varepsilon^{-1} r_0 + 2] + C\varepsilon^\eta H_{\varepsilon,\eta}(f) + C\, \|f\|_{L^\infty(\Omega)} r_0^\eta. \end{aligned}$$

This completes the proof. $\square$

Lemma 7.4.8. *Let Ω be a bounded $C^{2,\eta}$ domain for some $\eta \in (0,1)$. Suppose that $u_\varepsilon \in H^1(T_{3r})$ and $u_0 \in C^{2,\eta}(T_{3r})$ for some $0 < r < c_0 r_0$, where $r_0 = \text{diam}(\Omega)$. Assume that*

$$\mathcal{L}_\varepsilon(u_\varepsilon) = \mathcal{L}_0(u_0) \quad \text{in } T_{3r} \quad \text{and} \quad \frac{\partial u_\varepsilon}{\partial \nu_\varepsilon} = \frac{\partial u_0}{\partial \nu_0} \quad \text{on } I_{3r}.$$

Then, if $0 < \varepsilon < (r/2)$,

$$\begin{aligned} &\left\| \frac{\partial u_\varepsilon^\alpha}{\partial x_i} - \frac{\partial}{\partial x_i}\{\Psi_{\varepsilon,j}^{\alpha\beta}\} \cdot \frac{\partial u_0^\beta}{\partial x_j} \right\|_{L^\infty(T_r)} \\ &\quad \le \frac{C}{r} \fint_{T_{3r}} |u_\varepsilon - u_0| + C\varepsilon r^{-1} \ln[\varepsilon^{-1} r_0 + 2] \|\nabla u_0\|_{L^\infty(T_{3r})} \\ &\quad\quad + C\varepsilon^{1-\rho} r^\rho \ln[\varepsilon^{-1} r_0 + 2] \|\nabla^2 u_0\|_{L^\infty(T_{3r})} \\ &\quad\quad + C\varepsilon r^\rho \ln[\varepsilon^{-1} r_0 + 2] \|\nabla^2 u_0\|_{C^{0,\rho}(T_{3r})} \end{aligned} \tag{7.4.25}$$

for any $0 < \rho < \min(\eta, \tau)$.

Proof. By rescaling and translation, we may assume that $r = 1$ and $0 \in T_1$. Let

$$w_\varepsilon = u_\varepsilon - u_0 - \left\{\Psi_{\varepsilon,j}^\beta - P_j^\beta\right\} \cdot \frac{\partial u_0^\beta}{\partial x_j}.$$

Choose a $C^{2,\eta}$ domain $\widetilde{\Omega}$ such that $T_2 \subset \widetilde{\Omega} \subset T_3$. We now write

$$w_\varepsilon(x) = \int_{\widetilde{\Omega}} \widetilde{N}_\varepsilon(x,y) \mathcal{L}_\varepsilon(w_\varepsilon)\, dy + \int_{\partial\widetilde{\Omega}} \widetilde{N}_\varepsilon(x,y) \frac{\partial w_\varepsilon}{\partial \nu_\varepsilon}\, d\sigma(y) + \fint_{\partial\widetilde{\Omega}} w_\varepsilon$$

for $x \in T_2$, where $\widetilde{N}_\varepsilon(x,y)$ denotes the matrix of Neumann functions for $\mathcal{L}_\varepsilon$ in $\widetilde{\Omega}$. In view of the formula for $\mathcal{L}_\varepsilon(w_\varepsilon)$ in Lemma 7.1.2 and formula (7.4.7) for $\frac{\partial w_\varepsilon}{\partial \nu_\varepsilon}$, we have $w_\varepsilon = w_\varepsilon^{(1)} + w_\varepsilon^{(2)} + c$, where $c = \fint_{\partial\widetilde{\Omega}} w_\varepsilon$,

$$\begin{aligned} w_\varepsilon^{(1)}(x) = &-\varepsilon \int_{\widetilde{\Omega}} \frac{\partial}{\partial y_i}\{\widetilde{N}_\varepsilon(x,y)\} \cdot \{\phi_{jik}(y/\varepsilon)\} \cdot \frac{\partial^2 u_0}{\partial y_j \partial y_k}\, dy \\ &- \int_{\widetilde{\Omega}} \frac{\partial}{\partial y_i}\{\widetilde{N}_\varepsilon(x,y)\} \cdot a_{ij}(y/\varepsilon)\{\Psi_{\varepsilon,k}(y) - P_k(y)\} \cdot \frac{\partial^2 u_0}{\partial y_j \partial y_k}\, dy \\ &+ \int_{\widetilde{\Omega}} \widetilde{N}_\varepsilon(x,y) \cdot a_{ij}(y/\varepsilon) \frac{\partial}{\partial y_j}\{\Psi_{\varepsilon,k}(y) - P_k(y) - \varepsilon\chi_k(y/\varepsilon)\} \cdot \frac{\partial^2 u_0}{\partial y_i \partial y_k}\, dy \end{aligned} \tag{7.4.26}$$

and

$$\begin{aligned} w_\varepsilon^{(2)}(x) = \varepsilon &\int_{\partial\widetilde{\Omega}} \widetilde{N}_\varepsilon(x,y) \cdot n_i(y) \phi_{jik}(y/\varepsilon) \cdot \frac{\partial^2 u_0}{\partial y_j \partial y_k}\, d\sigma(y) \\ &+ \int_{\partial\widetilde{\Omega}} \widetilde{N}_\varepsilon(x,y) \cdot \left\{\frac{\partial u_\varepsilon}{\partial \nu_\varepsilon} - \frac{\partial u_0}{\partial \nu_0}\right\} d\sigma(y) \end{aligned} \tag{7.4.27}$$

(we have suppressed all superscripts for notational simplicity).

To estimate $w_\varepsilon^{(2)}$ in T_1, observe that $\mathcal{L}_\varepsilon(w_\varepsilon^{(2)}) = 0$ in $\widetilde{\Omega}$ and

$$\frac{\partial}{\partial \nu_\varepsilon}\{w_\varepsilon^{(2)}\} = \varepsilon n_i \phi_{jik}(x/\varepsilon)\frac{\partial^2 u_0}{\partial x_j \partial x_k} - \varepsilon \fint_{\partial\widetilde{\Omega}} n_i \phi_{jik}(x/\varepsilon)\frac{\partial^2 u_0}{\partial x_j \partial x_k} \qquad \text{on } I_2,$$

where we have used the fact that $\frac{\partial u_\varepsilon}{\partial \nu_\varepsilon} = \frac{\partial u_0}{\partial \nu_0}$ on I_2. Since

$$\Big\|\frac{\partial}{\partial \nu_\varepsilon}\{w_\varepsilon^{(2)}\}\Big\|_{L^\infty(I_2)} \le C\varepsilon\, \|\nabla^2 u_0\|_{L^\infty(T_3)}$$

and

$$\Big\|\frac{\partial}{\partial \nu_\varepsilon}\{w_\varepsilon^{(2)}\}\Big\|_{C^{0,\rho}(I_2)} \le C\varepsilon^{1-\rho}\|\nabla^2 u_0\|_{L^\infty T_2)} + C\varepsilon\, \|\nabla^2 u_0\|_{C^{0,\rho}(T_2)},$$

it follows from the boundary Lipschitz estimates in Section 6.5 that

$$\|\nabla w_\varepsilon^{(2)}\|_{L^\infty(T_1)} \le C\varepsilon^{1-\rho}\|\nabla^2 u_0\|_{L^\infty(T_2)} + C\varepsilon\, \|\nabla^2 u_0\|_{C^{0,\rho}(T_2)} + C\int_{T_2} |w_\varepsilon^{(2)} - c|$$

for any constant c. This leads to

$$\begin{aligned}
\|\nabla w_\varepsilon\|_{L^\infty(T_1)} &\le \|\nabla w_\varepsilon^{(1)}\|_{L^\infty(T_1)} + \|\nabla w_\varepsilon^{(2)}\|_{L^\infty(T_1)}\\
&\le C\int_{\Omega_2} |w_\varepsilon|\, dx + C\, \|\nabla w_\varepsilon^{(1)}\|_{L^\infty(T_2)} + C\varepsilon^{1-\rho}\|\nabla^2 u_0\|_{L^\infty(T_3)}\\
&\qquad + C\varepsilon\, \|\nabla^2 u_0\|_{C^{0,\rho}(T_3)}.
\end{aligned}$$

Since $|\Psi_{\varepsilon,j}^\beta - P_j^\beta| \le C\varepsilon \ln[\varepsilon^{-1}M + 2]$ by Lemma 7.4.5, we obtain

$$\begin{aligned}
&\Big\|\frac{\partial u_\varepsilon^\alpha}{\partial x_i} - \frac{\partial}{\partial x_i}\{\Psi_{\varepsilon,j}^{\alpha\beta}\}\cdot \frac{\partial u_0^\beta}{\partial x_j}\Big\|_{L^\infty(T_1)}\\
&\quad\le C\int_{\Omega_2} |u_\varepsilon - u_0| + C\varepsilon \ln[\varepsilon^{-1}M + 2]\|\nabla u_0\|_{L^\infty(T_3)}\\
&\quad + C\varepsilon \ln[\varepsilon^{-1}M + 2]\|\nabla^2 u_0\|_{L^\infty(T_3)} + C\varepsilon^{1-\rho}\|\nabla^2 u_0\|_{L^\infty(T_3)}\\
&\quad + C\varepsilon\, \|\nabla^2 u_0\|_{C^{0,\rho}(T_3)} + C\, \|\nabla w_\varepsilon^{(1)}\|_{L^\infty(T_2)}.
\end{aligned} \tag{7.4.28}$$

It remains to estimate $\nabla w_\varepsilon^{(1)}$ on T_2. The first two integrals in the RHS of (7.4.26) may be handled by applying Lemma 7.4.7 on $\widetilde{\Omega}$. Indeed, let

$$f(x) = -\varepsilon\phi_{jik}(x/\varepsilon)\cdot \frac{\partial^2 u_0}{\partial x_j \partial x_k} - a_{ij}(x/\varepsilon)\{\Psi_{\varepsilon,k}(x) - P_k(x)\}\frac{\partial^2 u_0}{\partial x_j \partial x_k}.$$

Note that $\|f\|_{L^\infty(\widetilde{\Omega})} \le C\varepsilon \ln[\varepsilon^{-1}r_0 + 2]\|\nabla^2 u_0\|_{L^\infty(T_3)}$ and

$$\begin{aligned}
H_{\varepsilon,\rho}(f) &\le C\varepsilon^{1-\rho} \ln[\varepsilon^{-1}r_0 + 2]\|\nabla^2 u\|_{L^\infty(T_3)}\\
&\quad + C\varepsilon \ln[\varepsilon^{-1}r_0 + 2]\|\nabla^2 u_0\|_{C^{0,\rho}(T_3)}.
\end{aligned}$$

It follows by Lemma 7.4.7 that the first two integrals in the RHS of (7.4.26) are bounded by

$$C\varepsilon \ln[\varepsilon^{-1} r_0 + 2]\big\{\varepsilon^{-\rho}\|\nabla^2 u_0\|_{L^\infty(T_3)} + \|\nabla^2 u_0\|_{C^{0,\rho}(T_3)}\big\}.$$

Finally, the third integral in (7.4.26) is bounded by

$$C\|\nabla^2 u_0\|_{L^\infty(T_3)} \int_{\widetilde{\Omega}} \frac{|\nabla_y\{\Psi_{\varepsilon,k}(y) - P_y(y) - \varepsilon\chi_k(y/\varepsilon)\}|}{|x-y|^{d-1}}\, dy. \tag{7.4.29}$$

Using the estimate

$$|\nabla_y\{\Psi_{\varepsilon,k}(y) - P_k(y) - \varepsilon\chi_k(y/\varepsilon)\}| \le C \min\big(1, \varepsilon[\text{dist}(y,\partial\Omega)]^{-1}\big),$$

one can show that the integral in (7.4.29) is bounded by $C\varepsilon\big[\ln(\varepsilon^{-1}+2)\big]^2$. As a result, we have proved that

$$\begin{aligned}\|w_\varepsilon^{(1)}\|_{L^\infty(T_2)} &\le C\varepsilon^{1-\rho}\ln[\varepsilon^{-1}M+2]\|\nabla^2 u_0\|_{L^\infty(T_3)}\\ &\quad + C\varepsilon\ln[\varepsilon^{-1}r_0+2]\|\nabla^2 u_0\|_{C^{0,\rho}(T_3)}.\end{aligned}$$

This, together with (7.4.28), yields the desired estimate. □

Proof of Theorem 7.4.6. Since $|\nabla_x N_\varepsilon(x,y)| \le C|x-y|^{1-d}$ and $|\nabla \Psi^\beta_{\varepsilon,j}| \le C$, we may assume that $\varepsilon < |x-y|$ and ρ is small. Fix $x_0, y_0 \in \Omega$, $1 \le \gamma \le d$ and let $r = |x_0 - y_0|/8$. Let

$$u_\varepsilon^\alpha(x) = N_\varepsilon^{\alpha\gamma}(x,y_0) \quad \text{and} \quad u_0^\alpha(x) = N_0^{\alpha\gamma}(x,y_0).$$

Observe that

$$\begin{cases} \mathcal{L}_\varepsilon(u_\varepsilon) = \mathcal{L}_0(u_0) = 0 & \text{in } T(x_0,r),\\ \left(\dfrac{\partial u_\varepsilon}{\partial \nu_\varepsilon}\right)^\alpha = \left(\dfrac{\partial u_0}{\partial \nu_0}\right)^\alpha = -|\partial\Omega|^{-1}\delta^{\alpha\gamma} & \text{on } I(x_0,r).\end{cases}$$

Also, note that $\|\nabla u_0\|_{L^\infty(T(x_0,r))} \le Cr^{1-d}$,

$$\|\nabla^2 u_0\|_{L^\infty(T(x_0,r))} \le Cr^{-d} \quad \text{and} \quad \|\nabla^2 u_0\|_{C^{0,\rho}(T(x_0,r))} \le Cr^{-d-\rho}.$$

Furthermore, it follows from Theorem 7.4.1 that

$$\|u_\varepsilon - u_0\|_{L^\infty(T(x_0,r))} \le C\varepsilon r^{1-d}\ln[\varepsilon^{-1}r+2].$$

Thus, by Lemma 7.4.8, we obtain

$$\left\|\frac{\partial u_\varepsilon^\alpha}{\partial x_i} - \frac{\partial}{\partial x_i}\{\Psi^{\alpha\beta}_{\varepsilon,j}\}\cdot\frac{\partial u_0^\beta}{\partial x_j}\right\|_{L^\infty(T(x_0,r/3))} \le C\varepsilon^{1-\rho}r^{\rho-d}\ln[\varepsilon^{-1}r_0+2].$$

This completes the proof. □

Remark 7.4.9. Using the relation $N_\varepsilon^{*\alpha\beta}(x,y) = N_\varepsilon^{\beta\alpha}(y,x)$, we deduce from Theorem 7.4.6 that for any $x, y \in \Omega$ and $x \neq y$,

$$\begin{aligned} \Big| \frac{\partial}{\partial y_j} \left\{ N_\varepsilon^{\alpha\beta}(x,y) \right\} - \frac{\partial}{\partial y_j} \left\{ \Psi_{\varepsilon,\ell}^{*\beta\sigma}(y) \right\} \cdot \frac{\partial}{\partial y_\ell} \left\{ N_0^{\alpha\sigma}(x,y) \right\} \Big| \\ \le \frac{C\varepsilon^{1-\rho} \ln[\varepsilon^{-1} r_0 + 2]}{|x-y|^{d-\rho}}, \end{aligned} \tag{7.4.30}$$

where $\Psi_{\varepsilon,j}^{*\alpha\beta}$ denotes the Neumann corrector for $\mathcal{L}_\varepsilon^*$ in Ω. Fix β and j. Let

$$u_\varepsilon^\alpha(x) = \frac{\partial}{\partial y_j} \left\{ N_\varepsilon^{\alpha\beta}(x,y) \right\}$$

and

$$u_0^\alpha(x) = \frac{\partial}{\partial y_j} \left\{ \Psi_{\varepsilon,\ell}^{*\beta\sigma}(y) \right\} \cdot \frac{\partial}{\partial y_\ell} \left\{ N_0^{\alpha\sigma}(x,y) \right\}.$$

Note that $\mathcal{L}_\varepsilon(u_\varepsilon) = \mathcal{L}_0(u_0) = 0$ in $\Omega \setminus \{y\}$ and $\frac{\partial u_\varepsilon}{\partial \nu_\varepsilon} = \frac{\partial u_0}{\partial \nu_0} = 0$ on $\partial\Omega$. We may use Lemma 7.4.8 and estimate (7.4.30) to deduce that if Ω is $C^{2,\eta}$ for some $\eta \in (0,1)$,

$$\begin{aligned} \Big| \frac{\partial^2}{\partial x_i \partial y_j} \left\{ N_\varepsilon^{\alpha\beta}(x,y) \right\} - \frac{\partial}{\partial x_i} \left\{ \Psi_{\varepsilon,k}^{\alpha\gamma}(x) \right\} \cdot \frac{\partial^2}{\partial x_k \partial y_\ell} \left\{ N_0^{\gamma\sigma}(x,y) \right\} \cdot \frac{\partial}{\partial y_j} \left\{ \Psi_{\varepsilon,\ell}^{*\beta\sigma}(y) \right\} \Big| \\ \le \frac{C_\rho \, \varepsilon^{1-\rho} \ln[\varepsilon^{-1} r_0 + 2]}{|x-y|^{d+1-\rho}} \end{aligned} \tag{7.4.31}$$

for any $x, y \in \Omega$ and $\rho \in (0,1)$, where C_ρ depends only on μ, λ, τ, ρ, and Ω.

7.5 Convergence rates in L^p and $W^{1,p}$

In this section we establish the convergence rates of solutions u_ε in $L^p(\Omega)$ and error estimates in $W^{1,p}$ for two-scale expansions, using the asymptotic expansions of Green and Neumann functions obtained in Sections 7.3 and 7.4.

We begin with the Dirichlet condition.

Theorem 7.5.1. *Suppose that A satisfies (2.1.2)–(2.1.3) and is 1-periodic. If $m \ge 2$, we also assume that A is Hölder continuous. Let Ω be a bounded $C^{1,1}$ domain. For $F \in L^2(\Omega; \mathbb{R}^m)$ and $\varepsilon \ge 0$, let $u_\varepsilon \in H_0^1(\Omega; \mathbb{R}^m)$ be the solution of $\mathcal{L}_\varepsilon(u_\varepsilon) = F$ in Ω. Then the estimate*

$$\|u_\varepsilon - u_0\|_{L^q(\Omega)} \le C_p \, \varepsilon \, \|F\|_{L^p(\Omega)} \tag{7.5.1}$$

holds if $1 < p < d$ and $\frac{1}{q} = \frac{1}{p} - \frac{1}{d}$, or $p > d$ and $q = \infty$. Moreover,

$$\|u_\varepsilon - u_0\|_{L^\infty(\Omega)} \le C\varepsilon \left[\ln(\varepsilon^{-1} r_0 + 2) \right]^{1-\frac{1}{d}} \|F\|_{L^d(\Omega)}, \tag{7.5.2}$$

where $r_0 = \mathrm{diam}(\Omega)$.

Proof. This theorem is a corollary of Theorem 7.3.1. Indeed, by the Green function representation and estimate (7.3.3),

$$|u_\varepsilon(x) - u_0(x)| \le C\varepsilon \int_\Omega \frac{|F(y)|}{|x-y|^{d-1}}\, dy, \qquad \text{for any } x \in \Omega.$$

This leads to (7.5.1) for $1 < p < d$ and $\frac{1}{q} = \frac{1}{p} - \frac{1}{d}$ by the estimate for fractional integrals in Proposition 3.1.9. The case of $p > d$ and $q = \infty$ follows readily from Hölder's inequality. To see (7.5.2) for $d \ge 3$, we bound $|G_\varepsilon(x,y) - G_0(x,y)|$ by $C|x-y|^{2-d}$ when $|x-y| < \varepsilon$, and by $C\varepsilon|x-y|^{1-d}$ when $|x-y| \ge \varepsilon$. By Hölder's inequality, this gives

$$\begin{aligned}
|u_\varepsilon(x) - u_0(x)| &\le C \int_{\Omega\cap B(x,\varepsilon)} \frac{|F(y)|}{|x-y|^{d-2}}\, dy + C\varepsilon \int_{\Omega\setminus B(x,\varepsilon)} \frac{|F(y)|}{|x-y|^{d-1}}\, dy \\
&\le C\varepsilon \, \|F\|_{L^d(\Omega)} + C\varepsilon \big[\ln\big(\varepsilon^{-1} r_0 + 2\big)\big]^{1-\frac{1}{d}} \|F\|_{L^d(\Omega)} \\
&\le C\varepsilon \big[\ln\big(\varepsilon^{-1} r_0 + 2\big)\big]^{1-\frac{1}{d}} \|F\|_{L^d(\Omega)}.
\end{aligned}$$

If $d = 2$, we bound $|G_\varepsilon(x,y) - G_0(x,y)|$ by $C(1 + |\ln|x-y||)$ when $|x-y| < \varepsilon$. The rest is the same as in the case $d \ge 3$. □

Theorem 7.5.2. *Suppose that A is 1-periodic and satisfies (2.1.2)–(2.1.3). Also assume that A satisfies (4.0.2). Let Ω be a bounded $C^{2,\eta}$ domain and $1 < p < \infty$. For $F \in L^p(\Omega; \mathbb{R}^m)$ and $\varepsilon \ge 0$, let $u_\varepsilon \in W_0^{1,p}(\Omega; \mathbb{R}^m)$ be the weak solution of $\mathcal{L}_\varepsilon(u_\varepsilon) = F$ in Ω. Then*

$$\Big\| u_\varepsilon - u_0 - \big\{\Phi_{\varepsilon,j}^\beta - P_j^\beta\big\} \frac{\partial u_0^\beta}{\partial x_j} \Big\|_{W_0^{1,p}(\Omega)} \le C_p\, \varepsilon \, \big\{ \ln[\varepsilon^{-1} r_0 + 2]\big\}^{4|\frac{1}{2} - \frac{1}{p}|} \|F\|_{L^p(\Omega)}, \tag{7.5.3}$$

where $r_0 = \operatorname{diam}(\Omega)$ and C_p depends only on p, μ, λ, τ, and Ω.

Proof. The case $p = 2$ is contained in Theorem 7.1.4. We will show that for any $1 \le p \le \infty$,

$$\Big\| \frac{\partial u_\varepsilon^\alpha}{\partial x_i} - \frac{\partial}{\partial x_i}\big\{\Phi_{\varepsilon,j}^{\alpha\beta}\big\} \cdot \frac{\partial u_0^\beta}{\partial x_j} \Big\|_{L^p(\Omega)} \le C\varepsilon \, \big\{\ln[\varepsilon^{-1} r_0 + 2]\big\}^{4|\frac{1}{2} - \frac{1}{p}|} \|F\|_{L^p(\Omega)}. \tag{7.5.4}$$

This, together with $\|\Phi_{\varepsilon,j}^\beta - P_j^\beta\|_\infty \le C\varepsilon$ and the estimate

$$\|\nabla^2 u_0\|_{L^p(\Omega)} \le C \|F\|_{L^p(\Omega)}$$

for $1 < p < \infty$, gives (7.5.3).

To see (7.5.4), we use Theorem 7.3.4 for $|x-y| \ge \varepsilon$ as well as estimates for $|\nabla_x G_\varepsilon(x,y)|$ and $|\nabla \Phi_\varepsilon|$ to deduce that

$$\Big| \frac{\partial u_\varepsilon^\alpha}{\partial x_i} - \frac{\partial}{\partial x_i}\big\{\Phi_{\varepsilon,j}^{\alpha\beta}\big\} \cdot \frac{\partial u_0^\beta}{\partial x_j} \Big| \le C \int_\Omega K_\varepsilon(x,y) |f(y)|\, dy,$$

where

$$K_\varepsilon(x,y) = \begin{cases} \varepsilon\, |x-y|^{-d} \ln\left[\varepsilon^{-1}|x-y|+2\right], & \text{if } |x-y| \ge \varepsilon, \\ |x-y|^{1-d}, & \text{if } |x-y| < \varepsilon. \end{cases} \tag{7.5.5}$$

It is not hard to show that

$$\sup_{x\in\Omega} \int_\Omega K_\varepsilon(x,y)\, dy + \sup_{y\in\Omega} \int_\Omega K_\varepsilon(x,y)\, dx \le C\varepsilon \big\{ \ln[\varepsilon^{-1} r_0 + 2] \big\}^2.$$

This gives (7.5.4) in the case $p = 1$ or ∞. Since the case $p = 2$ is contained in Theorem 7.1.4, the proof is finished by applying the Riesz–Thorin interpolation theorem [20]. □

Next we turn to the Neumann boundary conditions.

Theorem 7.5.3. *Suppose that A is 1-periodic and satisfies conditions* (2.1.20) *and* (4.0.2). *Let Ω a bounded $C^{1,1}$ domain and $1 < p < \infty$. For $\varepsilon \ge 0$ and $F \in L^p(\Omega; \mathbb{R}^m)$ with $\int_\Omega F\, dx = 0$, let $u_\varepsilon \in W^{1,p}(\Omega; \mathbb{R}^m)$ be the solution to the Neumann problem $\mathcal{L}_\varepsilon(u_\varepsilon) = F$ in Ω, $\frac{\partial u_\varepsilon}{\partial \nu_\varepsilon} = 0$ on $\partial\Omega$, and $\int_{\partial\Omega} u_\varepsilon\, d\sigma = 0$. Then*

$$\|u_\varepsilon - u_0\|_{L^q(\Omega)} \le C\varepsilon \ln[\varepsilon^{-1} r_0 + 2] \|F\|_{L^p(\Omega)} \tag{7.5.6}$$

holds if $1 < p < d$ and $\frac{1}{q} = \frac{1}{p} - \frac{1}{d}$, or $p > d$ and $q = \infty$, where $r_0 = \operatorname{diam}(\Omega)$. Moreover,

$$\|u_\varepsilon - u_0\|_{L^\infty(\Omega)} \le C\varepsilon \big[\ln(\varepsilon^{-1} r_0 + 2)\big]^{2-\frac{1}{d}} \|F\|_{L^d(\Omega)}. \tag{7.5.7}$$

Proof. This theorem is a corollary of Theorem 7.4.1. Note that by the estimate (7.4.4),

$$\begin{aligned} |u_\varepsilon(x) - u_0(x)| &\le \int_\Omega |N_\varepsilon(x,y) - N_0(x,y)| |F(y)|\, dy \\ &\le C\varepsilon \ln(\varepsilon^{-1} r_0 + 2) \int_\Omega \frac{|F(y)| dy}{|x-y|^{d-1}}. \end{aligned}$$

The rest of the proof is the same as that of Theorem 7.5.1. □

Theorem 7.5.4. *Suppose that A is 1-periodic and satisfies conditions* (2.1.20) *and* (4.0.2). *Let Ω be a bounded $C^{2,\alpha}$ domain and $1 < p < \infty$. For $\varepsilon \ge 0$ and $F \in L^p(\Omega; \mathbb{R}^m)$ with $\int_\Omega F = 0$, let $u_\varepsilon \in W^{1,p}(\Omega; \mathbb{R}^m)$ be the solution of the Neumann problem $\mathcal{L}_\varepsilon(u_\varepsilon) = F$ in Ω, $\frac{\partial u_\varepsilon}{\partial \nu_\varepsilon} = 0$ on $\partial\Omega$, and $\int_{\partial\Omega} u_\varepsilon\, d\sigma = 0$. Then*

$$\Big\| u_\varepsilon - u_0 - \big\{ \Psi^\beta_{\varepsilon,j} - P^\beta_j \big\} \frac{\partial u_0^\beta}{\partial x_j} \Big\|_{W^{1,p}(\Omega)} \le C_t\, \varepsilon^t \|F\|_{L^p(\Omega)} \tag{7.5.8}$$

for any $t \in (0,1)$, where $\big(\Psi^\beta_{\varepsilon,j}\big)$ is a matrix of Neumann correctors for $\mathcal{L}_\varepsilon$ in Ω such that $\Psi^\beta_{\varepsilon,j}(x_0) = P^\beta_j(x_0)$ for some $x_0 \in \Omega$, and C_t depends only on μ, λ, τ, t, p and Ω.

Proof. Recall that

$$\|\Psi_{\varepsilon,j}^{\beta} - P_k^{\beta}\|_{L^\infty(\Omega)} \le C\varepsilon \ln[\varepsilon^{-1} r_0 + 2].$$

Since $\|u_0\|_{W^{2,p}(\Omega)} \le C\|F\|_{L^p(\Omega)}$, in view of Theorem 7.5.3, it suffices to prove that

$$\left\| \frac{\partial u_\varepsilon^\alpha}{\partial x_i} - \frac{\partial}{\partial x_i}\{\Psi_{\varepsilon,j}^\beta\} \cdot \frac{\partial u_0^\beta}{\partial x_j} \right\|_{L^p(\Omega)} \le C_t\, \varepsilon^t \|F\|_{L^p(\Omega)}. \tag{7.5.9}$$

We will prove that the estimate (7.5.9) holds for any $1 \le p \le \infty$. To this end, observe that since $u_\varepsilon(x) = \int_\Omega N_\varepsilon(x,y) F(y)\, dy$, by Theorem 7.4.6,

$$\left| \frac{\partial u_\varepsilon^\alpha}{\partial x_i} - \frac{\partial}{\partial x_i}\{\Psi_{\varepsilon,j}^\beta\} \cdot \frac{\partial u_0^\beta}{\partial x_j} \right| \le C_\rho\, \varepsilon^{1-\rho} \ln[\varepsilon^{-1} r_0 + 2] \int_\Omega \frac{|F(y)|\, dy}{|x-y|^{d-\rho}} \tag{7.5.10}$$

for any $\rho \in (0,1)$. Note that if $\rho < 1-t$, then $\varepsilon^{1-\rho} \ln[\varepsilon^{-1} r_0 + 2] \le C\varepsilon^t$. Estimate (7.5.9) follows readily from (7.5.10). □

7.6 Notes

The material in Sections 7.1 and 7.2 is taken from [64] by C. Kenig, F. Lin, and Z. Shen.

The material in Sections 7.3, 7.4, and 7.5 is mostly taken from [66] by C. Kenig, F. Lin, and Z. Shen. Some modifications are made to cover the two-dimensional case and to remove the symmetry condition for the Neumann problems. Earlier work on asymptotic expansions for Green functions and Poisson kernels may be found in [14] by M. Avellaneda and F. Lin.

Higher-order convergence. It is proved in Chapter 3 that if Ω is a bounded $C^{1,1}$ domain in $\mathbb{R}^d$,

$$\|u_\varepsilon - u_0\|_{L^2(\Omega)} \le C\varepsilon\, \|u_0\|_{H^2(\Omega)},$$

where u_ε is the solution of the Dirichlet problem (3.0.1) or the Neumann problem (3.0.2), and u_0 the corresponding homogenized solution. It is natural to ask if there exists a function v_0, independent of ε, such that

$$\|u_\varepsilon - u_0 - \varepsilon(\chi(x/\varepsilon)\nabla u_0 + v_0)\|_{L^2(\Omega)} = O(\varepsilon^{1+\rho})$$

for some $\rho > 0$. In the case of the Dirichlet condition the problem may be reduced to the study of the Dirichlet problem with oscillating boundary data,

$$\begin{cases} \mathcal{L}_\varepsilon(w_\varepsilon) = 0 & \text{in } \Omega, \\ \quad w_\varepsilon = h(x, x/\varepsilon) & \text{on } \partial\Omega, \end{cases} \tag{7.6.1}$$

where the function $h(x,y)$ is 1-periodic in the variable y. This is mainly due to the fact that the boundary value of $u_\varepsilon - u_0 - \varepsilon\chi(x/\varepsilon)\nabla u_0$ on $\partial\Omega$ is given by

$-\varepsilon\chi(x/\varepsilon)\nabla u_0$, as $u_\varepsilon = u_0$ on $\partial\Omega$. Similarly, for the Neumann problem (3.0.2), by computing the conormal derivative

$$\frac{\partial}{\partial \nu_\varepsilon}\big(u_\varepsilon - u_0 - \varepsilon\chi(x/\varepsilon)\nabla u_0\big)$$

and with the help of the flux correctors, one is led to the study of the Neumann problem with first-order oscillating boundary data,

$$\begin{cases} \mathcal{L}_\varepsilon(w_\varepsilon) = 0 & \text{in } \Omega, \\ \dfrac{\partial w_\varepsilon}{\partial \nu_\varepsilon} = T_{ij}(x)\cdot\nabla\big(h_{ij}(x, x/\varepsilon)\big) & \text{on } \partial\Omega, \end{cases} \tag{7.6.2}$$

where $T_{ij}(x) = n_i(x)e_j - n_j(x)e_i = (0,\dots,n_i,\dots,-n_j,\dots,0)$ is a tangential vector field on $\partial\Omega$ and the functions $h_{ij}(x,y)$ are 1-periodic in y. We refer the reader to [1, 48, 49, 4] for the study of homogenization and boundary layers of the Dirichlet problem (7.6.1), and to [92] the Neumann problem (7.6.2). In particular, under the assumptions that A is elliptic, 1-periodic, and sufficiently smooth and that Ω is sufficiently smooth and strictly convex, it was proved in [49] that the solution w_ε of (7.6.1) converges to w_0 in $L^2(\Omega;\mathbb{R}^m)$. Moreover, the limit w_0 is the solution of the Dirichlet problem

$$\begin{cases} \mathcal{L}_0(w_0) = 0 & \text{in } \Omega, \\ w_0 = \overline{h} & \text{on } \partial\Omega, \end{cases} \tag{7.6.3}$$

where the homogenized data $\overline{h}$ at x depends on $h(x,\cdot)$, A and $n(x)$. The near optimal rates of convergence,

$$\|w_\varepsilon - w_0\|_{L^2(\Omega)} \le \begin{cases} C_\sigma\, \varepsilon^{\frac{1}{2}-\sigma} & \text{if } d \ge 3, \\ C_\sigma\, \varepsilon^{\frac{1}{4}-\sigma} & \text{if } d = 2, \end{cases}$$

were obtained in [4, 92] for any $\sigma \in (0, 1/4)$, where C_σ depends on σ, A, Ω, and the function h. Analogous results for the Neumann problem (7.6.2) were established in [92] for $d \ge 3$. As a result, it follows from [4, 92] that there exists v_0, independent of ε, such that

$$\|u_\varepsilon - u_0 - \varepsilon\chi_k(x/\varepsilon)\frac{\partial u_0}{\partial x_k} - \varepsilon v_0\|_{L^2(\Omega)} \le C_\sigma\, \varepsilon^{\frac{3}{2}-\sigma}\|u_0\|_{W^{3,\infty}(\Omega)} \tag{7.6.4}$$

for any $\sigma \in (0, 1/4)$, where C_σ depends only on σ, A and Ω. We mention that the starting points for the approaches used in [4, 92] are the asymptotic expansions of Green functions and Neumann functions established in Sections 7.3 and 7.4.

Chapter 8

L^2 Estimates in Lipschitz Domains

In this chapter we study L^2 boundary value problems for $\mathcal{L}_\varepsilon(u_\varepsilon) = 0$ in a bounded Lipschitz domain Ω. More precisely, we shall be interested in uniform estimates for the L^2 Dirichlet problem

$$\begin{cases} \mathcal{L}_\varepsilon(u_\varepsilon) = 0 & \text{in } \Omega, \\ u_\varepsilon = f \in L^2(\partial\Omega) & \text{on } \partial\Omega, \\ (u_\varepsilon)^* \in L^2(\partial\Omega), \end{cases} \tag{8.0.1}$$

the L^2 Neumann problem

$$\begin{cases} \mathcal{L}_\varepsilon(u_\varepsilon) = 0 & \text{in } \Omega, \\ \frac{\partial u_\varepsilon}{\partial \nu_\varepsilon} = g \in L^2(\partial\Omega) & \text{on } \partial\Omega, \\ (\nabla u_\varepsilon)^* \in L^2(\partial\Omega), \end{cases} \tag{8.0.2}$$

as well as the L^2 regularity problem

$$\begin{cases} \mathcal{L}_\varepsilon(u_\varepsilon) = 0 & \text{in } \Omega, \\ u_\varepsilon = f \in H^1(\partial\Omega) & \text{on } \partial\Omega, \\ (\nabla u_\varepsilon)^* \in L^2(\partial\Omega), \end{cases} \tag{8.0.3}$$

where $(u_\varepsilon)^*$ denotes the nontangential maximal function of u_ε, defined by (3.2.14). We will call the coefficient matrix $A \in \Lambda(\mu, \lambda, \tau)$ if A is 1-periodic and satisfies the Legendre ellipticity condition (2.1.20) and the Hölder continuity condition (4.0.2). Under the assumptions that $A \in \Lambda(\mu, \lambda, \tau)$ and $A^* = A$, we will show that the solutions to (8.0.1), (8.0.2) and (8.0.3) satisfy the estimates

$$\begin{aligned} \|(u_\varepsilon)^*\|_{L^2(\partial\Omega)} &\le C\, \|f\|_{L^2(\partial\Omega)}, \\ \|(\nabla u_\varepsilon)^*\|_{L^2(\partial\Omega)} &\le C\, \|g\|_{L^2(\partial\Omega)}, \\ \|(\nabla u_\varepsilon)^*\|_{L^2(\partial\Omega)} &\le C\, \|f\|_{H^1(\partial\Omega)}, \end{aligned} \tag{8.0.4}$$

Z. Shen, *Periodic Homogenization of Elliptic Systems*, Operator Theory: Advances and Applications 269, https://doi.org/10.1007/978-3-319-91214-1_8

respectively, where C depends only on μ, λ, τ, and the Lipschitz character of Ω. Since the Lipschitz character of Ω is scale-invariant, the constant C in (8.0.4) is independent of diam(Ω). Moreover, in view of the rescaling property (4.0.3) for $\mathcal{L}_\varepsilon$, these estimates are scale-invariant and as a result, it suffices to consider the case $\varepsilon = 1$.

The estimates in (8.0.4) are established by the method of layer potentials – the classical method of integral equations. In Section 8.1 we introduce the non-tangential convergence in Lipschitz domains and formulate the L^p boundary value problems for $\mathcal{L}_\varepsilon$. Sections 8.2–8.4 are devoted to the study of mapping properties of singular integral operators associated with the single- and double-layer potentials for $\mathcal{L}_\varepsilon$. The basic insight is to approximate the fundamental solution $\Gamma_\varepsilon(x, y)$ and its derivatives by freezing the coefficients when $|x - y| \le \varepsilon$. For $|x - y| > \varepsilon$, we use the asymptotic expansions of $\Gamma_\varepsilon(x, y)$ and its derivatives, established in Chapter 4, to bound the operators by the corresponding operators associated with $\mathcal{L}_0$.

The crucial step in the use of layer potentials to solve L^2 boundary value problems in Lipschitz domains is to establish the following Rellich estimates:

$$\begin{cases} \|\nabla u_\varepsilon\|_{L^2(\partial\Omega)} \le C \left\| \dfrac{\partial u_\varepsilon}{\partial \nu_\varepsilon} \right\|_{L^2(\partial\Omega)}, \\ \|\nabla u_\varepsilon\|_{L^2(\partial\Omega)} \le C \left\| \nabla_{\tan} u_\varepsilon \right\|_{L^2(\partial\Omega)}, \end{cases} \tag{8.0.5}$$

for (suitable) solutions of $\mathcal{L}_\varepsilon(u_\varepsilon) = 0$ in Ω, where $\nabla_{\tan} u_\varepsilon$ denotes the tangential gradient of u_ε on $\partial\Omega$. In Section 8.5 we give the proof of (8.0.5) and solve the L^2 Dirichlet, Neumann, and regularity problems in a Lipschitz domain in the case $\mathcal{L} = -\Delta$. These are classical results, due to B. Dahlberg, D. Jerison, C. Kenig, and G. Verchota [30, 31, 60, 104], which will be needed to deal with the general case by a method of continuity. In Section 8.6 we show that for a general operator $\mathcal{L}_\varepsilon$, the Rellich estimates in (8.0.5) are more or less equivalent to the well-posedness of (8.0.2) and (8.0.3).

In Section 8.7 we establish the estimates in (8.0.5) in the small-scale case, where diam(Ω) $\le \varepsilon$. This is a local result and the periodicity of A is not needed. If the coefficient matrix A is Lipschitz continuous, the estimates follow by using Rellich identities. If A is only Hölder continuous, the proof is more involved and uses a three-step approximate argument. The Rellich estimates in (8.0.5) for the large-scale case, where diam(Ω) $> \varepsilon$, are proved in Section 8.8. To do this we first use the small-scale estimates to reduce the problem to the L^2 estimate of ∇u_ε on a boundary layer $\{x \in \Omega : \text{dist}(x, \partial\Omega) < \varepsilon\}$. The latter is then handled by using the $O(\sqrt{\varepsilon})$ error estimates in $H^1(\Omega)$ we proved in Chapter 3 for a two-scale expansion.

The proof for (8.0.4) is given in Section 8.9, assuming $\partial\Omega$ is connected and $d \ge 3$. The case of arbitrary Lipschitz domains in $\mathbb{R}^d$, $d \ge 2$, is treated in Section 8.10. Finally, in Section 8.11, we prove the square function estimate as well as the $H^{1/2}$ estimate for solutions of Dirichlet problem (8.0.1).

8.1 Lipschitz domains and nontangential convergence

Recall that a bounded domain Ω in $\mathbb{R}^d$ is called a *Lipschitz domain* if there exists $r_0 > 0$ such that for each point $z \in \partial\Omega$, there is a new coordinate system of $\mathbb{R}^d$, obtained from the standard Euclidean coordinate system through translation and rotation, so that $z = (0,0)$ and

$$\begin{aligned}&B(z,r_0)\cap\Omega\\&=B(z,r_0)\cap\big\{(x',x_d)\in\mathbb{R}^d:\ x'\in\mathbb{R}^{d-1}\text{ and }x_d>\psi(x')\big\},\end{aligned}\tag{8.1.1}$$

where $\psi : \mathbb{R}^{d-1} \to \mathbb{R}$ is a Lipschitz function and $\psi(0) = 0$. Roughly speaking, the Lipschitz domains constitute a class of domains that satisfy the uniform interior and exterior cone conditions. This makes it possible to extend the notion of nontangential convergence and nontangential maximal functions from the upper half-space to a general Lipschitz domain.

Let Ω be a Lipschitz domain. Then $\partial\Omega$ possesses a tangent plane and a unit outward normal n at a.e. $z \in \partial\Omega$ with respect to the surface measure $d\sigma$ on $\partial\Omega$. In the local coordinate system for which (8.1.1) holds, one has

$$\begin{aligned}d\sigma&=\sqrt{|\nabla\psi(x')|^2+1}\,dx_1\cdots dx_{d-1},\\n&=\frac{(\nabla\psi(x'),-1)}{\sqrt{|\nabla\psi(x')|^2+1}}.\end{aligned}$$

Note that the Lipschitz function ψ in (8.1.1) may be taken to have compact support.

Let $r, h > 0$. We call Z a cylinder of radius r and height $2h$ if Z is obtained from the set $\{(x', x_d) \in \mathbb{R}^d : \ |x'| < r \text{ and } -h < x_d < h\}$ through translation and rotation. We will use tZ to denote the dilation of Z about its center by a factor of t. It follows that for each $z \in \partial\Omega$, we can find a Lipschitz function ψ on $\mathbb{R}^{d-1}$, a cylinder Z centered at z with radius r and height $2ar$, and a new coordinate system of $\mathbb{R}^d$ with the x_d axis containing the axis of Z, in which

$$10Z\cap\Omega=10Z\cap\big\{(x',x_d)\in\mathbb{R}^d:\ x'\in\mathbb{R}^{d-1}\text{ and }x_d>\psi(x')\big\}\tag{8.1.2}$$

and $z = (0, \psi(0)) = (0,0)$, where $a = 10(\|\nabla\psi\|_\infty + 1)$. We will call such cylinder Z a coordinate cylinder and the pair (Z, ψ) a coordinate pair. Since $\partial\Omega$ is compact, there exist $M > 0$ and a finite number of coordinate pairs,

$$\big\{(Z_i,\psi_i): i=1,2,\ldots,N\big\},$$

with the same radius r_0 such that $\|\nabla\psi_i\|_\infty \le M$ and $\partial\Omega$ is covered by Z_1, Z_2, $\ldots$, Z_N. Such Ω is said to belong to $\Xi(M,N)$ [4]. The constants M and N, which are translation and dilation invariant, describe the *Lipschitz character* of Ω. We

[4] Let Ω be a bounded C^1 domain. Then for any $\delta > 0$, there exists $N = N_\delta$ such that $\Omega \in \Xi(\delta, N)$.

shall say that a positive constant C depends on the Lipschitz character of Ω, if there exist M and N such that $\Omega \in \Xi(M,N)$ and the constant can be made uniform for all Lipschitz domains in $\Xi(M,N)$. For example, there is a constant C depending only on d and the Lipschitz character of Ω such that $|\partial\Omega| \le Cr_0^{d-1}$. By the isoperimetric inequality, we also have $|\Omega| \le Cr_0^d$, where C depends only on d and the Lipschitz character of Ω.

Let $h, \beta > 0$. We call Γ a (two-component) cone of height $2h$ and aperture β if Γ is obtained from the set

$$\Big\{(x', x_d) \in \mathbb{R}^d : \ |x'| < \beta x_d \text{ and } -h < x_d < h\Big\}$$

through translation and rotation. Let (Z, ψ) be a coordinate pair and r the radius of Z. For each $z \in 8Z \cap \partial\Omega$, it is not had to see that the cone with vertex at z, axis parallel to that of Z, height r and aperture $(\|\nabla\psi\|_\infty + 1)^{-1}$ has the property that one component is in Ω and the other in $\mathbb{R}^d \setminus \overline{\Omega}$. By a simple geometric observation, the cone with vertex at z, axis parallel to that of Z, height $r/2$ and aperture $\{2(\|\nabla\psi\|_\infty + 1)\}^{-1}$, is contained in the set

$$\gamma_\alpha(z) = \Big\{x \in \mathbb{R}^d \setminus \partial\Omega : \ |x - z| < \alpha \operatorname{dist}(x, \partial\Omega)\Big\}, \tag{8.1.3}$$

if $\alpha = \alpha(d, M) > 1$ is sufficiently large. By compactness, it is possible to choose $\alpha > 1$, depending only on d and the Lipschitz character of Ω, so that for any $z \in \partial\Omega$, $\gamma_\alpha(z)$ contains a cone of some fixed height and aperture, with vertex at z, one component in Ω and the other in $\mathbb{R}^d \setminus \overline{\Omega}$. Such $\{\gamma_\alpha(z) : z \in \partial\Omega\}$ will be called a family of *nontangential approach regions* for Ω.

Definition 8.1.1. Let Ω be a Lipschitz domain and $\{\gamma_\alpha(z) : z \in \partial\Omega\}$ a family of nontangential approach regions for Ω. For $u \in C(\Omega)$, the *nontangential maximal function* of u on $\partial\Omega$ is defined by

$$(u)^*_\alpha(z) = \sup\big\{|u(x)| : \ x \in \Omega \text{ and } x \in \gamma_\alpha(z)\big\}. \tag{8.1.4}$$

We say that $u = f$ on $\partial\Omega$ in the sense of nontangential convergence, written as $u = f$ n.t. on $\partial\Omega$, if

$$\lim_{\substack{x \to z \\ x \in \Omega \cap \gamma_\alpha(z)}} u(x) = f(z) \qquad \text{for a.e. } z \in \partial\Omega. \tag{8.1.5}$$

The nontangential maximal function $(u)^*_\alpha$ depends on the parameter α. However, the following proposition shows that the L^p norms of $(u)^*_\alpha$ are equivalent for different α's.

Proposition 8.1.2. *Let Ω be a Lipschitz domain and $1 < \alpha < \beta$. Then, for any $u \in C(\Omega)$ and any $t > 0$,*

$$\big|\{z \in \partial\Omega : \ (u)^*_\beta(z) > t\}\big| \le C\big|\{z \in \partial\Omega : \ (u)^*_\alpha(z) > t\}\big|, \tag{8.1.6}$$

where C depends only on α, β, and the Lipschitz character of Ω. Consequently,

$$\|(u)^*_\alpha\|_{L^p(\partial\Omega)} \le \|(u)^*_\beta\|_{L^p(\partial\Omega)} \le C\,\|(u)^*_\alpha\|_{L^p(\partial\Omega)}$$

for any $0 < p \le \infty$.

Proof. Fix $t > 0$ and let

$$E = \big\{z \in \partial\Omega:\ (u)^*_\alpha(z) > t\big\}.$$

We will show that there exists a constant $c > 0$, depending only on α, β, and the Lipschitz character of Ω, such that

$$\big\{z \in \partial\Omega:\ (u)^*_\beta(z) > t\big\} \subset \big\{z \in \partial\Omega:\ \mathcal{M}_{\partial\Omega}(\chi_E)(z) \ge c\big\}, \tag{8.1.7}$$

where $\mathcal{M}_{\partial\Omega}$ denotes the Hardy–Littlewood maximal operator on $\partial\Omega$. Since $\mathcal{M}_{\partial\Omega}$ is of weak type $(1, 1)$, the estimate (8.1.6) follows readily from (8.1.7).

To prove (8.1.7), we suppose that $z_0 \in \partial\Omega$ and $(u)^*_\beta(z_0) > t$. Then there exists $x \in \Omega$ such that $|u(x)| > t$ and $|x - z_0| < \beta\,\text{dist}(x, \partial\Omega)$. Choose $y_0 \in \partial\Omega$ so that $r = |x - y_0| = \text{dist}(x, \partial\Omega)$. Observe that if $y \in \partial\Omega$ and $|y - y_0| < (\alpha - 1)r$, we have

$$|x - y| \le |x - y_0| + |y - y_0| < \alpha r = \alpha\,\text{dist}(x, \partial\Omega).$$

This implies that $(u)^*_\alpha(y) \ge |u(x)| > t$. Thus,

$$B(y_0, (\alpha - 1)r) \cap \partial\Omega \subset E.$$

Finally, we note that $|z_0 - y_0| \le |z_0 - x| + |x - y_0| < (1 + \beta)r$. It follows that

$$B(z_0, (\alpha + \beta)r) \cap E \supset B(y_0, (\alpha - 1)r) \cap E.$$

Hence,

$$\frac{|B(z_0, (\alpha + \beta)r) \cap E|}{|B(z_0, (\alpha + \beta)r) \cap \partial\Omega|} \ge \frac{|B(y_0, (\alpha - 1)r) \cap \partial\Omega|}{|B(z_0, (\alpha + \beta)r) \cap \partial\Omega|} \ge c, \tag{8.1.8}$$

where $c > 0$ depends only on α, β, and the Lipschitz character of Ω. The last inequality in (8.1.8) follows from the fact that $|B(z, r) \cap \partial\Omega| \approx r^{d-1}$ for any $z \in \partial\Omega$ and $0 < r < \text{diam}(\partial\Omega)$. From (8.1.8) we conclude that $\mathcal{M}_{\partial\Omega}(\chi_E)(z_0) \ge c$. This finishes the proof of (8.1.7). □

Because of Proposition 8.1.2, from now on, we shall fix a family of nontangential approach regions for Ω and suppress the parameter α in the notation of nontangential maximal functions.

The next proposition will enable us to control $(u)^*$ by $(\nabla u)^*$.

Proposition 8.1.3. *Let Ω be a Lipschitz domain in $\mathbb{R}^d, d \ge 2$. Suppose that $u \in C^1(\Omega)$ and $(\nabla u)^* \in L^p(\partial\Omega)$ for some $p \ge 1$. Then $(\nabla u)^* \in L^q(\partial\Omega)$, where*

$$\begin{cases} 1 < q < \dfrac{d-1}{d-2} & \text{if } p = 1, \\ q = \dfrac{p(d-1)}{d-1-p} & \text{if } 1 < p < d-1, \\ 1 < q < \infty & \text{if } p = d-1, \\ q = \infty & \text{if } p > d-1. \end{cases}$$

Proof. Let $K = \{x \in \Omega : \operatorname{dist}(x, \partial\Omega) \ge \delta\, r_0\}$, where $\delta > 0$ is sufficiently small. We claim that for any $z \in \partial\Omega$,

$$\begin{aligned} (u)^*(z) &\le \sup_K |u| + C_\delta \int_{\partial\Omega} \frac{(\nabla u)^*(y)}{|z-y|^{d-2}}\, d\sigma(y) \qquad \text{if } d \ge 3, \\ (u)^*(z) &\le \sup_K |u| + C_{\delta,\eta} \int_{\partial\Omega} \frac{(\nabla u)^*(y)}{|z-y|^{\eta}}\, d\sigma(y) \qquad \text{if } d = 2 \end{aligned} \tag{8.1.9}$$

for any $\eta \in (0,1)$. The desired estimate for $(u)^*$ follows readily from the estimates for the fractional integral

$$I_t(f)(z) = \int_{\partial\Omega} \frac{f(y)}{|z-y|^{d-1-t}}\, d\sigma(y)$$

on $\partial\Omega$. Indeed, it is known that $\|I_t(f)\|_{L^q(\partial\Omega)} \le C\|f\|_{L^p(\partial\Omega)}$, where $1 < p < \frac{d-1}{t}$ and $\frac{1}{q} = \frac{1}{p} - \frac{t}{d-1}$. If $f \in L^1(\partial\Omega)$, we have $I_t(f) \in L^{q,\infty}(\partial\Omega) \subset L^{q_1}(\partial\Omega)$, where $q = \frac{d-1}{d-1-t}$ and $q_1 < q$. We refer the reader to [95, pp. 118–121] for a proof of the (L^p, L^q) estimates in the case of $\mathbb{R}^d$. The results extend readily to $\partial\Omega$ by a simple localization argument.

To prove (8.1.9), we fix $z \in \partial\Omega$. By translation and rotation we may assume that $z = 0$ and (8.1.1) holds. Let $x = (x', x_d) \in \gamma_\alpha(z)$ and $x \notin K$. Note that if $\widetilde{x} = (x', s)$, then

$$\begin{aligned} |\nabla u(\widetilde{x})| &\le \inf\Big\{(\nabla u)^*(y) : \ |\widetilde{x} - y| < \alpha \operatorname{dist}(\widetilde{x}, \partial\Omega)\Big\} \\ &\le \frac{C}{s^{d-1}} \int_{\substack{y \in \partial\Omega \\ |y| \le cs}} (\nabla u)^*(y)\, d\sigma(y). \end{aligned}$$

Choose $a \in \mathbb{R}$ so that $(x', a) \in K$. It follows that

$$\begin{aligned} |u(x)| &\le |u(x', a)| + \int_{x_d}^a |\nabla u(x', s)|\, ds \\ &\le \sup_K |u| + C \int_{x_d}^a \left\{ \int_{\substack{y \in \partial\Omega \\ |y| \le cs}} (\nabla u)^*(y)\, d\sigma(y) \right\} \frac{ds}{s^{d-1}} \\ &\le \sup_K |u| + C \int_{\partial\Omega} \frac{(\nabla u)^*(y)}{|y|^{d-2}}\, d\sigma(y), \end{aligned} \tag{8.1.10}$$

if $d \geq 3$. This gives the first estimate in (8.1.9). In the case $d = 2$, an inspection of the argument above shows that one needs to replace $\frac{1}{|y|^{d-2}}$ in (8.1.10) by $|\ln|y||+1$, which is bounded by $C_\eta |y|^{-\eta}$. □

Remark 8.1.4. Let $u \in C^1(\Omega)$. Suppose $(\nabla u)^* \in L^1(\partial\Omega)$. Then $(\nabla u)^*(z)$ is finite for a.e. $z \in \partial\Omega$. By the mean value theorem and the Cauchy criterion, it follows that u has nontangential limit a.e. on $\partial\Omega$.

It is often necessary to approximate a given Lipschitz domain Ω by a sequence of C^∞ domains $\{\Omega_j\}$ in such a manner that estimates on Ω_j with bounding constants depending on the Lipschitz characters may be extended to Ω by a limiting argument. The following theorem, whose proof may be found in [103] (also see [104]), serves this purpose.

Theorem 8.1.5. *Let Ω be a bounded Lipschitz domain in $\mathbb{R}^d$. Then there exist constants C and c, depending only on the Lipschitz character of Ω, and an increasing sequence of C^∞ domains Ω_j, $j = 1, 2, \ldots$ with the following properties.*

1. *For all j, $\overline{\Omega_j} \subset \Omega$.*
2. *There exists a sequence of homeomorphisms $\Lambda_j : \partial\Omega \to \partial\Omega_j$ such that*

$$\sup_{z\in\partial\Omega} |z - \Lambda_j(z)| \to 0 \text{ as } j \to \infty$$

and $\Lambda_j(z) \in \gamma_\alpha(z)$ for all j and all $z \in \partial\Omega$, where $\{\gamma_\alpha(z)\}$ is a family of nontangential approach regions.

3. *The unit outward normal to $\partial\Omega_j$, $n(\Lambda_j(z))$, converges to $n(z)$ for a.e. $z \in \partial\Omega$.*
4. *There are positive functions ω_j on $\partial\Omega$ such that $0 < c \leq \omega_j \leq C$ uniformly in j, $\omega_j \to 1$ a.e. as $j \to \infty$, and*

$$\int_E \omega_j \, d\sigma = \int_{\Lambda(E)} d\sigma_j \quad \text{for any measurable set } E \subset \partial\Omega.$$

5. *There exists a smooth vector field $h \in C_0^\infty(\mathbb{R}^d, \mathbb{R}^d)$ such that*

$$\langle h(z), n(z)\rangle \geq c > 0 \quad \text{for all } z \in \partial\Omega_j \text{ and all } j.$$

6. *There exists a finite covering of $\partial\Omega$ by coordinate cylinders so that for each coordinate cylinder (Z, ψ) in the covering, $10Z \cap \partial\Omega_j$ is given by the graph of a C^∞ function ψ_j,*

$$10Z \cap \partial\Omega_j = 10Z \cap \{(x', \psi_j(x')) : \ x' \in \mathbb{R}^{d-1}\}.$$

Furthermore, one has $\psi_j \to \psi$ uniformly, $\nabla\psi_j \to \nabla\psi$ a.e. as $j \to \infty$, and $\|\nabla\psi_j\|_\infty \leq \|\nabla\psi\|_\infty$.

We will use $\Omega_j \uparrow \Omega$ to denote an approximation sequence with properties 1–6. We may also approximate Ω by a decreasing sequence $\{\Omega_j\}$ from outside so that $\overline{\Omega} \subset \Omega_j$ and properties 2–6 hold. Such a sequence will be denoted by $\Omega_j \downarrow \Omega$.

Observe that if $(u)^* \in L^1(\partial\Omega)$ and $u = f$ n.t. on $\partial\Omega$, then

$$\int_{\partial\Omega} u(\Lambda_j(y))\, d\sigma(y) \to \int_{\partial\Omega} f\, d\sigma$$

by the Lebesgue dominated convergence theorem. This, in particular, allows us to extend the divergence theorem to Lipschitz domains for functions with nontangential limits.

Theorem 8.1.6. *Let Ω be a bounded Lipschitz domain in $\mathbb{R}^d$ and $u \in C^1(\Omega; \mathbb{R}^d)$. Suppose that* $\text{div}(u) \in L^1(\Omega)$ *and u has nontangential limit a.e. on $\partial\Omega$. Also assume that $(u)^* \in L^1(\partial\Omega)$. Then*

$$\int_{\Omega} \text{div}(u)\, dx = \int_{\partial\Omega} u \cdot n\, d\sigma. \tag{8.1.11}$$

Proof. Let Ω_j be a sequence of smooth domains such that $\Omega_j \uparrow \Omega$. Since $u \in C^1(\overline{\Omega_j})$, by the divergence theorem,

$$\int_{\Omega_j} \text{div}(u)\, dx = \int_{\partial\Omega_j} u \cdot n\, d\sigma. \tag{8.1.12}$$

Since $\text{div}(u) \in L^1(\Omega)$, the integral of $\text{div}(u)$ on Ω_j converges to its integral on Ω. Using a change of variables, one write the RHS of (8.1.12) as

$$\int_{\partial\Omega} u(\Lambda_j(y)) \cdot n(\Lambda_j(y)) \omega_j(y)\, d\sigma(y),$$

where Λ_j and ω_j are given by Theorem 8.1.5. Note that

$$u(\Lambda_j(y)) \cdot n(\Lambda_j(y)) \omega_j(y) \to u(y) \cdot n(y) \quad \text{for a.e. } y \in \partial\Omega$$

and

$$|u(\Lambda_j(y)) \cdot n(\Lambda_j(y)) \omega_j(y)| \le C\, (u)^*(y).$$

It follows by the dominated convergence theorem that the RHS of (8.1.12) converges to the integral in the RHS of (8.1.11). This completes the proof. □

Remark 8.1.7. Theorem 8.1.6 also holds for the unbounded domain $\Omega_- = \mathbb{R}^d \setminus \overline{\Omega}$, under the additional assumption that $u(x) = o(|x|^{1-d})$ as $|x| \to \infty$. Since n points away from Ω, in the case Ω_-, (8.1.11) is replaced by

$$\int_{\Omega_-} \text{div}(u)\, dx = -\int_{\partial\Omega} u \cdot n\, d\sigma. \tag{8.1.13}$$

We now formulate the L^p Dirichlet and Neumann problems for $\mathcal{L}_\varepsilon(u_\varepsilon) = 0$ in a Lipschitz domain Ω.

Definition 8.1.8 (L^p Dirichlet problem). Let Ω be a bounded Lipschitz domain. For $1 < p < \infty$, the L^p Dirichlet problem for $\mathcal{L}_\varepsilon(u_\varepsilon) = 0$ in Ω is said to be well posed, if for any $f \in L^p(\partial\Omega; \mathbb{R}^m)$, there exists a unique function u_ε in $W^{1,2}_{\text{loc}}(\Omega; \mathbb{R}^m)$ such that $\mathcal{L}_\varepsilon(u_\varepsilon) = 0$ in Ω, $(u_\varepsilon)^* \in L^p(\partial\Omega)$, and $u_\varepsilon = f$ n.t. on $\partial\Omega$. Moreover, the solution satisfies

$$\|(u_\varepsilon)^*\|_{L^p(\partial\Omega)} \le C\, \|f\|_{L^p(\partial\Omega)}.$$

Definition 8.1.9 (L^p Neumann problem). Let Ω be a bounded Lipschitz domain. For $1 < p < \infty$, the L^p Neumann problem for $\mathcal{L}_\varepsilon(u_\varepsilon) = 0$ in Ω is said to be well posed, if for any $g \in L^p(\partial\Omega; \mathbb{R}^m)$ with $\int_{\partial\Omega} g\, d\sigma = 0$, there exists a function u_ε in $W^{1,2}_{\text{loc}}(\Omega; \mathbb{R}^m)$, unique up to constants, such that $\mathcal{L}_\varepsilon(u_\varepsilon) = 0$ in Ω, $(\nabla u_\varepsilon)^* \in L^p(\partial\Omega)$, and $\frac{\partial u_\varepsilon}{\partial \nu_\varepsilon} = g$ on $\partial\Omega$ in the sense of nontangential convergence,

$$\lim_{\substack{x \to z \\ x \in \Omega \cap \gamma(z)}} n_i(z) a_{ij}^{\alpha\beta}(z/\varepsilon) \frac{\partial u_\varepsilon^\beta}{\partial x_j}(x) = g^\alpha(z) \qquad \text{for a.e. } z \in \partial\Omega. \tag{8.1.14}$$

Moreover, the solution satisfies

$$\|(\nabla u_\varepsilon)^*\|_{L^p(\partial\Omega)} \le C\, \|g\|_{L^p(\partial\Omega)}.$$

To formulate the L^p regularity problem, we first give the definition of $W^{1,p}(\partial\Omega)$.

Definition 8.1.10. For $1 \le p < \infty$, we say $f \in W^{1,p}(\partial\Omega)$ if $f \in L^p(\partial\Omega)$ and there exist functions $g_{jk} \in L^p(\partial\Omega)$ so that for all $\varphi \in C_0^\infty(\mathbb{R}^d)$ and $1 \le j, k \le d$,

$$\int_{\partial\Omega} f \left(n_j \frac{\partial}{\partial x_k} - n_k \frac{\partial}{\partial x_j} \right) \varphi\, d\sigma = - \int_{\partial\Omega} g_{jk} \varphi\, d\sigma.$$

By a partition of unity, one can show that $f \in W^{1,p}(\partial\Omega)$ if and only if $f\varphi \in W^{1,p}(\partial\Omega)$ for any $\varphi \in C_0^\infty(B(z, r_0))$ with $z \in \partial\Omega$. Thus, in a local coordinator system where (8.1.1) holds, $f \in W^{1,p}(\partial\Omega)$ means that $f(x', \psi(x'))$, as a function of x', is a $W^{1,p}$ function on the set $\{x' \in \mathbb{R}^{d-1} : |x'| < c r_0\}$. The space $W^{1,p}(\partial\Omega)$ is a Banach space with the (scale-invariant) norm

$$\|f\|_{W^{1,p}(\partial\Omega)} = |\partial\Omega|^{\frac{1}{1-d}} \|f\|_{L^p(\partial\Omega)} + \sum_{1 \le j,k \le d} \|g_{jk}\|_{L^p(\partial\Omega)}. \tag{8.1.15}$$

If $1 \le p < \infty$, the set $\{ f|_{\partial\Omega} : f \in C_0^\infty(\mathbb{R}^d) \}$ is dense in $W^{1,p}(\partial\Omega)$. Note that $g_{jk} = -g_{kj}$ and $g_{jj} = 0$. If $f \in C_0^1(\mathbb{R}^d)$ and $j \neq k$, then

$$g_{jk} = n_j \frac{\partial f}{\partial x_k} - n_k \frac{\partial f}{\partial x_j} \tag{8.1.16}$$

and

$$\sum_{1\le j<k\le d} |g_{jk}|^2 = |\nabla f|^2 - \Big|\frac{\partial f}{\partial n}\Big|^2 = |\nabla_{\tan} f|^2,$$

where $\nabla_{\tan} f = \nabla f - (n\cdot\nabla f)n$ is the tangential gradient of f on $\partial\Omega$. Moreover, from (8.1.16) we deduce that

$$n_\ell g_{jk} = n_k g_{jk} - n_j g_{kl} \qquad \text{for } 1\le j,k,\ell\le d. \tag{8.1.17}$$

By a simple density argument, the compatibility condition (8.1.17) holds for any $f\in W^{1,p}(\partial\Omega)$.

If the Dirichlet data f is taken from $W^{1,p}(\partial\Omega)$ instead of $L^p(\partial\Omega)$, one should expect the solution to have one order higher regularity. This is the so-called regularity problem (for the Dirichlet problem).

Definition 8.1.11 (L^p regularity problem)**.** Let Ω be a bounded Lipschitz domain. For $1<p<\infty$, the L^p regularity problem for $\mathcal{L}_\varepsilon(u_\varepsilon)=0$ in Ω is said to be well posed, if for any $f\in W^{1,p}(\partial\Omega;\mathbb{R}^m)$, there exists a unique function $u_\varepsilon\in W^{1,2}_{\text{loc}}(\Omega;\mathbb{R}^m)$ such that $\mathcal{L}_\varepsilon(u_\varepsilon)=0$ in Ω, $(\nabla u_\varepsilon)^*\in L^p(\partial\Omega)$, and $u_\varepsilon=f$ n.t. on $\partial\Omega$. Moreover, the solution satisfies

$$\|(\nabla u_\varepsilon)^*\|_{L^p(\partial\Omega)}\le C\,\|f\|_{W^{1,p}(\partial\Omega)}.$$

The rest of this chapter is devoted to the study of the L^2 Dirichlet, Neumann, and regularity problems for $\mathcal{L}_\varepsilon(u_\varepsilon)=0$ in a bounded Lipschitz domain Ω under the assumptions that A is elliptic, periodic, symmetric, and Hölder continuous. Our goal is to establish the uniform estimates in (8.0.4) with constants C depending at most on μ, λ, τ, and the Lipschitz character of Ω.

8.2 Estimates of fundamental solutions

Throughout Sections 8.2–8.9, with the exception of Section 8.5, we assume that $d\ge 3$. Let $A\in\Lambda(\mu,\lambda,\tau)$ and

$$\Gamma(x,y)=\Gamma(x,y;A)=\big(\Gamma^{\alpha\beta}(x,y;A)\big)_{m\times m}$$

denote the matrix of fundamental solutions for the operator

$$\mathcal{L}=\mathcal{L}_1=-\text{div}\big(A(x)\nabla\big)$$

in $\mathbb{R}^d$, with pole at y. Then

$$\begin{cases} |\Gamma(x,y)|\le C\,|x-y|^{2-d}, \\ |\nabla_x\Gamma(x,y)|+|\nabla_y\Gamma(x,y)|\le C\,|x-y|^{1-d}, \\ |\nabla_x\nabla_y\Gamma(x,y)|\le C\,|x-y|^{-d} \end{cases} \tag{8.2.1}$$

for any $x, y \in \mathbb{R}^d$ and $x \neq y$, where C depends only on μ, λ and τ (see Section 4.4; note that $\Gamma_1(x,y) = \Gamma(x,y;A)$ and $\Gamma_0(x,y) = \Gamma(x,y;\widehat{A})$). It follows from (4.4.19) and (4.4.25) that

$$\begin{aligned} \Big|\frac{\partial}{\partial x_i}\{\Gamma^{\alpha\beta}(x,y;A)\} - \Big\{\delta_{ij}\delta^{\alpha\gamma} + \frac{\partial}{\partial x_i}\{\chi_j^{\alpha\gamma}(x)\}\Big\}\frac{\partial}{\partial x_j}\Big\{\Gamma^{\gamma\beta}(x,y;\widehat{A})\Big\}\Big| \\ \le C\,|x-y|^{-d}\ln\big[|x-y|+2\big] \end{aligned} \tag{8.2.2}$$

and

$$\begin{aligned} \Big|\frac{\partial}{\partial y_i}\{\Gamma^{\alpha\beta}(x,y;A)\} - \Big\{\delta_{ij}\delta^{\beta\gamma} + \frac{\partial}{\partial y_i}\Big\{\chi_j^{*\beta\gamma}(y)\Big\}\Big\}\frac{\partial}{\partial y_j}\Big\{\Gamma^{\alpha\gamma}(x,y;\widehat{A})\Big\}\Big| \\ \le C\,|x-y|^{-d}\ln\big[|x-y|+2\big] \end{aligned} \tag{8.2.3}$$

for $x, y \in \mathbb{R}^d$ with $|x-y| \ge 1$, where $\widehat{A}$ is the (constant) matrix of homogenized coefficients. Let I denote the identity matrix. For brevity, we rewrite estimates (8.2.2) and (8.2.3) as

$$\begin{aligned} \big|\nabla_x\Gamma(x,y;A) - (I + \nabla\chi(x))\nabla_x\Gamma(x,y;\widehat{A})\big| \\ \le C\,|x-y|^{-d}\ln\big[|x-y|+2\big] \end{aligned} \tag{8.2.4}$$

and

$$\begin{aligned} \big|\nabla_y\big(\Gamma(x,y;A)\big)^T - \big(I + \nabla\chi^*(y)\big)\nabla_y\big(\Gamma(x,y;\widehat{A})\big)^T\big| \\ \le C\,|x-y|^{-d}\ln\big[|x-y|+2\big], \end{aligned} \tag{8.2.5}$$

where $(\Gamma(x,y;A))^T$ denotes the transpose of $\Gamma(x,y;A)$. These two inequalities give us the asymptotic behavior of $\nabla_x\Gamma(x,y;A)$ and $\nabla_y\Gamma(x,y;A)$ when $|x-y|$ is large.

For a function $F = F(x,y,z)$, we will use the notation

$$\nabla_1 F(x,y,z) = \nabla_x F(x,y,z) \quad \text{and} \quad \nabla_2 F(x,y,z) = \nabla_y F(x,y,z).$$

The next lemma describes the local behavior of $\Gamma(x,y;A)$. We remind the reader that $\Gamma(x,y;A(x))$ denotes the matrix of fundamental solutions for the operator $-\text{div}(E\nabla)$, where E is the constant matrix given by $A(x)$.

Lemma 8.2.1. *Let $A \in \Lambda(\mu,\lambda,\tau)$. Then*

$$\begin{aligned} |\Gamma(x,y;A) - \Gamma(x,y;A(x))| &\le C\,|x-y|^{2-d+\lambda}, \\ |\nabla_1\Gamma(x,y;A) - \nabla_1\Gamma(x,y;A(x))| &\le C\,|x-y|^{1-d+\lambda}, \\ |\nabla_1\Gamma(x,y;A) - \nabla_1\Gamma(x,y;A(y))| &\le C\,|x-y|^{1-d+\lambda} \end{aligned} \tag{8.2.6}$$

for any $x, y \in \mathbb{R}^d$ and $x \neq y$, where C depends only on μ, λ, and τ.

Proof. Let $B = (b_{ij}^{\alpha\beta}) \in \Lambda(\mu, \lambda, \tau)$. Then for any $x, y \in \mathbb{R}^d$,

$$\begin{aligned} &\Gamma^{\alpha\sigma}(x, y; A) - \Gamma^{\alpha\sigma}(x, y; B) \\ &= \int_{\mathbb{R}^d} \frac{\partial}{\partial z_i}\Big\{\Gamma^{\alpha\beta}(x, z; A)\Big\}\Big\{a_{ij}^{\beta\gamma}(z) - b_{ij}^{\beta\gamma}(z)\Big\}\frac{\partial}{\partial z_j}\Big\{\Gamma^{\gamma\sigma}(z, y; B)\Big\}\, dz. \end{aligned} \tag{8.2.7}$$

In view of (8.2.1) we obtain

$$|\Gamma(x, y; A) - \Gamma(x, y; B)| \le C \int_{\mathbb{R}^d} \frac{|A(z) - B(z)|}{|z - x|^{d-1}|z - y|^{d-1}}\, dz \tag{8.2.8}$$

and

$$|\nabla_1\Gamma(x, y; A) - \nabla_1\Gamma(x, y; B)| \le C \int_{\mathbb{R}^d} \frac{|A(z) - B(z)|}{|z - x|^{d}|z - y|^{d-1}}\, dz. \tag{8.2.9}$$

To show the first inequality in (8.2.6), we fix $x \in \mathbb{R}^d$ and let $B = A(x)$. Since

$$|A(z) - A(x)| \le \tau |z - x|^\lambda,$$

it follows from (8.2.8) that

$$\begin{aligned} |\Gamma(x, y; A) - \Gamma(x, y; A(x))| &\le C \int_{\mathbb{R}^d} \frac{dz}{|z - x|^{d-1-\lambda}|z - y|^{d-1}} \\ &\le C\, |x - y|^{2-d+\lambda}. \end{aligned}$$

The second inequality in (8.2.6) follows from (8.2.9) in the same manner.

To prove the third inequality in (8.2.6), note that if E_1, E_2 are two constant matrices satisfying the ellipticity condition (2.1.20), then

$$|\nabla_1^N \Gamma(x, y; E_1) - \nabla_1^N \Gamma(x, y; E_2)| \le C\, |E_1 - E_2||x - y|^{2-d-N}, \tag{8.2.10}$$

where C depends only on μ and N. This follows from (8.2.8). By taking $E_1 = A(x)$ and $E_2 = A(y)$, we obtain

$$|\nabla_1\Gamma(x, y; A(x)) - \nabla_1\Gamma(x, y; A(y))| \le C\, |x - y|^{1-d+\lambda}. \tag{8.2.11}$$

The third inequality in (8.2.6) follows from the second and (8.2.11). □

Remark 8.2.2. It follows from (8.2.8) that if $A, B \in \Lambda(\mu, \lambda, \tau)$,

$$|\Gamma(x, y; A) - \Gamma(x, y; B)| \le C\, \|A - B\|_\infty |x - y|^{2-d}. \tag{8.2.12}$$

Also, if we fix $y \in \mathbb{R}^d$ and let $B = A(y)$, the same argument as in the proof of Lemma 8.2.1 yields that

$$\begin{aligned} |\Gamma(x, y; A) - \Gamma(x, y; A(y))| &\le C\, |x - y|^{2-d+\lambda}, \\ |\nabla_2\Gamma(x, y; A) - \nabla_2\Gamma(x, y; A(y))| &\le C\, |x - y|^{1-d+\lambda}, \\ |\nabla_2\Gamma(x, y; A) - \nabla_2\Gamma(x, y; A(x))| &\le C\, |x - y|^{1-d+\lambda} \end{aligned} \tag{8.2.13}$$

for any $x, y \in \mathbb{R}^d$ with $x \ne y$.

In the rest of this section we will be concerned with the estimate of

$$\Gamma(x, y; A) - \Gamma(x, y; B),$$

when A is close to B in the Hölder space $C^\lambda(\mathbb{R}^d)$. The results are local and will be used in an approximation argument for a domains Ω with $\operatorname{diam}(\Omega) \le 1$ in Section 8.3.

Lemma 8.2.3. *Let $R \ge 1$ and $A, B \in \Lambda(\mu, \lambda, \tau)$. Then*

$$\begin{aligned} |\nabla_x \Gamma(x, y; A) - \nabla_x \Gamma(x, y; B)| &\le C_R \|A - B\|_{C^\lambda(\mathbb{R}^d)} |x - y|^{1-d}, \\ |\nabla_y \nabla_x \Gamma(x, y; A) - \nabla_y \nabla_x \Gamma(x, y; B)| &\le C_R \|A - B\|_{C^\lambda(\mathbb{R}^d)} |x - y|^{-d} \end{aligned} \tag{8.2.14}$$

for any $x, y \in \mathbb{R}^d$ with $0 < |x - y| \le R$, where C_R depends on μ, λ, τ, and R.

Proof. The estimates in (8.2.14) follow from (8.2.12) by (local) interior $C^{1,\lambda}$ estimates. Indeed, fix $x_0, y_0 \in \mathbb{R}^d$ with $r = |x_0 - y_0| \le R$, and consider $u(x) = \Gamma(x, y_0; A) - \Gamma(x, y_0; B)$ in $\Omega = B(x_0, r/2)$. Let $w(x) = -\Gamma(x, y_0; B)$. Then

$$\operatorname{div}(A\nabla u) = \operatorname{div}(A\nabla w) = \operatorname{div}\big((A - B)\nabla w\big) \quad \text{in } \Omega.$$

It follows that

$$\begin{aligned} |\nabla u(x_0)| &\le C r^{-1} \|u\|_{L^\infty(\Omega)} + C r^\lambda \|(A - B)\nabla w\|_{C^{0,\lambda}(\Omega)} \\ &\le C r^{1-d} \|A - B\|_{C^\lambda(\mathbb{R}^d)}, \end{aligned}$$

where C may depend on R. This gives the first inequality in (8.2.14). The second inequality follows in the same manner by considering

$$v(x) = \nabla_y \Gamma(x, y; A) - \nabla_y \Gamma(x, y; B). \qquad \square$$

For $A \in \Lambda(\mu, \lambda, \tau)$, define

$$\Pi(x, y; A) = \nabla_1 \Gamma(x, y; A) - \nabla_1 \Gamma(x, y; A(x)). \tag{8.2.15}$$

Theorem 8.2.4. *Let $A, B \in \Lambda(\mu, \lambda, \tau)$ and $R \ge 1$. Then*

$$|\Pi(x, y; A) - \Pi(x, y; B)| \le C_R \, \|A - B\|_{C^\lambda(\mathbb{R}^d)} |x - y|^{1-d+\lambda} \tag{8.2.16}$$

for any $x, y \in \mathbb{R}^d$ with $|x - y| < R$, where C_R depends on μ, λ, τ, and R.

Proof. Let $\Omega = B(x_0, 2R)$. It follows from integration by parts that for $x, y \in B(x_0, R)$,

$$\begin{aligned} &\Gamma^{\alpha\delta}(x, y; A) - \Gamma^{\alpha\delta}(x, y; A(P)) \\ &= \int_\Omega \frac{\partial}{\partial z_i}\Big\{\Gamma^{\alpha\beta}(x, z; A)\Big\}\Big\{a_{ij}^{\beta\gamma}(P) - a_{ij}^{\beta\gamma}(z)\Big\}\frac{\partial}{\partial z_j}\Big\{\Gamma^{\delta\gamma}(z, y; A(P))\Big\}\, dz \\ &\quad + \int_{\partial\Omega} \frac{\partial}{\partial z_i}\Big\{\Gamma^{\alpha\beta}(x, z; A)\Big\} a_{ij}^{\beta\gamma}(z) n_j(z) \Gamma^{\gamma\delta}(z, y; A(P))\, d\sigma(z) \\ &\quad - \int_{\partial\Omega} \Gamma^{\alpha\beta}(x, z; A) n_i(z) a_{ij}^{\beta\gamma}(P) \frac{\partial}{\partial z_j}\Big\{\Gamma^{\gamma\delta}(z, y; A(P))\Big\}\, d\sigma(z), \end{aligned} \tag{8.2.17}$$

where $P \in B(x_0, R)$ and $n = (n_1, \dots, n_d)$ denotes the unit outward normal to $\partial\Omega$. By taking the derivative with respect to x on both sides of (8.2.17) and subsequently setting $P = x$, we obtain

$$\begin{aligned}\Pi(x, y; A) = & \int_\Omega \nabla_x \nabla_z \Gamma(x, z; A)\{A(x) - A(z)\}\nabla_1 \Gamma(z, y; A(x))\, dz \\ & + \int_{\partial\Omega} \nabla_z \nabla_x \Gamma(x, z; A) A(z) n(z) \Gamma(z, y; A(x))\, d\sigma(z) \\ & - \int_{\partial\Omega} \nabla_x \Gamma(x, z; A) n(z) A(x) \nabla_z \Gamma(z, y; A(x))\, d\sigma(z).\end{aligned} \tag{8.2.18}$$

To estimate the difference $\Pi(x, y; A) - \Pi(x, y; B)$, we split its integrals over Ω as $I_1 + I_2 + I_3$, where

$$\begin{aligned} I_1 &= \int_\Omega \{\nabla_x \nabla_z \Gamma(x, z; A) - \nabla_x \nabla_z \Gamma(x, z; B)\}\{A(x) - A(z)\} \\ &\qquad\qquad \nabla_1 \Gamma(z, y; A(x))\, dz, \\ I_2 &= \int_\Omega \{\nabla_x \nabla_z \Gamma(x, z; B)\}\{A(x) - A(z) - (B(x) - B(z))\} \\ &\qquad\qquad \nabla_1 \Gamma(z, y; A(x))\, dz, \\ I_3 &= \int_\Omega \{\nabla_x \nabla_z \Gamma(x, z; B)\}\{B(x) - B(z)\} \\ &\qquad\qquad \{\nabla_1 \Gamma(z, y; A(x)) - \nabla_1 \Gamma(z, y; B(x))\}\, dz. \end{aligned}$$

It follows from (8.2.14) that

$$\begin{aligned} |I_1| &\le C_R \|A - B\|_{C^\lambda(\mathbb{R}^d)} \int_\Omega \frac{dz}{|x - z|^{d-\lambda} |z - y|^{d-1}} \\ &\le C_R \|A - B\|_{C^\lambda(\mathbb{R}^d)} |x - y|^{1-d+\lambda}. \end{aligned}$$

Similarly, by estimates (8.2.1) and (8.2.10), we obtain

$$\begin{aligned} |I_2| &\le C\, \|A - B\|_{C^{0,\lambda}(\mathbb{R}^d)} |x - y|^{1-d+\lambda}, \\ |I_3| &\le C\, \|A - B\|_\infty |x - y|^{1-d+\lambda}. \end{aligned}$$

Finally, we split the integrals over $\partial\Omega$ in $\Pi(x, y; A) - \Pi(x, y; B)$ in a similar manner. Using the fact that $|x - z| \ge R$ and $|y - z| \ge R$ for $z \in \partial\Omega$, we can show that they are bounded by $C_R \|A - B\|_{C^\lambda(\mathbb{R}^d)}$. □

Remark 8.2.5. For $A \in \Lambda(\mu, \lambda, \tau)$, define

$$\Theta(x, y; A) = \nabla_2 \Gamma(x, y; A) - \nabla_2 \Gamma(x, y; A(y)). \tag{8.2.19}$$

Let $A, B \in \Lambda(\mu, \lambda, \tau)$ and $R \ge 1$. Then

$$|\Theta(x, y; A) - \Theta(x, y; B)| \le C_R \|A - B\|_{C^\lambda(\mathbb{R}^d)} |x - y|^{1-d+\lambda} \tag{8.2.20}$$

for any $x, y \in \mathbb{R}^d$ with $|x - y| \le R$, where C_R depends only on μ, λ, τ, and R. The proof is similar to that of Theorem 8.2.4.

8.3 Estimates of singular integrals

Let Ω be a bounded Lipschitz domain. For $A \in \Lambda(\mu, \lambda, \tau)$, consider the two maximal singular integral operators $T_A^{1,*}$ and $T_A^{2,*}$ on $\partial\Omega$, defined by

$$\begin{aligned} T_A^{1,*}(f)(z) &= \sup_{r>0} \left| \int_{\substack{y\in\partial\Omega\\ |y-z|>r}} \nabla_1 \Gamma(z,y;A) f(y)\, d\sigma(y) \right|, \\ T_A^{2,*}(f)(z) &= \sup_{r>0} \left| \int_{\substack{y\in\partial\Omega\\ |y-z|>r}} \nabla_2 \Gamma(z,y;A) f(y)\, d\sigma(y) \right| \end{aligned} \tag{8.3.1}$$

for $z \in \partial\Omega$.

Theorem 8.3.1. *Let $f \in L^p(\partial\Omega;\mathbb{R}^m)$ for some $1 < p < \infty$. Then*

$$\|T_A^{1,*}(f)\|_{L^p(\partial\Omega)} + \|T_A^{2,*}(f)\|_{L^p(\partial\Omega)} \le C_p \, \|f\|_{L^p(\partial\Omega)}, \tag{8.3.2}$$

where C_p depends only on μ, λ, τ, p, and the Lipschitz character of Ω.

To establish the L^p boundedness of the operator $T_A^{1,*}$, we shall approximate its kernel $\nabla_1\Gamma(z,y;A)$ by $\nabla_1\Gamma(z,y;A(z))$ when $|z-y| \le 1$, and by $(I + \nabla\chi(z))\nabla_1\Gamma(z,y;\widehat{A})$ when $|z-y| \ge 1$. The operator $T_A^{2,*}$ can be handled in a similar manner.

Let $\mathcal{M}_{\partial\Omega}$ denote the Hardy–Littlewood maximal operator on $\partial\Omega$.

Lemma 8.3.2. *For each $z \in \partial\Omega$,*

$$\begin{aligned} T_A^{1,*}(f)(z) &\le C\mathcal{M}_{\partial\Omega}(f)(z) + 2\sup_{r>0} \left| \int_{\substack{y\in\partial\Omega\\ |y-z|>r}} \nabla_1 \Gamma(z,y;A(z)) f(y)\, d\sigma(y) \right| \\ &\quad + C\sup_{r>0} \left| \int_{\substack{y\in\partial\Omega\\ |y-z|>r}} \nabla_1 \Gamma(z,y;\widehat{A}) f(y)\, d\sigma(y) \right|, \\ T_A^{2,*}(f)(z) &\le C\mathcal{M}_{\partial\Omega}(f)(z) + 2\sup_{r>0} \left| \int_{\substack{y\in\partial\Omega\\ |y-z|>r}} \nabla_2 \Gamma(z,y;A(y)) f(y)\, d\sigma(y) \right| \\ &\quad + C\sup_{r>0} \left| \int_{\substack{y\in\partial\Omega\\ |y-z|>r}} \nabla_2 \Gamma(z,y;\widehat{A}) g(y)\, d\sigma(y) \right|, \end{aligned} \tag{8.3.3}$$

where g is a function satisfying $|g| \le C|f|$ on $\partial\Omega$, and C depends only on μ, λ, τ, and the Lipschitz character of Ω.

Proof. To prove (8.3.3), we fix $z \in \partial\Omega$ and $r > 0$. If $r \ge 1$, we use the estimate (8.2.4) and the boundedness of $\nabla\chi$ to obtain

$$\begin{aligned} &\left| \int_{|y-z|>r} \nabla_1 \Gamma(z,y;A) f(y)\, d\sigma(y) \right| \\ &\quad \le C \left| \int_{|y-z|>r} \nabla_1 \Gamma(z,y;\widehat{A}) f(y)\, d\sigma(y) \right| + C \int_{|y-z|>r} \frac{\ln[|z-y|+1]}{|z-y|^d} |f(y)|\, d\sigma(y) \end{aligned}$$

$$\leq C \sup_{t>0} \left| \int_{|y-z|>t} \nabla_1 \Gamma(z, y; \widehat{A}) f(y) \, d\sigma(y) \right| + C\mathcal{M}_{\partial\Omega}(f)(z).$$

If $0 < r < 1$, we split the set $\{y \in \partial\Omega : |y - z| > r\}$ as

$$\{y \in \partial\Omega : |y - z| > 1\} \cup \{y \in \partial\Omega : 1 \geq |y - z| > r\}.$$

The integral of $\nabla_1 \Gamma(z, y; A) f(y)$ on $\{y \in \partial\Omega : |y - z| > 1\}$ may be treated as above. To handle the integral on $\{y \in \partial\Omega : 1 \geq |y - z| > r\}$, we use the estimate (8.2.6) to obtain

$$\begin{aligned}
&\left| \int_{1\geq|y-z|>r} \nabla_1 \Gamma(z, y; A) f(y) \, d\sigma(y) \right| \\
&\leq \left| \int_{1\geq|y-z|>r} \nabla_1 \Gamma(z, y; A(z)) f(y) \, d\sigma(y) \right| + C \int_{|y-z|\leq 1} \frac{|f(y)|}{|y - z|^{d-1-\lambda}} \, d\sigma(y) \\
&\leq 2 \sup_{t>0} \left| \int_{|y-z|>t} \nabla_1 \Gamma(z, y; A(z)) f(y) \, d\sigma(y) \right| + C\mathcal{M}_{\partial\Omega}(f)(z).
\end{aligned}$$

This gives the desired estimates for $T_A^{1,*}(f)$. The estimate for $T_A^{2,*}(f)$ in (8.3.3) follows from (8.2.3) and (8.2.13) in the same manner. The details are left to the reader. □

Lemma 8.3.3. *Let $K(x, y)$ be odd in $x \in \mathbb{R}^d$ and homogenous of degree $1 - d$ in $x \in \mathbb{R}^d$, i.e.,*

$$K(-x, y) = -K(x, y) \quad \text{and} \quad K(tx, y) = t^{1-d} K(x, y)$$

for $x \in \mathbb{R}^d \setminus \{0\}$, $y \in \mathbb{R}^d$ and $t > 0$. Assume that for all $0 \leq N \leq N(d)$, where $N(d) > 1$ is sufficiently large, $\nabla_x^N K(x, y)$ is continuous on $\mathbb{S}^{d-1} \times \mathbb{R}^d$ and $|\nabla_x^N K(x, y)| \leq C_0$ for $x \in \mathbb{S}^{d-1}$ and $y \in \mathbb{R}^d$. Let $f \in L^p(\partial\Omega)$ for some $1 < p < \infty$. Define

$$\begin{aligned}
S^1(f)(x) &= \text{p.v.} \int_{\partial\Omega} K(x - y, x) f(y) \, d\sigma(y) \\
&:= \lim_{r\to 0} \int_{\substack{y\in\partial\Omega \\ |y-x|>r}} K(x - y, x) f(y) \, d\sigma(y), \\
S^2(f)(x) &= \text{p.v.} \int_{\partial\Omega} K(x - y, y) f(y) \, d\sigma(y), \\
S^{1,*}(f)(x) &= \sup_{r>0} \left| \int_{\substack{y\in\partial\Omega \\ |y-x|>r}} K(x - y, x) f(y) \, d\sigma(y) \right|, \\
S^{2,*}(f)(x) &= \sup_{r>0} \left| \int_{\substack{y\in\partial\Omega \\ |y-x|>r}} K(x - y, y) f(y) \, d\sigma(y) \right|.
\end{aligned}$$

Then $S^1(f)(x)$ and $S^2(f)(x)$ exist for a.e. $x \in \partial\Omega$, and

$$\|S^{1,*}(f)\|_{L^p(\partial\Omega)} + \|S^{2,*}(f)\|_{L^p(\partial\Omega)} \leq CC_0 \|f\|_{L^p(\partial\Omega)},$$

where C depends only on p and the Lipschitz character of Ω.

Proof. By considering $C_0^{-1}K(x,y)$, one may assume that $C_0 = 1$. In the special case where $K(x,y)$ is independent of y, the result is a consequence of the L^p boundedness of Cauchy integrals on Lipschitz curves [29]. The general case can be deduced from the special case by using the spherical harmonic decomposition (see, e.g., [74]). Note that only the continuity condition in the variable y is needed for $\nabla_x^N K(x,y)$. □

Proof of Theorem 8.3.1. This follows readily from the Lemmas 8.3.2 and 8.3.3. Note that if B is a constant matrix satisfying (2.2.28), the fundamental solution $\Gamma(x,y;B) = \Gamma(x-y,0;B)$ and $\Gamma(x,0;B)$ is even and homogeneous of degree $2-d$ in x. It follows that $\nabla_x\Gamma(x,0;B)$ is odd and homogeneous of degree $1-d$ in x. Also recall that $\mathcal{M}_{\partial\Omega}$ is bounded on $L^p(\partial\Omega)$ for $1 < p \le \infty$. □

Next we consider two singular integral operators:

$$\begin{aligned} T_A^1(f)(x) &= \text{p.v.} \int_{\partial\Omega} \nabla_1\Gamma(x,y;A)f(y)\,d\sigma(y), \\ T_A^2(f)(x) &= \text{p.v.} \int_{\partial\Omega} \nabla_2\Gamma(x,y;A)f(y)\,d\sigma(y). \end{aligned} \tag{8.3.4}$$

Theorem 8.3.4. *Let $A \in \Lambda(\mu,\lambda,\tau)$ and Ω be a bounded Lipschitz domain. Let $f \in L^p(\partial\Omega;\mathbb{R}^m)$ for some $1 < p < \infty$. Then $T_A^1(f)(x)$ and $T_A^2(f)(x)$ exist for a.e. $x \in \partial\Omega$, and*

$$\|T_A^1(f)\|_{L^p(\partial\Omega)} + \|T_A^2(f)\|_{L^p(\partial\Omega)} \le C_p\,\|f\|_{L^p(\partial\Omega)},$$

where C_p depends only on μ, λ, τ, p, and the Lipschitz character of Ω.

Proof. Let $f \in C_0^\infty(\mathbb{R}^d;\mathbb{R}^m)$. Write

$$\begin{aligned} &\text{p.v.} \int_{\partial\Omega} \nabla_1\Gamma(x,y;A)f(y)\,d\sigma(y) \\ &= \text{p.v.} \int_{\partial\Omega} \nabla_1\Gamma(x,y;A(x))f(y)\,d\sigma(y) \\ &\quad + \text{p.v} \int_{\partial\Omega} \big[\nabla_1\Gamma(x,y;A) - \nabla_1\Gamma(x,y;A(x))\big] f(y)\,d\sigma(y). \end{aligned} \tag{8.3.5}$$

Since

$$|\nabla_1\Gamma(x,y;A) - \nabla_1\Gamma(x,y;A(x))| \le C|x-y|^{1-d+\lambda},$$

the second term in the RHS of (8.3.5) exists for any $x \in \partial\Omega$. By Lemma 8.3.3, the first term in the RHS of (8.3.5) exists for a.e. $x \in \partial\Omega$. This shows that if $f \in C_0^\infty(\mathbb{R}^d;\mathbb{R}^m)$, $T_A^1(f)(x)$ exists for a.e. $x \in \partial\Omega$. Since $C_0^\infty(\mathbb{R}^d;\mathbb{R}^m)$ is dense in $L^p(\partial\Omega;\mathbb{R}^m)$ and the maximal singular integral operator $T_A^{1,*}$ is bounded on $L^p(\partial\Omega)$, we conclude that if $f \in L^p(\partial\Omega;\mathbb{R}^m)$, $T_A^1(f)(x)$ exists for a.e. $x \in \partial\Omega$. Since $|T_A^1(f)(P)| \le T_A^{1,*}(f)(x)$, it follows by Theorem 8.3.1 that $\|T_A^1(f)\|_{L^p(\partial\Omega)} \le C_p\,\|f\|_{L^p(\partial\Omega)}$. The case of $T_A^2(f)$ may be handled in a similar manner. □

Theorem 8.3.5. *Let T_A^1, T_A^2, T_B^1, and T_B^2 be defined by* (8.3.4)*, where $A, B \in \Lambda(\mu, \lambda, \tau)$. Let Ω be a bounded Lipschitz domain with* $\operatorname{diam}(\Omega) \le 10$*. Then, for $1 < p < \infty$,*

$$\begin{aligned} \|T_A^1(f) - T_B^1(f)\|_{L^p(\partial\Omega)} &\le C_p \, \|A - B\|_{C^\lambda(\mathbb{R}^d)} \|f\|_{L^p(\partial\Omega)}, \\ \|T_A^1(f) - T_B^1(f)\|_{L^p(\partial\Omega)} &\le C_p \, \|A - B\|_{C^\lambda(\mathbb{R}^d)} \|f\|_{L^p(\partial\Omega)}, \end{aligned} \tag{8.3.6}$$

where C_p depends only on μ, λ, τ, p, and the Lipschitz character of Ω.

Proof. Recall that $\Pi(x, y; A) = \nabla_1 \Gamma(x, y; A) - \nabla \Gamma(x, y, A(x))$. Write

$$\begin{aligned} &\nabla_1 \Gamma(x, y; A) - \nabla_1 \Gamma(x, y; B) \\ &\quad = \Pi(x, y; A) - \Pi(x, y; B) + \big\{ \nabla_1 \Gamma(x - y, 0; A(x)) - \nabla_1 \Gamma(x - y, 0; B(x)) \big\}. \end{aligned}$$

Since $\operatorname{diam}(\Omega) \le 10$, it follows from Theorem 8.2.4 that the norm of the operator with the integral kernel

$$\Pi(x, y; A) - \Pi(x, y; B)$$

on $L^p(\partial\Omega)$ is bounded by $C\|A - B\|_{C^\lambda(\mathbb{R}^d)}$ for $1 \le p \le \infty$. By Lemma 8.3.3, the norm of the operator with the integral kernel

$$\nabla_1 \Gamma(x - y, 0; A(x)) - \nabla_1 \Gamma(x - y, 0; B(x))$$

on $L^p(\partial\Omega)$ for $1 < p < \infty$ is bounded by $C\|A - B\|_\infty$. This gives the desired estimate for $\|T_A^1(f) - T_B^1(f)\|_{L^p(\partial\Omega)}$. The estimate for $\|T_A^2(f) - T_B^2(f)\|_{L^p(\partial\Omega)}$ follows from Remark 8.2.5 and Lemma 8.3.3 in the same manner. □

We end this section with some notation and a theorem on nontangential maximal functions. Since we will be dealing with functions on domains

$$\Omega_+ = \Omega \quad \text{and} \quad \Omega_- = \mathbb{R}^d \setminus \overline{\Omega}$$

simultaneously, we introduce the following notation. For a continuous function u in $\Omega_\pm$, the nontangential maximal function $(u)^*_\pm$ is defined by

$$(u)^*_\pm(z) = \sup \big\{ |u(x)| : \ x \in \Omega_\pm \quad \text{and} \quad |x - z| < \alpha \operatorname{dist}(x, \partial\Omega) \big\} \tag{8.3.7}$$

for $z \in \partial\Omega$. We will use $u_\pm(z)$ to denote the nontangential limit at z, if it exists, taken from $\Omega_\pm$ respectively. If u is continuous in $\mathbb{R}^d \setminus \partial\Omega$, we define

$$(u)^*(z) = \max \big\{ (u)^*_+(z), \, (u)^*_-(z) \big\}.$$

We also use $(u)^*$ to denote $(u)^*_+$ or $(u)^*_-$ if there is no danger of confusion.

For $f \in L^p(\partial\Omega; \mathbb{R}^m)$, consider the functions

$$\begin{aligned} v(x) &= \int_{\partial\Omega} \nabla_1 \Gamma(x, y; A) f(y) \, d\sigma(y), \\ w(x) &= \int_{\partial\Omega} \nabla_2 \Gamma(x, y; A) f(y) \, d\sigma(y), \end{aligned} \tag{8.3.8}$$

defined in $\mathbb{R}^d \setminus \partial\Omega$.

Theorem 8.3.6. *Let Ω be a bounded Lipschitz domain. Let v and w be defined by (8.3.8), where $A \in \Lambda(\mu, \lambda, \tau)$. Then, for $1 < p < \infty$,*

$$\|(v)^*\|_{L^p(\partial\Omega)} + \|(w)^*\|_{L^p(\partial\Omega)} \le C_p \|f\|_{L^p(\partial\Omega)}, \tag{8.3.9}$$

where C_p depends only on μ, λ, τ, p, and the Lipschitz character of Ω.

Proof. We first consider the case that $A = E$ is a constant matrix in $\Lambda(\mu, \lambda, \tau)$. Let

$$u(x) = \int_{\partial\Omega} \nabla_1 \Gamma(x, y; E) f(y)\, d\sigma(y).$$

Fix $z \in \partial\Omega$. Let $x \in \gamma(z) = \{y \in \mathbb{R}^d : |y - z| < \alpha\, \text{dist}(y, \partial\Omega)\}$ and $r = |x - z|$. Write $u(x) = I_1 + I_2 + I_3$, where

$$I_1 = \int_{\substack{y \in \partial\Omega \\ |y - z| \le 4r}} \nabla_1 \Gamma(x, y; E) f(y)\, d\sigma(y),$$

$$I_2 = \int_{\substack{y \in \partial\Omega \\ |y - z| > 4r}} \nabla_1 \Gamma(z, y; E) f(y)\, d\sigma(y),$$

$$I_3 = \int_{\substack{y \in \partial\Omega \\ |y - z| > 4r}} \{\nabla_1 \Gamma(x, y; E) - \nabla_1 \Gamma(z, y; E)\} f(y)\, d\sigma(y).$$

Using $|\nabla_x \Gamma(x, 0; E)| \le C|x|^{1-d}$ and $|\nabla_x^2 \Gamma(x, 0; E)| \le C|x|^{-d}$, we obtain

$$\begin{aligned} |I_1| + |I_3| &\le \frac{C}{r^{d-1}} \int_{\substack{y \in \partial\Omega \\ |y - z| \le 4r}} |f(y)|\, d\sigma(y) + Cr \int_{\substack{y \in \partial\Omega \\ |y - z| > 4r}} \frac{|f(y)|}{|y - z|^d}\, d\sigma(y) \\ &\le C \mathcal{M}_{\partial\Omega}(f)(z). \end{aligned}$$

It follows that

$$(u)^*(z) \le C \mathcal{M}_{\partial\Omega}(f)(z) + \sup_{r > 0} \left| \int_{\substack{y \in \partial\Omega \\ |y - z| > r}} \nabla_1 \Gamma(z, y; E) f(y)\, d\sigma(y) \right|. \tag{8.3.10}$$

In view of Theorem 8.3.1, this gives $\|(u)^*\|_{L^p(\partial\Omega)} \le C\|f\|_{L^p(\partial\Omega)}$.

We now return to the functions v and w for the general case $A \in \Lambda(\mu, \lambda, \tau)$. We claim that for any $z \in \partial\Omega$,

$$\begin{aligned} (v)^*(z) &\le C \mathcal{M}_{\partial\Omega}(f)(z) + C \sup_{t > 0} \left| \int_{\substack{y \in \partial\Omega \\ |y - z| > t}} \nabla_1 \Gamma(z, y, A(z)) f(y)\, d\sigma(y) \right| \\ &\quad + C \sup_{t > 0} \left| \int_{\substack{y \in \partial\Omega \\ |y - P| > t}} \nabla_1 \Gamma(z, y; \widehat{A}) f(y)\, d\sigma(y) \right|, \\ (w)^*(z) &\le C \mathcal{M}_{\partial\Omega}(f)(z) + C \sup_{t > 0} \left| \int_{\substack{y \in \partial\Omega \\ |y - z| > t}} \nabla_2 \Gamma(z, y, A(y)) f(y)\, d\sigma(y) \right| \\ &\quad + C \sup_{t > 0} \left| \int_{\substack{y \in \partial\Omega \\ |y - z| > t}} \nabla_2 \Gamma(z, y; \widehat{A}) g(y)\, d\sigma(y) \right|, \end{aligned} \tag{8.3.11}$$

where $|g(y)| \le C|f(y)|$, and C depends only on μ, λ, τ, and the Lipschitz character of Ω. Estimate (8.3.9) follows from (8.3.11) by Lemma 8.3.3. We will give the proof for $(v)^*$; the estimate for $(w)^*$ is obtained in the same way.

Fix $z \in \partial\Omega$. Let $x \in \gamma(z)$ and $r = |x - z|$. If $r \ge 1$, it follows from (8.2.4) that

$$\begin{aligned} |v(x) - (I + \nabla\chi(x))\nabla U(x)| &\le C \int_{\partial\Omega} \frac{\ln[|x-y|+2]}{|x-y|^d} |f(y)|\, d\sigma(y) \\ &\le C\mathcal{M}_{\partial\Omega}(f)(z), \end{aligned}$$

where

$$U(x) = \int_{\partial\Omega} \Gamma(x, y; \widehat{A}) f(y)\, d\sigma(y).$$

Hence,

$$\begin{aligned} |u(x)| &\le C\mathcal{M}_{\partial\Omega}(f)(z) + C(\nabla U)^*(z) \\ &\le C\mathcal{M}_{\partial\Omega}(f)(z) + C \sup_{t>0} \left| \int_{\substack{y\in\partial\Omega \\ |y-z|>t}} \nabla_1 \Gamma(z, y; \widehat{A}) f(y)\, d\sigma(y) \right|, \end{aligned}$$

where we have used (8.3.10).

Next suppose that $r = |x - z| < 1$. We write $u(x) = J_1 + J_2 + J_3$, where J_1, J_2, and J_3 denote the integral of $\nabla_1\Gamma(x, y; A) f(y)$ over

$$\begin{aligned} E_1 &= \{y \in \partial\Omega : \ |y - z| < r\}, \\ E_2 &= \{y \in \partial\Omega : \ r \le |y - z| \le 1\}, \\ E_3 &= \{y \in \partial\Omega : \ |y - z| > 1\}, \end{aligned}$$

respectively. Clearly, $|J_1| \le C\mathcal{M}_{\partial\Omega}(f)(P)$. For J_2, we use (8.2.6) to obtain

$$\begin{aligned} |J_2| &\le \left| \int_{E_2} \nabla_1\Gamma(x, y; A(y)) f(y)\, d\sigma(y) \right| + C \int_{E_2} \frac{|f(y)|}{|x-y|^{d-1-\lambda}}\, d\sigma(y) \\ &\le \left| \int_{E_2} \nabla_1\Gamma(z, y; A(z)) f(y)\, d\sigma(y) \right| + C\mathcal{M}_{\partial\Omega}(f)(z) + C \int_{E_2} \frac{|f(y)|}{|y-z|^{d-1-\lambda}}\, d\sigma(y) \\ &\le 2 \sup_{t>0} \left| \int_{\substack{y\in\partial\Omega \\ |y-z|>t}} \nabla_1\Gamma(z, y; A(z)) f(y)\, d\sigma(y) \right| + C\mathcal{M}_{\partial\Omega}(f)(z). \end{aligned}$$

In view of (8.2.4), we have

$$\begin{aligned} |J_3| &\le C \int_{E_3} \frac{\ln[|x-y|+2]}{|x-y|^d} |f(y)|\, d\sigma(y) + C \left| \int_{E_3} \nabla_1\Gamma(x, y; \widehat{A}) f(y)\, d\sigma(y) \right| \\ &\le C\mathcal{M}_{\partial\Omega}(f)(z) + C \sup_{t>0} \left| \int_{\substack{y\in\partial\Omega \\ |y-z|>t}} \nabla_1\Gamma(x, y; \widehat{A}) f(y)\, d\sigma(y) \right|. \end{aligned}$$

This, together with the estimates of J_1 and J_2, yields the desired estimate for $(v)^*(z)$. □

8.4 The method of layer potentials

In this section we fix $A = \big(a_{ij}^{\alpha\beta}(x)\big) \in \Lambda(\mu, \lambda, \tau)$ and let $\mathcal{L}_\varepsilon = -\text{div}(A(x/\varepsilon)\nabla)$.

Definition 8.4.1. Let Ω be a bounded Lipschitz domain and $f=(f^\alpha)\in L^p(\partial\Omega;\mathbb{R}^m)$ with $1 < p < \infty$. The single-layer potential $\mathcal{S}_\varepsilon(f) = (\mathcal{S}_\varepsilon^\alpha(f))$ is defined by

$$\mathcal{S}_\varepsilon^\alpha(f)(x) = \int_{\partial\Omega} \Gamma_\varepsilon^{\alpha\beta}(x,y) f^\beta(y)\, d\sigma(y), \tag{8.4.1}$$

where $\Gamma_\varepsilon(x,y) = \big(\Gamma_\varepsilon^{\alpha\beta}(x,y)\big)$ is the matrix of fundamental solutions for $\mathcal{L}_\varepsilon$ in $\mathbb{R}^d$. The double-layer potential $\mathcal{D}_\varepsilon(f) = (\mathcal{D}_\varepsilon^\alpha(f))$ is defined by

$$\mathcal{D}_\varepsilon^\alpha(f)(x) = \int_{\partial\Omega} n_j(y) a_{ij}^{\beta\gamma}(y/\varepsilon) \frac{\partial}{\partial y_i}\Big\{\Gamma_\varepsilon^{\alpha\beta}(x,y)\Big\} f^\gamma(y)\, d\sigma(y). \tag{8.4.2}$$

Observe that both $\mathcal{S}_\varepsilon(f)$ and $\mathcal{D}_\varepsilon(f)$ are solutions of $\mathcal{L}_\varepsilon(u) = 0$ in $\mathbb{R}^d \setminus \partial\Omega$. Since

$$\Gamma^{\alpha\beta}(x,y;A_\varepsilon^*) = \Gamma^{\beta\alpha}(y,x;A_\varepsilon) = \Gamma_\varepsilon^{\beta\alpha}(y,x), \tag{8.4.3}$$

where $A_\varepsilon(x) = A(x/\varepsilon)$, the double-layer potential may be written as

$$\mathcal{D}_\varepsilon^\alpha(f)(x) = \int_{\partial\Omega} \left(\frac{\partial}{\partial\nu_\varepsilon^*}\Big\{\Gamma^\alpha(y,x;A_\varepsilon^*)\Big\}\right)^\gamma f^\gamma(y)\, d\sigma(y), \tag{8.4.4}$$

where $\frac{\partial}{\partial\nu_\varepsilon^*}$ denotes the conormal derivative associated with $\mathcal{L}_\varepsilon^*$. The definitions of single- and double-layer potentials are motivated by the following Green representation formula.

Proposition 8.4.2. *Suppose that $\mathcal{L}_\varepsilon(u_\varepsilon) = F$ in Ω, where $F \in L^p(\Omega;\mathbb{R}^m)$ for some $p > d$. Also assume that $(\nabla u_\varepsilon)^* \in L^1(\partial\Omega)$ and $u_\varepsilon, \nabla u_\varepsilon$ have nontangential limits a.e. on $\partial\Omega$. Then for any $x \in \Omega$,*

$$u_\varepsilon(x) = \mathcal{S}_\varepsilon\left(\frac{\partial u_\varepsilon}{\partial\nu_\varepsilon}\right)(x) - \mathcal{D}_\varepsilon(u_\varepsilon)(x) + \int_\Omega \Gamma_\varepsilon(x,y)F(y)\, dy. \tag{8.4.5}$$

Proof. Fix $x \in \Omega$ and choose $r > 0$ so small that $B(x,4r) \subset \Omega$. Let $\varphi \in C_0^\infty(B(x,2r))$ be such that $\varphi = 1$ in $B(x,r)$. It follows from (4.4.2) that

$$\begin{aligned}
u_\varepsilon^\gamma(x) &= (u_\varepsilon\varphi)^\gamma(x)\\
&= \int_\Omega a_{ji}^{\beta\alpha}(y/\varepsilon)\frac{\partial}{\partial y_j}\Big\{\Gamma^{\beta\gamma}(y,x;A_\varepsilon^*)\Big\}\cdot\frac{\partial}{\partial y_i}(u_\varepsilon^\alpha\varphi)\, dy\\
&= \int_\Omega a_{ji}^{\beta\alpha}(y/\varepsilon)\frac{\partial}{\partial y_j}\Big\{\Gamma^{\beta\gamma}(y,x;A_\varepsilon^*)\Big\}\cdot\frac{\partial}{\partial y_i}\Big\{u_\varepsilon^\alpha(\varphi-1)\Big\}\, dy\\
&\quad + \int_\Omega a_{ji}^{\beta\alpha}(y/\varepsilon)\frac{\partial}{\partial y_j}\Big\{\Gamma^{\beta\gamma}(y,x;A_\varepsilon^*)\Big\}\cdot\frac{\partial u_\varepsilon^\alpha}{\partial y_i}\, dy\\
&= I_1^\gamma + I_2^\gamma.
\end{aligned}$$

Using the divergence theorem and $\mathcal{L}_\varepsilon^*\{\Gamma^\gamma(\cdot,x;A_\varepsilon^*)\}=0$ in $\mathbb{R}^d\setminus\{x\}$, we obtain

$$\begin{aligned}I_1^\gamma &= \int_{\partial\Omega} n_i(y) a_{ji}^{\beta\alpha}(y/\varepsilon)\frac{\partial}{\partial y_j}\Big\{\Gamma_\varepsilon^{\beta\gamma}(y,x;A_\varepsilon^*)\Big\}u_\varepsilon^\alpha(\varphi-1)\,d\sigma(y)\\ &= -\mathcal{D}_\varepsilon^\gamma(u_\varepsilon)(x),\end{aligned}$$

where we also used the fact that $\varphi-1=0$ in $B(x,r)$ and $\varphi-1=-1$ on $\partial\Omega$. Similarly, since $\mathcal{L}(u_\varepsilon)=F$ in Ω, it follows from the divergence theorem that

$$\begin{aligned}I_2^\gamma &= \int_{\partial\Omega}\Gamma_\varepsilon^{\beta\gamma}(y,x;A_\varepsilon^*)n_j(y)a_{ji}^{\beta\alpha}(y/\varepsilon)\frac{\partial u_\varepsilon^\alpha}{\partial y_i}\,d\sigma(y)+\int_\Omega \Gamma_\varepsilon^{\beta\gamma}(y,x;A_\varepsilon^*)F^\beta(y)\,dy\\ &= \mathcal{S}_\varepsilon^\gamma\left(\frac{\partial u_\varepsilon}{\partial\nu_\varepsilon}\right)(x)+\int_\Omega \Gamma_\varepsilon^{\gamma\beta}(x,y)F^\beta(y)\,dy.\end{aligned}$$

Hence,

$$\begin{aligned}u^\gamma(x) &= I_1^\gamma(x)+I_2^\gamma(x)\\ &= -\mathcal{D}_\varepsilon^\gamma(u_\varepsilon)(x)+\mathcal{S}_\varepsilon^\gamma\left(\frac{\partial u_\varepsilon}{\partial\nu_\varepsilon}\right)(x)+\int_\Omega\Gamma_\varepsilon^{\gamma\beta}(x,y)F^\beta(y)\,dy.\end{aligned}$$

We point out that since $F\in L^p(\Omega;\mathbb{R}^m)$ for some $p>d$, the solution $u_\varepsilon\in C^1(\Omega;\mathbb{R}^m)$. This, together with the assumptions that $(\nabla u_\varepsilon)^*\in L^1(\partial\Omega)$ and u_ε, ∇u_ε have nontangential limits, allows us to apply the divergence theorem on Ω. □

Theorem 8.4.3. *Let $A\in\Lambda(\mu,\lambda,\tau)$ and Ω be a bounded Lipschitz domain. Let $u_\varepsilon(x)=\mathcal{S}_\varepsilon(f)(x)$ and $v_\varepsilon=\mathcal{D}_\varepsilon(f)(x)$. Then for $1<p<\infty$,*

$$\|(\nabla u_\varepsilon)^*\|_{L^p(\partial\Omega)}+\|(v_\varepsilon)^*\|_{L^p(\partial\Omega)}\le C_p\,\|f\|_{L^p(\partial\Omega)},\tag{8.4.6}$$

where C_p depends only on μ, λ, τ, p, and the Lipschitz character of Ω.

Proof. Recall that

$$\Gamma_\varepsilon(x,y)=\Gamma(x,y;A_\varepsilon)=\varepsilon^{2-d}\Gamma(\varepsilon^{-1}x,\varepsilon^{-1}y;A),\tag{8.4.7}$$

where $A_\varepsilon(x)=A(x/\varepsilon)$. Thus, by rescaling, it suffices to prove the theorem for the case $\varepsilon=1$. However, in this case, the estimate (8.4.6) follows readily from Theorem 8.3.6. Here we have used the observation that the rescaled domain $\{x:\varepsilon x\in\Omega\}$ and Ω have the same Lipschitz characters. □

The next theorem gives the nontangential limits of $\nabla\mathcal{S}_\varepsilon(f)$ on $\partial\Omega$. Recall that for a function w in $\mathbb{R}^d\setminus\partial\Omega$, we use w_+ and w_- to denote its nontangential limits on $\partial\Omega$, taken from inside Ω and outside $\overline{\Omega}$, respectively.

Theorem 8.4.4. *Let $u_\varepsilon = \mathcal{S}_\varepsilon(f)$ for some $f \in L^p(\partial\Omega; \mathbb{R}^m)$ and $1 < p < \infty$. Then for a.e. $x \in \partial\Omega$,*

$$\left(\frac{\partial u_\varepsilon^\alpha}{\partial x_i}\right)_\pm (x) = \pm \frac{1}{2} n_i(x) b_\varepsilon^{\alpha\beta}(x) f^\beta(x) + \text{p.v.} \int_{\partial\Omega} \frac{\partial}{\partial x_i}\Big\{\Gamma_\varepsilon^{\alpha\beta}(x,y)\Big\} f^\beta(y)\, d\sigma(y), \tag{8.4.8}$$

where $\big(b_\varepsilon^{\alpha\beta}(x)\big)_{m\times m}$ is the inverse matrix of $\big(n_i(x)n_j(x)a_{ij}^{\alpha\beta}(x/\varepsilon)\big)_{m\times m}$.

Proof. By (8.4.7) and rescaling, it suffices to consider the case $\varepsilon = 1$. By Theorem 8.4.3, we may assume that $f \in C_0^\infty(\mathbb{R}^d; \mathbb{R}^m)$. We will also use the fact that if $f \in C_0^\infty(\mathbb{R}^d; \mathbb{R}^m)$, there exists a set $F \subset \partial\Omega$ such that $\sigma(\partial\Omega \setminus E) = 0$ and the trace formula (8.4.8) holds for any $x \in E$ and for any constant matrix A satisfying the ellipticity condition (2.2.28) (see [37, 43]).

Now fix $z \in E$ and consider

$$w^\alpha(x) = \int_{\partial\Omega} \Gamma^{\alpha\beta}(x, y; A(z)) f^\beta(y)\, d\sigma(y),$$

the single-layer potential for the (constant coefficient) operator $-\text{div}(A(z)\nabla)$. Note that by (8.2.6) and (8.2.11),

$$\begin{aligned}
&|\nabla_1\Gamma(x,y;A) - \nabla_1\Gamma(x,y;A(z))|\\
&\quad\le |\nabla_1\Gamma(x,y;A) - \nabla_1\Gamma(x,y;A(y))|\\
&\qquad + |\nabla_1\Gamma(x,y;A(y)) - \nabla_1\Gamma(x,y;A(z))|\\
&\quad\le C|x-y|^{1-d+\lambda} + C|x-y|^{1-d}|y-z|^\lambda\\
&\quad\le C|z-y|^{1-d+\lambda}
\end{aligned}$$

for $x \in \gamma(z)$ and $y \in \partial\Omega$. By the dominated convergence theorem, this implies that

$$\begin{aligned}
(\nabla u^\alpha)_\pm(z) &= (\nabla w^\alpha)_\pm(z) + \int_{\partial\Omega} \{\nabla_1\Gamma^{\alpha\beta}(z,y;A) - \nabla_1\Gamma^{\alpha\beta}(z,y;A(z))\} f^\beta(y)\, d\sigma(y)\\
&= \pm\frac{1}{2} n(z) b^{\alpha\beta}(z) f^\beta(z) + \text{p.v.} \int_{\partial\Omega} \nabla_1\Gamma^{\alpha\beta}(z,y;A) f^\beta(y)\, d\sigma(y).
\end{aligned}$$

The proof is complete. □

It follows from (8.4.8) that if $u_\varepsilon = \mathcal{S}_\varepsilon(f)$, then

$$n_j\left(\frac{\partial u_\varepsilon^\alpha}{\partial x_i}\right)_+ - n_i\left(\frac{\partial u_\varepsilon^\alpha}{\partial x_j}\right)_+ = n_j\left(\frac{\partial u_\varepsilon^\alpha}{\partial x_i}\right)_- - n_i\left(\frac{\partial u_\varepsilon^\alpha}{\partial x_j}\right)_-, \tag{8.4.9}$$

i.e., $(\nabla_{\tan} u_\varepsilon)_+ = (\nabla_{\tan} u_\varepsilon)_-$ on $\partial\Omega$. Moreover,

$$\left(\frac{\partial u_\varepsilon}{\partial \nu_\varepsilon}\right)_\pm = \left(\pm\frac{1}{2}I + \mathcal{K}_{\varepsilon,A}\right)(f), \tag{8.4.10}$$

where I denotes the identity operator and

$$(\mathcal{K}_{\varepsilon,A}(f)(x))^\alpha = \text{p.v.}\int_{\partial\Omega} K^{\alpha\beta}_{\varepsilon,A}(x,y) f^\beta(y)\, d\sigma(y), \tag{8.4.11}$$

with

$$K^{\alpha\beta}_{\varepsilon,A}(x,y) = n_i(x) a^{\alpha\gamma}_{ij}(x/\varepsilon)\frac{\partial}{\partial x_j}\left\{\Gamma^{\gamma\beta}_\varepsilon(x,y)\right\}. \tag{8.4.12}$$

In particular, we have the so-called *jump relation*

$$f = \left(\frac{\partial u_\varepsilon}{\partial \nu_\varepsilon}\right)_+ - \left(\frac{\partial u_\varepsilon}{\partial \nu_\varepsilon}\right)_- . \tag{8.4.13}$$

Note that by Theorems 8.4.3 and 8.4.4,

$$\|\mathcal{K}_{A,\varepsilon}(f)\|_{L^p(\partial\Omega)} \le C_p \|f\|_{L^p(\partial\Omega)} \qquad \text{for } 1 < p < \infty, \tag{8.4.14}$$

where C_p depends at most on μ, λ, τ, p, and the Lipschitz character of Ω. Let

$$L^p_0(\partial\Omega;\mathbb{R}^m) = \left\{ f \in L^p(\partial\Omega;\mathbb{R}^m) : \int_{\partial\Omega} f\, d\sigma = 0\right\}. \tag{8.4.15}$$

It follows from $\mathcal{L}_\varepsilon(u_\varepsilon) = 0$ in Ω that

$$\int_{\partial\Omega}\left(\frac{\partial u_\varepsilon}{\partial \nu_\varepsilon}\right)_+ d\sigma = 0.$$

Thus, for $1 < p < \infty$, the operator

$$(1/2)I + \mathcal{K}_{\varepsilon,A} : L^p_0(\partial\Omega;\mathbb{R}^m) \to L^p_0(\partial\Omega;\mathbb{R}^m) \tag{8.4.16}$$

is bounded; its operator norm is bounded by a constant independent of $\varepsilon > 0$. Also, by Theorem 8.4.3 and (8.4.9), if $f \in L^p(\partial\Omega;\mathbb{R}^m)$ for some $1 < p < \infty$, then $\mathcal{S}_\varepsilon(f) \in W^{1,p}(\partial\Omega;\mathbb{R}^m)$ and

$$\|\mathcal{S}_\varepsilon(f)\|_{W^{1,p}(\partial\Omega)} \le C_p \|f\|_{L^p(\partial\Omega)}, \tag{8.4.17}$$

where C_p depends only on μ, λ, τ, p, and the Lispchitz character of Ω.

The next theorem gives the trace of the double-layer potentials on $\partial\Omega$.

Theorem 8.4.5. *Let $w = \mathcal{D}_\varepsilon(f)$, where $f \in L^p(\partial\Omega;\mathbb{R}^m)$ and $1 < p < \infty$. Then*

$$w_\pm = \Big(\mp(1/2)I + \mathcal{K}^*_{\varepsilon,A^*}\Big)(f) \quad \text{on } \partial\Omega, \tag{8.4.18}$$

where $\mathcal{K}^*_{\varepsilon,A^*}$ is the adjoint operator of $\mathcal{K}_{\varepsilon,A^*}$, defined by (8.4.11) and (8.4.12) (with A replaced by A^*).

Proof. By rescaling we may assume $\varepsilon = 1$. Note that by (8.2.6) and (8.2.13),

$$|\nabla_1\Gamma(x,y;A) - \nabla_1\Gamma(x,y;A(y))| \le C\,|x-y|^{1-d+\lambda},$$
$$|\nabla_2\Gamma(x,y;A) - \nabla_2\Gamma(x,y;A(y))| \le C\,|x-y|^{1-d+\lambda}.$$

This, together with the observation $\nabla_1\Gamma(x,y;A(y)) = -\nabla_2\Gamma(x,y;A(y))$, gives

$$\Big|\frac{\partial}{\partial x_i}\Big\{\Gamma(x,y;A)\Big\} + \frac{\partial}{\partial y_i}\Big\{\Gamma(x,y;A)\Big\}\Big| \le C\,|x-y|^{1-d+\lambda}.$$

Hence, by the Lebesgue dominated convergence theorem,

$$\begin{aligned} w_\pm^\alpha(x) &= -v_\pm^\alpha(x)\\ &\quad + \int_{\partial\Omega} n_j(y)a_{ij}^{\beta\gamma}(y)\left\{\frac{\partial}{\partial y_i}\{\Gamma^{\alpha\beta}(x,y;A)\} + \frac{\partial}{\partial x_i}\{\Gamma^{\alpha\beta}(x,y;A)\}\right\} f^\gamma(y)\,d\sigma(y), \end{aligned}$$

where

$$v^\alpha(x) = \frac{\partial}{\partial x_i}\int_{\partial\Omega} n_j(y)a_{ij}^{\beta\gamma}(y)\Gamma^{\alpha\beta}(x,y;A)f^\gamma(y)\,d\sigma(y).$$

In view of the trace formula (8.4.8), we have

$$\begin{aligned} v_\pm^\alpha(x) &= \pm\,(1/2)n_i(x)b^{\alpha\beta}(x)\cdot n_j(x)a_{ij}^{\beta\gamma}(x)f^\gamma(x)\\ &\quad + \text{p.v.}\int_{\partial\Omega} n_j(y)a_{ij}^{\beta\gamma}(y)\frac{\partial}{\partial x_i}\Big\{\Gamma^{\alpha\beta}(x,y;A)\Big\} f^\gamma(y)\,d\sigma(y)\\ &= \pm\,(1/2)f^\alpha(x) + \text{p.v.}\int_{\partial\Omega} n_j(y)a_{ij}^{\beta\gamma}(y)\frac{\partial}{\partial x_i}\Big\{\Gamma^{\alpha\beta}(x,y;A)\Big\} f^\gamma(y)\,d\sigma(y). \end{aligned}$$

Hence,

$$\begin{aligned} w_\pm^\alpha(x) &= \mp\,(1/2)f^\alpha(x)\\ &\quad + \text{p.v.}\int_{\partial\Omega} n_j(y)a_{ij}^{\beta\gamma}(y)\frac{\partial}{\partial y_i}\Big\{\Gamma^{\alpha\beta}(x,y;A)\Big\} f^\gamma(y)\,d\sigma(y)\\ &= \mp\,(1/2)f^\alpha(x) + \text{p.v.}\int_{\partial\Omega} K_{1,A^*}^{\beta\alpha}(y,x)f^\beta(y)\,d\sigma(y), \end{aligned}$$

where $K_{1,A^*}^{\beta\alpha}(y,x)$ is defined by (8.4.12), but with A replaced by A^*. □

Remark 8.4.6. Let $f \in L^p(\partial\Omega;\mathbb{R}^m)$ and $u_\varepsilon = \mathcal{S}_\varepsilon(f)$. It follows from the Green representation formula (8.4.5) that

$$u_\varepsilon(x) = \mathcal{S}_\varepsilon(g)(x) - \mathcal{D}_\varepsilon(u_\varepsilon)(x)$$

for any $x \in \Omega$, where $g = (\partial u_\varepsilon/\partial\nu_\varepsilon)_+$. Letting $x \to z \in \partial\Omega$ nontangentially, we obtain

$$\mathcal{S}_\varepsilon(f) = \mathcal{S}_\varepsilon\left(((1/2)I + \mathcal{K}_{\varepsilon,A})(f)\right) - (-(1/2)I + \mathcal{K}_{\varepsilon,A^*}^*)\mathcal{S}_\varepsilon(f) \qquad \text{on } \partial\Omega.$$

This gives

$$\mathcal{S}_\varepsilon \mathcal{K}_{\varepsilon,A}(f) = \mathcal{K}^*_{\varepsilon,A^*} \mathcal{S}_\varepsilon(f) \tag{8.4.19}$$

for any $f \in L^p(\partial\Omega)$ with $1 < p < \infty$.

In summary, if $1 < p < \infty$ and $f \in L^p(\partial\Omega; \mathbb{R}^m)$, the single-layer potential $u_\varepsilon = \mathcal{S}_\varepsilon(f)$ is a solution to the L^p Neumann problem for $\mathcal{L}_\varepsilon(u_\varepsilon) = 0$ in Ω with the boundary data $((1/2)I + \mathcal{K}_{\varepsilon,A})f$, and the estimate

$$\|(\nabla u_\varepsilon)^*\|_{L^p(\partial\Omega)} \le C_p \|f\|_{L^p(\partial\Omega)}$$

holds. Similarly, the double-layer potential $v_\varepsilon = \mathcal{D}_\varepsilon(f)$ is a solution to the L^p Dirichlet problem in Ω with the boundary data $(-(1/2)I + \mathcal{K}^*_{\varepsilon,A^*})f$, and the estimate $\|(v_\varepsilon)^*\|_{L^p(\partial\Omega)} \le C_p \|f\|_{L^p(\partial\Omega)}$ holds. As a result, one may establish the existence of solutions as well as the nontangential-maximal-function estimates in the L^p Neumann and Dirichlet problems in Ω by showing that the operators $(1/2)I + \mathcal{K}_{\varepsilon,A}$ and $-(1/2)I + \mathcal{K}^*_{\varepsilon,A^*}$ are invertible on $L^p_0(\partial\Omega; \mathbb{R}^m)$ and $L^p(\partial\Omega; \mathbb{R}^m)$ (modulo possible finite-dimensional subspaces), respectively, and by proving that the operator norms of their inverses are bounded by constants independent of $\varepsilon > 0$. This approach to the boundary value problems is often referred as the method of layer potentials.

In the following sections we will show that if Ω is a bounded Lipschitz domain with connected boundary, the operators

$$\begin{aligned} (1/2)I + \mathcal{K}_{\varepsilon,A} &: L^2_0(\partial\Omega, \mathbb{R}^m) \to L^2_0(\partial\Omega, \mathbb{R}^m), \\ -(1/2)I + \mathcal{K}_{\varepsilon,A} &: L^2(\partial\Omega, \mathbb{R}^m) \to L^2(\partial\Omega, \mathbb{R}^m) \end{aligned} \tag{8.4.20}$$

are isomorphisms, under the assumptions that $A \in \Lambda(\mu, \lambda, \tau)$ and $A^* = A$. More importantly, we obtain the estimates

$$\begin{aligned} \|((1/2)I + \mathcal{K}_{\varepsilon,A})^{-1}\|_{L^2_0 \to L^2_0} &\le C, \\ \|(-(1/2)I + \mathcal{K}_{\varepsilon,A})^{-1}\|_{L^2 \to L^2} &\le C, \end{aligned} \tag{8.4.21}$$

where C depends only on μ, λ, τ, and the Lipschitz character of Ω.

We end this section with a simple observation.

Theorem 8.4.7. *Let Ω be a bounded Lipschitz domain with connected boundary and $A \in \Lambda(\mu, \lambda, \tau)$. Then the operators in* (8.4.20) *are injective.*

Proof. Suppose that $f \in L^2_0(\partial\Omega; \mathbb{R}^m)$ and $((1/2)I + \mathcal{K}_{\varepsilon,A})(f) = 0$. Let $u_\varepsilon = \mathcal{S}_\varepsilon(f)$. It follows via integration by parts that

$$\int_\Omega A(x/\varepsilon)\nabla u_\varepsilon \cdot \nabla u_\varepsilon \, dx = \int_{\partial\Omega} \left(\frac{\partial u_\varepsilon}{\partial \nu_\varepsilon}\right)_+ \cdot u_\varepsilon \, d\sigma. \tag{8.4.22}$$

Since $\left(\frac{\partial u_\varepsilon}{\partial \nu_\varepsilon}\right)_+ = ((1/2)I + \mathcal{K}_{\varepsilon,A})(f) = 0$ on $\partial\Omega$, we deduce from (8.4.22) that $\nabla u_\varepsilon = 0$ in Ω and hence $u_\varepsilon = b$ is constant in Ω. Recall that for $d \ge 3$, we have

$|\Gamma_\varepsilon(x,y)| \le C\,|x-y|^{2-d}$ and $|\nabla_x \Gamma_\varepsilon(x,y)| + |\nabla_y \Gamma_\varepsilon(x,y)| \le C\,|x-y|^{1-d}$. It follows that $u_\varepsilon(x) = O(|x|^{2-d})$ and $\nabla u_\varepsilon(x) = O(|x|^{1-d})$, as $|x| \to \infty$. Hence, we may use integration by parts in Ω_- to show that

$$\int_{\Omega_-} A(x/\varepsilon)\nabla u_\varepsilon \cdot \nabla u_\varepsilon \, dx = -\int_{\partial\Omega} \left(\frac{\partial u_\varepsilon}{\partial \nu_\varepsilon}\right)_- \cdot u_\varepsilon \, d\sigma, \tag{8.4.23}$$

where $\Omega_- = \mathbb{R}^d \setminus \overline{\Omega}$. Note that

$$\begin{aligned}
\int_{\partial\Omega} \left(\frac{\partial u_\varepsilon}{\partial \nu_\varepsilon}\right)_- \cdot u_\varepsilon \, d\sigma &= b \cdot \int_{\partial\Omega} \left(\frac{\partial u_\varepsilon}{\partial \nu_\varepsilon}\right)_- d\sigma \\
&= -b \cdot \int_{\partial\Omega} f \, d\sigma = 0,
\end{aligned}$$

where we have used the jump relation (8.4.13) and $\int_{\partial\Omega} f\, d\sigma = 0$. In view of (8.4.23) this implies that $\nabla u_\varepsilon = 0$ in Ω_-. Since Ω_- is connected and $u_\varepsilon(x) = O(|x|^{2-d})$ as $|x| \to \infty$, u_ε must be zero in Ω_-. It follows that $\left(\frac{\partial u_\varepsilon}{\partial \nu_\varepsilon}\right)_- = 0$ on $\partial\Omega$. By the jump relation, we obtain $f = 0$. Thus we have proved that $(1/2)I + \mathcal{K}_{\varepsilon,A}$ is injective on $L^2_0(\partial\Omega; \mathbb{R}^m)$. That $-(1/2)I + \mathcal{K}_{\varepsilon,A}$ is injective on $L^2(\partial\Omega; \mathbb{R}^m)$ is established in a similar manner. □

8.5 Laplace's equation

In this section we establish the estimates in (8.4.21) and solve the L^2 Dirichlet, Neumann, and regularity problems in Lipschitz domains in the case $\mathcal{L}_\varepsilon = -\Delta$. This not only illustrates the crucial role of Rellich identities in the study of boundary value problems in nonsmooth domains in the simplest setting, but the results of this section will be used to handle the operator $\mathcal{L}_\varepsilon$ in the general case. Furthermore, the argument presented in this section extends readily to the case of second-order elliptic systems with constant coefficients satisfying (2.1.20).

Throughout this section we will assume that Ω is a bounded Lipschitz domain in $\mathbb{R}^d$, $d \ge 2$, with connected boundary. By rescaling we may also assume $\operatorname{diam}(\Omega) = 1$.

Lemma 8.5.1. *Suppose that $\Delta u \in L^2(\Omega)$ and $(\nabla u)^* \in L^2(\partial\Omega)$. Also assume that ∇u has nontangential limit a.e. on $\partial\Omega$. Then*

$$\begin{aligned}
\int_{\partial\Omega} h_i n_i |\nabla u|^2 \, d\sigma = 2\int_{\partial\Omega} h_i \frac{\partial u}{\partial x_i} \cdot \frac{\partial u}{\partial n}\, d\sigma + \int_\Omega \operatorname{div}(h)|\nabla u|^2 \, dx \\
- 2\int_\Omega \frac{\partial h_i}{\partial x_j} \cdot \frac{\partial u}{\partial x_i} \cdot \frac{\partial u}{\partial x_j}\, dx - 2\int_\Omega h_i \frac{\partial u}{\partial x_i}\, \Delta u \, dx,
\end{aligned} \tag{8.5.1}$$

where $h = (h_1, \dots, h_d) \in C^1_0(\mathbb{R}^d; \mathbb{R}^d)$, n denotes the outward unit normal to $\partial\Omega$ and $\frac{\partial u}{\partial n} = \nabla u \cdot n$.

Proof. We begin by choosing a sequence of smooth domains $\{\Omega_\ell\}$ so that $\Omega_\ell \uparrow \Omega$. By the divergence theorem we have

$$\begin{aligned}\int_{\Omega_\ell} h_i n_i |\nabla u|^2 \, d\sigma &= \int_{\Omega\ell} \frac{\partial}{\partial x_i} \left\{ h_i \frac{\partial u}{\partial x_j} \frac{\partial u}{\partial x_j} \right\} dx \\ &= \int_{\Omega_\ell} \operatorname{div}(h) |\nabla u|^2 \, dx + 2 \int_{\Omega_\ell} h_i \frac{\partial^2 u}{\partial x_i \partial x_j} \cdot \frac{\partial u}{\partial x_j} \, dx.\end{aligned} \tag{8.5.2}$$

Using integration by parts, we see that the last term in (8.5.2) is equal to

$$-2 \int_{\Omega_\ell} h_i \frac{\partial u}{\partial x_i} \Delta u \, dx - 2 \int_{\Omega_\ell} \frac{\partial h_i}{\partial x_j} \cdot \frac{\partial u}{\partial x_i} \cdot \frac{\partial u}{\partial x_j} \, dx + 2 \int_{\partial\Omega_\ell} h_i \frac{\partial u}{\partial x_i} \cdot \frac{\partial u}{\partial n} \, d\sigma.$$

This gives the identity (8.5.1), but with Ω replaced by Ω_ℓ. Finally we let $\ell \to \infty$. Since $(\nabla u)^* \in L^2(\partial\Omega)$ and ∇u has nontangential limit a.e. on $\partial\Omega$, the identity for Ω follows by the Lebesgue dominated convergence theorem. ☐

Lemma 8.5.2. *Under the same assumptions as in Lemma* 8.5.1, *we have*

$$\begin{aligned}\int_{\partial\Omega} h_i n_i |\nabla u|^2 \, d\sigma = 2 \int_{\partial\Omega} & h_i \frac{\partial u}{\partial x_j} \left\{ n_i \frac{\partial u}{\partial x_j} - n_j \frac{\partial u}{\partial x_i} \right\} d\sigma \\ & - \int_{\Omega} \operatorname{div}(h) |\nabla u|^2 \, dx + 2 \int_{\Omega} \frac{\partial h_i}{\partial x_j} \cdot \frac{\partial u}{\partial x_i} \cdot \frac{\partial u}{\partial x_j} \, dx \\ & - 2 \int_{\Omega} h_i \frac{\partial u}{\partial x_i} \Delta u \, dx.\end{aligned} \tag{8.5.3}$$

Proof. Let I and J denote the left- and right-hand sides of (8.5.1) respectively. Identity (8.5.3) follows from (8.5.1) by writing J as $2I - J$. ☐

Formulas (8.5.1) and (8.5.3) are referred to as the Rellich identities for Laplace's equation. They also hold on $\Omega_- = \mathbb{R}^d \setminus \overline{\Omega}$ under the assumptions that $\Delta u \in L^2(\Omega_-)$, $(\nabla u)^* \in L^2(\partial\Omega)$ and ∇u has nontangential limit a.e. on $\partial\Omega$. The use of the divergence theorem on the unbounded domain Ω_- is justified, as the vector field h has compact support. We should point out that in the case of Ω_-, all solid integrals in (8.5.1) and (8.5.3) need to change signs. This is because n always denotes the outward unit normal to $\partial\Omega$.

The following lemma will be used to handle solid integrals in (8.5.1) and (8.5.3). Recall that $(\nabla u)^* \in L^1(\partial\Omega)$ implies that u has nontangential limit a.e. on $\partial\Omega$.

Lemma 8.5.3.

1. *Suppose that* $\Delta u = 0$ *in* Ω *and* $(\nabla u)^* \in L^2(\partial\Omega)$. *Also assume that* ∇u *has nontangential limit a.e. on* $\partial\Omega$. *Then*

$$\int_{\Omega} |\nabla u|^2 \, dx \le C \left\| \frac{\partial u}{\partial n} \right\|_{L^2(\partial\Omega)} \|\nabla_{\tan} u\|_{L^2(\partial\Omega)}. \tag{8.5.4}$$

2. *Suppose that* $\Delta u = 0$ *in* Ω_- *and* $(\nabla u)^* \in L^2(\partial\Omega)$. *Also assume that* ∇u *has nontangential limit a.e. on* $\partial\Omega$ *and that as* $|x| \to \infty$, $|u(x)| = O(|x|^{2-d})$ *for* $d \geq 3$ *and* $|u(x)| = o(1)$ *for* $d = 2$. *Then*

$$\int_{\Omega_-} |\nabla u|^2 \, dx \leq C \left\| \frac{\partial u}{\partial n} \right\|_{L^2(\partial\Omega)} \|\nabla_{\tan} u\|_{L^2(\partial\Omega)} + \left| \int_{\partial\Omega} \frac{\partial u}{\partial n} \, d\sigma \right| \left| \fint_{\partial\Omega} u \right|. \quad (8.5.5)$$

Proof. For part (1) we use the divergence theorem and $\int_{\partial\Omega} u \, d\sigma = 0$ to obtain

$$\int_\Omega |\nabla u|^2 \, dx = \int_{\partial\Omega} \frac{\partial u}{\partial n} u \, d\sigma = \int_{\partial\Omega} \frac{\partial u}{\partial n} \left(u - \fint_{\partial\Omega} u \right) d\sigma.$$

Estimate (8.5.4) follows from this by applying the Cauchy inequality and then the Poincaré inequality. Part (2) may be proved in a similar manner. By interior estimates for harmonic functions and the decay assumption at infinity, we see that $|\nabla u(x)| = O(|x|^{1-d})$ as $|x| \to \infty$. Hence,

$$\int_{|x|=R} |\nabla u| \, |u| \, d\sigma \to 0 \quad \text{as } R \to \infty.$$

This is enough to justify the integration by parts in Ω_-. □

Theorem 8.5.4. *Suppose that* $\Delta u = 0$ *in* Ω *and* $(\nabla u)^* \in L^2(\partial\Omega)$. *Also assume that* ∇u *has nontangential limit a.e. on* $\partial\Omega$. *Then*

$$\begin{cases} \|\nabla u\|_{L^2(\partial\Omega)} \leq C \left\| \dfrac{\partial u}{\partial n} \right\|_{L^2(\partial\Omega)}, \\ \|\nabla u\|_{L^2(\partial\Omega)} \leq C \, \|\nabla_{\tan} u\|_{L^2(\partial\Omega)}, \end{cases} \quad (8.5.6)$$

where C *depends only on the Lipschitz character of* Ω.

Proof. Let $h \in C_0^\infty(\mathbb{R}^d; \mathbb{R}^d)$ be a vector field such that $h \cdot n \geq c_0 > 0$ on $\partial\Omega$. It follows from (8.5.1) that

$$\begin{aligned} \int_{\partial\Omega} |\nabla u|^2 \, d\sigma &\leq C \, \|\nabla u\|_{L^2(\partial\Omega)} \left\| \frac{\partial u}{\partial n} \right\|_{L^2(\partial\Omega)} + C \int_\Omega |\nabla u|^2 \, dx \\ &\leq C \, \|\nabla u\|_{L^2(\partial\Omega)} \left\| \frac{\partial u}{\partial n} \right\|_{L^2(\partial\Omega)} + C \left\| \frac{\partial u}{\partial n} \right\|_{L^2(\partial\Omega)} \|\nabla_{\tan} u\|_{L^2(\partial\Omega)}, \end{aligned}$$

where we have used (8.5.4) for the second inequality. Since $|\nabla_{\tan} u| \leq |\nabla u|$, this gives the first estimate in (8.5.6). The second estimate in (8.5.6) follows from the formula (8.5.3) and (8.5.4) in the same manner. □

Theorem 8.5.4 shows that for (suitable) harmonic functions in a Lipschitz domain Ω, the L^2 norms of the normal derivative $\frac{\partial u}{\partial n}$ and the tangential derivatives $\nabla_{\tan} u$ on $\partial\Omega$ are equivalent. This is also true for harmonic functions in Ω_-, modulo some linear functionals. We leave the proof of the following theorem to the reader.

Theorem 8.5.5. *Suppose that $\Delta u = 0$ in Ω_- and $(\nabla u)^* \in L^2(\partial\Omega)$. Also assume that ∇u has nontangential limit a.e. on $\partial\Omega$ and that as $|x| \to \infty$, $|u(x)| = O(|x|^{2-d})$ for $d \ge 3$ and $|u(x)| = o(1)$ for $d = 2$. Then*

$$\begin{aligned} \|\nabla u\|^2_{L^2(\partial\Omega)} &\le C \left\| \frac{\partial u}{\partial n} \right\|^2_{L^2(\partial\Omega)} + C \left| \int_{\partial\Omega} \frac{\partial u}{\partial n}\, d\sigma \right| \left| \fint_{\partial\Omega} u \right|, \\ \|\nabla u\|^2_{L^2(\partial\Omega)} &\le C \, \|\nabla_{\tan} u\|^2_{L^2(\partial\Omega)} + C \left| \int_{\partial\Omega} \frac{\partial u}{\partial n}\, d\sigma \right| \left| \fint_{\partial\Omega} u \right|, \end{aligned} \tag{8.5.7}$$

where C depends only on the Lipschitz character of Ω.

Let ω_d denote the surface area of the unit sphere in $\mathbb{R}^d$. The fundamental solution for $\mathcal{L} = -\Delta$ with pole at the origin is given by

$$\begin{cases} \Gamma(x) = \dfrac{1}{(d-2)w_d |x|^{d-2}} & \text{for } d \ge 3, \\ \Gamma(x) = -\dfrac{1}{2\pi} \ln |x| & \text{for } d = 2. \end{cases} \tag{8.5.8}$$

For $f \in L^p(\partial\Omega)$ with $1 < p < \infty$, let

$$u(x) = \mathcal{S}(f)(x) = \int_{\partial\Omega} \Gamma(x-y) f(y)\, d\sigma(y) \tag{8.5.9}$$

be the single-layer potential for $\mathcal{L} = -\Delta$. It follows from Section 7.5 that

$$\|(\nabla u)^*\|_{L^p(\partial\Omega)} \le C_p \, \|f\|_{L^p(\partial\Omega)}$$

and

$$\left(\frac{\partial u}{\partial n} \right)_+ = \big((1/2)I + \mathcal{K}\big) f \quad \text{and} \quad \left(\frac{\partial u}{\partial n} \right)_- = \big(-(1/2)I + \mathcal{K}\big) f \tag{8.5.10}$$

on $\partial\Omega$, where $\mathcal{K}$ is a bounded singular integral operator on $L^p(\partial\Omega)$. Also recall that $(\nabla_{\tan} u)_+ = (\nabla_{\tan} u)_-$ on $\partial\Omega$. As indicated at the beginning of this section, we will show that $(1/2) + \mathcal{K}$ and $-(1/2)I + \mathcal{K}$ are isomorphisms on $L^2_0(\partial\Omega)$ and $L^2(\partial\Omega)$, respectively.

Lemma 8.5.6. *Let $f \in L^2(\partial\Omega)$. Then*

$$\begin{aligned} \|f\|_{L^2(\partial\Omega)} &\le C \left\{ \|((1/2)I + \mathcal{K}) f\|_{L^2(\partial\Omega)} + \left| \int_{\partial\Omega} f \, d\sigma \right| \right\}, \\ \|f\|_{L^2(\partial\Omega)} &\le C \, \|(-(1/2)I + \mathcal{K}) f\|_{L^2(\partial\Omega)}, \end{aligned} \tag{8.5.11}$$

where C depends only on the Lipschitz character of Ω.

Proof. We first consider the case $f \in L_0^2(\partial\Omega)$. Let $u = \mathcal{S}(f)$. The additional condition $\int_{\partial\Omega} f\, d\sigma = 0$ implies that $u(x) = O(|x|^{1-d})$ as $|x| \to \infty$, which assures that estimates (8.5.6) and (8.5.7) hold for all $d \geq 2$. By the jump relation (8.4.13), it also implies that the mean value of $\left(\frac{\partial u}{\partial n}\right)_-$ is zero. Thus we may deduce from (8.5.7) that

$$\begin{aligned}\|(\nabla u)_-\|_{L^2(\partial\Omega)} &\leq C\,\|(\nabla_{\tan} u)_-\|_{L^2(\partial\Omega)} = C\,\|(\nabla_{\tan} u)_+\|_{L^2(\partial\Omega)}\\ &\leq C\left\|\left(\frac{\partial u}{\partial n}\right)_+\right\|_{L^2(\partial\Omega)}.\end{aligned}$$

By the jump relation it follows that

$$\begin{aligned}\|f\|_{L^2(\partial\Omega)} &\leq \left\|\left(\frac{\partial u}{\partial n}\right)_+\right\|_{L^2(\partial\Omega)} + \left\|\left(\frac{\partial u}{\partial n}\right)_-\right\|_{L^2(\partial\Omega)}\\ &\leq C\left\|\left(\frac{\partial u}{\partial n}\right)_+\right\|_{L^2(\partial\Omega)}\\ &= C\,\|((1/2)I + \mathcal{K})f\|_{L^2(\partial\Omega)}.\end{aligned}$$

This gives the first inequality in (8.5.11) for the case $f \in L_0^2(\partial\Omega)$. The general case follows by considering $f - E$, where E is the mean value of f on $\partial\Omega$.

Similarly, to establish the second inequality in (8.5.11), we use (8.5.6) and (8.5.7) to obtain

$$\begin{aligned}\|(\nabla u)_+\|_{L^2(\partial\Omega)} &\leq C\,\|(\nabla_{\tan} u)_+\|_{L^2(\partial\Omega)} = C\,\|(\nabla_{\tan} u)_-\|_{L^2(\partial\Omega)}\\ &\leq C\left\|\left(\frac{\partial u}{\partial n}\right)_-\right\|_{L^2(\partial\Omega)},\end{aligned}$$

where we also used $\int_{\partial\Omega}(\partial u/\partial n)_-\, d\sigma = 0$. By the jump relation this gives the second inequality in (8.5.11) for the case $f \in L_0^2(\partial\Omega)$. As before, it follows that for any $f \in L^2(\partial\Omega)$,

$$\|f\|_{L^2(\partial\Omega)} \leq C\,\|(-(1/2)I + \mathcal{K})f\|_{L^2(\partial\Omega)} + C|E|,$$

where E is the mean value of f on $\partial\Omega$. To finish the proof we simply observe that by the jump relation, E is also the mean value of $(-(1/2)I + \mathcal{K})f$ on $\partial\Omega$. Hence,

$$|E| \leq C\,\|(-(1/2)I + \mathcal{K})f\|_{L^2(\partial\Omega)},$$

which completes the proof. □

It follows readily from Lemma 8.5.6 that $(1/2)I + \mathcal{K}$ and $-(1/2)I + \mathcal{K}$ are injective on $L_0^2(\partial\Omega)$ and $L^2(\partial\Omega)$, respectively. The lemma also implies that the ranges of the operators are closed in $L^2(\partial\Omega)$. This is a consequence of the next theorem, whose proof is left to the reader.

Theorem 8.5.7. *Let $T : X \to Y$ be a bounded linear operator, where X, Y are normed linear spaces. Suppose that*

1. *X is Banach;*
2. *the dimension of the null space $\{f \in X : T(f) = 0\}$ is finite;*
3. *for any $f \in X$,*
$$\|f\|_X \le C\,\|Tf\|_Y + C\sum_{j=1}^{\ell} \|S_j f\|_{Y_j},$$
where $S_j : X \to Y_j$, $j = 1, \ldots, \ell$ are compact operators.

Then the range of T is closed in Y.

We will use a continuity method to show that $\pm(1/2)I + \mathcal{K}$ are surjective on $L^2_0(\partial\Omega)$ and $L^2(\partial\Omega)$, respectively. To this end we consider a family of operators

$$T_\lambda = \lambda I + \mathcal{K}, \tag{8.5.12}$$

where $\lambda \in \mathbb{R}$.

Lemma 8.5.8. *If $|\lambda| > 1/2$, the operator $T_\lambda : L^2(\partial\Omega) \to L^2(\partial\Omega)$ is injective.*

Proof. Let $f \in L^2(\partial\Omega)$ and $u = \mathcal{S}(f)$. Suppose that $T_\lambda(f) = 0$. Then

$$\left(\frac{\partial u}{\partial n}\right)_+ = ((1/2) - \lambda)f \quad \text{and} \quad \left(\frac{\partial u}{\partial n}\right)_+ = (-(1/2) - \lambda)f.$$

Note that the first equation above implies $f \in L^2_0(\partial\Omega)$. It follows that $u(x) = O(|x|^{1-d})$ as $|x| \to \infty$. By the divergence theorem,

$$\int_{\Omega_\pm} |\nabla u|^2\, dx = \pm \int_{\partial\Omega} u \left(\frac{\partial u}{\partial n}\right)_\pm d\sigma = \left(\frac{1}{2} \mp \lambda\right) \int_{\partial\Omega} uf\, d\sigma. \tag{8.5.13}$$

It follows that

$$\begin{aligned}
\int_{\Omega_+} |\nabla u|^2\, dx + \int_{\Omega_-} |\nabla u|^2\, dx &= \int_{\partial\Omega} uf\, d\sigma,\\
\int_{\Omega_+} |\nabla u|^2\, dx - \int_{\Omega_-} |\nabla u|^2\, dx &= -2\lambda \int_{\partial\Omega} uf\, d\sigma.
\end{aligned}$$

This implies that

$$2|\lambda| \left| \int_{\partial\Omega} uf\, d\sigma \right| \le \left| \int_{\partial\Omega} uf\, d\sigma \right|.$$

Since $2|\lambda| > 1$, we obtain $\int_{\partial\Omega} uf\, d\sigma = 0$. Hence, by (8.5.13), $|\nabla u| = 0$ in $\Omega_\pm$. Consequently, by the jump relation (8.4.13), we get $f = 0$. □

Let $h = (h_1, \dots, h_d) \in C_0^1(\mathbb{R}^d; \mathbb{R}^d)$. It follows from the trace formula (8.4.8) that if $u = \mathcal{S}(f)$,

$$h_i \left(\frac{\partial u}{\partial x_i}\right)_\pm = \pm\frac{1}{2} h \cdot n + \mathcal{K}_h(f),$$

where

$$\mathcal{K}_h(f)(x) = \text{p.v.} \int_{\partial\Omega} \frac{(y-x)\cdot h(x)}{\omega_d |x-y|^d} f(y)\, d\sigma(y).$$

Observe that

$$(\mathcal{K}_h + \mathcal{K}_h^*)(f)(x) = \text{p.v.} \int_{\partial\Omega} \frac{(y-x)\cdot (h(x)-h(y))}{\omega_d |x-y|^d} f(y)\, d\sigma(y).$$

Since $|h(x) - h(y)| \le C\,|x-y|$, the integral kernel of $\mathcal{K}_h + \mathcal{K}_h^*$ is bounded by $C|y-x|^{2-d}$. This implies that the operator $\mathcal{K}_h + \mathcal{K}_h^*$ is compact on $L^p(\partial\Omega)$ for $1 \le p \le \infty$.

Lemma 8.5.9. *Let $|\lambda| > 1/2$. Then for any $f \in L^2(\partial\Omega)$,*

$$\|f\|_{L^2(\partial\Omega)} \le C_\lambda \left\{ \|T_\lambda f\|_{L^2(\partial\Omega)} + \|(\mathcal{K}_h + \mathcal{K}_h^*) f\|_{L^2(\partial\Omega)} + \|\mathcal{S}(f)\|_{L^2(\partial\Omega)} \right\},$$

where C_λ depends on λ.

Proof. Let $f \in L^2(\partial\Omega)$ and $u = \mathcal{S}(f)$. Then

$$\begin{aligned}
|(\nabla u)_+|^2 &\ge \left|\left(\frac{\partial u}{\partial n}\right)_+\right|^2 = \left|\left(\frac{1}{2} I + \mathcal{K}\right) f\right|^2 = \left|\left(\frac{1}{2} - \lambda\right) f + T_\lambda(f)\right|^2 \\
&\ge \left(\frac{1}{2} - \lambda\right)^2 |f|^2 + (1 - 2\lambda) f\, T_\lambda(f).
\end{aligned}$$

This, together with the Rellich identity (8.5.1), gives

$$\begin{aligned}
&\left(\frac{1}{2} - \lambda\right) \int_{\partial\Omega} (h\cdot n)|f|^2\, d\sigma \\
&\qquad \le (2\lambda - 1) \int_{\partial\Omega} (h\cdot n) f\, T_\lambda(f)\, d\sigma + C \int_\Omega |\nabla u|^2\, dx \\
&\qquad\quad + 2 \int_{\partial\Omega} \left\{ \frac{1}{2}(h\cdot n) f + \mathcal{K}_h(f) \right\} \left\{ \left(\frac{1}{2} - \lambda\right) f + T_\lambda(f) \right\} d\sigma.
\end{aligned}$$

It follows that

$$\begin{aligned}
&\left(\lambda^2 - \frac{1}{4}\right) \int_{\partial\Omega} (h\cdot n)|f|^2\, d\sigma \\
&\quad \le C_\lambda \left\{ \|f\|_{L^2(\partial\Omega)} \|T_\lambda f\|_{L^2(\partial\Omega)} + \left| \int_{\partial\Omega} \mathcal{K}_h(f)\cdot f\, d\sigma \right| + \int_\Omega |\nabla u|^2\, dx \right\}.
\end{aligned}$$

The desired estimate follows from this and the observation that

$$2\int_{\partial\Omega} \mathcal{K}_h(f) \cdot f \, d\sigma = \int_{\partial\Omega} (\mathcal{K}_h + \mathcal{K}_h^*)(f) \cdot f \, d\sigma$$

and

$$\int_\Omega |\nabla u|^2 \, dx = \int_{\partial\Omega} \frac{\partial u}{\partial n} u \, d\sigma \le C \, \|f\|_{L^2(\partial\Omega)} \|\mathcal{S}(f)\|_{L^2(\partial\Omega)}. \qquad \square$$

Lemma 8.5.10. *If $|\lambda| \ge 1/2$, the operator $T_\lambda : L^2(\partial\Omega) \to L^2(\partial\Omega)$ has a closed range.*

Proof. The case $|\lambda| = 1/2$ follows from Theorem 8.5.7 and Lemma 8.5.6. To deal with the case $|\lambda| > 1/2$, we recall that $\mathcal{S} : L^2(\partial\Omega) \to W^{1,2}(\partial\Omega)$ is bounded. Since the embedding $W^{1,2}(\partial\Omega) \subset L^2(\partial\Omega)$ is compact, it follows that the operator $\mathcal{S}$ is compact on $L^2(\partial\Omega)$. Since $\mathcal{K}_h + \mathcal{K}_h^*$ is also compact on $L^2(\partial\Omega)$, in view of Lemma 8.5.9 and Theorem 8.5.7, we conclude that $T_\lambda : L^2(\partial\Omega) \to L^2(\partial\Omega)$ has a closed range. $\square$

Lemma 8.5.11. *Let $\{E(\lambda) : \lambda \in I\}$ be a continuous family of bounded linear operators from $X \to Y$, where X, Y are Banach spaces and $I \subset \mathbb{C}$ is connected. Suppose that*

1. *for each $\lambda \in I$, $E(\lambda)$ is injective and its range is closed;*
2. *$E(\lambda_0)$ is an isomorphism for some $\lambda_0 \in I$.*

Then $E(\lambda)$ is an isomorphism for all $\lambda \in I$.

Proof. Let

$$J = \{\lambda \in I : E(\lambda) \text{ is an isomorphism}\}.$$

Since $E(\lambda)$ is continuous, it is easy to see that $J \ne \emptyset$ is relative open in I. Since I is connected, we only need to show that J is also relative closed in I.

To this end, let

$$C(\lambda) = \sup \left\{ \frac{\|g\|}{\|E(\lambda) g\|} : \, g \in X \text{ and } g \ne 0 \right\}.$$

By the uniform boundedness theorem, $C(\lambda)$ is finite for all $\lambda \in I$. Suppose that $\lambda_j \in J$ and $\lambda_j \to \lambda \in I$. We will show that $\lambda \in J$. Since

$$\begin{aligned} \|g\| \le C(\lambda)\|E(\lambda)g\| &\le C(\lambda)\{\|E(\lambda)g - E(\lambda_j)g\| + \|E(\lambda_j)g\|\} \\ &\le C(\lambda)\|E(\lambda_j) - E(\lambda)\|\|g\| + C(\lambda)\|E(\lambda_j)g\|, \end{aligned}$$

we obtain

$$\big(1 - C(\lambda)\|E(\lambda) - E(\lambda_j)\|\big)\|g\| \le C(\lambda)\|E(\lambda_j)g\|.$$

It follows that if $1 - C(\lambda)\|E(\lambda_j) - E(\lambda)\| < 1$,

$$C(\lambda_j) \le \frac{C(\lambda_0)}{1 - C(\lambda)\|E(\lambda_j) - E(\lambda)\|}.$$

This shows that $\{C(\lambda_j)\}$ is bounded in $\mathbb{R}$. Now let $f \in Y$. Since $E(\lambda_j)$ is an isomorphism, there exists $g_j \in X$ such that $E(\lambda_j)g_j = f$. Note that $\|g_j\| \le C(\lambda_j)\|f\|$ and thus $\{\|g_j\|\}$ is bounded. Also observe that

$$\begin{aligned}\|g_i - g_j\| &\le C(\lambda_j)\|E(\lambda_j)g_i - E(\lambda_j)g_j\| \\ &\le C(\lambda_j)\|E(\lambda_j)g_i - E(\lambda_i)g_i\| \\ &\le C(\lambda)\|E(\lambda_j) - E(\lambda_i)\|\|g_i\|.\end{aligned}$$

Hence, $\{g_j\}$ is a Cauchy sequence in X. Suppose that $g_j \to g \in X$. It is not hard to see that $E(\lambda)g = f$. This shows that $E(\lambda)$ is surjective and thus an isomorphism. □

We are now ready to prove the main theorem of this section.

Theorem 8.5.12. *Let Ω be a bounded Lipschitz domain in $\mathbb{R}^d$, $d \ge 2$, with connected boundary. Then operators $(1/2)I + \mathcal{K} : L^2_0(\partial\Omega) \to L^2_0(\partial\Omega)$ and $-(1/2)I + \mathcal{K} : L^2(\partial\Omega) \to L^2(\partial\Omega)$ are isomorphisms. Moreover,*

$$\begin{aligned}&\|f\|_{L^2(\partial\Omega)} \le C\|((1/2)I + \mathcal{K})f\|_{L^2(\partial\Omega)} \qquad \text{for } f \in L^2_0(\partial\Omega),\\ &\|f\|_{L^2(\partial\Omega)} \le C\|(-(1/2)I + \mathcal{K})f\|_{L^2(\partial\Omega)} \qquad \text{for } f \in L^2(\partial\Omega),\end{aligned} \tag{8.5.14}$$

where C depends only on the Lipschitz character of Ω.

Proof. By rescaling we may assume $\operatorname{diam}(\Omega) = 1$. Note that the estimates in (8.5.14) follow from Lemma 8.5.6. To show $(1/2)+\mathcal{K}$ is an isomorphism on $L^2_0(\partial\Omega)$, we apply Lemma 8.5.11 to $E(\lambda) = \lambda I + \mathcal{K}$ for $\lambda \in I = [1/2, \infty)$. Observe that $E(\lambda)$ is a bounded operator on $L^2_0(\partial\Omega)$ for any $\lambda \in \mathbb{R}$. Clearly, $E(\lambda)$ is an isomorphism on $L^2_0(\partial\Omega)$ if λ is greater than the operator norm of $\mathcal{K}$ on $L^2_0(\partial\Omega)$. By Lemma 8.5.8 (for $\lambda > 1/2$) and Lemma 8.5.6 (for $\lambda = 1/2$), $\lambda I + \mathcal{K}$ is injective for all $\lambda \in I$. That the range of $\lambda I + \mathcal{K}$ is closed was proved in Lemma 8.5.10. It now follows from Lemma 8.5.11 that $\lambda I + \mathcal{K}$ is an isomorphism on $L^2_0(\partial\Omega)$ for all $\lambda \ge 1/2$.

To show $-(1/2)I + \mathcal{K}$ is an isomorphism on $L^2(\partial\Omega)$, we consider $E(\lambda) = \lambda I + \mathcal{K}$ on $L^2(\partial\Omega)$ for $\lambda \in (-\infty, -1/2]$. The argument is similar to that for the case $(1/2)I + \mathcal{K}$. The details are left to the reader. □

As we mentioned before, the invertibility of $\pm(1/2)I + \mathcal{K}$ on L^2 leads to the existence of solutions to the L^2 Neumann and Dirichlet problems.

Theorem 8.5.13 (L^2 Neumann problem)**.** *Let Ω be a bounded Lipschitz domain in $\mathbb{R}^d$, $d \ge 2$, with connected boundary. Then, given any $g \in L^2_0(\partial\Omega)$, there exists a unique (up to constants) harmonic function in Ω such that $\frac{\partial u}{\partial n} = g$ n.t. on $\partial\Omega$.*

Moreover, the solution may be represented by a single-layer potential $\mathcal{S}(h)$ with $\|h\|_{L^2(\partial\Omega)} \le C \|g\|_{L^2(\partial\Omega)}$ and satisfies the estimate $\|(\nabla u)^*\|_{L^2(\partial\Omega)} \le C \|g\|_{L^2(\partial\Omega)}$, where C depends only on the Lipschitz character of Ω.

Proof. Let $g \in L^2_0(\partial\Omega)$. By Theorem 8.5.12, there exists $h \in L^2_0(\partial\Omega)$ such that $((1/2)I + \mathcal{K})h = g$ on $\partial\Omega$ and $\|h\|_{L^2(\partial\Omega)} \le C \|g\|_{L^2(\partial\Omega)}$. Then $u = \mathcal{S}(h)$ is a solution to the L^2 Neumann problem for $\Delta u = 0$ in Ω with boundary data g. Moreover,

$$\|(\nabla u)^*\|_{L^2(\partial\Omega)} \le C \|h\|_{L^2(\partial\Omega)} \le C \|g\|_{L^2(\partial\Omega)},$$

where C depends only on the Lipschitz character of Ω.

The uniqueness follows from Green's identity. Indeed, suppose that $\Delta u = 0$ in Ω, $(\nabla u)^* \in L^2(\partial\Omega)$ and $\frac{\partial u}{\partial n} = 0$ n.t. on $\partial\Omega$. Note that by Proposition 8.1.3, $(\nabla u)^* \in L^2(\partial\Omega)$ implies that $(u)^* \in L^2(\partial\Omega)$. Let $\Omega_j \uparrow \Omega$. By using the Green identity in Ω_j and then the dominated convergence theorem, we deduce that

$$\int_\Omega |\nabla u|^2\, dx = \int_{\partial\Omega} \frac{\partial u}{\partial n}\, u\, d\sigma = 0.$$

Hence u is constant in Ω. □

Theorem 8.5.14 (L^2 Dirichlet problem). *Let Ω be a bounded Lipschitz domain in $\mathbb{R}^d$, $d \ge 2$, with connected boundary. Then, given any $f \in L^2(\partial\Omega)$, there exists a unique harmonic function u in Ω such that $u = f$ n.t. on $\partial\Omega$ and $(u)^* \in L^2(\partial\Omega)$. Moreover, the solution may be represented by a double-layer potential $\mathcal{D}(h)$ with $\|h\|_{L^2(\partial\Omega)} \le C \|f\|_{L^2(\partial\Omega)}$ and satisfies the estimate $\|(u)^*\|_{L^2(\partial\Omega)} \le C\|f\|_{L^2(\partial\Omega)}$, where C depends only on the Lipschitz character of Ω.*

Proof. By Theorem 8.5.12, the operator $-(1/2)I+\mathcal{K}$ is an isomorphism on $L^2(\partial\Omega)$. By duality, the operator $-(1/2)I + \mathcal{K}^*$ is also an isomorphism on $L^2(\partial\Omega)$. Thus, given any $f \in L^2(\partial\Omega)$, there exists $h \in L^2(\partial\Omega)$ such that $(-(1/2)I+\mathcal{K}^*)h = f$ on $\partial\Omega$. It follows that $u = \mathcal{D}(h)$ is a solution to the L^2 Dirichlet problem in Ω with boundary data f. Moreover,

$$\|(u)^*\|_{L^2(\partial\Omega)} \le C \|h\|_{L^2(\partial\Omega)} \le C \|f\|_{L^2(\partial\Omega)},$$

where C depends only on the Lipschitz character of Ω.

To establish the uniqueness, we will show that if $\Delta u = 0$ in Ω, $(u)^* \in L^2(\partial\Omega)$ and $u = f$ n.t. on $\partial\Omega$, then

$$\int_\Omega |u|^2\, dx \le C \int_{\partial\Omega} |f|^2\, d\sigma. \tag{8.5.15}$$

Consequently, $f = 0$ on $\partial\Omega$ impies that $u = 0$ in Ω.

To prove (8.5.15), we let $\{\Omega_j\}$ be a sequence of smooth domains such that $\Omega_j \uparrow \Omega$. Let $F \in C_0^\infty(\Omega_j)$ and w be the solution to the Dirichlet problem $\Delta w = F$

in Ω_j and $w = 0$ on $\partial\Omega_j$. Since Ω_j and F are smooth, we have $w \in C^\infty(\overline{\Omega_j})$. Note that by the Cauchy and Poincaré inequalities,

$$\begin{aligned}\int_{\Omega_j} |\nabla w|^2\, dx &= -\int_{\Omega_j} F w\, dx \le \|F\|_{L^2(\Omega_j)} \|w\|_{L^2(\Omega_j)}\\ &\le C\, \|F\|_{L^2(\Omega_j)} \|\nabla w\|_{L^2(\Omega_j)},\end{aligned}$$

where C does not depend on j. It follows that $\|\nabla w\|_{L^2(\Omega_j)} \le C \|F\|_{L^2(\Omega_j)}$.

Next we observe that, by Green's identity,

$$\int_{\Omega_j} u\, F\, dx = \int_{\Omega_j} u\, \Delta w\, dx = \int_{\partial\Omega_j} u\, \frac{\partial w}{\partial n}\, d\sigma, \tag{8.5.16}$$

where we have used the assumption $\Delta u = 0$ in Ω. Also, since $w = 0$ on $\partial\Omega_j$, we may use the Rellich identity (8.5.3) to obtain

$$\begin{aligned}\int_{\partial\Omega_j} |\nabla w|^2\, d\sigma &\le C \left\{ \int_{\Omega_j} |\nabla w|\, |F|\, dx + \int_{\Omega_j} |\nabla w|^2\, dx \right\}\\ &\le C \int_{\Omega_j} |F|^2\, dx.\end{aligned}$$

This, together with (8.5.16), yields

$$\begin{aligned}\left| \int_\Omega u\, F\, dx \right| &\le \|u\|_{L^2(\partial\Omega_j)} \|\nabla w\|_{L^2(\partial\Omega_j)}\\ &\le C\, \|u\|_{L^2(\partial\Omega_j)} \|F\|_{L^2(\Omega_j)}.\end{aligned}$$

It follows by duality that

$$\int_{\Omega_j} |u|^2\, dx \le C \int_{\partial\Omega_j} |u|^2\, d\sigma. \tag{8.5.17}$$

This gives the estimate (8.5.15) by letting $j \to \infty$. □

Finally we consider the L^2 regularity problem.

Theorem 8.5.15 (L^2 regularity problem). *Let Ω be a bounded Lipschitz domain in $\mathbb{R}^d$, $d \ge 2$, with connected boundary. Given any $f \in W^{1,2}(\partial\Omega)$, there exists a unique harmonic function u in Ω such that $(\nabla u)^* \in L^2(\partial\Omega)$ and $u = f$ n.t. on $\partial\Omega$. Furthermore, the solution satisfies*

$$\|(\nabla u)^*\|_{L^2(\partial\Omega)} + |\partial\Omega|^{\frac{1}{d-1}} \|(u)^*\|_{L^2(\partial\Omega)} \le C\, \|f\|_{W^{1,2}(\partial\Omega)}, \tag{8.5.18}$$

where C depends only on the Lipschitz character of Ω.

Proof. Since $(\nabla u)^* \in L^2(\partial\Omega)$ implies $(u)^* \in L^2(\partial\Omega)$, the uniqueness follows from that of the L^2 Dirichlet problem. One may also prove the uniqueness by using Green's identity, as in the case of the L^2 Neumann problem.

To establish the existence, we normalize Ω so that $|\partial\Omega| = 1$. First, let us consider the case where $f = F|_{\partial\Omega}$, where $F \in C_0^\infty(\mathbb{R}^d)$. Choose a sequence of smooth domains $\{\Omega_j\}$ such that $\Omega_j \downarrow \Omega$ with homomorphism $\Lambda_j : \partial\Omega \to \partial\Omega_j$, given by Theorem 8.1.5. Let w_j be the solution of the Dirichlet problem $\Delta w_j = 0$ in Ω_j and $w_j = F$ on $\partial\Omega_j$. It follows from Theorem 8.5.14 that

$$
\begin{aligned}
&\|(\nabla w_j)^*\|_{L^2(\partial\Omega_j)} + \|(w_j)^*\|_{L^2(\partial\Omega_j)} \\
&\qquad \le C\left\{\|\nabla w_j\|_{L^2(\partial\Omega_j)} + \|w_j\|_{L^2(\partial\Omega_j)}\right\} \\
&\qquad \le C\,\|F\|_{W^{1,2}(\partial\Omega_j)},
\end{aligned}
\tag{8.5.19}
$$

where we have used Theorem 8.5.4 for the second inequality. We emphasize that the constant C in (8.5.19) depends only on the Lipschitz character of Ω.

Next we observe that, by (8.5.19), the sequence $\{w_j\}$ is bounded in $W^{1,2}(\Omega)$. Thus, by passing to a subsequence, we may assume that w_j converges to w in $L^2(\Omega)$. It follows from the mean value property and interior estimates for harmonic functions that $w_j \to w$ and $\nabla w_j \to \nabla w$ uniformly on any compact subset of Ω. Moreover, w is harmonic in Ω. For $z \in \partial\Omega$ and $\delta > 0$, let

$$
\mathcal{M}_\delta^2(u)(z) = \sup\left\{|u(x)| : \ x \in \gamma_\alpha(z) \text{ and } \operatorname{dist}(x, \partial\Omega) \ge \delta\right\}.
$$

Note that by (8.5.19), if j is large, then

$$
\|\mathcal{M}_\delta^2(\nabla w_j)\|_{L^2(\partial\Omega)} \le C\,\|(\nabla w_j)^*\|_{L^2(\partial\Omega_j)} \le C\,\|F\|_{W^{1,2}(\partial\Omega_j)}. \tag{8.5.20}
$$

Since $\nabla w_j \to \nabla w$ uniformly on any compact subset of Ω, we see that $\mathcal{M}_\delta^2(\nabla w_j)$ converges to $\mathcal{M}_\delta^2(\nabla w)$ uniformly on $\partial\Omega$. Thus, by letting $j \to \infty$ in (8.5.20), we obtain

$$
\|\mathcal{M}_\delta^2(\nabla w)\|_{L^2(\partial\Omega)} \le C\,\|f\|_{W^{1,2}(\partial\Omega)}.
$$

We now let $\delta \to 0$. By the monotone convergence theorem, this gives

$$
\|(\nabla w)^*\|_{L^2(\partial\Omega)} \le C\,\|f\|_{W^{1,2}(\partial\Omega)}.
$$

Let u be the solution of the L^2 Dirichlet problem in Ω with boundary data f. Using

$$
\begin{aligned}
\|(w_j - u)^*\|_{L^2(\partial\Omega)} &\le C\,\|w_j - f\|_{L^2(\partial\Omega)} \\
&\le C\,\|w_j - F_j\|_{L^2(\partial\Omega)} + C\,\|F_j - F\|_{L^2(\partial\Omega)} \to 0,
\end{aligned}
$$

where $F_j(z) = F(\Lambda_j(z))$, we see that $w_j \to u$ in $L^2(\Omega)$. As a result, $u = w$ in Ω. This shows that $w = f$ n.t. on $\partial\Omega$.

Finally, suppose that $f \in W^{1,2}(\partial\Omega)$. We choose sequences of smooth functions $\{f_j\}$ in $\mathbb{R}^d$ such that $\|f_j - f\|_{W^{1,2}(\partial\Omega)} \to 0$ as $j \to \infty$. Let u_j be the solution of the L^2 Dirichlet problem in Ω with boundary data f_j. We have proved above that

$$\|(u_j - u_k)^*\|_{L^2(\partial\Omega)} + \|(\nabla u_j - \nabla u_k)^*\|_{L^2(\partial\Omega)} \le C\,\|f_j - f_k\|_{W^{1,2}(\partial\Omega)}. \quad (8.5.21)$$

It follows that u_j converges to u uniformly on any compact subset of Ω and that u is harmonic in Ω. Let $k \to \infty$ in (8.5.21). We obtain

$$\|\mathcal{M}^2_\delta(u_j - u)\|_{L^2(\partial\Omega)} + \|\mathcal{M}^2_\delta(\nabla u_j - \nabla u)\|_{L^2(\partial\Omega)} \le C\,\|f_j - f\|_{W^{1,2}(\partial\Omega)}.$$

As before, this leads to

$$\|(u_j - u)^*\|_{L^2(\partial\Omega)} + \|(\nabla u_j - \nabla u)^*\|_{L^2(\partial\Omega)} \le C\,\|f_j - f\|_{W^{1,2}(\partial\Omega)}, \quad (8.5.22)$$

by the monotone convergence theorem. Using (8.5.22) and

$$\limsup_{\substack{x\to z\\ x\in\Omega\cap\gamma_\alpha(z)}} |u(x) - f(z)| \le (u - u_j)^*(z) + |f_j(z) - f(z)|,$$

we may conclude that $u = f$ n.t. on $\partial\Omega$. To finish the proof, we observe that $\|(\nabla u)^*\|_{L^2(\partial\Omega)} \le C\,\|f\|_{W^{1,2}(\partial\Omega)}$. This follows from the estimate $\|(\nabla u_j)^*\|_{L^2(\partial\Omega)} \le C\,\|f_j\|_{L^2(\partial\Omega)}$ by the same argument as in the proof of (8.5.22). □

The next theorem shows that if u is a solution of the L^p regularity problem in Ω, then ∇u has nontangential limit a.e. on $\partial\Omega$.

Theorem 8.5.16. *Let u be harmonic in a bounded Lipschitz domain Ω. Suppose that $(\nabla u)^* \in L^p(\partial\Omega)$ for some $1 < p < \infty$. Then u and ∇u have nontangential limits a.e. on $\partial\Omega$. Furthermore, one has $u|_{\partial\Omega} \in W^{1,p}(\partial\Omega)$ and*

$$\|(\nabla u)^*\|_{L^p(\partial\Omega)} \le C_p\,\|\nabla u\|_{L^p(\partial\Omega)},$$

where C_p depends only on p and the Lipschitz character of Ω.

Proof. Recall that $(\nabla u)^* \in L^p(\partial\Omega)$ implies that $(u)^* \in L^p(\partial\Omega)$ and u has nontangential limit a.e. on $\partial\Omega$. Let $\{\Omega_\ell\}$ be a sequence of smooth domains such that $\Omega_\ell \uparrow \Omega$ and $\Lambda_\ell : \partial\Omega \to \partial\Omega_\ell$ the homomorphisms, given by Theorem 8.1.5. Since $(\nabla u)^* \in L^p(\partial\Omega)$, it follows that $\{\frac{\partial u}{\partial x_j} \circ \Lambda_\ell\}$ is bounded in $L^p(\partial\Omega)$. Hence, by passing to a subsequence, we may assume that $\frac{\partial u}{\partial x_j} \circ \Lambda_\ell$ converges weakly to g_j in $L^p(\partial\Omega)$ as $\ell \to \infty$. It follows by passing to the limit argument that

$$\int_{\partial\Omega} u\left(n_j \frac{\partial\varphi}{\partial x_k} - n_k \frac{\partial\varphi}{\partial x_j}\right) d\sigma = -\int_{\partial\Omega} (n_j g_k - n_k g_j)\varphi\, d\sigma$$

for any $\varphi \in C^1_0(\mathbb{R}^d)$. By definition this implies that $u|_{\partial\Omega} \in W^{1,p}(\partial\Omega)$ and

$$\|\nabla_{\tan} u\|_{L^p(\partial\Omega)} \le C \sum_j \|g_j\|_{L^p(\partial\Omega)} \le C\,\|(\nabla u)^*\|_{L^p(\partial\Omega)}.$$

Next, to show that ∇u has nontangential limit on $\partial\Omega$ and

$$\|(\nabla u)^*\|_{L^p(\partial\Omega)} \le C_p \|\nabla u\|_{L^p(\partial\Omega)},$$

we use Green's representation formula in Ω_ℓ:

$$u(x) = \int_{\partial\Omega_\ell} \Gamma(x-y)\frac{\partial u}{\partial n}\, d\sigma(y) - \int_{\partial\Omega_\ell} \frac{\partial}{\partial n(y)}\Big\{\Gamma(x-y)\Big\} u(y)\, d\sigma(y). \tag{8.5.23}$$

It is easy to see that the second integral in the RHS of (8.5.23) converges to $\mathcal{D}(u)(x)$ as $\ell \to \infty$. To handle the first integral, we write it as

$$\int_{\partial\Omega_\ell} \Gamma(x - \Lambda_\ell(y)) h_\ell(y)\, d\sigma(y),$$

where $h_\ell(y) = \nabla u(\Lambda_\ell(y)) \cdot n(\Lambda_\ell(y))\omega_\ell(y)$. Since $\{h_\ell\}$ is bounded in $L^p(\partial\Omega)$, by passing to a subsequence, we may assume that $h_\ell \to h$ weakly in $L^p(\partial\Omega)$. It follows that the first integral in the RHS of (8.5.23) converges to $\mathcal{S}(h)(x)$ and hence, for $x \in \Omega$

$$u(x) = \mathcal{S}(h)(x) - \mathcal{D}(u)(x). \tag{8.5.24}$$

We claim that if $f \in W^{1,p}(\partial\Omega)$ and $1 < p < \infty$, then $\nabla\mathcal{D}(f)$ has nontangential limit a.e. on $\partial\Omega$ and

$$\|(\nabla\mathcal{D}(f))^*\|_{L^p(\partial\Omega)} \le C_p \|\nabla_{\tan} f\|_{L^p(\partial\Omega)}, \tag{8.5.25}$$

where C_p depends only on p and the Lipschitz character of Ω. Assume the claim for a moment. We may deduce from (8.5.24) that ∇u has nontangential limit a.e. on $\partial\Omega$ and thus $h = n \cdot \nabla u$. Moreover,

$$\begin{aligned}\|(\nabla u)^*\|_{L^p(\partial\Omega)} &\le C_p \left\{\|h\|_{L^p(\partial\Omega)} + \|\nabla_{\tan} u\|_{L^p(\partial\Omega)}\right\}\\ &\le C_p \|\nabla u\|_{L^p(\partial\Omega)}.\end{aligned}$$

It remains to prove the claim. Let $f \in W^{1,p}(\partial\Omega)$ and $w = \mathcal{S}(f)$. Observe that

$$\begin{aligned} w(x) &= \int_{\partial\Omega} \frac{\partial}{\partial n(y)}\Big\{\Gamma(x-y)\Big\} f(y)\, d\sigma(y)\\ &= -\frac{\partial}{\partial x_k}\int_{\partial\Omega} \Gamma(x-y) n_k(y) f(y)\, d\sigma(y).\end{aligned}$$

It follows that for $x \in \Omega$,

$$\begin{aligned}\frac{\partial w}{\partial x_j} &= -\frac{\partial^2}{\partial x_j \partial x_k}\int_{\partial\Omega} \Gamma(x-y) n_k(y) f(y)\, d\sigma(y)\\ &= \frac{\partial}{\partial x_k}\int_{\partial\Omega} \frac{\partial}{\partial y_j}\Big\{\Gamma(x-y)\Big\} n_k(y) f(y)\, d\sigma(y)\\ &= \frac{\partial}{\partial x_k}\int_{\partial\Omega}\left\{n_k\frac{\partial}{\partial y_j} - n_j\frac{\partial}{\partial y_k}\right\}\Big\{\Gamma(x-y)\Big\} f(y)\, d\sigma(y)\\ &= \frac{\partial}{\partial x_k}\int_{\partial\Omega} \Gamma(x-y) g_{jk}(y)\, d\sigma(y),\end{aligned}$$

where $g_{jk} = \frac{\partial f}{\partial t_{jk}}$ and we have used the fact that $\Delta_y\{\Gamma(x-y)\} = 0$ for $y \neq x$ in the third equality. By Theorem 8.4.4, we conclude that ∇w has nontangential limit a.e. on $\partial\Omega$ and that

$$\|(\nabla w)^*\|_{L^p(\partial\Omega)} \le C_p \sum_{j,k} \|g_{jk}\|_{L^p(\partial\Omega)} \le C_p \|\nabla_{\tan} f\|_{L^p(\partial\Omega)}.$$

This completes the proof. □

Recall that $\mathcal{S} : L^p(\partial\Omega) \to W^{1,p}(\partial\Omega)$ is bounded for $1 < p < \infty$.

Theorem 8.5.17. *Let Ω be a bounded Lipschitz domain in $\mathbb{R}^d$, $d \ge 3$, with connected boundary. Then $\mathcal{S} : L^2(\partial\Omega) \to W^{1,2}(\partial\Omega)$ is an isomorphism. Furthermore, we have*

$$\|g\|_{L^2(\partial\Omega)} \le C\, \|\mathcal{S}(g)\|_{W^{1,2}(\partial\Omega)}$$

for any $g \in L^2(\partial\Omega)$, where C depends only on the Lipschitz character of Ω. Consequently, the unique solution of the L^2 regularity problem in Ω with data $f \in W^{1,2}(\partial\Omega)$ may be represented by a single-layer potential $\mathcal{S}(g)$, where $g \in L^2(\partial\Omega)$ and $\|g\|_{L^2(\partial\Omega)} \le C\|f\|_{W^{1,2}(\partial\Omega)}$.

Proof. By dilation we may assume that $|\partial\Omega| = 1$. Let $g \in L^2(\partial\Omega)$ and $u = \mathcal{S}(g)$. Then $u(x) = O(|x|^{2-d})$ as $|x| \to \infty$. By the jump relation (8.4.13) and Theorems 8.5.4 and 8.5.5,

$$\begin{aligned}
\|g\|_{L^2(\partial\Omega)} &\le \left\| \left(\frac{\partial u}{\partial n}\right)_+ \right\|_{L^2(\partial\Omega)} + \left\| \left(\frac{\partial u}{\partial n}\right)_- \right\|_{L^2(\partial\Omega)} \\
&\le C \left\{ \|\nabla_{\tan} u\|_{L^2(\partial\Omega)} + \|u\|_{L^2(\partial\Omega)} \right\} \\
&\le C\, \|\mathcal{S}(g)\|_{W^{1,2}(\partial\Omega)},
\end{aligned}$$

where C depends only on the Lipschitz character of Ω.

It remains to show that $\mathcal{S} : L^2(\partial\Omega) \to W^{1,2}(\partial\Omega)$ is surjective. To this end, let $f \in W^{1,2}(\partial\Omega)$ and u be the unique solution of the L^2 regularity problem in Ω with data f, given by Theorem 8.5.15. Since $(\nabla u)^* \in L^2(\partial\Omega)$, by Theorem 8.5.16, ∇u has nontangential limit a.e. on $\partial\Omega$. Thus u may be regarded as a solution to the L^2 Neumann problem with data $n \cdot \nabla u$. It follows from Theorem 8.5.13 that $u = \mathcal{S}(h_1) + \beta$ for some $h_1 \in L^2_0(\partial\Omega)$ and $\beta \in \mathbb{R}$.

Finally, we recall that the range of the operator $(1/2)I + \mathcal{K}$ on $L^2(\partial\Omega)$ is $L^2_0(\partial\Omega)$. It follows that there exists $h_2 \in L^2(\partial\Omega)$ such that $h_2 \neq 0$ and $\big((1/2)I + \mathcal{K}\big)(h_2) = 0$. Let $v = \mathcal{S}(h_2)$. Then v is a nonzero constant in Ω. Hence there exists $\alpha \in \mathbb{R}$ such that $\beta = \mathcal{S}(\alpha h_2)$ in Ω. As a result we obtain $u = \mathcal{S}(h_1 + \alpha h_2)$ in $\overline{\Omega}$. This finishes the proof. □

8.6 The Rellich property

Definition 8.6.1. Let $\mathcal{L} = -\text{div}\big(A(x)\nabla\big)$ and Ω be a bounded Lipschitz domain with connected boundary. We say that the operator $\mathcal{L}$ has the *Rellich property in Ω with constant $C = C(\Omega)$* if

$$\begin{cases} \|\nabla u\|_{L^2(\partial\Omega)} \le C \left\| \dfrac{\partial u}{\partial \nu} \right\|_{L^2(\partial\Omega)}, \\ \|\nabla u\|_{L^2(\partial\Omega)} \le C \, \|\nabla_{\tan} u\|_{L^2(\partial\Omega)}, \end{cases} \tag{8.6.1}$$

whenever u is a solution of $\mathcal{L}(u) = 0$ in Ω such that $(\nabla u)^* \in L^2(\partial\Omega)$ and ∇u exists n.t. on $\partial\Omega$.

In the previous section we proved that the Laplace operator $\mathcal{L} = -\Delta$ has the Rellich property in any Lipschitz domain Ω with constant $C(\Omega)$ depending only on the Lipschitz character of Ω. This was used to establish the invertibility of $\pm(1/2)I + \mathcal{K}$ on $L^2(\partial\Omega)$ and consequently solve the L^2 Dirichlet and Neumann problems. We will see in this section that for second-order elliptic operators with variable coefficients, the well-posedness of the L^2 Neumann, Dirichlet, and regularity problems in Lipschitz domains can be reduced to the boundary Rellich estimates in (8.6.1) by the method of layer potentials and a localization argument.

The following two theorems are the main results of this section. The first theorem treats the well-posedness for the small scale; the constants C in the nontangential-maximal-function estimates in (8.6.2) may depend on $\text{diam}(\Omega)$, if $\text{diam}(\Omega) \ge 1$. The assumptions and conclusions in the second theorem are scale-invariant. As a result, by rescaling, they lead to uniform L^2 estimates in a Lipschitz domain for the family of elliptic operators $\{\mathcal{L}_\varepsilon\}$.

Theorem 8.6.2. *Let $d \ge 3$ and $\mathcal{L} = -\text{div}(A(x)\nabla)$ with $A \in \Lambda(\mu, \lambda, \tau)$. Let $R \ge 1$. Suppose that for any Lipschitz domain Ω with $\text{diam}(\Omega) \le (1/4)$ and connected boundary, there exists $C(\Omega)$ depending on the Lipschitz character of Ω such that for each $s \in (0, 1]$, the operator*

$$\mathcal{L}^s = -\text{div}\big((sA + (1-s)I)\nabla\big)$$

has the Rellich property in Ω with constant $C(\Omega)$. Then for any Lipschitz domain Ω with $\text{diam}(\Omega) \le R$ and connected boundary, the L^2 Neumann and regularity problems for $\mathcal{L}(u) = 0$ in Ω are well posed and the solutions satisfy the estimates

$$\begin{cases} \|(\nabla u)^*\|_{L^2(\partial\Omega)} \le C \left\| \dfrac{\partial u}{\partial \nu} \right\|_{L^2(\partial\Omega)}, \\ \|(\nabla u)^*\|_{L^2(\partial\Omega)} \le C \, \|\nabla_{\tan} u\|_{L^2(\partial\Omega)}, \end{cases} \tag{8.6.2}$$

where C depends only on μ, λ, τ, the Lipschitz character of Ω, and R (if $\text{diam}(\Omega) \ge 1$). Furthermore, the L^2 Dirichlet problem for $\mathcal{L}^*(u) = 0$ in Ω is well posed with the estimate $\|(u)^*\|_{L^2(\partial\Omega)} \le C\|u\|_{L^2(\partial\Omega)}$.

Theorem 8.6.3. *Let $d \geq 3$ and $\mathcal{L} = -\text{div}(A(x)\nabla)$ with $A \in \Lambda(\mu, \lambda, \tau)$. Suppose that for any Lipschitz domain Ω with connected boundary, there exists $C(\Omega)$ depending on the Lipschitz character of Ω such that for each $s \in (0,1]$, the operator $\mathcal{L}^s = -\text{div}\big((sA+(1-s)I)\nabla\big)$ has the Rellich property in Ω with constant $C(\Omega)$. Then for any Lipschitz domain Ω with connected boundary, the L^2 Neumann and regularity problems for $\mathcal{L}(u) = 0$ in Ω are well posed and the solutions satisfy the estimates in (8.6.2) with a constant C depending only on μ, λ, τ, and the Lipschitz character of Ω. Furthermore, the L^2 Dirichlet problem for $\mathcal{L}^*(u) = 0$ in Ω is well posed with the estimate $\|(u)^*\|_{L^2(\partial\Omega)} \leq C\|u\|_{L^2(\partial\Omega)}$.*

The uniqueness for the L^2 Neumann and regularity problems follows readily from the Green identity

$$\int_\Omega A(x)\nabla u \cdot \nabla u \, dx = \int_{\partial\Omega} \frac{\partial u}{\partial \nu} \cdot u \, d\sigma. \tag{8.6.3}$$

We will use the method of layer potentials to establish the existence of solutions in Theorems 8.6.2 and 8.6.3.

Recall that $\Omega_+ = \Omega$ and $\Omega_- = \mathbb{R}^d \setminus \overline{\Omega}$.

Lemma 8.6.4. *Let Ω be a bounded Lipschitz domain with $r_0 = \text{diam}(\Omega) \leq R$. Suppose that $\mathcal{L}(u) = 0$ in $\Omega_\pm$, $(\nabla u)^* \in L^2(\partial\Omega)$, and $(\nabla u)_\pm$ exists n.t. on $\partial\Omega$. Under the same assumptions on A as in Theorem 8.6.2, we have*

$$\begin{aligned} \int_{\partial\Omega} |(\nabla u)_\pm|^2 \, d\sigma &\leq C \int_{\partial\Omega} \left| \left(\frac{\partial u}{\partial \nu}\right)_\pm \right|^2 d\sigma + \frac{C}{r_0} \int_{N_\pm} |\nabla u|^2 \, dx, \\ \int_{\partial\Omega} |(\nabla u)_\pm|^2 \, d\sigma &\leq C \int_{\partial\Omega} \left| (\nabla_{\tan} u)_\pm \right|^2 d\sigma + \frac{C}{r_0} \int_{N_\pm} |\nabla u|^2 \, dx, \end{aligned} \tag{8.6.4}$$

where

$$N_\pm = \Big\{ x \in \Omega_\pm : \ \text{dist}(x, \partial\Omega) \leq r_0 \Big\},$$

and C depends on the Lipschitz character of Ω and R (if $r_0 > 1$).

Proof. The proof uses a localization argument. Let $\psi : \mathbb{R}^{d-1} \to \mathbb{R}$ be a Lipschitz function such that $\psi(0) = 0$ and $\|\nabla\psi\|_\infty \leq M$. Let

$$\begin{aligned} D_r &= \big\{ (x', x_d) \in \mathbb{R}^d : \ |x'| < r \text{ and } \psi(x') < x_d < 10(M+1)r \big\}, \\ \Delta_r &= \big\{ (x', \psi(x')) \in \mathbb{R}^d : \ |x'| < r \big\}. \end{aligned} \tag{8.6.5}$$

Suppose that $\mathcal{L}(u) = 0$ in $\Omega_0 = D_{3r}$, $(\nabla u)^* \in L^2(\partial\Omega_0)$, and ∇u exists n.t. on $\partial\Omega_0$. Assume that $\text{diam}(D_{2r}) \leq (1/4)$. Then for any $t \in (1, 2)$, $\mathcal{L}$ has the Rellich property in the Lipschitz domain D_{tr} with constant $C_0(D_{tr})$ depending only on

M. It follows that

$$\begin{aligned}\int_{\Delta_r} |\nabla u|^2 \, d\sigma &\le \int_{\partial D_{tr}} |\nabla u|^2 \, d\sigma \\ &\le C_0 \int_{\Delta_{2r}} \left|\frac{\partial u}{\partial \nu}\right|^2 d\sigma + CC_0 \int_{\partial D_{tr}\setminus \Delta_{tr}} |\nabla u|^2 \, d\sigma,\end{aligned} \tag{8.6.6}$$

where C depends only on μ. We now integrate both sides of (8.6.6) with respect to t over the interval $(1,2)$ to obtain

$$\int_{\Delta_r} |\nabla u|^2 \, d\sigma \le C \int_{\Delta_{2r}} \left|\frac{\partial u}{\partial \nu}\right|^2 d\sigma + \frac{C}{r} \int_{D_{2r}} |\nabla u|^2 \, dx. \tag{8.6.7}$$

Finally, we choose $r = c(M)r_0$ if $r_0 \le 1$ and $r = c(M)$ if $r_0 > 1$. The first inequality in (8.6.2) follows by covering $\partial\Omega$ with sets $\{\Delta_i\}$, each of which is obtained from Δ_r through translation and rotation. The proof for the second inequality in (8.6.4) is similar. □

Remark 8.6.5. Under the same conditions on A as in Theorem 8.6.3, the estimates in (8.6.4) hold with constant C independent of R. This is because we may choose $r = c(M)r_0$ for any Ω.

We will use $\mathcal{S}$, $\mathcal{D}$ and $\mathcal{K}_A$ to denote $\mathcal{S}_\varepsilon$, $\mathcal{D}_\varepsilon$ and $\mathcal{K}_{A,\varepsilon}$, respectively, when $\varepsilon = 1$.

Lemma 8.6.6. *Let $R \ge 1$ and Ω be a bounded Lipschitz domain with connected boundary and* $\operatorname{diam}(\Omega) \le R$. *Under the same conditions on A as in Theorem* 8.6.2, *the operators $(1/2)I + \mathcal{K}_A$ and $-(1/2)I + \mathcal{K}_A$ are isomorphisms on $L^2_0(\partial\Omega;\mathbb{R}^m)$ and $L^2(\partial\Omega;\mathbb{R}^m)$, respectively. Moreover,*

$$\begin{cases} \|\big((1/2)I + \mathcal{K}_A\big)^{-1}\|_{L^2_0 \to L^2_0} \le C, \\ \|\big(-(1/2)I + \mathcal{K}_A\big)^{-1}\|_{L^2 \to L^2} \le C, \end{cases} \tag{8.6.8}$$

where C depends on the Lipschitz character of Ω and R (if $\operatorname{diam}(\Omega) \ge 1$*).*

Proof. Let $f \in L^2_0(\partial\Omega;\mathbb{R}^m)$ and $u = \mathcal{S}(f)$. Then $\mathcal{L}(u) = 0$ in $\mathbb{R}^d \setminus \partial\Omega$, $(\nabla u)^* \in L^2(\partial\Omega)$, and $(\nabla u)_\pm$ exists n.t. on $\partial\Omega$. Also recall that $(\nabla_{\tan} u)_+ = (\nabla_{\tan} u)_-$ on $\partial\Omega$. Note that

$$|u(x)| + |x||\nabla u(x)| = O(|x|^{2-d})$$

as $|x| \to \infty$, for $d \ge 3$. It follows via integration by parts that

$$\int_{\Omega_-} A(x)\nabla u \cdot \nabla u \, dx = -\int_{\partial\Omega} \left(\frac{\partial u}{\partial \nu}\right)_- \cdot u \, d\sigma. \tag{8.6.9}$$

By the jump relation (8.4.13),

$$\int_{\partial\Omega} \left(\frac{\partial u}{\partial \nu}\right)_- d\sigma = -\int_{\partial\Omega} f \, d\sigma = 0.$$

As in the case of the Laplacian, using the Poincaré inequality on $\partial\Omega$ and the Green identities (8.6.9) and (8.6.3), we obtain

$$\int_{\Omega_\pm} |\nabla u|^2 \, dx \le C r_0 \left\| \left(\frac{\partial u}{\partial \nu}\right)_\pm \right\|_{L^2(\partial\Omega)} \|\nabla_{\tan} u\|_{L^2(\partial\Omega)}. \tag{8.6.10}$$

Combining (8.6.4) with (8.6.10) and then using the Cauchy inequality, we see that

$$\begin{cases} \|(\nabla u)_\pm\|_{L^2(\partial\Omega)} \le C \, \|\nabla_{\tan} u\|_{L^2(\partial\Omega)}, \\ \|(\nabla u)_\pm\|_{L^2(\partial\Omega)} \le C \left\| \left(\frac{\partial u}{\partial \nu}\right)_\pm \right\|_{L^2(\partial\Omega)}. \end{cases} \tag{8.6.11}$$

It follows that

$$\begin{aligned} \left\| \left(\frac{\partial u}{\partial \nu}\right)_\pm \right\|_{L^2(\partial\Omega)} &\le C \, \|\nabla_{\tan} u\|_{L^2(\partial\Omega)} \le C \, \|(\nabla u)_\mp\|_{L^2(\partial\Omega)} \\ &\le C \left\| \left(\frac{\partial u}{\partial \nu}\right)_\mp \right\|_{L^2(\partial\Omega)}. \end{aligned}$$

By the jump relation, this gives

$$\begin{aligned} \|f\|_{L^2(\partial\Omega)} &\le C \left\| \left(\frac{\partial u}{\partial \nu}\right)_\pm \right\|_{L^2(\partial\Omega)} \\ &= C \, \|\big(\pm (1/2)I + \mathcal{K}_A\big) f\|_{L^2(\partial\Omega)} \end{aligned} \tag{8.6.12}$$

for any $f \in L^2_0(\partial\Omega; \mathbb{R}^m)$. By considering $f - \fint_{\partial\Omega} f$, as in the case of Laplace's equation, we deduce from (8.6.12) that

$$\|f\|_{L^2(\partial\Omega)} \le C \, \|\big(-(1/2)I + \mathcal{K}_A\big) f\|_{L^2(\partial\Omega)} \tag{8.6.13}$$

for any $f \in L^2(\partial\Omega; \mathbb{R}^m)$.

Thus, to complete the proof, it suffices to show that the operators $(1/2)I + \mathcal{K}_A : L^2_0(\partial\Omega; \mathbb{R}^m) \to L^2_0(\partial\Omega; \mathbb{R}^m)$ and $-(1/2)I + \mathcal{K}_A : L^2(\partial\Omega; \mathbb{R}^m) \to L^2(\partial\Omega; \mathbb{R}^m)$ are surjective. This is done by a continuity method. Indeed, let us consider a family of matrices $A^s = sA + (1-s)I$ in $\Lambda(\mu, \lambda, \tau)$, where $0 \le s \le 1$. Note that by Theorem 8.5.12, $\pm(1/2)I + \mathcal{K}_{A^0}$ are isomorphisms on $L^2_0(\partial\Omega; \mathbb{R}^m)$ and $L^2(\partial\Omega; \mathbb{R}^m)$, respectively. Also observe that for each $s \in [0, 1]$, the matrix A^s satisfies the same conditions as A. Hence,

$$\begin{aligned} \|f\|_{L^2(\partial\Omega)} &\le C \, \|\big((1/2)I + \mathcal{K}_{A^s}\big) f\|_{L^2(\partial\Omega)} \quad \text{for any } f \in L^2_0(\partial\Omega; \mathbb{R}^m), \\ \|f\|_{L^2(\partial\Omega)} &\le C \, \|\big(-(1/2)I + \mathcal{K}_{A^s}\big) f\|_{L^2(\partial\Omega)} \quad \text{for any } f \in L^2(\partial\Omega; \mathbb{R}^m), \end{aligned}$$

where C is independent of s. Since

$$\|A^{s_1} - A^{s_2}\|_{C^\lambda(\mathbb{R}^d)} \le C|s_1 - s_2|\|A\|_{C^\lambda(\mathbb{R}^d)},$$

it follows from Theorem 8.3.5 that

$$\big\{(1/2)I + \mathcal{K}_{A^s} : 0 \le s \le 1\big\} \quad \text{and} \quad \big\{ - (1/2)I + \mathcal{K}_{A^s} : 0 \le s \le 1\big\}$$

are continuous families of bounded operators on $L^2_0(\partial\Omega; \mathbb{R}^m)$ and $L^2(\partial\Omega; \mathbb{R}^m)$, respectively. In view of Lemma 8.5.11, we conclude that $\pm(1/2) + \mathcal{K}_A$ are isomorphisms on $L^2_0(\partial\Omega; \mathbb{R}^m)$ and $L^2(\partial\Omega; \mathbb{R}^m)$, respectively. This finishes the proof. □

Remark 8.6.7. Suppose $d \ge 3$. Under the same assumptions on A and Ω as in Lemma 8.6.6, the operator $\mathcal{S} : L^2(\partial\Omega; \mathbb{R}^m) \to W^{1,2}(\partial\Omega; \mathbb{R}^m)$ is an isomorphism and

$$\|\mathcal{S}^{-1}\|_{W^{1,2} \to L^2} \le C. \tag{8.6.14}$$

To see this, we let $f \in W^{1,2}(\partial\Omega; \mathbb{R}^m)$ and $u = \mathcal{S}(f)$. Note that for $d \ge 3$ we have $|u(x)| = O(|x|^{2-d})$ and $|\nabla u(x)| = O(|x|^{1-d})$, as $|x| \to \infty$, which allow us to use the Green identity on Ω_-. It follows from the proof of Lemma 8.6.6 that

$$\|(\nabla u)_-\|_{L^2(\partial\Omega)} \le C \|\nabla_{\tan} u\|_{L^2(\partial\Omega)} + C r_0^{-1} \|u\|_{L^2(\partial\Omega)}.$$

This, together with $\|(\nabla u)_+\|_{L^2(\partial\Omega)} \le C \|\nabla_{\tan} u\|_{L^2(\partial\Omega)}$ and the jump relation, gives

$$\begin{aligned} \|f\|_{L^2(\partial\Omega)} &\le C \left\{ \|\nabla_{\tan} \mathcal{S}(f)\|_{L^2(\partial\Omega)} + r_0^{-1} \|\mathcal{S}(f)\|_{L^2(\partial\Omega)} \right\} \\ &\le C \|\mathcal{S}(f)\|_{W^{1,2}(\partial\Omega)}. \end{aligned} \tag{8.6.15}$$

Hence, $\mathcal{S} : L^2(\partial\Omega; \mathbb{R}^m) \to W^{1,2}(\partial\Omega; \mathbb{R}^m)$ is injective. A continuity argument similar to that in the proof of Lemma 8.6.6 shows that the operator is in fact an isomorphism.

Remark 8.6.8. Under the same assumptions on A as in Theorem 8.6.3, the estimates in (8.6.8) and (8.6.14) (for $d \ge 3$) hold with a constant depending on the Lipschitz character of Ω.

We are now in a position to give the proof of Theorems 8.6.2 and 8.6.3.

Proof of Theorems 8.6.2 *and* 8.6.3. The existence of solutions to the L^2 Neumann and regularity problems for $\mathcal{L}(u) = 0$ in Ω is a direct consequence of the invertibility of $(1/2)I + \mathcal{K}_A$ on $L^2_0(\partial\Omega; \mathbb{R}^m)$ and that of $\mathcal{S} : L^2(\partial\Omega; \mathbb{R}^m) \to W^{1,2}(\partial\Omega; \mathbb{R}^m)$, respectively. Since $-(1/2)I + \mathcal{K}_A$ is invertible on $L^2(\partial\Omega; \mathbb{R}^m)$, it follows by duality that $-(1/2)I + \mathcal{K}_A^*$ is also invertible on $L^2(\partial\Omega; \mathbb{R}^m)$ and

$$\|\big(- (1/2)I + \mathcal{K}_A\big)^{-1}\|_{L^2 \to L^2} = \|\big(- (1/2)I + \mathcal{K}_A^*\big)^{-1}\|_{L^2 \to L^2}.$$

This gives the existence of solutions to the L^2 Dirichlet problem for $\mathcal{L}^*(u) = 0$ in Ω. Now observe that under the conditions in Theorem 8.6.3, the norms of the operators $\big(\pm(1/2)I+\mathcal{K}_A\big)^{-1}$ and $\mathcal{S}^{-1}$ are bounded by constants depending on the Lipschitz character of Ω. As a result, the constants C in (8.6.2) and in the estimate $\|(u)^*\|_{L^2(\partial\Omega)} \le C\,\|u\|_{L^2(\partial\Omega)}$ depend on the Lipschitz character of Ω, not on $\operatorname{diam}(\Omega)$.

As we mentioned earlier, the uniqueness for the L^2 Neumann and regularity problems follows readily from Green's identity (8.6.3). To establish the uniqueness for the L^2 Dirichlet problem for $\mathcal{L}^*(u)=0$ in Ω, we construct a matrix of Green functions $\big(G^{\alpha\beta}(x,y)\big)$ for Ω, where

$$G^{\alpha\beta}(x,y)=\Gamma^{\alpha\beta}(x,y;A)-W^{\alpha\beta}(x,y),$$

and for each β and $y\in\Omega$, $W^\beta(\cdot,y)=\big(W^{1\beta}(\cdot,y),\dots,W^{m\beta}(\cdot,y)\big)$ is the solution to the L^2 regularity problem for $\mathcal{L}(u)=0$ in Ω with boundary data

$$\Gamma^\beta(\cdot,y)=\big(\Gamma^{1\beta}(\cdot,y),\dots,\Gamma^{m\beta}(\cdot,y)\big) \quad \text{on } \partial\Omega.$$

Since $|\nabla_x\Gamma(x,y;A)|\le C\,|x-y|^{1-d}$, by the well-posedness of the L^2 regularity problem,

$$\big(\nabla_x W^\beta(\cdot,y)\big)^*\in L^2(\partial\Omega). \tag{8.6.16}$$

Suppose now that $\mathcal{L}^*(u)=0$ in Ω, $(u)^*\in L^2(\partial\Omega)$ and $u=0$ n.t. on $\partial\Omega$. For $\rho>0$ small, choose $\varphi=\varphi_\rho$ so that $\varphi=1$ in $\{x\in\Omega:\operatorname{dist}(x,\partial\Omega)\ge 2\rho\}$, $\varphi=0$ in $\{x\in\Omega:\operatorname{dist}(x,\partial\Omega)\le\rho\}$, and $|\nabla\varphi|\le C\rho^{-1}$. Fix $y\in\Omega$ so that $\operatorname{dist}(y,\partial\Omega)\ge 2\rho$. Then

$$\begin{aligned} u^\gamma(y)=u^\gamma(y)\varphi(y)&=\int_\Omega a_{ij}^{\alpha\beta}(x)\frac{\partial}{\partial x_j}\Big\{G^{\beta\gamma}(x,y)\Big\}\frac{\partial}{\partial x_i}(u^\alpha\varphi)\,dx\\ &=-\int_\Omega a_{ij}^{\alpha\beta}(x)G^{\beta\gamma}(x,y)\frac{\partial u^\alpha}{\partial x_i}\cdot\frac{\partial\varphi}{\partial x_j}\,dx\\ &\quad+\int_\Omega a_{ij}^{\alpha\beta}(x)\frac{\partial}{\partial x_j}\Big\{G^{\beta\gamma}(x,y)\Big\}u^\alpha\frac{\partial\varphi}{\partial x_i}\,dx, \end{aligned}$$

where we have used integration by parts and $\mathcal{L}^*(u)=0$ in Ω for the last equality. This gives

$$|u(y)|\le\frac{C}{\rho}\int_{E_\rho}|G(x,y)|\,|\nabla u(x)|\,dx+\frac{C}{\rho}\int_{E_\rho}|\nabla_x G(x,y)|\,|u(x)|\,dx, \tag{8.6.17}$$

where $E_\rho=\{x\in\mathbb{R}^d:\rho\le\operatorname{dist}(x,\partial\Omega)\le 2\rho\}$. Using the fact that $G(\cdot,y)=u=0$ on $\partial\Omega$ n.t. on $\partial\Omega$ and the interior Lipschitz estimate of u, we deduce from (8.6.17) that

$$|u(y)|\le C\int_{\partial\Omega}\mathcal{M}^1_{3\rho}(\nabla_x G(\cdot,y))\cdot\mathcal{M}^1_{3\rho}(u)\,d\sigma, \tag{8.6.18}$$

where

$$\mathcal{M}^1_t(u)(z)=\sup\Big\{|u(x)|:\ x\in\gamma(z)\cap\Omega\text{ and }\operatorname{dist}(x,\partial\Omega)<t\Big\}. \tag{8.6.19}$$

Finally, we note that by (8.6.16), $\mathcal{M}^1_\delta(\nabla_x G(\cdot,y))\in L^2(\partial\Omega)$ if $\delta=\operatorname{dist}(y,\partial\Omega)/9$.

This, together with the assumption $(u)^* \in L^2(\partial\Omega)$, shows that

$$\mathcal{M}^1_\delta(\nabla_x G(\cdot, y)) \cdot (u)^* \in L^1(\partial\Omega).$$

Also, observe that as $\rho \to 0$, $\mathcal{M}^1_\rho(u)(z) \to 0$ for a.e. $z \in \partial\Omega$. It follows by the dominated convergence theorem that the RHS of (8.6.18) goes to zero as $\rho \to 0$. This yields that $u(y) = 0$ for any $y \in \Omega$, and completes the proof. □

8.7 The well-posedness for the small scale

In this section we establish the well-posedness of the L^2 Dirichlet, Neumann, and regularity problems for $\mathcal{L}(u) = 0$ in a Lipschitz domain Ω in $\mathbb{R}^d$, $d \geq 3$, under the ellipticity condition (2.1.20), the smoothness condition (4.0.2) and the symmetry condition $A^* = A$. The periodicity condition is not needed. However, the constants C may depend on $\text{diam}(\Omega)$ if $\text{diam}(\Omega)$ is large.

Theorem 8.7.1. *Let $d \geq 3$ and $\mathcal{L} = -\text{div}(A(x)\nabla)$. Assume that $A = A(x)$ satisfies* (2.1.20), (4.0.2), *and the symmetry condition $A^* = A$. Let $R \geq 1$. Then for any bounded Lipschitz domain Ω with connected boundary and* $\text{diam}(\Omega) \leq R$, *the L^2 Neumann and regularity problems for $\mathcal{L}(u) = 0$ in Ω are well posed and the solutions satisfy the estimates in* (8.6.2) *with constant C depending only on μ, λ, τ, the Lipschitz character of Ω, and R (if* $\text{diam}(\Omega) > 1$*). Furthermore, the L^2 Dirichlet problem for $\mathcal{L}(u) = 0$ in Ω is well posed with the estimate* $\|(u)^*\|_2 \leq C\|u\|_2$.

Remark 8.7.2. Note that the periodicity of A is not needed in Theorem 8.7.1. This is because we can reduce the general case to the case of periodic coefficients. Indeed, by translation, we may assume that $0 \in \Omega$. If $\text{diam}(\Omega) \leq (1/4)$, we construct $\widetilde{A} \in \Lambda(\mu, \lambda, \tau_0)$ so that $\widetilde{A} = A$ on $[-3/8, 3/8]^d$, where τ_0 depends on μ and τ. The boundary value problems for $\text{div}(\widetilde{A}\nabla u) = 0$ in Ω are the same as those for $\text{div}(A\nabla u) = 0$ in Ω. Suppose now that $r_0 = \text{diam}(\Omega) > (1/4)$. By rescaling, the boundary value problems for $\text{div}(A\nabla u) = 0$ in Ω are equivalent to those for $\text{div}(B\nabla u) = 0$ in Ω_1, where $B(x) = A(4r_0 x)$ and $\Omega_1 = \{x \in \mathbb{R}^d : 4r_0 x \in \Omega\}$. Since $\text{diam}(\Omega_1) = (1/4)$, we have reduced the case to the previous one. Note that in this case the bounding constants C may depend on r_0.

By Remark 8.7.2 it is enough to prove Theorem 8.7.1 under the additional assumption that $\text{diam}(\Omega) \leq (1/4)$ and A is 1-periodic (thus $A \in \Lambda(\mu, \lambda, \tau)$). Furthermore, in view of Theorem 8.6.2, it suffices to show that if $\mathcal{L} = -\text{div}(A\nabla)$ with $A \in \Lambda(\mu, \lambda, \tau)$ and $A^* = A$ and if Ω is a Lipschitz domain with $\text{diam}(\Omega) \leq (1/4)$, then $\mathcal{L}$ has the Rellich property in Ω with constant $C(\Omega)$ depending only on μ, λ, τ, and the Lipschitz character of Ω. We point out that if A is Lipschitz continuous, the Rellich property follows readily from the Rellich type identities (see Lemma 8.7.5), as in the case of Laplace's equation. However, the proof for operators with Hölder continuous coefficients is more involved.

As we pointed out above, Theorem 8.7.1 is a consequence of the following.

Theorem 8.7.3. *Let $\mathcal{L} = -\text{div}(A\nabla)$ with $A \in \Lambda(\mu, \lambda, \tau)$ and $A^* = A$. Let Ω be a bounded Lipschitz domain with* $\text{diam}(\Omega) \le (1/4)$ *and connected boundary. Then $\mathcal{L}$ has the Rellich property in Ω with constant $C(\Omega)$ depending only on μ, λ, τ, and the Lipschitz character of Ω.*

By translation we may assume that $0 \in \Omega$ and thus $\Omega \subset [-1/4, 1/4]^d$. We divide the proof of Theorem 8.7.3 into three steps.

Step 1. Establish the invertibility of $\pm(1/2)I + \mathcal{K}_A$ under the additional assumption that

$$\begin{cases} A \in C^1([-1/2, 1/2]^d \setminus \partial\Omega), \\ |\nabla A(x)| \le C_0\{\text{dist}(x, \partial\Omega)\}^{\lambda_0 - 1} \quad \text{for any } x \in [-1/2, 1/2]^d \setminus \partial\Omega, \end{cases} \tag{8.7.1}$$

where $\lambda_0 \in (0, 1]$.

Clearly, if $A \in C^1([-1/2, 1/2]^d)$, then it satisfies (8.7.1).

We start out with two Rellich type identities for the system $\mathcal{L}(u) = 0$.

Lemma 8.7.4. *Let Ω be a bounded Lipschitz domain. Let $A \in \Lambda(\mu, \lambda, \tau)$ be such that $A^* = A$ and the condition (8.7.1) holds. Suppose that $u \in C^1(\overline{\Omega})$ and $\mathcal{L}(u) = 0$ in Ω. Then*

$$\begin{aligned} \int_{\partial\Omega} (h \cdot n)\, A\nabla u \cdot \nabla u \, d\sigma = 2 \int_{\partial\Omega} \left(\frac{\partial u}{\partial \nu}\right)^{\alpha} \cdot \frac{\partial u^\alpha}{\partial x_k} h_k \, d\sigma \\ + \int_\Omega \text{div}(h) \cdot a_{ij}^{\alpha\beta} \frac{\partial u^\alpha}{\partial x_i} \cdot \frac{\partial u^\beta}{\partial x_j} \, dx \\ + \int_\Omega h_k \frac{\partial}{\partial x_k}\left\{a_{ij}^{\alpha\beta}\right\} \cdot \frac{\partial u^\alpha}{\partial x_i} \cdot \frac{\partial u^\beta}{\partial x_j} \, dx \\ - 2 \int_\Omega \frac{\partial h_k}{\partial x_i} \cdot a_{ij}^{\alpha\beta} \frac{\partial u^\alpha}{\partial x_k} \cdot \frac{\partial u^\beta}{\partial x_j} \, dx \end{aligned} \tag{8.7.2}$$

and

$$\begin{aligned} \int_{\partial\Omega} (h \cdot n)\, A\nabla u \cdot \nabla u \, d\sigma = 2 \int_{\partial\Omega} \left\{ n_k \frac{\partial}{\partial x_i} - n_i \frac{\partial}{\partial x_k} \right\} u^\alpha \cdot a_{ij}^{\alpha\beta} \frac{\partial u^\beta}{\partial x_j} h_k \, d\sigma \\ - \int_\Omega \text{div}(h) \cdot a_{ij}^{\alpha\beta} \frac{\partial u^\alpha}{\partial x_i} \cdot \frac{\partial u^\beta}{\partial x_j} \, dx \\ - \int_\Omega h_k \frac{\partial}{\partial x_k}\left\{a_{ij}^{\alpha\beta}\right\} \cdot \frac{\partial u^\alpha}{\partial x_i} \cdot \frac{\partial u^\beta}{\partial x_j} \, dx \\ + 2 \int_\Omega \frac{\partial h_k}{\partial x_i} \cdot a_{ij}^{\alpha\beta} \frac{\partial u^\alpha}{\partial x_k} \cdot \frac{\partial u^\beta}{\partial x_j} \, dx, \end{aligned} \tag{8.7.3}$$

where $h = (h_1, \ldots, h_d) \in C_0^1(\mathbb{R}^d; \mathbb{R}^d)$.

Proof. Using the assumptions that $\mathcal{L}(u) = 0$ in Ω and $A^* = A$, one may verify that

$$
\begin{aligned}
&\frac{\partial}{\partial x_k}\left\{h_k a_{ij}^{\alpha\beta}\frac{\partial u^\alpha}{\partial x_i}\cdot\frac{\partial u^\beta}{\partial x_j}\right\} - 2\frac{\partial}{\partial x_i}\left\{h_k a_{ij}^{\alpha\beta}\frac{\partial u^\alpha}{\partial x_k}\cdot\frac{\partial u^\beta}{\partial x_j}\right\}\\
&\quad= \operatorname{div}(h)\cdot a_{ij}^{\alpha\beta}\frac{\partial u^\alpha}{\partial x_i}\cdot\frac{\partial u^\beta}{\partial x_j} + h_k\frac{\partial}{\partial x_k}\left\{a_{ij}^{\alpha\beta}\right\}\cdot\frac{\partial u^\alpha}{\partial x_i}\cdot\frac{\partial u^\beta}{\partial x_j}\\
&\qquad - 2\frac{\partial h_k}{\partial x_i}\cdot a_{ij}^{\alpha\beta}\frac{\partial u^\alpha}{\partial x_k}\cdot\frac{\partial u^\beta}{\partial x_j}.
\end{aligned}
$$

This gives (8.7.2) by the divergence theorem. Let I and J denote the left- and right-hand sides of (8.7.2), respectively. The identity (8.7.3) follows by rewriting $I = J$ as $I = 2I - J$. $\square$

By an approximation argument, Rellich identities (8.7.2) and (8.7.3) continue to hold under the assumptions that $\mathcal{L}(u) = 0$ in Ω, $(\nabla u)^* \in L^2(\partial\Omega)$, and ∇u exists n.t. on $\partial\Omega$.

Lemma 8.7.5. *Let Ω be a bounded Lipschitz domain with connected boundary. Suppose that $0 \in \Omega$ and $r_0 = \operatorname{diam}(\Omega) \le (1/4)$. Let $A \in \Lambda(\mu, \lambda, \tau)$ be such that $A^* = A$ and the condition (8.7.1) holds. Assume that $\mathcal{L}(u) = 0$ in Ω, $(\nabla u)^* \in L^2(\partial\Omega)$ and ∇u exists n.t. on $\partial\Omega$. Then*

$$
\begin{aligned}
\int_{\partial\Omega}|\nabla u|^2\,d\sigma &\le C\int_{\partial\Omega}\left|\frac{\partial u}{\partial\nu}\right|^2 d\sigma + C\int_\Omega (|\nabla A| + r_0^{-1})|\nabla u|^2\,dx,\\
\int_{\partial\Omega}|\nabla u|^2\,d\sigma &\le C\int_{\partial\Omega}|\nabla_{\tan} u|^2\,d\sigma + C\int_\Omega (|\nabla A| + r_0^{-1})|\nabla u|^2\,dx,
\end{aligned}
\tag{8.7.4}
$$

where C depends only on μ and the Lipschitz character of Ω.

Proof. Let $h \in C_0^1(\mathbb{R}^d; \mathbb{R}^d)$ be a vector field such that

$$
\operatorname{supp}(h) \subset \{x : \operatorname{dist}(x, \partial\Omega) < cr_0\},
$$

$|\nabla h| \le Cr_0^{-1}$, and $h \cdot n \ge c > 0$ on $\partial\Omega$. It follows from (8.7.2) and (8.7.3) that

$$
\begin{aligned}
\int_{\partial\Omega}(h\cdot n)A\nabla u\cdot\nabla u\,d\sigma &= 2\int_{\partial\Omega}(h\cdot\nabla u^\alpha)\left(\frac{\partial u}{\partial\nu}\right)^\alpha d\sigma + I_1,\\
\int_{\partial\Omega}(h\cdot n)A\nabla u\cdot\nabla u\,d\sigma &= 2\int_{\partial\Omega}h_k a_{ij}^{\alpha\beta}\frac{\partial u^\beta}{\partial x_j}\left(n_k\frac{\partial}{\partial x_i} - n_i\frac{\partial}{\partial x_k}\right)u^\alpha\,d\sigma + I_2,
\end{aligned}
\tag{8.7.5}
$$

where

$$
|I_1| + |I_2| \le C\int_\Omega \{|\nabla h| + |h||\nabla A|\}|\nabla u|^2\,dx,
$$

and C depends only on μ. The estimates (8.7.4) follow from (8.7.5) by using the inequality (2.1.10). $\square$

Remark 8.7.6. Let $\mathcal{L}(u) = 0$ in $(-1/2, 1/2)^d \setminus \overline{\Omega}$. Suppose that $(\nabla u)^* \in L^2(\partial\Omega)$ and ∇u exists n.t. on $\partial\Omega$. Under the same conditions on Ω and A as in Lemma 8.7.5, we have

$$\begin{aligned}
\int_{\partial\Omega} |(\nabla u)_-|^2 \, d\sigma &\le C \int_{\partial\Omega} \left| \left(\frac{\partial u}{\partial \nu}\right)_- \right|^2 d\sigma \\
&\quad + C \int_{\Omega_- \cap [-1/2, 1/2]^d} (|\nabla A| + r_0^{-1}) |\nabla u|^2 \, dx, \\
\int_{\partial\Omega} |(\nabla u)_-|^2 \, d\sigma &\le C \int_{\partial\Omega} |(\nabla_{\tan} u)_-|^2 \, d\sigma \\
&\quad + C \int_{\Omega_- \cap [-(1/2), 1/2]^d} (|\nabla A| + r_0^{-1}) |\nabla u|^2 \, dx,
\end{aligned} \tag{8.7.6}$$

where C depends only on μ and the Lipschitz character of Ω. The proof is similar to that of Lemma 8.7.5.

Lemma 8.7.7. *Under the same assumptions as in Lemma* 8.7.5, *we have*

$$\begin{aligned}
\int_{\partial\Omega} |\nabla u|^2 \, d\sigma &\le C \Big\{ 1 + r_0^{2\lambda_0} \rho^{2\lambda_0 - 2} \Big\} \int_{\partial\Omega} \left| \frac{\partial u}{\partial \nu} \right|^2 d\sigma \\
&\quad + C(\rho r_0)^{\lambda_0} \int_{\partial\Omega} |(\nabla u)^*|^2 \, d\sigma, \\
\int_{\partial\Omega} |\nabla u|^2 \, d\sigma &\le C \Big\{ 1 + r_0^{2\lambda_0} \rho^{2\lambda_0 - 2} \Big\} \int_{\partial\Omega} |\nabla_{\tan} u|^2 \, d\sigma \\
&\quad + C(\rho r_0)^{\lambda_0} \int_{\partial\Omega} |(\nabla u)^*|^2 \, d\sigma,
\end{aligned} \tag{8.7.7}$$

where $0 < \rho < 1$ *and* C *depends only on* μ, *the Lipschitz character of* Ω, *and* (λ_0, C_0) *in* (8.7.1).

Proof. Write $\Omega = E_1 \cup E_2$, where

$$\begin{aligned}
E_1 &= \{x \in \Omega : \text{dist}(x, \partial\Omega) \le \rho r_0\}, \\
E_2 &= \{x \in \Omega : \text{dist}(x, \partial\Omega) > \rho r_0\}.
\end{aligned}$$

Using the condition (8.7.1), we obtain

$$\begin{aligned}
\int_\Omega |\nabla A| |\nabla u|^2 \, dx &\le C_0 \int_{E_1} \big\{ \text{dist}(x, \partial\Omega) \big\}^{\lambda_0 - 1} |\nabla u|^2 \, dx \\
&\quad + C_0 (\rho r_0)^{\lambda_0 - 1} \int_{E_2} |\nabla u|^2 \, dx \\
&\le C(\rho r_0)^{\lambda_0} \int_{\partial\Omega} |(\nabla u)^*|^2 \, d\sigma + C_0 (\rho r_0)^{\lambda_0 - 1} \int_\Omega |\nabla u|^2 \, dx.
\end{aligned} \tag{8.7.8}$$

This, together with (8.7.4) and (8.6.10) for Ω_+, gives

$$\begin{aligned}\|\nabla u\|^2_{L^2(\partial\Omega)} \le C\left\|\frac{\partial u}{\partial\nu}\right\|^2_{L^2(\partial\Omega)}& \\ + C(1+r_0^{\lambda_0}\rho^{\lambda_0-1})&\left\|\frac{\partial u}{\partial\nu}\right\|_{L^2(\partial\Omega)}\|\nabla_{\tan}u\|_{L^2(\partial\Omega)} \\ + C(\rho r_0)^{\lambda_0}&\|(\nabla u)^*\|^2_{L^2(\partial\Omega)}.\end{aligned} \tag{8.7.9}$$

The first inequality in (8.7.7) follows from (8.7.9) by applying the inequality (2.1.10). The proof of the second inequality in (8.7.7) is similar. □

Remark 8.7.8. Let $\mathcal{L}(u) = 0$ in Ω_-. Suppose that $(\nabla u)^* \in L^2(\partial\Omega)$, $(\nabla u)_-$ exists n.t. on $\partial\Omega$, and $|u(x)| = O(|x|^{2-d})$ as $|x| \to \infty$. In view of Remark 8.7.6 and (8.6.10) for Ω_-, the same argument as in the proof of Lemma 8.7.7 shows that

$$\begin{aligned}\int_{\partial\Omega}|(\nabla u)_-|^2\,d\sigma &\le C\left\{1+r_0^{2\lambda_0}\rho^{2\lambda_0-2}\right\}\left\|\left(\frac{\partial u}{\partial\nu}\right)_-\right\|_2^2 \\ &\quad + C(\rho r_0)^{\lambda_0}\|(\nabla u)^*\|_2^2 \\ &\quad + C(\rho r_0)^{\lambda_0-1}\left|\fint_{\partial\Omega}u\right|\left|\int_{\partial\Omega}\left(\frac{\partial u}{\partial\nu}\right)_- d\sigma\right|, \\ \int_{\partial\Omega}|(\nabla u)_-|^2\,d\sigma &\le C\left\{1+r_0^{2\lambda_0}\rho^{2\lambda_0-2}\right\}\|(\nabla_{\tan}u)_-\|_2^2 \\ &\quad + C(\rho r_0)^{\lambda_0}\|(\nabla u)^*\|_2^2 \\ &\quad + C(\rho r_0)^{\lambda_0-1}\left|\fint_{\partial\Omega}u\right|\left|\int_{\partial\Omega}\left(\frac{\partial u}{\partial\nu}\right)_- d\sigma\right|,\end{aligned} \tag{8.7.10}$$

for any $0 < \rho < 1$.

The following theorem completes Step 1.

Theorem 8.7.9. *Suppose that Ω and A satisfy the conditions in Theorem 8.7.3. Assume further that $0 \in \Omega$ and A satisfies (8.7.1). Then $(1/2)I+\mathcal{K}_A$ and $-(1/2)I+\mathcal{K}_A$ are invertible on $L^2_0(\partial\Omega;\mathbb{R}^m)$ and $L^2(\partial\Omega;\mathbb{R}^d)$, respectively, and the estimates in (8.6.8) hold with a constant C depending only on μ, λ, τ, the Lipschitz character of Ω, and (C_0, λ_0) in (8.7.1).*

Proof. Let $u = \mathcal{S}(f)$ for some $f \in L^2_0(\partial\Omega;\mathbb{R}^m)$. Recall that $\int_{\partial\Omega}\left(\frac{\partial u}{\partial\nu}\right)_- d\sigma = 0$, $(\nabla_{\tan}u)_- = (\nabla_{\tan}u)_+$, and $\|(\nabla u)^*\|_{L^2(\partial\Omega)} \le C\,\|f\|_{L^2(\partial\Omega)}$. Thus, by the second inequality in (8.7.10), we obtain

$$\|(\nabla u)_-\|_{L^2(\partial\Omega)} \le C\rho_1^{\lambda_0-1}\|(\nabla u)_+\|_{L^2(\partial\Omega)} + C\rho_1^{\lambda_0/2}\|f\|_{L^2(\partial\Omega)} \tag{8.7.11}$$

for any $0 < \rho_1 < 1$. Similarly, by the first inequality in (8.7.7),

$$\|(\nabla u)_+\|_{L^2(\partial\Omega)} \le C\rho_2^{\lambda_0-1}\Big\|\Big(\frac{\partial u}{\partial \nu}\Big)_+\Big\|_{L^2(\partial\Omega)} + C\rho_2^{\lambda_0/2}\|f\|_{L^2(\partial\Omega)} \tag{8.7.12}$$

for any $0 < \rho_2 < 1$. It follows from the jump relation (8.4.13), (8.7.11) and (8.7.12) that

$$\begin{aligned}\|f\|_{L^2(\partial\Omega)} &\le \Big\|\Big(\frac{\partial u}{\partial \nu}\Big)_+\Big\|_{L^2(\partial\Omega)} + \Big\|\Big(\frac{\partial u}{\partial \nu}\Big)_-\Big\|_{L^2(\partial\Omega)}\\ &\le C\rho_1^{\lambda_0-1}\rho_2^{\lambda_0-1}\Big\|\Big(\frac{\partial u}{\partial \nu}\Big)_+\Big\|_{L^2(\partial\Omega)}\\ &\quad + C\Big\{\rho_1^{\lambda_0-1}\rho_2^{\lambda_0/2} + \rho_1^{\lambda_0/2}\Big\}\|f\|_{L^2(\partial\Omega)}.\end{aligned} \tag{8.7.13}$$

We now choose $\rho_1 \in (0,1)$ and then $\rho_2 \in (0,1)$ so that

$$C\left\{\rho_1^{\lambda_0-1}\rho_2^{\lambda_0/2} + \rho_1^{\lambda_2/2}\right\} \le (1/2).$$

This gives

$$\|f\|_2 \le C\Big\|\Big(\frac{\partial u}{\partial \nu}\Big)_+\Big\|_{L^2(\partial\Omega)} = C\,\|((1/2)I + \mathcal{K}_A)f\|_{L^2(\partial\Omega)} \tag{8.7.14}$$

for any $f \in L^2_0(\partial\Omega;\mathbb{R}^m)$. The same argument also shows that for $f \in L^2_0(\partial\Omega,\mathbb{R}^m)$,

$$\|f\|_{L^2(\partial\Omega)} \le C\Big\|\Big(\frac{\partial u}{\partial \nu}\Big)_-\Big\|_{L^2(\partial\Omega)} = C\,\|(-(1/2)I + \mathcal{K}_A)f\|_{L^2(\partial\Omega)}. \tag{8.7.15}$$

The rest of the proof is the same as that of Lemma 8.6.6. □

Remark 8.7.10. Let $f \in L^2(\partial\Omega;\mathbb{R}^m)$ and $u = \mathcal{S}(f)$. It follows from (8.7.7) and (8.7.10) that

$$\begin{aligned}\|(\nabla u)_+\|_{L^2(\partial\Omega)} &\le C\rho_1^{\lambda_0-1}\|\nabla_{\tan} u\|_{L^2(\partial\Omega)} + C\rho_1^{\lambda_0/2}\|f\|_{L^2(\partial\Omega)},\\ \|(\nabla u)_-\|_{L^2(\partial\Omega)} &\le C\rho_2^{\lambda_0-1}\|\nabla_{\tan} u\|_{L^2(\partial\Omega)} + C\rho_2^{\lambda_0/2}\|f\|_{L^2(\partial\Omega)}\\ &\quad + Cr_0^{-1}\|u\|_{L^2(\partial\Omega)}\end{aligned} \tag{8.7.16}$$

for any $\rho_1, \rho_2 \in (0,1)$. This, together with the jump relation, implies that

$$\begin{aligned}\|f\|_{L^2(\partial\Omega)} &\le C\,\|\nabla_{\tan}\mathcal{S}(f)\|_{L^2(\partial\Omega)} + Cr_0^{-1}\|\mathcal{S}(f)\|_{L^2(\partial\Omega)}\\ &\le C\,\|\mathcal{S}(f)\|_{W^{1,2}(\partial\Omega)}.\end{aligned} \tag{8.7.17}$$

Thus $\mathcal{S} : L^2(\partial\Omega;\mathbb{R}^m) \to W^{1,2}(\partial\Omega;\mathbb{R}^m)$ is one-to-one. A continuity argument similar to that in the proof of Lemma 8.6.6 shows that the operator is in fact invertible.

Step 2. Given any $A \in \Lambda(\mu, \lambda, \tau)$ and Ω such that $A^* = A$, $0 \in \Omega$ and $r_0 = \text{diam}(\Omega) \leq (1/4)$, construct $\widetilde{A} \in \Lambda(\mu, \lambda_0, \tau_0)$ with λ_0 and τ_0 depending only on μ, λ, τ, and the Lipschitz character of Ω, such that

$$\widetilde{A}(x) = A(x) \qquad \text{if } \text{dist}(x, \partial\Omega) \leq c r_0, \tag{8.7.18}$$

and such that the operators

$$\begin{aligned} (1/2)I + \mathcal{K}_{\widetilde{A}} &: L^2_0(\partial\Omega; \mathbb{R}^m) \to L^2_0(\partial\Omega; \mathbb{R}^m), \\ -(1/2)I + \mathcal{K}_{\widetilde{A}} &: L^2(\partial\Omega; \mathbb{R}^m) \to L^2(\partial\Omega; \mathbb{R}^m), \\ \mathcal{S}_{\widetilde{A}} &: L^2(\partial\Omega; \mathbb{R}^m) \to W^{1,2}(\partial\Omega; \mathbb{R}^m) \end{aligned} \tag{8.7.19}$$

are isomorphisms and the operator norms of their inverses are bounded by constants depending only on μ, λ, τ, and the Lipschitz character of Ω.

Lemma 8.7.11. *Given $A \in \Lambda(\mu, \lambda, \tau)$ and a Lipschitz domain Ω such that $\text{diam}(\Omega) \leq (1/4)$ and $0 \in \Omega$. There exists $\bar{A} \in \Lambda(\mu, \lambda_0, \tau_0)$ such that $\bar{A} = A$ on $\partial\Omega$ and $\bar{A}$ satisfies the condition* (8.7.1), *where $\lambda_0 \in (0, \lambda]$, τ_0, and C_0 in* (8.7.1) *depend only on μ, λ, τ and the Lipschitz character of Ω. In addition, $(\bar{A})^* = \bar{A}$ if $A^* = A$.*

Proof. By periodicity, it suffices to define $\bar{A} = (\bar{a}_{ij}^{\alpha\beta})$ on $[-1/2, 1/2]^d$. This is done as follows. On Ω we define $\bar{A}$ to be the Poisson extension of A on $\partial\Omega$, i.e., $\bar{a}_{ij}^{\alpha\beta}$ is harmonic in Ω and $\bar{a}_{ij}^{\alpha\beta} = a_{ij}^{\alpha\beta}$ on $\partial\Omega$, for each i, j, α, β. On $[-1/2, 1/2]^d \setminus \overline{\Omega}$, we define $\bar{A}$ to be the harmonic function in $(-1/2, 1/2)^d \setminus \overline{\Omega}$ with boundary data $\bar{A} = A$ on $\partial\Omega$ and $\bar{A} = I$ on $\partial[-1/2, 1/2]^d$. Note that the latter boundary condition allows us to extend $\bar{A}$ to $\mathbb{R}^d$ by periodicity.

Since $\bar{a}_{ij}^{\alpha\beta}\xi_i^\alpha\xi_j^\beta$ is harmonic in $(-1/2, 1/2)^d \setminus \partial\Omega$, the ellipticity condition (2.1.20) for $\bar{A}$ follows readily from the maximum principle. By the solvability of Laplace's equation in Lipschitz domains with Hölder continuous data, there exists $\lambda_1 \in (0, 1)$, depending only on the Lipschitz character of Ω, such that $\bar{A} \in C^{\lambda_0}(\overline{\Omega})$ and $\bar{A} \in C^{\lambda_0}([-1/2, 1/2]^d \setminus \Omega)$, where $\lambda_0 = \lambda$ if $\lambda < \lambda_1$, and $\lambda_0 = \lambda_1$ if $\lambda \geq \lambda_1$. It follows that $\bar{A} \in C^{\lambda_0}(\mathbb{R}^d)$. Using the interior estimates for harmonic functions, one can also show that

$$|\nabla \bar{A}(x)| \leq C_0 \big\{\text{dist}(x, \partial\Omega)\big\}^{\lambda_0 - 1} \qquad \text{for } x \in [-3/4, 3/4]^d \setminus \partial\Omega,$$

where C_0 depends only on μ, λ, τ, and the Lipschitz character of Ω. Thus we have proved that $\bar{A} \in \Lambda(\mu, \lambda_0, \tau)$ and satisfies the condition (8.7.1). Clearly, $(\bar{A})^* = \bar{A}$ if $A^* = A$. □

Let $\eta \in C_0^\infty(-1/2, 1/2)$ such that $0 \leq \eta \leq 1$ and $\eta = 1$ on $(-1/4, 1/4)$. Given $A \in \Lambda(\mu, \lambda, \tau)$ with $A^* = A$, define

$$A^\rho(x) = \eta\left(\frac{\delta(x)}{\rho}\right) A(x) + \left[1 - \eta\left(\frac{\delta(x)}{\rho}\right)\right] \bar{A}(x) \tag{8.7.20}$$

for $x \in [-1/2, 1/2]^d$, where $\rho \in (0, 1/8)$, $\delta(x) = \text{dist}(x, \partial\Omega)$ and $\bar{A}(x)$ is the matrix constructed in Lemma 8.7.11. Extend A^ρ to $\mathbb{R}^d$ by periodicity. Clearly, A^ρ satisfies the ellipticity condition (2.1.20) and $(A^\rho)^* = A^\rho$.

Lemma 8.7.12. *Let A^ρ be defined by (8.7.20). Then*

$$\|A^\rho - \bar{A}\|_\infty \le C\rho^{\lambda_0} \quad \textit{and} \quad \|A^\rho - \bar{A}\|_{C^{0,\lambda_0}(\mathbb{R}^d)} \le C, \tag{8.7.21}$$

where C depends only on μ, λ, τ, and the Lipschitz character of Ω.

Proof. Let $H^\rho = A^\rho - \bar{A}$. Given $x \in [-1/2, 1/2]^d$, let $z \in \partial\Omega$ such that $|x - z| = \delta(x)$. Since $A(z) = \bar{A}(z)$, we have

$$\begin{aligned} |A(x) - \bar{A}(x)| &\le |A(x) - A(z)| + |\bar{A}(z) - \bar{A}(x)| \\ &\le C|x - z|^{\lambda_0} = C\{\delta(x)\}^{\lambda_0}. \end{aligned}$$

It follows that

$$\begin{aligned} |H^\rho(x)| &\le C\theta(\rho^{-1}\delta(x))\{\delta(x)\}^{\lambda_0} \\ &= C\theta(\rho^{-1}\delta(x))\{\rho^{-1}\delta(x)\}^{\lambda_0}\rho^{\lambda_0} \le C\rho^{\lambda_0}. \end{aligned}$$

This gives $\|A^\rho - A\|_\infty \le C\rho^{\lambda_0}$.

Next we show $|H^\rho(x) - H^\rho(y)| \le C|x - y|^{\lambda_0}$ for any $x, y \in \mathbb{R}^d$. Since $\|H^\rho\|_\infty \le C\rho^{\lambda_0}$, we may assume that $|x - y| \le \rho$. Note that $H^\rho = 0$ on $[-1/2, 1/2]^d \setminus [-3/8, 3/8]^d$. Thus it is enough to consider the case where $x, y \in [-1/2, 1/2]^d$. We may further assume that $\delta(x) \le \rho$ or $\delta(y) \le \rho$. For otherwise, $H^\rho(x) = H^\rho(y) = 0$ and there is nothing to show. Finally, suppose that $\delta(y) \le \rho$. Then

$$\begin{aligned} |H^\rho(x) - H^\rho(y)| &\le \theta(\rho^{-1}\delta(x))|(A(x) - \bar{A}(x)) - (A(y) - \bar{A}(y))| \\ &\qquad + |A(y) - \bar{A}(y)||\theta(\rho^{-1}\delta(x)) - \theta(\rho^{-1}\delta(y))| \\ &\le C|x - y|^{\lambda_0} + C\{\delta(y)\}^{\lambda_0}|x - y| \cdot \rho^{-1} \\ &\le C|x - y|^{\lambda_0} + C\rho^{\lambda_0 - 1}|x - y| \\ &\le C|x - y|^{\lambda_0}. \end{aligned}$$

The proof for the case $\delta(x) \le \rho$ is the same. □

It follows from Lemma 8.7.12 that for $\rho \in (0, 1/4)$,

$$\|A^\rho - \bar{A}\|_{C^{\lambda_0/2}(\mathbb{R}^d)} \le C\rho^{\lambda_0/2}. \tag{8.7.22}$$

Since $A^\rho = A = \bar{A}$ on $\partial\Omega$, we may deduce from Theorem 8.3.5 that

$$\|\mathcal{K}_{A^\rho} - \mathcal{K}_{\bar{A}}\|_{L^2 \to L^2} \le C\|A^\rho - \bar{A}\|_{C^{\lambda_0/2}(\mathbb{R}^d)} \le C\rho^{\lambda_0/2} \tag{8.7.23}$$

for any $\rho \in (0, 1/4)$. Note that by Lemma 8.7.11 and Theorem 8.7.9, the operator $(1/2)I + \mathcal{K}_{\bar{A}}$ is invertible on $L^2_0(\partial\Omega; \mathbb{R}^m)$ and

$$\|\big((1/2)I + \mathcal{K}_{\bar{A}}\big)^{-1}\|_{L^2_0 \to L^2_0} \le C.$$

Write

$$(1/2)I + \mathcal{K}_{A^\rho} = (1/2)I + \mathcal{K}_{\bar{A}} + (\mathcal{K}_{A^\rho} - \mathcal{K}_{\bar{A}}).$$

In view of (8.7.23), one may choose $\rho > 0$, depending only on μ, λ, τ, and the Lipschitz character of Ω, so that

$$\|\big((1/2)I + \mathcal{K}_{\bar{A}}\big)^{-1}(\mathcal{K}_{A^\rho} - \mathcal{K}_{\bar{A}})\|_{L^2_0 \to L^2_0} \le 1/2.$$

It follows that $(1/2)I + \mathcal{K}_{A^\rho}$ is an isomorphism on $L^2_0(\partial\Omega; \mathbb{R}^m)$ and

$$\begin{aligned}\|\big((1/2)I + \mathcal{K}_{A^\rho}\big)^{-1}\|_{L^2_0 \to L^2_0} &\le 2\|\big((1/2)I + \mathcal{K}_{\bar{A}}\big)^{-1}\|_{L^2_0 \to L^2_0} \\ &\le 2C.\end{aligned}$$

Similar arguments show that it is possible to choose ρ, depending only on μ, λ, τ, and the Lipschitz character of Ω, such that $-(1/2)I + \mathcal{K}^*_{A^\rho} : L^2(\partial\Omega; \mathbb{R}^m) \to L^2(\partial\Omega; \mathbb{R}^m)$ and $\mathcal{S}_{A^\rho} : L^2(\partial\Omega; \mathbb{R}^m) \to W^{1,2}(\partial\Omega; \mathbb{R}^m)$ are isomorphisms and the operator norms of their inverses are bounded by constants depending only on μ, λ, τ, and the Lipschitz character of Ω. Let $\widetilde{A} = A^\rho$. Note that if $\text{dist}(x, \partial\Omega) \le (1/4)\rho$, $A^\rho(x) = A(x)$. This completes Step 2.

Step 3. Use a perturbation argument to complete the proof.

Lemma 8.7.13. *Let $A = \big(a_{ij}^{\alpha\beta}\big)$ and $B = \big(b_{ij}^{\alpha\beta}\big) \in \Lambda(\mu, \lambda, \tau)$. Let Ω be a bounded Lipschitz domain with connected boundary. Suppose that*

$$A = B \quad \textit{in } \big\{x \in \Omega : \text{dist}(x, \partial\Omega) \le c_0 r_0\big\}$$

for some $c_0 > 0$, where $r_0 = \text{diam}(\Omega)$. Assume that $\mathcal{L}^A = -\text{div}(A\nabla)$ has the Rellich property in Ω with constant C_0. Then $\mathcal{L}^B = -\text{div}(B\nabla)$ has the Rellich property in Ω with constant C_1, where C_1 depends only on μ, λ, τ, c_0, C_0, and the Lipschitz character of Ω.

Proof. Suppose that $\mathcal{L}^B(u) = 0$ in Ω, $(\nabla u)^* \in L^2(\partial\Omega)$ and ∇u exists n.t. on $\partial\Omega$. Let $\varphi \in C_0^\infty(\mathbb{R}^d)$ such that $|\nabla\varphi| \le C r_0^{-1}$,

$$\begin{aligned}\varphi &= 1 \qquad \text{on } \{x \in \mathbb{R}^d : \text{dist}(x, \partial\Omega) \le (1/4)c_0 r_0\},\\ \varphi &= 0 \qquad \text{on } \{x \in \mathbb{R}^d : \text{dist}(x, \partial\Omega) \ge (1/2)c_0 r_0\}.\end{aligned}$$

Let $\bar{u} = \varphi(u - E)$, where E is the mean value of u over Ω. Note that

$$\big(\mathcal{L}^A(\bar{u})\big)^\alpha = -\partial_i\Big\{a_{ij}^{\alpha\beta}(\partial_j\varphi)(u - E)^\beta\Big\} - a_{ij}^{\alpha\beta}(\partial_i\varphi)(\partial_j u^\beta),$$

where we have used the fact that

$$\mathcal{L}^A(u) = \mathcal{L}^B(u) = 0 \quad \text{in } \{x \in \Omega : \text{dist}(x, \partial\Omega) < c_0 r_0\}.$$

It follows from the proof of (8.4.5) that

$$\begin{aligned} \bar{u}(x) &= \mathcal{S}_A\left(\frac{\partial \bar{u}}{\partial \nu_A}\right) - \mathcal{D}_A(\bar{u}) + v(x) \\ &= w(x) + v(x), \end{aligned} \tag{8.7.24}$$

where v satisfies

$$\begin{aligned} |\nabla v(x)| \le C &\int_\Omega |\nabla_x \nabla_y \Gamma(x, y; A)| \, |\nabla \varphi| \, |u - E| \, dy \\ &+ C \int_\Omega |\nabla_x \Gamma(x, y; A)| \, |\nabla \varphi| \, |\nabla u| \, dy. \end{aligned} \tag{8.7.25}$$

This, together with estimates

$$|\nabla_x \Gamma(x, y; A)| \le C \, |x - y|^{1-d} \quad \text{and} \quad |\nabla_x \nabla_y \Gamma(x, y; A)| \le C \, |x - y|^{-d},$$

implies that if $x \in \Omega$ and $\text{dist}(x, \partial\Omega) \le (1/5) c_0 r_0$,

$$\begin{aligned} |\nabla v(x)|^2 &\le \frac{C}{r_0^d} \int_\Omega |\nabla u|^2 \, dy \\ &\le C r_0^{1-d} \Big\| \frac{\partial u}{\partial \nu_B} \Big\|_{L^2(\partial\Omega)} \|\nabla_{\tan} u\|_{L^2(\partial\Omega)}, \end{aligned} \tag{8.7.26}$$

where we have used (8.6.10) for the last inequality.

Next, note that $\mathcal{L}^A(w) = 0$ in Ω, where $w = \bar{u} - v$. Using (8.7.26) and the assumption $(\nabla u)^* \in L^2(\partial\Omega)$, we deduce that $(\nabla w)^* \in L^2(\partial\Omega)$ and ∇w exists n.t. on $\partial\Omega$. Since $\mathcal{L}^A$ has the Rellich property, this implies that

$$\begin{aligned} \|\nabla w\|_{L^2(\partial\Omega)} &\le C_0 \Big\| \frac{\partial w}{\partial \nu_A} \Big\|_{L^2(\partial\Omega)} \\ &\le C \left\{ \Big\| \frac{\partial u}{\partial \nu_B} \Big\|_{L^2(\partial\Omega)} + \|\nabla v\|_{L^2(\partial\Omega)} \right\} \\ &\le C \left\{ \Big\| \frac{\partial u}{\partial \nu_B} \Big\|_{L^2(\partial\Omega)} + \Big\| \frac{\partial u}{\partial \nu_B} \Big\|_{L^2(\partial\Omega)}^{1/2} \|\nabla_{\tan} u\|_{L^2(\partial\Omega)}^{1/2} \right\}, \end{aligned} \tag{8.7.27}$$

where we used (8.7.26) in the last inequality. Using (8.7.26) again, we obtain

$$\begin{aligned} \|\nabla u\|_{L^2(\partial\Omega)} &\le C \Big\{ \|\nabla w\|_{L^2(\partial\Omega)} + \|\nabla v\|_{L^2(\partial\Omega)} \Big\} \\ &\le C \left\{ \Big\| \frac{\partial u}{\partial \nu_B} \Big\|_{L^2(\partial\Omega)} + \Big\| \frac{\partial u}{\partial \nu_B} \Big\|_{L^2(\partial\Omega)}^{1/2} \|\nabla_{\tan} u\|_{L^2(\partial\Omega)}^{1/2} \right\}. \end{aligned} \tag{8.7.28}$$

The desired estimate $\|\nabla u\|_{L^2(\partial\Omega)} \le C \| \frac{\partial u}{\partial \nu_B} \|_{L^2(\partial\Omega)}$ follows readily from (8.7.28) by applying the inequality (2.1.10). The proof of $\|\nabla u\|_{L^2(\partial\Omega)} \le C \|\nabla_{\tan} u\|_{L^2(\partial\Omega)}$ is similar and left to the reader. □

Finally we give the proof of Theorem 8.7.3.

Proof of Theorem 8.7.3. By Step 2, there exists $\widetilde{A} \in \Lambda(\mu, \lambda_0, \tau_0)$ such that

$$\widetilde{A} = A \quad \text{in } \{x \in \mathbb{R}^d : \text{dist}(x, \partial\Omega) \le cr_0\}$$

and $(1/2)I + \mathcal{K}_{\widetilde{A}} : L_0^2(\partial\Omega, \mathbb{R}^m) \to L_0^2(\partial\Omega, \mathbb{R}^m)$, $\mathcal{S}_{\widetilde{A}} : L^2(\partial\Omega, \mathbb{R}^m) \to W^{1,2}(\partial\Omega, \mathbb{R}^m)$ are isomorphisms. Moreover, the operator norms of their inverses are bounded by constants depending only on μ, λ, τ, and the Lipschitz character of Ω. It follows that the L^2 Neumann and regularity problems for $\widetilde{\mathcal{L}}(u) = \text{div}(\widetilde{A}\nabla u) = 0$ in Ω are well posed and the solutions satisfy

$$\|(\nabla u)^*\|_{L^2(\partial\Omega)} \le C \left\| \frac{\partial u}{\partial \nu} \right\|_{L^2(\partial\Omega)} \quad \text{and} \quad \|(\nabla u)^*\|_{L^2(\partial\Omega} \le C \, \|\nabla_{\tan} u\|_{L^2(\partial\Omega)},$$

with a constant C depending only on μ, λ, τ, and the Lipschitz character of Ω. In particular, the operator $\widetilde{\mathcal{L}}$ has the Rellich property in Ω with constant $C(\Omega)$ depending only on μ, λ, τ, and the Lipschitz character of Ω. By Lemma 8.7.13, this implies that $\mathcal{L}$ has the Rellich property in Ω with a constant C depending only on μ, λ, τ, and the Lipschitz character of Ω. The proof is complete. □

8.8 Rellich estimates for the large scale

Let $\psi : \mathbb{R}^{d-1} \to \mathbb{R}$ be a Lipschitz function such that $\psi(0) = 0$ and $\|\nabla\psi\|_\infty \le M$. Recall that

$$D_r = D(r, \psi) = \big\{(x', x_d) \in \mathbb{R}^d : \ |x'| < r \ \text{ and } \ \psi(x') < x_d < 10(M+1)r\big\},$$
$$\Delta_r = \Delta(\psi, r) = \big\{(x', \psi(x')) \in \mathbb{R}^d : \ |x'| < r\big\}.$$

It follows from Theorem 8.7.3 that if A satisfies the ellipticity condition (2.1.20), the smoothness condition (4.0.2), and the symmetry condition $A^* = A$, and if $0 < r \le 10$, then the operator $\mathcal{L} = -\text{div}(A\nabla)$ has the Rellich property in $\Omega = D_r$ with a constant $C(\Omega)$ depending only on μ, λ, τ, and M. In this section we show that the same conclusion in fact holds for any $0 < r < \infty$, with $C(\Omega)$ independent of r, if A also satisfies the periodicity condition.

For $x = (x', \psi(x'))$, let

$$N_\rho(w)(x) = \sup\big\{|w(x', \psi(x') + t)| : \ 0 < t < \rho\big\}.$$

Theorem 8.8.1. *Let $\mathcal{L} = -\text{div}(A\nabla)$ with $A \in \Lambda(\mu, \lambda, \tau)$ and $A^* = A$. Suppose that $\mathcal{L}(u) = 0$ in D_{8r} for some $r > 0$, where $u \in C^1(D_{8r})$, $N_r(\nabla u) \in L^2(\Delta_{6r})$ and ∇u exists n.t. on Δ_{6r}. Then*

$$\begin{aligned}
\int_{\Delta_r} |\nabla u|^2 \, d\sigma &\le C \int_{\Delta_{4r}} \left|\frac{\partial u}{\partial \nu}\right|^2 d\sigma + \frac{C}{r} \int_{D_{4r}} |\nabla u|^2 \, dx, \\
\int_{\Delta_r} |\nabla u|^2 \, d\sigma &\le C \int_{\Delta_{4r}} |\nabla_{\tan} u|^2 \, d\sigma + \frac{C}{r} \int_{D_{4r}} |\nabla u|^2 \, dx,
\end{aligned} \tag{8.8.1}$$

where C depends only on μ, λ, τ, and M.

As we mentioned above, the estimates in (8.8.1) hold for $0 < r \le 10$ with $C = C(\mu, \lambda, \tau, M) > 0$. To treat the case $r > 10$, we shall apply the small-scale estimates for $r = 1$ and reduce the problem to the control of the integral of $|\nabla u|^2$ on a boundary layer

$$\Big\{(x', x_d) \in \mathbb{R}^d : |x'| < r \text{ and } \psi(x') < x_d < \psi(x') + 1\Big\}.$$

We then use the error estimates in H^1 for a two-scale expansion obtained in Chapter 3 to handle ∇u over the boundary layer.

Lemma 8.8.2. *Suppose that A is 1-periodic and satisfies* (2.1.20). *Also assume $A^* = A$. Let $u \in H^1(D_{3r})$ be a weak solution of $\text{div}(A\nabla u) = 0$ in D_{3r} with $\frac{\partial u}{\partial \nu} = g$ on Δ_{3r}, where $r > 3$. Then*

$$\int_{\substack{|x'|<r \\ \psi(x')<x_d<\psi(x')+1}} |\nabla u|^2\, dx \le C \int_{\Delta_{2r}} |g|^2\, d\sigma + \frac{C}{r} \int_{D_{2r}} |\nabla u|^2\, dx, \tag{8.8.2}$$

where C depends only on μ and M.

Proof. The lemma follows from the error estimate (3.3.5) in Section 3.3. To see this, we consider the function $u_\varepsilon(x) = \varepsilon u(x/\varepsilon)$ in Ω, where $\varepsilon = r^{-1} < 1$,

$$\Omega = D(t, \psi_\varepsilon),$$

$t \in (1, 2)$ and $\psi_\varepsilon(x') = \varepsilon\psi(x'/\varepsilon)$. Observe that $\psi_\varepsilon(0) = 0$ and $\|\nabla\psi_\varepsilon\|_\infty = \|\nabla\psi\|_\infty \le M$. It follows that the Lipschitz character of Ω depends only on M.

Since $\mathcal{L}_\varepsilon(u_\varepsilon) = 0$ in Ω, in view of (3.3.5), we obtain

$$\|\nabla u_\varepsilon - \nabla u_0\|_{L^2(\Omega_{4\varepsilon})} \le \|w_\varepsilon\|_{H^1(\Omega)} \le C\sqrt{\varepsilon}\, \|g_\varepsilon\|_{L^2(\partial\Omega)},$$

where w_ε is defined by (3.2.2), $g_\varepsilon(x) = g(x/\varepsilon)$, and

$$\Omega_{4\varepsilon} = \{x \in \Omega : \text{dist}(x, \partial\Omega) < 4\varepsilon\}.$$

Moreover, since $\mathcal{L}_0(u_0) = 0$ in Ω, we can use the nontangential-maximal-function estimate for the operator $\mathcal{L}_0$ to show that

$$\begin{aligned} \|\nabla u_0\|_{L^2(\Omega_{4\varepsilon})} &\le C\sqrt{\varepsilon}\, \|(\nabla u_0)^*\|_{L^2(\partial\Omega)} \\ &\le C\sqrt{\varepsilon}\, \|g_\varepsilon\|_{L^2(\partial\Omega)}, \end{aligned} \tag{8.8.3}$$

where C depends only on μ and M. It follows that

$$\|\nabla u_\varepsilon\|_{L^2(\Omega_{4\varepsilon})} \le C\sqrt{\varepsilon}\, \|g_\varepsilon\|_{L^2(\partial\Omega)}.$$

By a change of variables this yields

$$\begin{aligned} \int_{\substack{|x'|<r \\ \psi(x')<x_d<\psi(x')+1}} |\nabla u|^2\, dx &\le C \int_{\partial D(tr,\psi)} \left|\frac{\partial u}{\partial \nu}\right|^2 d\sigma \\ &\le C \int_{\Delta_{2r}} |g|^2\, d\sigma + C \int_{\partial D_{tr} \setminus \Delta_{tr}} |\nabla u|^2\, d\sigma \end{aligned}$$

for any $t \in (1,2)$. Integrating the inequality above in t over the interval $(1,2)$ we obtain (8.8.2). □

Lemma 8.8.3. *Let $r > 1$. Suppose that A satisfies the same conditions as in Lemma 8.8.2. Let $u \in H^1(D_{3r})$ be a weak solution of* $\text{div}(A\nabla u) = 0$ *in D_{3r} with $u = f$ on Δ_{3r}, where $r > 3$. Then*

$$\int_{\substack{|x'|<r \\ \psi(x')<x_d<\psi(x')+1}} |\nabla u|^2\, dx \le C \int_{\Delta_{2r}} |\nabla_{\tan} f|^2\, d\sigma + \frac{C}{r} \int_{D_{2r}} |\nabla u|^2\, dx, \qquad (8.8.4)$$

where C depends only on μ and M.

Proof. As in the case of Lemma 8.8.2, the estimate follows from the error estimate (3.2.13) by a rescaling argument. Indeed, let u_ε and Ω be defined as in the proof of Lemma 8.8.2. The same argument shows that

$$\|\nabla u_\varepsilon\|_{L^2(\Omega_{4\varepsilon})} \le C\sqrt{\varepsilon}\, \|u_\varepsilon\|_{H^1(\partial\Omega)}.$$

Since $\partial\Omega$ is connected, by subtracting a constant from u_ε and using the Poincaré inequality, we obtain

$$\|\nabla u_\varepsilon\|_{L^2(\Omega_{4\varepsilon})} \le C\sqrt{\varepsilon}\, \|\nabla_{\tan} u_\varepsilon\|_{L^2(\partial\Omega)}.$$

By a change of variables this gives

$$\begin{aligned}\int_{\substack{|x'|<r \\ \psi(x')<x_d<\psi(x')+1}} |\nabla u|^2\, dx &\le C \int_{\partial D(tr,\psi)} |\nabla_{\tan} u|^2\, d\sigma \\ &\le C \int_{\Delta_{2r}} |\nabla_{\tan} f|^2\, d\sigma + C \int_{\partial D_{tr}\setminus \Delta_{tr}} |\nabla u|^2\, d\sigma\end{aligned}$$

for any $t \in (1,2)$. Integrating the inequality above in t over the interval $(1,2)$ we obtain (8.8.4). □

Proof of Theorem 8.8.1. By considering $v_\delta(x', x_d) = u(x', x_d + \delta)$, we may assume $u \in C^1(\overline{D_{7r}})$. We may also assume $r > 3$. Covering Δ_r with surface balls of small radius $c(M)$ on $\{(x', \psi(x')) : x' \in \mathbb{R}^{d-1}\}$ and using the first inequality in (8.8.1) on each surface ball, we obtain

$$\int_{\Delta_r} |\nabla u|^2\, d\sigma \le C \int_{\Delta_{2r}} \left|\frac{\partial u}{\partial \nu}\right|^2 d\sigma + C \int_{\substack{|x'|<r \\ \psi(x')<x_d<\psi(x')+1}} |\nabla u|^2\, dx.$$

This, together with Lemma 8.8.2, gives the first inequality in (8.8.1). The second inequality in (8.8.1) follows from Lemma 8.8.3 in a similar fashion. □

8.9 L^2 boundary value problems

Let Ω be a bounded Lipschitz domain in $\mathbb{R}^d$, $d \geq 3$, with connected boundary. In this section we establish the well-posedness of the L^2 Dirichlet, Neumann, and regularity problems with uniform nontangential-maximal-function estimates for $\mathcal{L}_\varepsilon(u_\varepsilon) = 0$ in Ω.

Theorem 8.9.1. *Suppose that $A \in \Lambda(\mu, \lambda, \tau)$ and $A^* = A$. Let $\mathcal{L} = -\text{div}(A\nabla)$ and Ω be a bounded Lipschitz domain in $\mathbb{R}^d$, $d \geq 3$ with connected boundary. Then the operators*

$$\begin{aligned}(1/2)I + \mathcal{K}_A &: L^2_0(\partial\Omega; \mathbb{R}^m) \to L^2_0(\partial\Omega; \mathbb{R}^m),\\ -(1/2)I + \mathcal{K}_A &: L^2(\partial\Omega; \mathbb{R}^m) \to L^2(\partial\Omega, \mathbb{R}^m),\\ \mathcal{S}_A &: L^2(\partial\Omega, \mathbb{R}^m) \to H^1(\partial\Omega, \mathbb{R}^m)\end{aligned}$$

are isomorphisms and the operator norms of their inverses are bounded by constants depending only on μ, λ, τ, and the Lipschitz character of Ω.

Proof. Suppose that $\mathcal{L}(u) = 0$ in Ω, $(\nabla u)^* \in L^2(\partial\Omega)$, and ∇u exists n.t. on $\partial\Omega$. Let $z \in \partial\Omega$. It follows from Theorem 8.8.1 by a change of the coordinate system that

$$\int_{B(z,cr_0)\cap\partial\Omega} |\nabla u|^2 \, d\sigma \leq C \int_{\partial\Omega} \left|\frac{\partial u}{\partial \nu}\right|^2 d\sigma + \frac{C}{r_0} \int_\Omega |\nabla u|^2 \, dx, \tag{8.9.1}$$

where $r_0 = \text{diam}(\Omega)$ and C depends at most on μ, λ, τ and the Lipschitz character of Ω. By covering $\partial\Omega$ with a finite number of balls $\{B(z_k, cr_0) : k = 1, \ldots, N\}$, where $z_k \in \partial\Omega$, we obtain

$$\begin{aligned}\int_{\partial\Omega} |\nabla u|^2 \, d\sigma &\leq C \int_{\partial\Omega} \left|\frac{\partial u}{\partial \nu}\right|^2 d\sigma + \frac{C}{r_0} \int_\Omega |\nabla u|^2 \, dx\\ &\leq C \int_{\partial\Omega} \left|\frac{\partial u}{\partial \nu}\right|^2 d\sigma + C \left\|\frac{\partial u}{\partial \nu}\right\|_{L^2(\partial\Omega)} \|\nabla_{\tan} u\|_{L^2(\partial\Omega)},\end{aligned}$$

where we have used (8.6.10) for the second inequality. By using the inequality (2.1.10), this gives

$$\|\nabla u\|_{L^2(\partial\Omega)} \leq C \left\|\frac{\partial u}{\partial \nu}\right\|_{L^2(\partial\Omega)}.$$

The estimate

$$\|\nabla u\|_{L^2(\partial\Omega)} \leq C \, \|\nabla_{\tan} u\|_{L^2(\partial\Omega)}$$

may be proved in a similar manner. Thus we have proved that $\mathcal{L}$ has the Rellich property in any Lipschitz domain Ω with connected boundary, and that the constant $C = C(\Omega)$ in (8.6.1) depends only on μ, λ, τ, and the Lipschitz character of Ω. Clearly, the same is true when A is replaced by $sA + (1-s)I$ for $0 \leq s \leq 1$. In view of Remark 8.6.8, this implies that the operator norms of $\big((1/2)I + \mathcal{K}_A\big)^{-1}$ on $L^2_0(\partial\Omega; \mathbb{R}^m)$, $\big(-(1/2)I + \mathcal{K}_A\big)^{-1}$ on $L^2(\partial\Omega; \mathbb{R}^m)$, and $\mathcal{S}_A^{-1} : H^1(\partial\Omega; \mathbb{R}^m) \to L^2(\partial\Omega; \mathbb{R}^m)$ are bounded by constants depending only on μ, λ, τ, and the Lipschitz character of Ω. □

Theorem 8.9.2 (L^2 Neumann problem). *Let $\mathcal{L}_\varepsilon = -\text{div}(A(x/\varepsilon)\nabla)$ with $A \in \Lambda(\mu, \lambda, \tau)$ and $A^* = A$. Let Ω be a bounded Lipschitz domain in $\mathbb{R}^d$, $d \geq 3$ with connected boundary. Then for any $g \in L^2_0(\partial\Omega; \mathbb{R}^m)$, there exists $u_\varepsilon \in C^1(\Omega; \mathbb{R}^m)$, unique up to constants, such that $\mathcal{L}_\varepsilon(u_\varepsilon) = 0$ in Ω, $(\nabla u_\varepsilon)^* \in L^2(\partial\Omega)$ and $\frac{\partial u_\varepsilon}{\partial \nu_\varepsilon} = g$ n.t. on $\partial\Omega$. Moreover, the solution u_ε satisfies the estimate*

$$\|(\nabla u_\varepsilon)^*\|_{L^2(\partial\Omega)} \leq C \, \|g\|_{L^2(\partial\Omega)},$$

and can be represented by a single-layer potential $\mathcal{S}_\varepsilon(h_\varepsilon)$ with $h_\varepsilon \in L^2_0(\partial\Omega; \mathbb{R}^m)$ and

$$\|h_\varepsilon\|_{L^2(\partial\Omega)} \leq C \, \|g\|_{L^2(\partial\Omega)}.$$

The constant C depends only on μ, λ, τ, and the Lipschitz character of Ω.

Theorem 8.9.3 (L^2 Dirichlet problem). *Assume A and Ω satisfy the same conditions as in Theorem 8.9.2. Then for any $f \in L^2(\partial\Omega; \mathbb{R}^m)$, there exists a unique $u_\varepsilon \in C^1(\Omega; \mathbb{R}^m)$ such that $\mathcal{L}_\varepsilon(u_\varepsilon) = 0$ in Ω, $(u_\varepsilon)^* \in L^2(\partial\Omega)$ and $u_\varepsilon = f$ n.t. on $\partial\Omega$. Moreover, the solution u_ε satisfies the estimate*

$$\|(u_\varepsilon)^*\|_{L^2(\partial\Omega)} \leq C \, \|f\|_{L^2(\partial\Omega)},$$

and can be represented by a double-layer potential $\mathcal{D}_\varepsilon(h_\varepsilon)$ with

$$\|h_\varepsilon\|_{L^2(\partial\Omega)} \leq C \, \|f\|_{L^2(\partial\Omega)}.$$

The constant C depends only on μ, λ, τ, and the Lipschitz character of Ω.

Theorem 8.9.4 (L^2 regularity problem). *Assume A and Ω satisfy the same conditions as in Theorem 8.9.2. Then for any $f \in H^1(\partial\Omega; \mathbb{R}^m)$, there exists a unique $u_\varepsilon \in C^1(\Omega; \mathbb{R}^m)$ such that $\mathcal{L}_\varepsilon(u_\varepsilon) = 0$ in Ω, $(\nabla u_\varepsilon)^* \in L^2(\partial\Omega)$ and $u_\varepsilon = f$ n.t. on $\partial\Omega$. Moreover, the solution u_ε satisfies the estimate*

$$\|(\nabla u_\varepsilon)^*\|_{L^2(\partial\Omega)} \leq C \, \|\nabla_{\tan} f\|_{L^2(\partial\Omega)},$$

and can be represented by a single-layer potential $\mathcal{S}_\varepsilon(h_\varepsilon)$ with

$$\|h_\varepsilon\|_{L^2(\partial\Omega)} \leq C \, \|f\|_{W^{1,2}(\partial\Omega)}.$$

The constant C depends only on μ, λ, τ, and the Lipschitz character of Ω.

Proof of Theorems 8.9.2, 8.9.3, *and* 8.9.4. By a simple rescaling we may assume that $\varepsilon = 1$. In this case the existence of solutions in Theorems 8.9.2, 8.9.3, and 8.9.4 is a direct consequence of the invertibility of $(1/2)I + \mathcal{K}_A$, $-(1/2)I + \mathcal{K}_A$, and $\mathcal{S}_A$ as well as estimates of the operator norms of their inverses, established in Theorem 8.9.1. The uniqueness for the L^2 Neumann and regularity problems follows readily from Green's identity (8.6.3), while the uniqueness for the L^2 Dirichlet problem may be proved by constructing a matrix of Green functions, as in the proof of Theorems 8.6.2 and 8.6.3. $\square$

Remark 8.9.5. The method of layer potentials also applies to Lipschitz domains Ω whose boundaries may not be connected. In this case, under the assumptions that $A \in \Lambda(\mu, \lambda, \tau)$ and $A^* = A$, the operator $(1/2)I + \mathcal{K}_A$ continues to be an isomorphism on $L^2_0(\partial\Omega, \mathbb{R}^m)$, while $-(1/2)I + \mathcal{K}_A$ is an isomorphism on the subspace

$$L^2_{0'}(\partial\Omega; \mathbb{R}^m) = \left\{ f \in L^2_{0'}(\partial\Omega; \mathbb{R}^m) : \int_{\partial\Omega'_j} f \, d\sigma = 0 \text{ for } j = 1, \dots, \ell \right\},$$

where Ω'_j, $j = 1, \dots, \ell$ are bounded connected components of $\Omega_- = \mathbb{R}^d \setminus \overline{\Omega}$. Also, $\mathcal{S}_A : L^2(\partial\Omega; \mathbb{R}^m) \to H^1(\partial\Omega; \mathbb{R}^m)$ is an isomorphism.

We end this section with some L^2 estimates for local solutions with Dirichlet or Neumann conditions. Recall that if Ω is a bounded Lipschitz domain and $u \in C(\Omega)$, the nontangential maximal function $(u)^*$ is defined by

$$(u)^*(z) = \sup \left\{ |u(x)| : x \in \Omega \text{ and } |x - z| < C_0 \operatorname{dist}(x, \partial\Omega) \right\}$$

for $z \in \partial\Omega$, where $C_0 = C(\Omega) > 1$ is sufficiently large. Let

$$\begin{aligned} \mathcal{M}^1_r(u)(z) &= \sup \left\{ |u(x)| : x \in \Omega, \ |x - z| \le r \text{ and } |x - z| < C_0 \operatorname{dist}(x, \partial\Omega) \right\}, \\ \mathcal{M}^2_r(u)(z) &= \sup \left\{ |u(x)| : x \in \Omega, \ |x - z| > r \text{ and } |x - z| < C_0 \operatorname{dist}(x, \partial\Omega) \right\}, \end{aligned} \tag{8.9.2}$$

where $0 < r < c_0 \operatorname{diam}(\Omega)$. Clearly,

$$(u)^*(z) = \max \left\{ \mathcal{M}^1_r(u)(z), \mathcal{M}^2_r(u)(z) \right\}. \tag{8.9.3}$$

Theorem 8.9.6. *Let $\mathcal{L}_\varepsilon = -\operatorname{div}(A(x/\varepsilon)\nabla)$ with $A \in \Lambda(\mu, \lambda, \tau)$ and $A^* = A$. Let Ω be a bounded Lipschitz domain in $\mathbb{R}^d$, $d \ge 3$. Assume that $\mathcal{L}_\varepsilon(u_\varepsilon) = 0$ in $B(z, 4r) \cap \Omega$ for some $z \in \partial\Omega$ and $0 < r < c_0 \operatorname{diam}(\Omega)$.*

1. *Suppose that $\mathcal{M}^1_r(\nabla u_\varepsilon) \in L^2(B(z, 3r) \cap \partial\Omega; \mathbb{R}^m)$ and $\frac{\partial u_\varepsilon}{\partial \nu_\varepsilon} = g$ n.t. on $B(z, 4r) \cap \partial\Omega$. Then*

$$\begin{aligned} &\int_{B(z,r)\cap\partial\Omega} |\mathcal{M}^1_r(\nabla u_\varepsilon)|^2 \, d\sigma \\ &\le C \int_{B(z,3r)\cap\partial\Omega} |g|^2 \, d\sigma + \frac{C}{r} \int_{B(z,4r)\cap\Omega} |\nabla u_\varepsilon|^2 \, dx. \end{aligned} \tag{8.9.4}$$

2. *Suppose that $\mathcal{M}^1_r(\nabla u_\varepsilon) \in L^2(B(z, 3r) \cap \partial\Omega; \mathbb{R}^m)$ and $u_\varepsilon = f$ n.t. on $B(z, 4r) \cap \partial\Omega$. Then*

$$\begin{aligned} &\int_{B(z,r)\cap\partial\Omega} |\mathcal{M}^1_r(\nabla u_\varepsilon)|^2 \, d\sigma \\ &\le C \int_{B(z,3r)\cap\partial\Omega} |\nabla_{\tan} f|^2 \, d\sigma + \frac{C}{r} \int_{B(z,4r)\cap\Omega} |\nabla u_\varepsilon|^2 \, dx. \end{aligned} \tag{8.9.5}$$

3. *Suppose that* $\mathcal{M}_r^1(u_\varepsilon) \in L^2(B(z,3r)\cap\partial\Omega;\mathbb{R}^m)$ *and* $u_\varepsilon = f$ *n.t. on* $B(z,4r)\cap\partial\Omega$. *Then*

$$\begin{aligned} &\int_{B(z,r)\cap\partial\Omega} |\mathcal{M}_r^1(u_\varepsilon)|^2\, d\sigma \\ &\quad \le C\int_{B(z,3r)\cap\partial\Omega} |f|^2\, d\sigma + \frac{C}{r}\int_{B(z,4r)\cap\Omega} |u_\varepsilon|^2\, dx. \end{aligned} \tag{8.9.6}$$

The constants C *in* (8.9.4), (8.9.5) *and* (8.9.6) *depend only on* μ, λ, τ, *and the Lipschitz character of* Ω.

Proof. We give the proof of (8.9.4). The estimates (8.9.5) and (8.9.6) may be proved in a similar manner.

Let

$$\begin{aligned} D_r &= \big\{(x',x_d)\in\mathbb{R}^d:\ |x'|<r \text{ and } \psi(x')<x_d<10(M+1)r\big\},\\ \Delta_r &= \big\{(x',\psi(x'))\in\mathbb{R}^d:\ |x'|<r\big\}, \end{aligned}$$

where $\psi:\mathbb{R}^{d-1}\to\mathbb{R}$ is a Lipschitz function with $\psi(0)=0$ and $\|\nabla\psi\|_\infty\le M$. Suppose that $\mathcal{L}_\varepsilon(u_\varepsilon)=0$ in D_{5r}, $u_\varepsilon\in H^1(D_{5r};\mathbb{R}^m)$, and $\frac{\partial u_\varepsilon}{\partial\nu_\varepsilon}=g$ on Δ_{3r}. We will show that

$$\int_{\Delta_r} |\mathcal{M}_r^1(\nabla u_\varepsilon)|^2\, d\sigma \le C\int_{\Delta_{3r}} |g|^2\, d\sigma + \frac{C}{r}\int_{D_{3r}} |\nabla u_\varepsilon|^2\, dx. \tag{8.9.7}$$

By translation and rotation as well as a simple covering argument, this implies the estimate (8.9.4). To see (8.9.7), we first assume $u_\varepsilon\in C^1(\overline{D_{3r}};\mathbb{R}^m)$. Let $(w)^*_D$ denote the nontangential maximal function of w with respect to the Lipschitz domain D_{sr}, where $s\in(2,3)$. By applying the L^2 estimate in Theorem 8.9.2 in D_{sr}, we obtain

$$\begin{aligned} \int_{\partial D_{sr}} |(\nabla u_\varepsilon)^*_D|^2\, d\sigma &\le C\int_{\partial D_{sr}} \left|\frac{\partial u_\varepsilon}{\partial\nu_\varepsilon}\right|^2 d\sigma\\ &\le C\int_{\Delta_{3r}} |g|^2\, d\sigma + C\int_{\partial D_{sr}\setminus\Delta_{3r}} |\nabla u_\varepsilon|^2\, d\sigma, \end{aligned}$$

where C depends only on μ, λ, τ, and M. This leads to

$$\int_{\Delta_r} |\mathcal{M}_r^1(\nabla u_\varepsilon)|^2\, d\sigma \le C\int_{\Delta_{3r}} |g|^2\, d\sigma + C\int_{\partial D_{sr}\setminus\Delta_{3r}} |\nabla u_\varepsilon|^2\, d\sigma. \tag{8.9.8}$$

The estimate (8.9.7) now follows by integrating both sides of (8.9.8) in s over the interval $(2,3)$. Finally, to get rid of the assumption $u_\varepsilon\in C^1(\overline{D_{3r}})$, we apply the estimate (8.9.7) to the function $v_\varepsilon(x',x_d)=u_\varepsilon(x',x_d+\delta)$ and then let $\delta\to 0^+$. The proof is finished by a simple limiting argument. □

8.10 L^2 estimates in arbitrary Lipschitz domains

In Theorems 8.9.2, 8.9.3, and 8.9.4, we solve the L^2 boundary value problems for $\mathcal{L}_\varepsilon(u_\varepsilon) = 0$ in Ω, assuming that $d \ge 3$ and $\partial\Omega$ is a bounded Lipschitz domain with *connected* boundary. Although the method of layer potentials may be applied to a general Lipschitz domain in $\mathbb{R}^d$, $d \ge 2$, we will show in this section that the general case follows from the localized L^2 estimates in Theorem 8.9.6 by some approximation argument. The descent method is used to handle the case $d = 2$.

Theorem 8.10.1 (L^2 Neumann problem). *Suppose that $A \in \Lambda(\mu, \lambda, \tau)$ and $A^* = A$. Let Ω be a bounded Lipschitz domain in $\mathbb{R}^d$, $d \ge 2$. Given $g \in L^2_0(\partial\Omega; \mathbb{R}^m)$, let $u_\varepsilon \in H^1(\Omega; \mathbb{R}^m)$ be the weak solution to the Neumann problem $\mathcal{L}_\varepsilon(u_\varepsilon) = 0$ in Ω and $\frac{\partial u_\varepsilon}{\partial \nu_\varepsilon} = g$ on $\partial\Omega$, given by Theorem 2.4.4. Then the nontangential limits of ∇u_ε exist a.e. on $\partial\Omega$, and*

$$\|(\nabla u_\varepsilon)^*\|_{L^2(\partial\Omega)} \le C\, \|g\|_{L^2(\partial\Omega)},$$

where C depends only on μ, λ, τ, and the Lipschitz character of Ω.

Proof. By dilation we may assume that $\operatorname{diam}(\Omega) = 1$. We may also assume that $\int_\Omega u_\varepsilon\, dx = 0$. Suppose that $d \ge 3$. To establish the estimate $\|(\nabla u_\varepsilon)^*\|_{L^2(\partial\Omega)} \le C\|g\|_{L^2(\partial\Omega)}$, we first consider the case $u_\varepsilon \in C^1(\overline{\Omega})$. By covering $\partial\Omega$ with coordinate cylinders we deduce from the estimate (8.9.4) and interior Lipschitz estimates that

$$\begin{aligned}\int_{\partial\Omega} |(\nabla u_\varepsilon)^*|^2\, d\sigma &\le C \int_{\partial\Omega} |g|^2\, d\sigma + C \int_\Omega |\nabla u_\varepsilon|^2\, dx \\ &\le C \int_{\partial\Omega} |g|^2\, d\sigma,\end{aligned} \tag{8.10.1}$$

where we have used the energy estimate $\|\nabla u_\varepsilon\|_{L^2(\Omega)} \le C\|g\|_{L^2(\partial\Omega)}$.

Next we consider the case $g \in C_0^\infty(\mathbb{R}^d; \mathbb{R}^m)$. We choose a sequence of smooth domains $\{\Omega_\ell\}$ such that $\Omega_\ell \downarrow \Omega$. Let $u_{\varepsilon,\ell} \in H^1(\Omega_\ell; \mathbb{R}^m)$ be a weak solution to the Neumann problem for $\mathcal{L}_\varepsilon(u_{\varepsilon,\ell}) = 0$ in Ω_ℓ with boundary data $g - \fint_{\partial\Omega_\ell} g$. Since Ω_ℓ and g are smooth, we have $u_{\varepsilon,\ell} \in C^1(\overline{\Omega_\ell})$. It follows from (8.10.1) that

$$\|(\nabla u_{\varepsilon,\ell})^*\|_{L^2(\partial\Omega)} \le C\, \|(\nabla u_{\varepsilon,\ell})^*\|_{L^2(\partial\Omega_\ell)} \le C\, \|g\|_{L^2(\partial\Omega_\ell)}. \tag{8.10.2}$$

Assume that $\int_{\Omega_\ell} u_{\varepsilon,\ell} = 0$. Since $\|u_{\varepsilon,\ell}\|_{H^1(\Omega_\ell)} \le C\|g\|_{L^2(\partial\Omega_\ell)}$, the sequence $\{u_{\varepsilon,\ell}\}$ is bounded in $H^1(\Omega; \mathbb{R}^m)$. As a result, by passing to a subsequence, we may assume that $u_{\varepsilon,\ell} \to w_\varepsilon$ weakly in $H^1(\Omega; \mathbb{R}^m)$, as $\ell \to \infty$. Note that for any $\varphi \in C_0^\infty(\mathbb{R}^d; \mathbb{R}^m)$,

$$\begin{aligned}&\int_{\partial\Omega_\ell} \left(g - \fint_{\partial\Omega_\ell} g\right) \cdot \varphi\, d\sigma \to \int_{\partial\Omega} g \cdot \varphi\, d\sigma,\\ &\int_{\Omega_\ell \setminus \Omega} A(x/\varepsilon)\nabla u_{\varepsilon,\ell} \cdot \nabla\varphi\, dx \to 0,\end{aligned}$$

$$\int_\Omega A(x/\varepsilon)\nabla u_{\varepsilon,\ell}\cdot\nabla\varphi\,dx \to \int_\Omega A(x/\varepsilon)\nabla w_\varepsilon\cdot\nabla\varphi\,dx,$$

as $\ell\to\infty$. This implies that $w_\varepsilon\in H^1(\Omega;\mathbb{R}^m)$ is a weak solution of the Neumann problem for $\mathcal{L}_\varepsilon(w_\varepsilon)=0$ in Ω with boundary data g. Since $\int_\Omega w_\varepsilon\,dx=0$, we obtain $u_\varepsilon=w_\varepsilon$ in Ω. Using interior Lipschitz estimates and the observation that $u_{\varepsilon,\ell}\to u_\varepsilon$ strongly in $L^2(\Omega;\mathbb{R}^m)$, we may deduce that $\nabla u_{\varepsilon,\ell}\to\nabla u_\varepsilon$ uniformly on any compact subset of Ω. In view of (8.10.2) this yields

$$\|\mathcal{M}^2_\delta(\nabla u_\varepsilon)\|_{L^2(\partial\Omega)}\le C\,\|g\|_{L^2(\partial\Omega)}. \tag{8.10.3}$$

Letting $\delta\to 0$ in (8.10.3) and using Fatou's lemma, we obtain

$$\|(\nabla u_\varepsilon)^*\|_{L^2(\partial\Omega)}\le C\,\|g\|_{L^2(\partial\Omega)}.$$

Suppose now that $g\in L^2_0(\partial\Omega;\mathbb{R}^m)$. We choose a sequence of functions $\{g_\ell\}$ in $C_0^\infty(\mathbb{R}^d;\mathbb{R}^m)$ such that $g_\ell\to g$ in $L^2(\partial\Omega;\mathbb{R}^m)$ and $\int_{\partial\Omega} g_\ell\,d\sigma=0$. Let $v_{\varepsilon,\ell}$ be the weak solution of the Neumann problem for $\mathcal{L}_\varepsilon(v_{\varepsilon,\ell})=0$ in Ω with boundary data g_ℓ and $\int_\Omega v_{\varepsilon,\ell}\,dx=0$. Since

$$\|v_{\varepsilon,\ell}-u_\varepsilon\|_{H^1(\Omega)}\le C\,\|g_\ell-g\|_{L^2(\partial\Omega)}\to 0,\qquad \text{as } \ell\to\infty,$$

by interior Lipschitz estimates, we see that $v_{\varepsilon,\ell}\to u_\varepsilon$ uniformly on any compact subset of Ω. Note that

$$\|(\nabla v_{\varepsilon,\ell})^*\|_{L^2(\partial\Omega)}\le C\,\|g_\ell\|_{L^2(\partial\Omega)}. \tag{8.10.4}$$

As before, by a simple limiting argument, this leads to the estimate (8.10.3) and hence to

$$\|(\nabla u_\varepsilon)^*\|_{L^2(\partial\Omega)}\le C\,\|g\|_{L^2(\partial\Omega)}.$$

To show that the nontangential limits of ∇u_ε exist a.e. on $\partial\Omega$, consider the set

$$E(T)=\Big\{z\in\partial\Omega : (\nabla u_\varepsilon)^*(z)\le T\Big\}.$$

Since $(\nabla u_\varepsilon)^*(z)<\infty$ for a.e. $z\in\partial\Omega$, it suffices to show that ∇u_ε exist a.e. on $E(T)$ for each fixed $T>1$. Fix $z_0\in E(T)$ and $\rho>0$ small. We construct a bounded Lipschitz domain $\widetilde{\Omega}$ with connected boundary, such that

$$\widetilde{\Omega}\subset\Big\{x\in\Omega : x\in\gamma(z) \text{ for some } z\in E(T)\Big\}$$

and

$$\partial\widetilde{\Omega}\cap\partial\Omega\cap B(z_0,\rho)=E(T)\cap B(z_0,\rho),$$

where $\gamma(z)=\big\{x\in\Omega : |x-z|<C_0\,\text{dist}(x,\partial\Omega)\big\}$. Note that $|\nabla u_\varepsilon|\le T$ on $\widetilde{\Omega}$. Let v_ε be the solution of the L^2 Neumann problem for $\mathcal{L}_\varepsilon(v_\varepsilon)=0$ in $\widetilde{\Omega}$ with boundary data $h=\frac{\partial u_\varepsilon}{\partial\nu_\varepsilon}$, given by Theorem 8.9.2. By the uniqueness of weak

solutions of the Neumann problem in $H^1(\Omega;\mathbb{R}^m)$, $u_\varepsilon - v_\varepsilon$ is constant in $\widetilde{\Omega}$. Since the nontangential limits of ∇v_ε exist a.e. on $\partial\widetilde{\Omega}$, we may conclude that ∇u_ε exist a.e. on $E(T)\cap B(Q_0,\rho)$ and hence a.e. on $\partial\Omega$.

Finally, we use the descent method to treat the case $d=2$. Consider the function $v_\varepsilon(x,t)=u_\varepsilon(x)$ in $\Omega_1=\Omega\times(0,1)$, a bounded Lipschitz domain in $\mathbb{R}^3$. Observe that $v_\varepsilon \in H^1(\Omega_1;\mathbb{R}^m)$ is a weak solution of the Neumann problem for

$$\left(\mathcal{L}_\varepsilon - \frac{\partial^2}{\partial t^2}\right) v_\varepsilon = 0 \quad \text{in } \Omega_1,$$

with boundary data $\frac{\partial v_\varepsilon}{\partial \nu_\varepsilon} = g$ on $\partial\Omega\times(0,1)$, and $\frac{\partial v_\varepsilon}{\partial \nu_\varepsilon}=0$ on $\Omega\times\{t=0\}$ and on $\Omega\times\{t=1\}$. It is not hard to see that the desired results for u_ε in Ω follow from the corresponding results for v_ε in Ω_1. This completes the proof. □

Theorem 8.10.2 (L^2 regularity problem). *Suppose that $A\in\Lambda(\mu,\lambda,\tau)$ and $A^*=A$. Let Ω be a bounded Lipschitz domain in $\mathbb{R}^d$, $d\geq 2$. Given $f\in H^1(\partial\Omega;\mathbb{R}^m)$, let $u_\varepsilon\in H^1(\Omega;\mathbb{R}^m)$ be the weak solution to the Dirichlet problem $\mathcal{L}_\varepsilon(u_\varepsilon)=0$ in Ω and $u_\varepsilon=f$ on $\partial\Omega$, given by Theorem 2.1.5. Then the nontangential limits of u_ε and ∇u_ε exist a.e. on $\partial\Omega$, $u_\varepsilon=f$ n.t. on $\partial\Omega$, and*

$$\begin{aligned} &\|(\nabla u_\varepsilon)^*\|_{L^2(\partial\Omega)} + r_0^{-1}\|(u_\varepsilon)^*\|_{L^2(\partial\Omega)} \\ &\qquad \leq C\Big\{\|\nabla_{\tan} f\|_{L^2(\partial\Omega)} + r_0^{-1}\|f\|_{L^2(\partial\Omega)}\Big\}, \end{aligned} \tag{8.10.5}$$

where r_0 =diam(Ω) and C depends only on μ, λ, τ, and the Lipschitz character of Ω.

Proof. By dilation we may assume that $r_0=1$. The case $d=2$ can be handled by the descent method, as in the proof of Theorem 8.10.1. We will assume $d\geq 3$. Furthermore, by the approximation argument as well as the argument for the existence of nontangential limits in the proof of Theorem 8.10.1, it suffices to prove the estimate (8.10.5) for $f\in C_0^\infty(\mathbb{R}^d;\mathbb{R}^m)$.

Let $\{\Omega_\ell\}$ be a sequence of smooth domains such that $\Omega_\ell\downarrow\Omega$. Let $u_{\varepsilon,\ell}\in H^1(\Omega_\ell;\mathbb{R}^m)$ be the weak solution of $\mathcal{L}_\varepsilon(u_{\varepsilon,\ell})=0$ in Ω with boundary data $u_{\varepsilon,\ell}=f$ on $\partial\Omega$. Since Ω_ℓ and f are smooth, $u_{\varepsilon,\ell}\in C^1(\overline{\Omega_\ell})$. It follows from Theorem 8.9.6 that

$$\begin{aligned} &\|(\nabla u_{\varepsilon,\ell})^*\|_{L^2(\partial\Omega_\ell)} + \|(u_{\varepsilon,\ell})^*\|_{L^2(\partial\Omega_\ell)} \\ &\qquad \leq C\left\{\|f\|_{H^1(\partial\Omega_\ell)} + \|\nabla u_{\varepsilon,\ell}\|_{L^2(\Omega_\ell)} + \|u_{\varepsilon,\ell}\|_{L^2(\Omega_\ell)}\right\} \\ &\qquad \leq C\,\|f\|_{H^1(\partial\Omega_\ell)}, \end{aligned} \tag{8.10.6}$$

where we have used the bound $\|u_{\varepsilon,\ell}\|_{H^1(\Omega_\ell)}\leq C\|f\|_{H^1(\partial\Omega_\ell)}$. Hence,

$$\|(\nabla u_{\varepsilon,\ell})^*\|_{L^2(\partial\Omega)} + \|(u_{\varepsilon,\ell})^*\|_{L^2(\partial\Omega)} \leq C\,\|f\|_{H^1(\partial\Omega_\ell)}. \tag{8.10.7}$$

Note that

$$\|u_{\varepsilon,\ell}\|_{H^1(\Omega)} \leq \|u_{\varepsilon,\ell}\|_{H^1(\Omega_\ell)} \leq C\,\|f\|_{H^1(\partial\Omega_\ell)} \leq C_f < \infty.$$

By passing to a subsequence we may assume that $u_{\varepsilon,\ell} \to w_\varepsilon$ weakly in $H^1(\Omega)$, as $\ell \to \infty$. Clearly, $\mathcal{L}_\varepsilon(w_\varepsilon) = 0$ in Ω. Since $u_{\varepsilon,\ell} \to w_\varepsilon$ strongly in $L^2(\Omega)$, by interior Lipschitz estimates, $u_{\varepsilon,\ell}$ converges to w_ε uniformly on any compact subset of Ω. This, together with (8.10.7), implies that

$$\|(\nabla w_\varepsilon)^*\|_{L^2(\partial\Omega)} + \|(w_\varepsilon)^*\|_{L^2(\partial\Omega)} \le C\,\|f\|_{H^1(\partial\Omega)}, \tag{8.10.8}$$

by a limiting argument.

Finally, note that the trace operator from $H^1(\Omega;\mathbb{R}^m)$ to $L^2(\partial\Omega;\mathbb{R}^m)$ is compact. It follows that $u_{\varepsilon,\ell}|_{\partial\Omega} \to w_\varepsilon|_{\partial\Omega}$ strongly in $L^2(\partial\Omega;\mathbb{R}^m)$, as $\ell\to\infty$. However, using (8.10.6), it is not hard to verify that $u_{\varepsilon,\ell}|_{\partial\Omega} \to f|_{\partial\Omega}$ a.e. on $\partial\Omega$. Hence, $w_\varepsilon = f$ on $\partial\Omega$. As a result, by the uniqueness of the Dirichlet problem, we obtain $w_\varepsilon = u_\varepsilon$ in Ω. In view of (8.10.8) this completes the proof. □

To handle the L^2 Dirichlet problem we first establish two lemmas.

Lemma 8.10.3. *Let $F \in C_0^\infty(\mathbb{R}^d)$ and*

$$u(x) = \int_{\mathbb{R}^d} \frac{F(y)}{|x-y|^{d-1}}\, dy.$$

Let Ω be a bounded Lipschitz domain in $\mathbb{R}^d$, $d \ge 2$. Then $\|u\|_{L^2(\partial\Omega)} \le C\|F\|_{L^p(\mathbb{R}^d)}$, where $p = \frac{2d}{d+1}$ and C depends only on the Lipschitz character of Ω.

Proof. By dilation we may assume that $\operatorname{diam}(\Omega) = 1$. It follows from (3.1.15) that

$$\|u\|_{L^2(\partial\Omega)} \le C\,\|u\|_{W^{1,p}(\Omega)}. \tag{8.10.9}$$

By the fractional integral estimates and singular integral estimates,

$$\|u\|_{L^p(\Omega)} + \|\nabla u\|_{L^p(\Omega)} \le C\,\|F\|_{L^p(\mathbb{R}^d)}.$$

which together with (8.10.9) gives $\|u\|_{L^2(\partial\Omega)} \le C\|F\|_{L^p(\mathbb{R}^d)}$. □

Lemma 8.10.4. *Suppose that $A \in \Lambda(\mu,\lambda,\tau)$ and $A^* = A$. Let Ω be a bounded Lipschitz domain in $\mathbb{R}^d$, $d \ge 2$. Let $u_\varepsilon \in H^1(\Omega;\mathbb{R}^m)$ be a weak solution of the Dirichlet problem $\mathcal{L}_\varepsilon(u_\varepsilon) = 0$ in Ω and $u_\varepsilon = f$ on $\partial\Omega$, where $f \in H^1(\partial\Omega;\mathbb{R}^m)$. Then*

$$\int_\Omega |u_\varepsilon|^2\, dx \le C r_0 \int_{\partial\Omega} |f|^2\, d\sigma, \tag{8.10.10}$$

where $r_0 = \operatorname{diam}(\Omega)$ and C depends only on μ, λ, τ, and the Lipschitz character of Ω.

Proof. By dilation we may assume that $r_0 = 1$. Choose a ball B of radius 2 such that $\Omega \subset B$. Let $G_\varepsilon(x,y)$ denote the matrix of Green functions for $\mathcal{L}_\varepsilon$ in $2B$. Let $F \in C_0^\infty(\Omega;\mathbb{R}^m)$ and

$$v_\varepsilon(x) = \int_\Omega G_\varepsilon(x,y) F(y)\, dy.$$

Then $v_\varepsilon \in C^1(2B; \mathbb{R}^m)$ and $\mathcal{L}_\varepsilon(v_\varepsilon) = F$ in Ω. Recall that

$$|G_\varepsilon(x,y)| \le C|x-y|^{2-d} \text{ for } d \ge 3, \text{ and } |\nabla_x G_\varepsilon(x,y)| \le C|x-y|^{1-d} \text{ for } d \ge 2.$$

Also, $|G_\varepsilon(x,y)| \le C\left(1 + \big|\ln|x-y|\big|\right)$ if $d=2$. Hence, by Lemma 8.10.3,

$$\|v_\varepsilon\|_{L^2(\partial\Omega)} \le C\,\|F\|_{L^2(\Omega)} \quad \text{and} \quad \|\nabla v_\varepsilon\|_{L^2(\partial\Omega)} \le C\,\|F\|_{L^2(\Omega)}.$$

In particular, we see that $\|v_\varepsilon\|_{H^1(\partial\Omega)} \le C\|F\|_{L^2(\Omega)}$.

Let $w_\varepsilon \in H^1(\Omega; \mathbb{R}^m)$ be the weak solution of $\mathcal{L}_\varepsilon(w_\varepsilon) = 0$ in Ω with $w_\varepsilon = v_\varepsilon$ on $\partial\Omega$. By Theorem 8.10.2, ∇w_ε has nontangential limit a.e. on $\partial\Omega$, and

$$\|\nabla w_\varepsilon\|_{L^2(\partial\Omega)} \le C\,\|v_\varepsilon\|_{H^1(\partial\Omega)} \le C\,\|F\|_{L^2(\Omega)}. \tag{8.10.11}$$

Let $z_\varepsilon = v_\varepsilon - w_\varepsilon$ in Ω. Then $z_\varepsilon \in H_0^1(\Omega; \mathbb{R}^m)$ and $\mathcal{L}_\varepsilon(z_\varepsilon) = F$ in Ω. It follows by the divergence theorem that

$$\begin{aligned} \left| \int_\Omega u_\varepsilon \cdot F\, dx \right| &= \left| \int_{\partial\Omega} \frac{\partial z_\varepsilon}{\partial \nu_\varepsilon} \cdot u_\varepsilon \, d\sigma \right| \\ &\le C \left\|\nabla z_\varepsilon\right\|_{L^2(\partial\Omega)} \|f\|_{L^2(\partial\Omega)} \\ &\le C\,\|F\|_{L^2(\Omega)} \|f\|_{L^2(\partial\Omega)}. \end{aligned} \tag{8.10.12}$$

By duality this gives $\|u_\varepsilon\|_{L^2(\Omega)} \le C\|f\|_{L^2(\partial\Omega)}$. □

Theorem 8.10.5 (L^2 Dirichlet problem). *Let Ω be a bounded Lipschitz domain in $\mathbb{R}^d$, $d \ge 2$. Given $f \in L^2(\partial\Omega; \mathbb{R}^m)$, there exists a unique $u_\varepsilon \in C^1(\Omega; \mathbb{R}^m)$ such that $\mathcal{L}_\varepsilon(u_\varepsilon) = 0$ in Ω, $u_\varepsilon = f$ n.t. on $\partial\Omega$, and $(u_\varepsilon)^* \in L^2(\partial\Omega)$. Moreover, the solution satisfies*

$$\|(u_\varepsilon)^*\|_{L^2(\partial\Omega)} \le C\,\|f\|_{L^2(\partial\Omega)},$$

where C depends only on μ, λ, τ, and the Lipschitz character of Ω.

Proof. By dilation we may assume that $\operatorname{diam}(\Omega) = 1$. The uniqueness follows readily from Lemma 8.10.4 by choosing a sequence of smooth domains $\{\Omega_\ell\}$ such that $\Omega_\ell \uparrow \Omega$. To establish the existence as well as the nontangential-maximal-function estimate, we first consider the case $d \ge 3$. Choose a sequence of functions $\{f_\ell\}$ in $C_0^\infty(\mathbb{R}^d; \mathbb{R}^m)$ such that $f_\ell \to f$ in $L^2(\partial\Omega; \mathbb{R}^m)$. Let $u_{\varepsilon,\ell}$ be the solution to the L^2 regularity problem for $\mathcal{L}_\varepsilon$ in Ω with Dirichlet data f_ℓ, given by Theorem 8.10.2. By the localized L^2 estimate (8.9.6) we obtain

$$\begin{aligned} \int_{\partial\Omega} |(u_{\varepsilon,\ell})^*|^2\, d\sigma &\le C \int_{\partial\Omega} |f_\ell|^2\, d\sigma + C \int_\Omega |u_{\varepsilon,\ell}|^2\, dx \\ &\le C \int_{\partial\Omega} |f_\ell|^2\, d\sigma, \end{aligned} \tag{8.10.13}$$

where we have used Lemma 8.10.4. Similarly, we have

$$\|(u_{\varepsilon,\ell} - u_{\varepsilon,k})^*\|_{L^2(\partial\Omega)} \le C\,\|f_\ell - f_k\|_{L^2(\partial\Omega)} \tag{8.10.14}$$

for any $k, \ell \geq 1$. By interior Lipschitz estimates, this implies that $u_{\varepsilon,\ell} \to u_\varepsilon$ and $\nabla u_{\varepsilon,\ell} \to \nabla u_\varepsilon$ uniformly on any compact subset of Ω. Clearly, $u_\varepsilon \in H^1_{\text{loc}}(\Omega; \mathbb{R}^m)$ and $\mathcal{L}_\varepsilon(u_\varepsilon) = 0$ in Ω. By a limiting argument we also deduce from (8.10.13) and (8.10.14) that $\|(u_\varepsilon)^*\|_{L^2(\partial\Omega)} \leq C\|f\|_{L^2(\partial\Omega)}$ and

$$\|(u_{\varepsilon,\ell} - u_\varepsilon)^*\|_{L^2(\partial\Omega)} \leq C\,\|f_\ell - f\|_{L^2(\partial\Omega)}. \tag{8.10.15}$$

Next, to show $u_\varepsilon = f$ n.t. on $\partial\Omega$, we observe that

$$\begin{aligned}
&\limsup_{x\to Q} |u_\varepsilon(x) - f(Q)| \\
&\quad\leq \limsup_{x\to Q} |u_\varepsilon(x) - u_{\varepsilon,\ell}(x)| + \limsup_{x\to Q} |u_{\varepsilon,\ell}(x) - f_\ell(Q)| + |f_\ell(Q) - f(Q)| \\
&\quad\leq (u_\varepsilon - u_{\varepsilon,\ell})^*(Q) + |f_\ell(Q) - f(Q)|,
\end{aligned}$$

where the limits are taken nontangentially from Ω. This, together with (8.10.15), implies that $u_\varepsilon = f$ n.t. on $\partial\Omega$.

Finally, we consider the case $d = 2$. By the approximation argument above, it suffices to show that if u_ε is the solution of the L^2 regularity problem for $\mathcal{L}_\varepsilon$ in Ω with $f \in H^1(\partial\Omega; \mathbb{R}^m)$, given by Theorem 8.10.2, then $\|(u_\varepsilon)^*\|_{L^2(\partial\Omega)} \leq C\|f\|_{L^2(\partial\Omega)}$. This may be done by the descent method. Indeed, consider

$$v_\varepsilon(x_1, x_2, x_3) = u_\varepsilon(x_1, x_2) \quad \text{in } \Omega_1 = \Omega \times (0, 1).$$

Since $\|(v_\varepsilon)^*\|_{L^2(\partial\Omega_1)} \leq C\|v_\varepsilon\|_{L^2(\partial\Omega_1)}$, it follows that

$$\begin{aligned}
\|(u_\varepsilon)^*\|_{L^2(\partial\Omega)} &\leq C\left\{\|f\|_{L^2(\partial\Omega)} + \|u_\varepsilon\|_{L^2(\Omega)}\right\} \\
&\leq C\,\|f\|_{L^2(\partial\Omega)},
\end{aligned}$$

where we have used Lemma 8.10.4 for the last inequality. □

Remark 8.10.6. Let $A \in \Lambda(\mu, \lambda, \tau)$, $A^* = A$, and Ω be a bounded Lipschitz domain. As in the proof of Theorem 8.6.3, one may construct a matrix of Green functions $G_\varepsilon(x, y) = \big(G_\varepsilon^{\alpha\beta}(x, y)\big)_{m\times m}$ for $\mathcal{L}_\varepsilon$ in Ω, where for $d \geq 3$,

$$G_\varepsilon^{\alpha\beta}(x, y) = \Gamma_\varepsilon^{\alpha\beta}(x, y) - W_\varepsilon^{\alpha\beta}(x, y) \tag{8.10.16}$$

and for $\beta \in \{1, \ldots, m\}$ and $y \in \Omega$, $W^\beta(\cdot, y) = \big(W^{1\beta}(\cdot, y), \ldots, W^{m\beta}(\cdot, y)\big)$ is the unique solution of the L^2 regularity problem for $\mathcal{L}_\varepsilon(u_\varepsilon) = 0$ in Ω with Dirichlet data $\big(\Gamma_\varepsilon^{1\beta}(\cdot, y), \ldots, \Gamma_\varepsilon^{m\beta}(\cdot, y)\big)$ on $\partial\Omega$. If $d = 2$, we replace the matrix of fundamental solutions in (8.10.16) by the matrix of Green functions in a ball B such that $\Omega \subset (1/2)B$. Note that $G_\varepsilon(\cdot, y) = 0$ n.t. on $\partial\Omega$ and if $r < \text{dist}(y, \partial\Omega)/4$,

$$\mathcal{M}_r^1(\nabla_x G_\varepsilon(\cdot, y)) \in L^2(\partial\Omega), \tag{8.10.17}$$

where the operator $\mathcal{M}_r^1$ is defined in (8.9.2). We claim that if u_ε is a solution of the L^2 Dirichlet problem for $\mathcal{L}_\varepsilon(u_\varepsilon) = 0$ in Ω, it may be represented by the Poisson integral,

$$u_\varepsilon(x) = -\int_{\partial\Omega} \frac{\partial}{\partial\nu_\varepsilon(y)}\{G_\varepsilon(y,x)\}u_\varepsilon(y)\, d\sigma(y) \tag{8.10.18}$$

for any $x \in \Omega$.

To see (8.10.18), we first assume that u_ε is a solution of the L^2 regularity problem in Ω. Let Ω_ℓ be a sequence of smooth domains such that $\Omega_\ell \uparrow \Omega$. By the Green representation formula,

$$\begin{aligned} u_\varepsilon(x) = &\int_{\partial\Omega_\ell} G_\varepsilon(y,x)\frac{\partial u_\varepsilon}{\partial\nu_\varepsilon}\, d\sigma(y) \\ &-\int_{\partial\Omega_\ell} \frac{\partial}{\partial\nu_\varepsilon(y)}\{G_\varepsilon(y,x)\}u_\varepsilon(y)\, d\sigma(y), \end{aligned} \tag{8.10.19}$$

where we have used the symmetry condition $A^* = A$. Letting $\ell \to \infty$ in (8.10.19) and using $(\nabla u_\varepsilon)^* \in L^2(\partial\Omega)$ and the dominated convergence theorem, we see that the first integral in (8.10.19) converges to zero and the second converges to the RHS of (8.10.18). In general, if u_ε is the solution of the L^2 Dirichlet problem with data $f \in L^2(\partial\Omega;\mathbb{R}^m)$, we choose $f_\ell \in C_0^\infty(\mathbb{R}^d;\mathbb{R}^m)$ such that $f_\ell \to f$ in $L^2(\partial\Omega;\mathbb{R}^m)$. Let $u_{\varepsilon,\ell}$ be the solution of the L^2 Dirichlet problem in Ω with data f_ℓ. Then $(\nabla u_{\varepsilon,\ell})^* \in L^2(\partial\Omega)$ and thus (8.10.18) holds for $u_{\varepsilon,\ell}$. Since

$$\|(u_{\varepsilon,\ell} - u_\varepsilon)^*\|_{L^2(\partial\Omega)} \le C\, \|f_\ell - f\|_{L^2(\partial\Omega)},$$

it follows that $u_{\varepsilon,\ell}$ converges to u_ε uniformly on any compact subset of Ω. By letting $\ell \to \infty$, we conclude that (8.10.18) continues to hold for u_ε.

8.11 Square function and $H^{1/2}$ estimates

Let Ω be a bounded Lipschitz domain in $\mathbb{R}^d$. For a function $u \in H^1_{\text{loc}}(\Omega)$, the square function $S(u)$ on $\partial\Omega$ is defined by

$$S(u)(x) = \left(\int_{\gamma(x)} |\nabla u(y)|^2 |y-x|^{2-d}\, dy\right)^{1/2}$$

for $x \in \partial\Omega$, where $\gamma(x) = \{y \in \Omega : |y-x| < (1+\alpha)\text{dist}(y,\partial\Omega)\}$ and $\alpha = \alpha(\Omega) > 1$ is sufficiently large. It is not hard to see that

$$\int_{\partial\Omega} |S(u)|^2\, d\sigma \approx \int_\Omega |\nabla u(x)|^2\, \text{dist}(x,\partial\Omega)\, dx.$$

In this section we show that solutions to the L^2 Dirichlet problem for $\mathcal{L}_\varepsilon$ satisfy uniform square function estimates and $H^{1/2}$ estimates.

Theorem 8.11.1. *Suppose that $A \in \Lambda(\mu, \lambda, \tau)$ and $A^* = A$. Let Ω be a bounded Lipschitz domain and $f \in L^2(\partial\Omega; \mathbb{R}^m)$. Let u_ε be the solution of the Dirichlet problem $\mathcal{L}_\varepsilon(u_\varepsilon) = 0$ in Ω and $u_\varepsilon = f$ n.t. on $\partial\Omega$ with $(u_\varepsilon)^* \in L^2(\partial\Omega)$. Then*

$$\begin{aligned} \left(\int_\Omega |\nabla u_\varepsilon(x)|^2 \operatorname{dist}(x, \partial\Omega)\, dx \right)^{1/2} &\le C\, \|f\|_{L^2(\partial\Omega)}, \\ \left(\int_\Omega \int_\Omega \frac{|u_\varepsilon(x) - u_\varepsilon(y)|^2}{|x-y|^{d+1}}\, dxdy \right)^{1/2} &\le C\, \|f\|_{L^2(\partial\Omega)}, \end{aligned} \tag{8.11.1}$$

where C depends only on μ, λ, τ, and the Lipschitz character of Ω.

Let D be the region above a Lipschitz graph, i.e.,

$$D = \big\{ (x', x_d) \in \mathbb{R}^d : \ x' \in \mathbb{R}^{d-1} \text{ and } x_d > \psi(x') \big\}, \tag{8.11.2}$$

where ψ is a Lipschitz function in $\mathbb{R}^{d-1}$ such that $\psi(0) = 0$ and $\|\nabla\psi\|_\infty \le M_0$.

Lemma 8.11.2. *Let $g \in L^2(\partial D; \mathbb{R}^m)$ and*

$$u_\varepsilon(x) = \int_{\partial D} \frac{\partial}{\partial y_k} \{\Gamma_\varepsilon(x, y)\} g(y)\, d\sigma(y),$$

where D is given by (8.11.2). Then

$$\left(\int_D |\nabla u_\varepsilon(x)|^2 \operatorname{dist}(x, \partial D)\, dx \right)^{1/2} \le C\, \|g\|_{L^2(\partial D)}, \tag{8.11.3}$$

where C depends only on μ, λ, τ, and M_0.

Proof. By rescaling we may assume that $\varepsilon = 1$. We first estimate the integral of $|\nabla u_1(x)|^2 \operatorname{dist}(x, \partial D)$ on the set

$$H = D + (0, \dots, 0, 1) = \big\{ (x', x_d) : \ x_d > \psi(x') + 1 \big\}.$$

This is done by using the asymptotic estimates of $\nabla_x \nabla_y \Gamma_1(x, y)$ for $|x - y| \ge 1$. Indeed, by Theorem 4.4.7,

$$\Big| \nabla_x \nabla_y \Gamma_1(x, y) - \big(I + \nabla\chi(x)\big) \nabla_x \nabla_y \Gamma_0(x, y) \big(I + \nabla\chi^*(y)\big)^T \Big| \le \frac{C \ln[|x-y| + 2]}{|x-y|^{d+1}}$$

for any $x, y \in \mathbb{R}^d$ and $x \ne y$, where $\Gamma_0(x, y)$ is the matrix of fundamental solutions for $\mathcal{L}_0$. It follows that

$$|\nabla u_1(x) - W(x)| \le C \int_{\partial D} \frac{\ln[|x-y| + 2]}{|x-y|^{d+1}}\, |g(y)|\, d\sigma(y), \tag{8.11.4}$$

where $W(x)$ is a finite sum of functions of the form

$$e_{ij}(x) \int_{\partial D} \frac{\partial^2}{\partial x_i \partial x_j} \{\Gamma_0(x, y)\} h(y)\, d\sigma(y),$$

with $|e_{ij}(x)| \le C$, and $|h(y)| \le C|g(y)|$. As $\mathcal{L}_0$ is a second-order elliptic operator with constant coefficients, it is known that

$$\int_D |W(x)|^2 \operatorname{dist}(x, \partial D)\, dx \le C \int_{\partial D} |g|^2\, d\sigma \tag{8.11.5}$$

(see, e.g., [35]). Let $I(x)$ denote the integral in the RHS of (8.11.4). By the Cauchy inequality,

$$|I(x)|^2 \le \frac{C \ln[\operatorname{dist}(x, \partial D) + 2]}{[\operatorname{dist}(x, \partial D)]^2} \int_{\partial D} \frac{\ln[|x-y|+2]}{|x-y|^{d+1}} |g(y)|^2\, d\sigma(y).$$

This gives

$$\int_H |I(x)|^2 \operatorname{dist}(x, \partial D)\, dx \le C \int_{\partial D} |g|^2\, d\sigma. \tag{8.11.6}$$

In view of (8.11.5) and (8.11.6), we have therefore proved that

$$\int_H |\nabla u_1(x)|^2 \operatorname{dist}(x, \partial D)\, dx \le C\, \|g\|^2_{L^2(\partial D)}.$$

To handle $|\nabla u_1|$ in $D \setminus H$, we let

$$\Delta_r = \big\{ (x', \psi(x')) \in \mathbb{R}^d : \ |x'| < r \big\},$$
$$D_r = \big\{ (x', x_d) \in \mathbb{R}^d : \ |x'| < r \text{ and } \psi(x') < x_d < 10(M_0+1)r \big\}.$$

We will show that if $\mathcal{L}_1(u) = 0$ in the Lipschitz domain D_2 and $(u)^* \in L^2(\partial D_2)$, then

$$\int_{D_1} |\nabla u(x)|^2 |x_d - \psi(x')|\, dx \le C \int_{\Delta_2} |u|^2\, d\sigma + C \int_{D_2} |u|^2\, dx, \tag{8.11.7}$$

which is bounded by $C \int_{\Delta_2} |(u)^*|^2\, d\sigma$. By a simple covering argument one can deduce from (8.11.7) that

$$\begin{aligned} \int_{D\setminus H} |\nabla u_1(x)|^2 \operatorname{dist}(x, \partial D)\, dx &\le C \int_{\partial D} |(u_1)^*|^2\, d\sigma \\ &\le C \int_{\partial D} |g|^2\, d\sigma, \end{aligned} \tag{8.11.8}$$

where we have used Theorem 8.3.6 for the last inequality.

Finally, to see (8.11.7), we use the square function estimate for solutions of $\mathcal{L}_1(u) = 0$ in the domain D_r for $(3/2) < r < 2$,

$$\int_{D_r} |\nabla u(x)|^2 \operatorname{dist}(x, \partial D_r)\, dx \le C \int_{\partial D_r} |u|^2\, d\sigma, \tag{8.11.9}$$

to obtain

$$\int_{D_1} |\nabla u(x)|^2 \, |x_d - \psi(x')| \, dx \le C \int_{\Delta_2} |u|^2 \, d\sigma + C \int_{\partial D_r \setminus \Delta_2} |u|^2 \, d\sigma. \tag{8.11.10}$$

The estimate (8.11.7) follows by integrating both sides of (8.11.10) with respect to r over the interval $(3/2, 2)$. Note that (8.11.9) is a special case of (8.11.1) with $\varepsilon = 1$ and $\operatorname{diam}(\Omega) \le C$. Under the conditions that $A \in \Lambda(\mu, \lambda, \tau)$ and $A^* = A$, the square function estimate (8.11.9) follows from the double-layer-potential representation obtained in Theorem 8.9.3 for solutions of the L^2 Dirichlet problem for $\mathcal{L}_1$, by a $T(b)$-theorem argument (see, e.g., [73, pp. 9–11]). This completes the proof. □

The next lemma shows that for solutions of $\mathcal{L}_\varepsilon(u_\varepsilon) = 0$ in a Lipschitz domain Ω,

$$\|u_\varepsilon\|^2_{H^{1/2}(\Omega)} \approx \int_\Omega |\nabla u_\varepsilon(x)|^2 \operatorname{dist}(x, \partial\Omega) \, dx + \int_\Omega |u_\varepsilon(x)|^2 \, dx.$$

Lemma 8.11.3. *Let Ω be a bounded Lipschitz domain with* $\operatorname{diam}(\Omega) = 1$. *Then*

$$\|u\|^2_{H^{1/2}(\Omega)} \le C \left\{ \int_\Omega |\nabla u(x)|^2 \operatorname{dist}(x, \partial\Omega) \, dx + \int_\Omega |u(x)|^2 \, dx \right\}, \tag{8.11.11}$$

where C depends only on the Lipschitz character of Ω. Moreover, if $A \in \Lambda(\mu, \lambda, \tau)$ and $\mathcal{L}_\varepsilon(u_\varepsilon) = 0$ in Ω, then

$$\int_\Omega |\nabla u_\varepsilon(x)|^2 \operatorname{dist}(x, \partial\Omega) \, dx \le C \, \|u_\varepsilon\|^2_{H^{1/2}(\Omega)}, \tag{8.11.12}$$

where C depends only on μ, λ, τ, and the Lipschitz character of Ω.

Proof. We first prove (8.11.11). By a partition of unity we may assume that $\operatorname{supp}(u) \subset B(z, c) \cap \Omega$ for some $z \in \partial\Omega$ (it is easy to see that $\|u\|_{H^1(\Omega')}$ is bounded by the RHS of (8.11.11), if $\Omega' \subset \overline{\Omega'} \subset \Omega$). By a change of the coordinate system we may further assume that $z = 0$ and

$$\begin{aligned} &\Omega \cap \big\{(x', x_d) : \ |x'| < 4c \text{ and } -4c < x_d < 4c\big\} \\ &\quad = \big\{(x', x_d) : \ |x'| < 4c \text{ and } \psi(x') < x_d < 4c\big\}, \end{aligned}$$

where ψ is a Lipschitz function in $\mathbb{R}^{d-1}$ such that $\psi(0) = 0$ and $\|\nabla\psi\|_\infty \le M_0$. We will use the fact that

$$H^{1/2}(\Omega) = \big[L^2(\Omega), H^1(\Omega)\big]_{1/2,2},$$

known from real interpolation theory. Thus, $\|u\|_{H^{1/2}(\Omega)}$ is comparable to the infimum over all functions $f : [0, \infty) \to L^2(\Omega) + H^1(\Omega)$ with $f(0) = u$ of

$$\left(\int_0^\infty \|t^{1/2} f(t)\|^2_{H^1(\Omega)} \frac{dt}{t} \right)^{1/2} + \left(\int_0^\infty \|t^{1/2} f'(t)\|^2_{L^2(\Omega)} \frac{dt}{t} \right)^{1/2}. \tag{8.11.13}$$

See [61]. Let $f(t) = u(x', x_d + t)\theta(t)$, where $\theta \in C_0^\infty(\mathbb{R})$, $\theta(t) = 1$ for $|t| \le c$, and $\theta(t) = 0$ for $|t| \ge 2c$. Clearly, $f(0) = u$. Also, it is not hard to see that the expression (8.11.13) is bounded by

$$C\left(\int_0^{2c}\int_\Omega \Big\{|\nabla u(x', x_d + t)|^2 + |u(x', x_d+t)|^2\Big\}\Big\{|\theta(t)|^2 + |\theta'(t)|^2\Big\}\, dxdt\right)^{1/2}.$$

This implies that

$$\begin{aligned}
\|u\|^2_{H^{1/2}(\Omega)} &\le C\int_0^{2c}\int_{|x'|<c}\int_{\psi(x')}^{2c} \{|\nabla u(x', x_d+t)|^2 + |u(x', x_d+t)|^2\}\, dx_d dx' dt\\
&\le C\int_{|x'|<c}\int_{\psi(x')}^{4c} \{|\nabla u(x', x_d)|^2 + |u(x', x_d)|^2\}\, |x_d - \psi(x')|\, dx_d dx'\\
&\le C\left\{\int_\Omega |\nabla u(x)|^2 \operatorname{dist}(x, \partial\Omega)\, dx + \int_\Omega |u(x)|^2\, dx\right\}.
\end{aligned}$$

Next, let $u_\varepsilon \in H^1_{\text{loc}}(\Omega; \mathbb{R}^m)$ be a solution to $\mathcal{L}_\varepsilon(u_\varepsilon) = 0$ in Ω. To prove (8.11.12), we use the interior estimate (4.1.2) to obtain

$$\begin{aligned}
|\nabla u_\varepsilon(x)| &\le \frac{C}{r}\left(\fint_{B(x,r/2)} |u_\varepsilon(y) - u_\varepsilon(x)|^2\, dy\right)^{1/2}\\
&\le C r^{\frac{d-1}{2}}\left(\fint_{B(x,r/2)} \frac{|u_\varepsilon(y) - u_\varepsilon(x)|^2}{|y-x|^{d+1}}\, dy\right)^{1/2},
\end{aligned}$$

where $r = \operatorname{dist}(x, \partial\Omega)$. It follows that

$$\begin{aligned}
|\nabla u_\varepsilon(x)|^2 \operatorname{dist}(x, \partial\Omega) &\le C\int_{B(x,r/2)} \frac{|u_\varepsilon(y) - u_\varepsilon(x)|^2}{|y-x|^{d+1}}\, dy\\
&\le C\int_\Omega \frac{|u_\varepsilon(y) - u_\varepsilon(x)|^2}{|y-x|^{d+1}}\, dy.
\end{aligned}$$

Hence,

$$\begin{aligned}
\int_\Omega |\nabla u_\varepsilon(x)|^2 \operatorname{dist}(x, \partial\Omega)\, dx &\le C\int_\Omega\int_\Omega \frac{|u_\varepsilon(y) - u_\varepsilon(x)|^2}{|y-x|^{d+1}}\, dydx\\
&\le C\, \|u_\varepsilon\|^2_{H^{1/2}(\Omega)}.
\end{aligned}$$

∎

Proof of Theorem 8.11.1. It suffices to show that

$$\left(\int_\Omega |\nabla u_\varepsilon(x)|^2 \operatorname{dist}(x, \partial\Omega)\, dx\right)^{1/2} \le C\, \|f\|_{L^2(\partial\Omega)}. \tag{8.11.14}$$

By Lemma 8.11.3, the estimate $\|u\|_{H^{1/2}(\Omega)} \le C\|f\|_{L^2(\partial\Omega)}$ follows from (8.11.14) and the observation that $\|u\|_{L^2(\Omega)} \le C\|(u)^*\|_{L^2(\partial\Omega)} \le C\|f\|_{L^2(\partial\Omega)}$.

By Theorem 8.9.3, the solution u_ε is given by a double-layer potential with a density function g_ε and $\|g_\varepsilon\|_{L^2(\partial\Omega)} \le C\|f\|_{L^2(\partial\Omega)}$. As a result, it suffices to show that the LHS of (8.11.14) is bounded by $C\|g\|_{L^2(\partial\Omega)}$, if u_ε is given by

$$u_\varepsilon(x) = \int_{\partial\Omega} \frac{\partial}{\partial y_k}\{\Gamma_\varepsilon(x,y)\} g(y)\, d\sigma(y) \tag{8.11.15}$$

for some $g \in L^2(\partial\Omega)$. Using the estimate $|\nabla_x \nabla_y \Gamma_\varepsilon(x,y)| \le C|x-y|^{-d}$ and a partition of unity, we may further reduce the problem to the estimate

$$\int_{\Omega\cap B(z,cr_0)} |\nabla u_\varepsilon(x)|^2 \operatorname{dist}(x,\partial\Omega)\, dx \le C\int_{\partial\Omega} |g|^2\, d\sigma, \tag{8.11.16}$$

where u_ε is given by (8.11.15) and $\operatorname{supp}(g) \subset B(z,cr_0)\cap\partial\Omega$ for $r_0 = \operatorname{diam}(\Omega)$ and some $z \in \partial\Omega$. However, the estimate (8.11.16) follows readily from Lemma 8.11.2 by a change of the coordinate system. This completes the proof. □

We end this section with an error estimate in $H^{1/2}$.

Theorem 8.11.4. *Suppose that $A \in \Lambda(\mu,\lambda,\tau)$ and $A^* = A$. Let Ω be a bounded Lipschitz domain in $\mathbb{R}^d$, $d \ge 2$. Let u_ε ($\varepsilon \ge 0$) be the solution to the Dirichlet problem $\mathcal{L}_\varepsilon(u_\varepsilon) = F$ in Ω and $u_\varepsilon = f$ on $\partial\Omega$, where $F \in L^2(\Omega;\mathbb{R}^d)$ and $f \in H^1(\partial\Omega;\mathbb{R}^m)$. Then*

$$\|u_\varepsilon - u_0 - \varepsilon\chi(x/\varepsilon)\nabla u_0\|_{H^{1/2}(\Omega)} \le C\varepsilon\,\|u_0\|_{H^2(\Omega)}, \tag{8.11.17}$$

where C depends only on μ, λ, τ and Ω.

Proof. Let

$$\begin{aligned} w_\varepsilon &= u_\varepsilon - u_0 - \varepsilon\chi(x/\varepsilon)\nabla u_0 \\ &= w_\varepsilon^{(1)} + w_\varepsilon^{(2)}, \end{aligned}$$

where

$$\mathcal{L}_\varepsilon(w_\varepsilon^{(1)}) = \mathcal{L}_\varepsilon(w_\varepsilon) \quad \text{in } \Omega \quad \text{and} \quad w_\varepsilon^{(1)} = 0 \quad \text{on } \partial\Omega,$$

and

$$\mathcal{L}_\varepsilon(w_\varepsilon^{(2)}) = 0 \quad \text{in } \Omega \quad \text{and} \quad w_\varepsilon^{(2)} = w_\varepsilon \quad \text{on } \partial\Omega.$$

By (4.4.14) and the energy estimate,

$$\|w_\varepsilon^{(1)}\|_{H^1(\Omega)} \le C\varepsilon\,\|\nabla^2 u_0\|_{L^2(\Omega)}. \tag{8.11.18}$$

Note that $w_\varepsilon = -\chi(x/\varepsilon)\nabla u_0$ on $\partial\Omega$. Hence, by Theorem 8.11.1,

$$\|w_\varepsilon^{(2)}\|_{H^{1/2}(\Omega)} \le C\varepsilon\,\|\nabla u_0\|_{L^2(\partial\Omega)},$$

which, together with (8.11.18), gives

$$\begin{aligned} \|w_\varepsilon\|_{H^{1/2}(\Omega)} &\le C\varepsilon\big\{\|\nabla^2 u_0\|_{L^2(\Omega)} + \|\nabla u_0\|_{L^2(\partial\Omega)}\big\} \\ &\le C\varepsilon\,\|u_0\|_{H^2(\Omega)}. \end{aligned}$$

□

8.12 Notes

The main results in this chapter were proved by C. Kenig and Z. Shen in [68]. In particular, the material in Sections 8.2–8.4, 8.6–8.7, and 8.9 is taken from [68]. The treatment for the large-scale Rellich estimates in Section 8.8, which is different from that in [68], follows an approach used in [90]. The square function estimate in Section 8.11 was proved in [63]. Early work on square function estimates in Lipschitz domains may be found in [32, 35].

There exists an extensive literature on L^p boundary value problems for elliptic equations and systems in nonsmooth domains. Classical results for Laplace's equation in Lipschitz domains, which are presented in Section 8.5 in the case $p = 2$, may be found in [30, 31, 60, 104, 33]. We refer the reader to [62, 73, 86, 87] for further references. For elliptic operators with periodic coefficients, the L^2 Dirichlet problem in Lipschitz domains for the scalar case ($m = 1$) was solved by B. Dahlberg, using the method of harmonic measure (unpublished; a proof may be found in the Appendix of [67]). This extends an earlier work in [12, 11] for $C^{1,\alpha}$ domains by M. Avellaneda and F. Lin. The well-posedness of the L^2 Neumann and regularity problems in the periodic setting was first obtained by C. Kenig and Z. Shen in [67] in the case $m = 1$. In [67, 68] the large-scale Rellich estimates were established by using integration by parts.

The estimates in (8.0.4) continue to hold for elliptic systems of elasticity. For the L^2 Dirichlet problem (8.0.1) and regularity problem (8.0.3), this follows readily from the observation that the system of elasticity may be rewritten in such a way that the new coefficient matrix $\widetilde{A} \in \Lambda(\mu, \lambda, \tau)$ and is symmetric (see the end of Section 2.4). For the L^2 Neumann problem, the results were proved in [47] by J. Geng, Z. Shen, and L. Song.

For the L^p estimates in the periodic setting, the results are known for $1 < p < \infty$, if Ω is $C^{1,\alpha}$ [11, 65]. For a Lipschitz domain Ω, if $m = 1$ or $d = 2, 3$, one can establish the L^p estimate for $2 - \delta < p < \infty$ in the case of the Dirichlet problem, and for $1 < p < 2+\delta$ in the case of the Neumann and regularity problems. With minor modifications, the methods used for the case of constant coefficients in [33, 34] work equally well in the periodic setting. If $m \geq 2$ and $d \geq 4$, partial results may be obtained using the approaches in [86, 87].

Bibliography

[1] G. Allaire and M. Amar, *Boundary layer tails in periodic homogenization*, ESAIM Control Optim. Calc. Var. **4** (1999), 209–243.

[2] S.N. Armstrong, T. Kuusi, and J.-C. Mourrat, *Mesoscopic higher regularity and subadditivity in elliptic homogenization*, Comm. Math. Phys. **347** (2016), no. 2, 315–361.

[3] ———, *The additive structure of elliptic homogenization*, Invent. Math. **208** (2017), no. 3, 999–1154.

[4] S.N. Armstrong, T. Kuusi, J.-C. Mourrat, and C. Prange, *Quantitative analysis of boundary layers in periodic homogenization*, Arch. Ration. Mech. Anal. **226** (2017), no. 2, 695–741.

[5] S.N. Armstrong and J.-C. Mourrat, *Lipschitz regularity for elliptic equations with random coefficients*, Arch. Ration. Mech. Anal. **219** (2016), no. 1, 255–348.

[6] S.N. Armstrong and Z. Shen, *Lipschitz estimates in almost-periodic homogenization*, Comm. Pure Appl. Math. **69** (2016), no. 10, 1882–1923.

[7] S.N. Armstrong and C. Smart, *Quantitative stochastic homogenization of convex integral functionals*, Ann. Sci. Éc. Norm. Supér. (4) **49** (2016), no. 2, 423–481.

[8] P. Auscher, *On necessary and sufficient conditions for L^p-estimates of Riesz transforms associated to elliptic operators on $\mathbb{R}^n$ and related estimates*, Mem. Amer. Math. Soc. **186** (2007).

[9] P. Auscher, T. Coulhon, X.T. Duong, and S. Hofmann, *Riesz transform on manifolds and heat kernel regularity*, Ann. Sci. École Norm. Sup. (4) **37** (2004), no. 6, 911–957.

[10] P. Auscher and M. Qafsaoui, *Observations on $W^{1,p}$ estimates for divergence elliptic equations with VMO coefficients*, Boll. Unione Mat. Ital. Sez. B Artic. Ric. Mat. (8) **5** (2002), no. 2, 487–509.

[11] M. Avellaneda and F. Lin, *Compactness methods in the theory of homogenization*, Comm. Pure Appl. Math. **40** (1987), no. 6, 803–847.

Z. Shen, *Periodic Homogenization of Elliptic Systems*, Operator Theory: Advances and Applications 269, https://doi.org/10.1007/978-3-319-91214-1

[12] ———, *Homogenization of elliptic problems with L^p boundary data*, Appl. Math. Optim. **15** (1987), no. 2, 93–107.

[13] ———, *Compactness methods in the theory of homogenization. II. Equations in nondivergence form*, Comm. Pure Appl. Math. **42** (1989), no. 2, 139–172.

[14] ———, *Homogenization of Poisson's kernel and applications to boundary control*, J. Math. Pures Appl. (9) **68** (1989), no. 1, 1–29.

[15] ———, *Un théorème de Liouville pour des équations elliptiques à coefficients périodiques*, C. R. Acad. Sci. Paris Sér. I Math. **309** (1989), no. 5, 245–250.

[16] ———, *L^p bounds on singular integrals in homogenization*, Comm. Pure Appl. Math. **44** (1991), no. 8-9, 897–910.

[17] N.S. Bahvalov, *Averaged characteristics of bodies with periodic structure*, Dokl. Akad. Nauk SSSR **218** (1974), 1046–1048.

[18] ———, *Averaging of partial differential equations with rapidly oscillating coefficients*, Dokl. Akad. Nauk SSSR **221** (1975), no. 3, 516–519.

[19] A. Bensoussan, J.-L. Lions, and G. Papanicolaou, *Asymptotic Analysis for Periodic Structures*, Studies in Mathematics and its Applications, vol. 5, North-Holland Publishing Co., Amsterdam-New York, 1978.

[20] J. Bergh and J. Löfström, *Interpolation Spaces. An Introduction*, Springer-Verlag, Berlin-New York, 1976, Grundlehren der Mathematischen Wissenschaften, No. 223.

[21] S. Byun, *Elliptic equations with BMO coefficients in Lipschitz domains*, Trans. Amer. Math. Soc. **357** (2005), no. 3, 1025–1046.

[22] S. Byun and L. Wang, *Elliptic equations with BMO coefficients in Reifenberg domains*, Comm. Pure Appl. Math. **57** (2004), no. 10, 1283–1310.

[23] ———, *The conormal derivative problem for elliptic equations with BMO coefficients on Reifenberg flat domains*, Proc. London Math. Soc. (3) **90** (2005), no. 1, 245–272.

[24] L. Caffarelli and I. Peral, *On $W^{1,p}$ estimates for elliptic equations in divergence form*, Comm. Pure Appl. Math. **51** (1998), no. 1, 1–21.

[25] G.A. Chechkin, A.L. Piatnitski, and A.S. Shamaev, *Homogenization*, Translations of Mathematical Monographs, vol. 234, American Mathematical Society, Providence, RI, 2007, Methods and applications, Translated from the 2007 Russian original by Tamara Rozhkovskaya.

[26] Y.Z. Chen and L.C. Wu, *Second Order Elliptic Equations and Elliptic Systems*, Translations of Mathematical Monographs, vol. 174, American Mathematical Society, Providence, RI, 1998, Translated from the 1991 Chinese original by Bei Hu.

[27] J. Choi and S. Kim, *Neumann functions for second order elliptic systems with measurable coefficients*, Trans. Amer. Math. Soc. **365** (2013), no. 12, 6283–6307.

[28] D. Cioranescu and P. Donato, *An Introduction to Homogenization*, Oxford Lecture Series in Mathematics and its Applications, vol. 17, The Clarendon Press, Oxford University Press, New York, 1999.

[29] R. Coifman, A. McIntosh, and Y. Meyer, *L'intégrale de Cauchy définit un opérateur borné sur L^2 pour les courbes lipschitziennes*, Ann. of Math. (2) **116** (1982), no. 2, 361–387.

[30] B. Dahlberg, *Estimates of harmonic measure*, Arch. Rational Mech. Anal. **65** (1977), no. 3, 275–288.

[31] ———, *On the Poisson integral for Lipschitz and C^1-domains*, Studia Math. **66** (1979), no. 1, 13–24.

[32] ———, *Weighted norm inequalities for the Lusin area integral and the non-tangential maximal functions for functions harmonic in a Lipschitz domain*, Studia Math. **67** (1980), no. 3, 297–314.

[33] B. Dahlberg and C. Kenig, *Hardy spaces and the Neumann problem in L^p for Laplace's equation in Lipschitz domains*, Ann. of Math. (2) **125** (1987), no. 3, 437–465.

[34] ———, *L^p estimates for the three-dimensional systems of elastostatics on Lipschitz domains*, Analysis and partial differential equations, Lecture Notes in Pure and Appl. Math., vol. 122, Dekker, New York, 1990, pp. 621–634.

[35] B. Dahlberg, C. Kenig, J. Pipher, and G. Verchota, *Area integral estimates for higher order elliptic equations and systems*, Ann. Inst. Fourier (Grenoble) **47** (1997), no. 5, 1425–1461.

[36] B. Dahlberg, C. Kenig, and G. Verchota, *Boundary value problems for the systems of elastostatics in Lipschitz domains*, Duke Math. J. **57** (1988), no. 3, 795–818.

[37] B. Dahlberg and G. Verchota, *Galerkin methods for the boundary integral equations of elliptic equations in nonsmooth domains*, Harmonic analysis and partial differential equations (Boca Raton, FL, 1988), Contemp. Math., vol. 107, Amer. Math. Soc., Providence, RI, 1990, pp. 39–60.

[38] E. De Giorgi, *Sulla convergenza di alcune successioni d'integrali del tipo dell'area*, Rend. Mat. (6) **8** (1975), 277–294, Collection of articles dedicated to Mauro Picone on the occasion of his ninetieth birthday.

[39] ———, *Convergence problems for functionals and operators*, Proceedings of the International Meeting on Recent Methods in Nonlinear Analysis (Rome, 1978), Pitagora, Bologna, 1979, pp. 131–188.

[40] E. De Giorgi and S. Spagnolo, *Sulla convergenza degli integrali dell'energia per operatori ellittici del secondo ordine*, Boll. Un. Mat. Ital. (4) **8** (1973), 391–411.

[41] H. Dong and D. Kim, *Elliptic equations in divergence form with partially BMO coefficients*, Arch. Ration. Mech. Anal. **196** (2010), no. 1, 25–70.

[42] E. Fabes, *Layer potential methods for boundary value problems on Lipschitz domains*, Potential theory – surveys and problems (Prague, 1987), Lecture Notes in Math., vol. 1344, Springer, Berlin, 1988, pp. 55–80.

[43] W. Gao, *Layer potentials and boundary value problems for elliptic systems in Lipschitz domains*, J. Funct. Anal. **95** (1991), no. 2, 377–399.

[44] J. Geng, *$W^{1,p}$ estimates for elliptic problems with Neumann boundary conditions in Lipschitz domains*, Adv. Math. **229** (2012), no. 4, 2427–2448.

[45] J. Geng and Z. Shen, *Uniform regularity estimates in parabolic homogenization*, Indiana Univ. Math. J. **64** (2015), no. 3, 697–733.

[46] J. Geng, Z. Shen, and L. Song, *Uniform $W^{1,p}$ estimates for systems of linear elasticity in a periodic medium*, J. Funct. Anal. **262** (2012), no. 4, 1742–1758.

[47] ———, *Boundary Korn inequality and Neumann problems in homogenization of systems of elasticity*, Arch. Ration. Mech. Anal. **224** (2017), no. 3, 1205–1236.

[48] D. Gérard-Varet and N. Masmoudi, *Homogenization in polygonal domains*, J. Eur. Math. Soc. (JEMS) **13** (2011), no. 5, 1477–1503.

[49] ———, *Homogenization and boundary layers*, Acta Math. **209** (2012), no. 1, 133–178.

[50] M. Giaquinta and L. Martinazzi, *An Introduction to the Regularity Theory for Elliptic Systems, harmonic maps and minimal graphs*, second ed., Appunti. Scuola Normale Superiore di Pisa (Nuova Serie) [Lecture Notes. Scuola Normale Superiore di Pisa (New Series)], vol. 11, Edizioni della Normale, Pisa, 2012.

[51] A. Gloria, S. Neukamm, and F. Otto, *Quantification of ergodicity in stochastic homogenization: optimal bounds via spectral gap on Glauber dynamics*, Invent. Math. **199** (2015), no. 2, 455–515.

[52] A. Gloria and F. Otto, *Quantitative results on the corrector equation in stochastic homogenization*, J. Eur. Math. Soc. (JEMS) **19** (2017), no. 11, 3489–3548.

[53] G. Griso, *Error estimate and unfolding for periodic homogenization*, Asymptot. Anal. **40** (2004), no. 3-4, 269–286.

[54] ———, *Interior error estimate for periodic homogenization*, Anal. Appl. (Singap.) **4** (2006), no. 1, 61–79.

[55] M. Grüter and K.-O. Widman, *The Green function for uniformly elliptic equations*, Manuscripta Math. **37** (1982), no. 3, 303–342.

[56] S. Gu, *Convergence rates in homogenization of Stokes systems*, J. Differential Equations **260** (2016), no. 7, 5796–5815.

[57] ———, *Convergence rates of Neumann problems for Stokes systems*, J. Math. Anal. Appl. **457** (2018), no. 1, 305–321.

[58] S. Gu and Z. Shen, *Homogenization of Stokes systems and uniform regularity estimates*, SIAM J. Math. Anal. **47** (2015), no. 5, 4025–4057.

[59] S. Hofmann and S. Kim, *The Green function estimates for strongly elliptic systems of second order*, Manuscripta Math. **124** (2007), no. 2, 139–172.

[60] D. Jerison and C. Kenig, *The Neumann problem on Lipschitz domains*, Bull. Amer. Math. Soc. (N.S.) **4** (1981), no. 2, 203–207.

[61] ———, *The inhomogeneous Dirichlet problem in Lipschitz domains*, J. Funct. Anal. **130** (1995), no. 1, 161–219.

[62] C. Kenig, *Harmonic Analysis Techniques for Second Order Elliptic Boundary Value Problems*, CBMS Regional Conference Series in Math., vol. 83, AMS, Providence, RI, 1994.

[63] C. Kenig, F. Lin, and Z. Shen, *Convergence rates in L^2 for elliptic homogenization problems*, Arch. Ration. Mech. Anal. **203** (2012), no. 3, 1009–1036.

[64] ———, *Estimates of eigenvalues and eigenfunctions in periodic homogenization*, J. Eur. Math. Soc. (JEMS) **15** (2013), no. 5, 1901–1925.

[65] ———, *Homogenization of elliptic systems with Neumann boundary conditions*, J. Amer. Math. Soc. **26** (2013), no. 4, 901–937.

[66] ———, *Periodic homogenization of Green and Neumann functions*, Comm. Pure Appl. Math. **67** (2014), no. 8, 1219–1262.

[67] C. Kenig and Z. Shen, *Homogenization of elliptic boundary value problems in Lipschitz domains*, Math. Ann. **350** (2011), no. 4, 867–917.

[68] ———, *Layer potential methods for elliptic homogenization problems*, Comm. Pure Appl. Math. **64** (2011), no. 1, 1–44.

[69] D. Kim and N.V. Krylov, *Elliptic differential equations with coefficients measurable with respect to one variable and VMO with respect to the others*, SIAM J. Math. Anal. **39** (2007), no. 2, 489–506.

[70] S.M. Kozlov, *Asymptotic behavior of fundamental solutions of divergence second-order differential equations*, Mat. Sb. (N.S.) **113(155)** (1980), no. 2(10), 302–323, 351.

[71] V.A. Marchenko and E.Ya. Khruslov, *Homogenization of Partial Differential Equations*, Progress in Mathematical Physics, vol. 46, Birkhäuser Boston, Inc., Boston, MA, 2006, Translated from the 2005 Russian original by M. Goncharenko and D. Shepelsky.

[72] N. Meyers, *An L^p e-estimate for the gradient of solutions of second order elliptic divergence equations*, Ann. Scuola Norm. Sup. Pisa (3) **17** (1963), 189–206.

[73] D. Mitrea, M. Mitrea, and M. Taylor, *Layer Potentials, the Hodge Laplacian, and Global Boundary Problems in Nonsmooth Riemannian Manifolds*, Mem. Amer. Math. Soc. **150** (2001), no. 713, x+120.

[74] M. Mitrea and M. Taylor, *Boundary layer methods for Lipschitz domains in Riemannian manifolds*, J. Funct. Anal. **163** (1999), no. 2, 181–251.

[75] F. Murat, *Compacité par compensation. II*, Proceedings of the International Meeting on Recent Methods in Nonlinear Analysis (Rome, 1978), Pitagora, Bologna, 1979, pp. 245–256.

[76] J. Necas, *Direct Methods in the Theory of Elliptic Equations*, Springer Monographs in Mathematics, Springer, Heidelberg, 2012, Translated from the 1967 French original by Gerard Tronel and Alois Kufner, Editorial coordination and preface by Šárka Nečasová and a contribution by Christian G. Simader.

[77] N. Niu, Z. Shen, and Y. Xu, *Convergence rates and interior estimates in homogenization of higher order elliptic systems*, J. Funct. Anal. **274** (2018), no. 8, 2356–2398.

[78] O.A. Oleĭnik, A.S. Shamaev, and G.A. Yosifian, *Mathematical Problems in Elasticity and Homogenization*, Studies in Mathematics and its Applications, vol. 26, North-Holland Publishing Co., Amsterdam, 1992.

[79] D. Onofrei and B. Vernescu, *Error estimates for periodic homogenization with non-smooth coefficients*, Asymptot. Anal. **54** (2007), no. 1-2, 103–123.

[80] M.A. Pakhnin and T.A. Suslina, *Operator error estimates for the homogenization of the elliptic Dirichlet problem in a bounded domain*, Algebra i Analiz **24** (2012), no. 6, 139–177.

[81] S.E. Pastukhova, *On some estimates from the homogenization of problems in plasticity theory*, Dokl. Akad. Nauk **406** (2006), no. 5, 604–608.

[82] B.C. Russell, *Homogenization in perforated domains and interior Lipschitz estimates*, J. Differential Equations **263** (2017), no. 6, 3396–3418.

[83] E. Schmutz, *Rational points on the unit sphere*, Cent. Eur. J. Math. **6** (2008), no. 3, 482–487.

[84] E.V. Sevostyanova, *Asymptotic expansion of the solution of a second-order elliptic equation with periodic rapidly oscillating coefficients*, Mat. Sb. (N.S.) **115(157)** (1981), no. 2, 204–222, 319.

[85] Z. Shen, *Bounds of Riesz transforms on L^p spaces for second order elliptic operators*, Ann. Inst. Fourier (Grenoble) **55** (2005), no. 1, 173–197.

[86] ______, *Necessary and sufficient conditions for the solvability of the L^p Dirichlet problem on Lipschitz domains*, Math. Ann. **336** (2006), no. 3, 697–725.

[87] ______, *The L^p boundary value problems on Lipschitz domains*, Adv. Math. **216** (2007), no. 1, 212–254.

[88] ______, *$W^{1,p}$ estimates for elliptic homogenization problems in nonsmooth domains*, Indiana Univ. Math. J. **57** (2008), no. 5, 2283–2298.

[89] ______, *Convergence rates and Hölder estimates in almost-periodic homogenization of elliptic systems*, Anal. PDE **8** (2015), no. 7, 1565–1601.

[90] ______, *Boundary estimates in elliptic homogenization*, Anal. PDE **10** (2017), no. 3, 653–694.

[91] Z. Shen and J. Zhuge, *Convergence rates in periodic homogenization of systems of elasticity*, Proc. Amer. Math. Soc. **145** (2017), no. 3, 1187–1202.

[92] ———, *Boundary layers in periodic homogenization of Neumann problems*, Comm. Pure Appl. Math. (to appear) (2018).

[93] S. Spagnolo, *Sul limite delle soluzioni di problemi di Cauchy relativi all'equazione del calore*, Ann. Scuola Norm. Sup. Pisa (3) **21** (1967), 657–699.

[94] ———, *Sulla convergenza di soluzioni di equazioni paraboliche ed ellittiche*, Ann. Scuola Norm. Sup. Pisa (3) 22 (1968), 571–597; errata, ibid. (3) **22** (1968), 673.

[95] E.M. Stein, *Singular Integrals and Differentiability Properties of Functions*, Princeton Mathematical Series, No. 30, Princeton University Press, Princeton, N.J., 1970.

[96] ———, *Harmonic Analysis: Real-variable Methods, Orthogonality, and Oscillatory Integrals*, Princeton Mathematical Series, vol. 43, Princeton University Press, Princeton, NJ, 1993, With the assistance of Timothy S. Murphy, Monographs in Harmonic Analysis, III.

[97] T.A. Suslina, *Homogenization of the Dirichlet problem for elliptic systems: L_2-operator error estimates*, Mathematatika **59** (2013), no. 2, 463–476.

[98] ———, *Homogenization of the Neumann problem for elliptic systems with periodic coefficients*, SIAM J. Math. Anal. **45** (2013), no. 6, 3453–3493.

[99] L. Tartar, *Problèmes d'homogénéisation dans les équations aux dérivées partielles, Cours Peccot Collége de France. H-Convergence*, F. Murat, ed., Séminaire d'Analyse Fonctionnelle et Numérique, 1977/78, Université d'Alger, 1978.

[100] ———, *The General Theory of Homogenization, a personalized Introduction*, Lecture Notes of the Unione Matematica Italiana, vol. 7, Springer-Verlag, Berlin; UMI, Bologna, 2009, A personalized introduction.

[101] J. Taylor, S. Kim, and R. Brown, *The Green function for elliptic systems in two dimensions*, Comm. Partial Differential Equations **38** (2013), no. 9, 1574–1600.

[102] A.F.M. ter Elst, D.W. Robinson, and A. Sikora, *On second-order periodic elliptic operators in divergence form*, Math. Z. **238** (2001), no. 3, 569–637.

[103] G. Verchota, *Layer potentials and boundary value problems for Laplace's equation on Lipschitz domains*, Thesis, University of Minnesota (1982).

[104] ———, *Layer potentials and regularity for the Dirichlet problem for Laplace's equation in Lipschitz domains*, J. Funct. Anal. **59** (1984), no. 3, 572–611.

[105] ———, *Remarks on 2nd order elliptic systems in Lipschitz domains*, Miniconference on operator theory and partial differential equations (North Ryde, 1986), Proc. Centre Math. Anal. Austral. Nat. Univ., vol. 14, Austral. Nat. Univ., Canberra, 1986, pp. 303–325.

[106] L. Wang, *A geometric approach to the Calderón–Zygmund estimates*, Acta Math. Sin. (Engl. Ser.) **19** (2003), no. 2, 381–396.

[107] Q. Xu, *Convergence rates for general elliptic homogenization problems in Lipschitz domains*, SIAM J. Math. Anal. **48** (2016), no. 6, 3742–3788.

[108] L.-M. Yeh, *L^p gradient estimate for elliptic equations with high-contrast conductivities in $\mathbb{R}^n$*, J. Differential Equations **261** (2016), no. 2, 925–966.

[109] V.V. Zhikov, *On operator estimates in homogenization theory*, Dokl. Akad. Nauk **403** (2005), no. 3, 305–308.

[110] ———, *On some estimates from homogenization theory*, Dokl. Akad. Nauk **406** (2006), no. 5, 597–601.

[111] V.V. Zhikov, S.M. Kozlov, and O.A. Oleĭnik, *Homogenization of Differential Operators and Integral Functionals*, Springer-Verlag, 1994, Translated from the Russian by G.A. Yosifian.

[112] V.V. Zhikov and S.E. Pastukhova, *On operator estimates for some problems in homogenization theory*, Russ. J. Math. Phys. **12** (2005), no. 4, 515–524.

Index

A^*, 20
$C^{0,\alpha}(E)$, 100
D_r, 6
$E(\kappa_1, \kappa_2)$, 26
$H^k_{\text{per}}(Y; \mathbb{R}^m)$, 15
$H^{1/2}(\partial\Omega)$, 12
S_ε, 37
V-ellipticity, 9
Δ_r, 6
$\Lambda(\mu, \lambda, \tau)$, 205
Ω_+, 222
Ω_-, 222
χ_j^β, 17
$\chi_j^{\alpha\beta}$, 16
$\fint_E u$, 5
$\frac{\partial u_0}{\partial \nu_0}$, 26
$\frac{\partial u_\varepsilon}{\partial \nu_\varepsilon}$, 12
$\mathcal{L}_0$, 16
$\mathcal{L}_\varepsilon$, 7
$\mathcal{L}_\varepsilon^*$, 20
$\mathcal{M}_{\partial\Omega}$, 132
$\mathcal{R}$, 29
$\phi_{kij}^{\alpha\beta}$, 35
$\widehat{A}$, 16
$\widehat{a}_{ij}^{\alpha\beta}$, 16
$b_{ij}^{\alpha\beta}$, 34
1-periodic, 7

Caccioppoli's inequality, 11, 102, 137, 161
Calderón–Zygmund decomposition, 75
conormal derivative, 12
corrector, 17

Dirichlet corrector, 118
Div-Curl Lemma, 21
divergence theorem, 212
double-layer potential, 225

effective coefficient, 17
elasticity condition, 26
ellipticity condition, 8

first Korn inequality, 26
flux corrector, 34, 36

Hardy–Littlewood maximal function, 74
homogenized coefficient, 17

jump relation, 228

Legendre condition, 13
Legendre–Hadamard condition, 10
Liouville property, 73, 109
Lipschitz character, 207
Lipschitz function, 100

Neumann corrector, 193
nontangential
  approach region, 208
  convergence, 208
  maximal function, 6, 208

Rellich identity, 231, 253
Rellich property, 246
reverse Hölder inequality, 11, 12

second Korn inequality, 29
single-layer potential, 225
square function, 275

weak solution, 8, 144
  of the Neumann problem, 13
well posed, 213, 214

Z. Shen, *Periodic Homogenization of Elliptic Systems*, Operator Theory: Advances and Applications 269, https://doi.org/10.1007/978-3-319-91214-1